SUPERTANKERS
Anatomy and Operation

by

Dr Raymond Solly

First Published 2001

© Dr Raymond Solly, 2001
ISBN 1 85609 181 3

WITHERBY

PUBLISHERS

All rights reserved

British Library Cataloguing in Publication Data

Dr Raymond Solly
Supertankers: Anatomy and Operation
First Edition
1. Title

ISBN 1 85609 181 3

NOTICE OF TERMS OF USE AND COPYRIGHT

It should be understood that this work is not intended as a professional manual, but as a technical interest book. It is considered necessary to stress, therefore, that whilst every effort has been made to ensure that the information contained therein is accurate, no responsibility can be accepted by the author and publishers for any errors which may have occurred.

All rights reserved. No part of this publication may be reproduced, stored in a retrieval system, or transmitted in any form or by any means, electronic, mechanical, photocopying, recording or otherwise, without the prior permission of the publisher and copyright owner.

While the principles discussed and the details given in this book are the product of careful study, the author and the publisher cannot in any way guarantee the suitability of recommendations made in this book for individual problems, and they shall not be under any legal liability of any kind in respect of or arising out of the form of contents of this book or any error therein, or the reliance of any person thereon.

Printed in Great Britain by
Witherby & Co Ltd
32–36 Aylesbury Street
London EC1R 0ET, England

Cover Photographs: ULCC "FRONT CHIEF"
(By kind permission of Ian Cochran, London Correspondent of Tradewinds Magazine)

SUPERTANKERS
Anatomy and Operation

by

Dr Raymond Solly

Published by
Witherby & Company Limited
32-36 Aylesbury Street, London EC1R 0ET
Tel No. 020-7251 5341   Fax No. 020-7251 1296
International Tel No. +44 207 251 5341   Fax No. +44 207 251 1296
E-mail:books @ witherbys.co.uk
www.witherbys.com

DEDICATION

This book is dedicated with affectionate respect
to the memory of:

Captain A. "Tommy" AGNEW

(Ex-Commodore of ESSO TANKER COMPANY)

From amongst the many Ships-Masters, under whom I had the privilege to serve during my years at sea as a deck officer aboard supertankers, dry cargo ships, ferries and coasters, it was truly an honour to have sailed with a Master of his calibre.

THE AUTHOR

Dr. Raymond Solly served for a number of years in the Merchant Navy aboard Supertankers following extended experience on deep-sea dry-cargo ships, coasters and ferries. He rose through the ranks from deck cadet to chief officer, achieving command on one occasion following the untimely demise of the ship's master.

After coming ashore he attended a number of university courses enabling him to qualify as a schoolmaster at an East Kent HMC independent school. He retains active interest in the sea as a marine author and Lieutenant-Commander in the Royal Naval Reserve. Amongst other naval duties, he organises and teaches coastal navigation and radar courses to potential merchant navy deck cadets, and Royal Naval midshipmen, through the School of Maritime Operations in HMS. "DRYAD", and on the Naval Proficiency courses at HMS "RALEIGH", as well as practical sea training. He also lectures extensively on 'matters maritime' to a number of shipping organisations.

CONTENTS

	Page
Dedication	iv
The Author	iv
Foreword	vii
Preface	ix
Acknowledgements	xix
Glossary	xxiii

Part One — Introduction — 1
Chapter One — Background — 3

Part Two — The 'First' Generation ULCC/VLCC — 21
Chapter Two — Construction — 21
Chapter Three — Hull and Tank Protection — 53
Chapter Four — Deck Gear — 81
Chapter Five — Section Analysis — 119
Chapter Six — Ship Stresses and Stability — 157

Part Three — Operational Techniques — 187
Chapter Seven — Cargo Voyages — 187
Chapter Eight — Tank Cleaning Procedures — 207
Chapter Nine — Deck Duties — 267
Chapter Ten — Bridge Watches and Ship Handling — 293
Chapter Eleven — Anchoring/Berthing/Lightening — 331
Chapter Twelve — Training Deck Cadets — 355
Chapter Thirteen — Life Aboard Supertankers — 381

Part Four — The 'Second' Generation VLCC — 401
Chapter Fourteen — Double Hull Construction — 401
Chapter Fifteen — Investigating Computers at Sea — 435
Chapter Sixteen — Use of Computers in Navigation — 443
Chapter Seventeen — Other Bridge Computerisation — 477
Chapter Eighteen — Computerised Cargo Control — 529

Part Five — Conclusion — 555
Chapter Nineteen — Supertankers — The Future? — 555

Appendix — Organisations — 569
Bibliography — 577
Index — 581

Inside back cover – General Arrangement and Capacity Plans for each of Single and Double-hulled Supertankers

FOREWORD

Doctor Solly's love affair with the supertanker shines from every page of this majestic volume. He has produced a tour d'horizon which is also a tour de force. Aimed at a diverse readership "Supertankers" traces a logical path, in easily-understood anecdotal style, through the history of these great ships, their development, design, operation, manning and the economies of scale they represent.

The term "supertanker" has been used, to my knowledge, since the 1950's to describe classes of oil tanker starting at twenty-eight thousand deadweight tons and rising progressively to the five hundred thousand-ton behemoths which figure in this book. The history of their development, over four decades, embraces, technical innovation, tragic accident, human enterprise and human frailty, economic pressure, geopolitics and a vastly altered world scene.

Quite apart from the physical growth in the size of crude carrying oil tankers there have been enormous advances in the detail, the techniques and the human involvement, all superbly handled by Raymond Solly. Going back to near the beginning, who now remembers the great "bridge aft" debate of forty years ago which changed forever the silhouette of all tankers — and made them safer by getting the living quarters off the top of the cargo compartments? Before that, modular construction under cover and improved welding techniques meant bigger ships could be built faster, better and cheaper. But then, once the bigger and better ships were in service, there were the disasters and near-disasters which occurred to ships like the MACTRA (subject of what was claimed to be the longest inquiry since the TITANIC), the MARPESSA and the KONG HAAKON, all within a short time of each other. These incidents alerted the industry to the dangers of static generation when washing gas-filled tanks with sea water and one important result was adoption of the now universal application of inert gas blanketting of cargo tanks, although the technique had been invented in the United States by the Sun Oil Company before the Second World War.

Notorious navigational errors typified by the TORREY CANYON, the AMOCO CADIZ and too many others, including the EXXON VALDEZ, resulted, inevitably, in deafening condemnation from politicians, press and public. The industry was pilloried for the sins of the few and "supertanker" almost became a term of opprobrium. The TORREY CANYON incident was the seminal event and led to fundamental examinations into the way ships were designed and operated. One outcome was the creation of the Oil Companies' International Marine Forum and of two anti-pollution insurance funds later to be superseded by more powerful international agreements set up under the auspices of IMO (the International Maritime Organization).

The Arab-Israeli War of 1967 closed the Suez Canal as had the earlier war in 1956. Middle East crude had to be moved round the Cape of Good Hope emphasising the benefits of scale already apparent to those operating larger ships. The oil crisis of 1973, when the Gulf states wrested complete control over Middle East oil and the price rose three-fold, led to a world recession, a massive drop in demand for oil and a surplus of one hundred million deadweight tons of VLCC shipping by 1975. So many ships had been ordered by optimistic opportunists in the happy years of high freight rates, cheap money and the increase in the number of shipyards worldwide able to build VLCCs quickly and at low cost.

Supertankers – Anatomy and Operation

Over the years the ships have become so much bigger but, equally important, techniques are greatly improved. Cathodic protection and later coating of cargo tanks, advances in mechanical tank-cleaning equipment, the retention on board of water-washing slops and subsequent load-on-top methods, introduction of crude oil washing of cargo tanks, all became possible when the inert gas system provided a safe atmosphere. And, of course, there have been major changes in structural design, firstly by the inclusion of segregated ballast tanks and subsequently, the introduction of double hull construction. Certainly, four decades of the supertanker era have their own complex and fascinating history.

Much of that part of this book dealing with navigation, watch-keeping, ship-handling and on-board routines applies to all types of ship, and is none the worse for that. As a small aside I was pleased to see, in Chapter 10, some excellent diagrams of stopping and turning distances for VLCC's which I remember my former colleagues producing twenty-five years ago, putting realistic figures to some of the horrendous allegations of the time. "They have no brakes! They take seventeen miles to stop!" The author also deals honestly with the social side of seafaring and with training. I endorse his comments on the serious gap in professional recruitment and training which occurred in major maritime countries and particularly the United Kingdom in the 1980's, the consequences of which will be resolved neither quickly nor easily.

I am sure that non-seagoing readers of these pages will learn a great deal more from them than they bargained for. Those of us who are, or have been, seagoing participants will be able to identify with much of the author's thinking. The fascination that such great ships have for him is understandable, his own perception all-embracing. Seamen sometimes claim that all ships are the same size after two days. I have some sympathy with this contention — except in the case of the supertanker. After years of experience I am still acutely aware of the sheer scale of the ship, the solidity of the deck underfoot, the distance from one end to the other, whether arriving over the gangway or landing by helicopter and that awareness never leaves me. (I am reminded of the American lady, travelling in the old QUEEN MARY, who asked, "What time does this place get to Southampton?")

Raymond Solly's knowledge of his profession, and particularly that part of it covered in these six hundred pages, is encyclopaedic and he shares it with his readers lucidly and effectively. This book is a meticulously researched and clearly expressed academic work of major importance to the maritime community.

G.A.B. King

PREFACE

(General comments on Supertankers and their Refineries – Some definitions – A brief historical look into tonnage developments – Comparative deck officer rank styles – Chief Engineer's status – An "Incident" at Oxford University – Outline of aims, intentions and approaches – Intended levels and status of readership).

It is perfectly true to say that not everyone has an interest in ships but, equally, the evidence accumulated over twenty-five years indicates that a surprising number of people do have more than a passing curiosity about those particular ships which come into the Supertanker category. The interest appears to be concerned as much with the vessels' clear-cut lines as with the more obvious factor regarding their sheer size. The opportunity of seeing Supertankers at close quarters, with the chance to examine at leisure their massive solidity does not occur very often. This is because the refineries where they berth are invariably in remote parts of the country, surrounded by rigid security measures, whilst their anchorage's are often far from land. The Supertanker is more frequently observed when it is under way. In this condition the craft shows a demeanour which makes it appear ponderous, and yet graceful, with an inevitability to its progress that seems unstoppable. (Photos. 1, 2 and 3). Of course, such observations are undoubtedly romantic, as indeed are some of the views held about the vessels and their operation but, notwithstanding this, once a Supertanker is seen there can be little doubt of its identity. In south-east England, at Folkestone or the Dover cliffs for instance, the centre of the outward bound shipping lanes is only six or so miles off-shore, whilst on a day of good visibility, it is frequently possible to see fully-laden ships inward bound off the French coast some twenty miles distance. It will be a very rare occasion indeed if a Supertanker is not seen on either passage and I have overheard comments, that have been uttered almost in wonder, as even the most land-bound of people have followed-down the progress of an outward-bound Supertanker in the near shipping lanes.

Supertankers are engaged in the carrying only of crude oil, a cargo that comprises around 80% of all oil transported by sea. A Shipping Company, in the 1980's, did try an experiment carrying in tanks, 'crude cargo', with a part parcel of 'gas oil' in other tanks, for one voyage. The results were not sufficiently encouraging, so the practice was discontinued and it has not since been attempted. A homogeneous carriage of crude is generally considered the most practical arrangement and, certainly, it is a more straightforward method for the officers engaged on the operations of loading and discharging. 'Crude' is oil in its most basic form as it comes out of the ground so that, before it can be used for anything else, carriage to a refinery for processing into a product is necessary. The next stage from crude is fuel oil, which is used for ships' bunkers to run virtually all commercial power driven craft. Later stages of processing include diesel, paraffin, petrol and the other commonly known refined fuels. The carrying of crude is a very complex business. It is more than merely filling up the vessel until the tanks are full. A colleague on a dry-cargo ship once remarked that a cargo plan for a Supertanker would consist of only an outline of the vessel with the words "oil-oil-oil-oil-oil- …" in the tank spaces, instead of the complexities of the loading plan which used to beset the officers aboard many dry-cargo ships. He was (I trusted) being humorous, but I did wonder … .

Supertankers – Anatomy and Operation

1. VLCC "RANIA CHANDRIS".
(By kind permission of SKYFOTOS.)
"Supertankers under way show a demeanour which make them ...

2. VLCC "LATIA".
(By kind permission of Shell International Trading and Shipping).
... seem ponderous, yet graceful, ...

3. VLCC "RANIA CHANDRIS".
(By kind permission of SKYFOTOS.)
... with an inevitability to their progress that seems unstoppable.

There can be no doubt that environmental protection is very much a major policy of all reputable shipowners, their captains and officers. They all share genuine concern for hull integrity and the practices of safe navigation comforming to the legal requirements of IMO's SOLAS 1974/78 Regulations. It is lamentable therefore that the few accidents involving these ships over past decades, with such sadly appalling effects on the environment, has directed adverse attention into the minds of a wider general non-seafaring public. This publicity is understandable, but it is also unfair for such serious catastrophes are the exception rather than the norm, bearing in mind that the movements of around 450 Supertankers on any single day are recorded by Lloyd's of London, each following its lawful trade in complete safety, and certainly without major incident. Invariably, there exists also a media 'hype' in reporting, such that even tankers of less capacity than the smallest VLCC involved in any kind of oil spillage, are pounced upon and immediately referred to as 'Supertankers'. The grounding of the 92,000 sdwt "BRAER" in Scotland and the 130,000 sdwt "SEA EMPRESS" in Milford Haven are two noticeable examples in this respect.

Perhaps it is an appropriate place here to define a few terms. There was never a revelation such that early one morning the maritime industry awoke to find that a new breed of ship, like the dry-cargo container vessels or specialised gas carriers, had appeared overnight. The development of the Supertanker was a natural progression and the word is generic because it has been used to describe all tankers as they increased in size over the years. As an innocent and 'starry eyed' sixteen year old in 1957, about to start pre-sea training in a nautical college, I recall the Shell Group launching their new "Z" class tankers "ZENATIA", (Photo. 4), and "ZATHON", each of some 38,000 tons summer deadweight, (sdwt), with a draught of 10.9 metres enabling them to make a fully laden transit of the Suez Canal. The magnificent lines and size of these tankers made a lasting impression probably because they were my first clear impressions of a very large ship even though, as much as four years earlier, the Onassis Group had launched their 45,437 sdwt "TINA ONASSIS". (Photo. 5). Just a couple of years later, in 1959, Universe Tankships' mt "UNIVERSE APOLLO" (Photo. 6), at 106,416 sdwt, broke new tonnage records. It was from this date that even larger

4. The mt "ZENATIA" launched in 1957. The ship that started the author's life-long interest in tankers. (By kind permission of World Ship Society Library).

tankers competed for the world's similarly increasing demand for oil. The first 300,000 tonner was introduced in 1968, again by Universe Tankships of America, with their 326,585 sdwt "UNIVERSE IRELAND". (Photo. 7). I caught first sighting of the "UNIVERSE APOLLO" in 1960, whilst serving on a dry cargo ship, when our tracks crossed in Capetown. I can still see "in my mind's eye" not only the impressive dimensions, but more particularly the lines of the bridge structure. I decided then that, at some stage in my seagoing career, I should like to serve as a deck officer aboard such a ship. The ULCC's "GLOBTIK LONDON" and "GLOBTIK TOKYO", (Photo. 8), were built in 1973, each over 476,000 sdwt and, although plans were devised for a one million tonned Supertanker, the largest ship ever built is the ULCC "JAHRE VIKING" her current name, (Photo. 9), which is a theoretical 564,763 sdwt and continues in regular trading. Following an enlargement to her hull, this magnificent tanker is now around 460 metres in length overall with a beam of 69 metres. The largest tanker which I can recall seeing, and identifying, is the ULCC "BURMAH ENDEAVOUR", (Photo. 10), taken when she was laid up in Southampton Docks whilst I was on the way to Warsash Maritime Centre to take an ARPA Radar refresher course. Of 457,841 sdwt this Supertanker was built in 1977 and continues trading as the ULCC "STENA QUEEN", owned by Concordia Shipping Company.

It was the London tanker freight brokers who also in 1968, working through their Average Freight Rate Assessment (AFRA) panel which acts for the chartering of tankers, divided into categories by tonnage the various sizes of tankers. It was this panel who later, in January 1974, introduced the term "Very Large Crude Carrier" (VLCC) for those ships between 160,000 and 320,000 summer deadweight tons. In May/June of the same year, they devised the expression 'Ultra Large Crude Carrier' (ULCC) for tankers in excess of 320,000 sdwt.

By far the most popular question asked by professionals working ashore is quite specific and pointed: 'What exactly does a navigating officer do aboard a Supertanker at sea?' The implication in the mind of the questioner seems to pre-suppose that the officer's time is spent between "rather vague and ill-defined duties on the bridge" and afternoons sun-bathing languidly beside the pool. To an extent on occasions, of course, both functions are true. It is rather a case of what happens in between pool and bridge, and the nature and extent of those duties performed in and out of the wheelhouse. In practice today, although it has not always been the case, all officers duties are laid down in a job specification which varies from company to company, and even ship to ship, largely at the discretion of the captain.

5. At 236m in length and 29m in beam, the mt "TINA ONASSIS" (1953), demonstrates the increasing size of tankers.
(By kind permission of SKYFOTOS).

There exists also some confusion concerning why he is simultaneously a 'mate' or a 'deck officer' and even 'a navigating officer', a question which is soon answered because the reasons are largely traditional. In formal Merchant Navy administration of the maritime based civil service kind which governs the issue of 'Certificates of Competency', ship's deck officers have always been called Master and Mates and it is very much an honorary title for them to be known as captain and officers. These terms originate from sailing ship days when, after the commissioned captain, the sailing master and his mate were responsible for all aspects of navigation aboard a Royal Naval ship. About 30% of a mate's duties on a Supertanker are concerned with deck and cargo work, hence the 'deck' contribution has developed into a title. The remainder of his time is spent performing bridge duties, and engaged on anchor watch or keeping the ship to its course-line, during which occasions he is termed a 'navigating officer'. All of these references, however, have become acceptably interchangeable and, to provide variety, will be used as such in this book. The title 'The Mate' is a specific rank that refers only to the chief officer who is 'On articles' aboard any ship and is in charge of the deck department. His next step on the promotion ladder would be to Master, and it is he, or she, who is the person that would normally be directed by the company to assume command in any event which led to an incapacity of his serving captain. All future references here to the 'chief officer', will refer only to that specific officer, 'The Mate'.

6. mt. "UNIVERSE APOLLO" broke the 100,000 ton barrier when she was launched in 1959, and was one of the last large tankers to be built with a centre castle.
(By kind permission of World Ship Society Library.)

There is an additional distinction that might be worthy of clarification. The chief engineering officer, whilst nominally of equivalent rank to the captain, has total control of the department governing all aspects of a ship's engine and associated machinery. Under present manning structures, however, he lacks 'on line' bridge/deck experience and so, in the event that the captain was removed from the ship for any reason (i.e.: illness or death), the chief engineering officer would not be directed to command. This situation is likely to alter only when one of the comparatively new dual certificated deck/engineer officers attains chief engineer officer rank, a situation, that although certainly conceivable, lies in the future.

Discussions with adults, and verbal engagements with the uncomplicated and enthusiastically enquiring minds of Naval cadets, or young people at Fifth and Sixth Form levels, raise just about every conceivable maritime issue. Usually the next line of popular questions, after dimensions, concerns ship handling and construction characteristics. Factors affecting stopping distances of very large ships, with comments citing the oft-quoted, but invariably misquoted, ranges that cover any distance between five and twenty miles, are frequently posed. Enquiries regarding cargo come a ready fourth with more than one view being expressed that VLCC's are full to the main deck with petroleum or even chemical cargoes. This creates, not surprisingly, an intimation of unease about the true function of Supertankers and there is conveyed an accompanying, almost sub-conscious, quest for re-assurance involving safety factors concerning human life and environmental protection.

Perhaps, however, the most bizarre conversation ever experienced came with a university don, who really should have known better, during a formal dinner at a *very* traditional Oxford college. Having made extensive enquiries on the subject, he asked whether there was always sufficient wind to enable a Supertanker to sail. I have on many occasions in retrospect made an equally serious effort to comprehend how the image of a Supertanker fitted out with a complete set of sails, was built up in the mind of this redoubtable don as he assimilated my replies to his pressing questions. Possibly, he had been reading experiments that had occurred with small dry cargo ships some years previously but, having explained to him the dimensions of Supertankers, I remain unconvinced and feel that so far, I still have not found the answer ....

The 'Oxford incident', as I have come to label it, reinforced in my mind, however, that there is perhaps a place in the market for a book which would appeal to a range of professional and academic people who show interest in ships of this class. A work that would almost, as it were, anchor the Supertanker for a while in order to find answers to a range of questions at a

7. The ULCC "UNIVERSE IRELAND" (326,585 sdwt) was the first ultra large crude carrier in the world when she entered service in 1968.
(By kind permission of A. Duncan).

number of different levels. I want to use the 'early' ships, that were built in the years prior to the double hulled Supertankers of the 1990's, as a norm from which to launch investigations into aspects specifically of a deck officer's interests. In Part One, some introductory comments will be made showing comparisons between these very large ships and a range of everyday familiar objects, together with details of the ports visited by VLCC's during one day in 2000. Tracks are shown also of some typical voyages of VLCC's undertaken during that day. This part will include also a brief examination of the regulations covering tonnage measurement and load lines, some knowledge of which is essential in order to understand the implications contained in subsequent chapters.

The Second Part of the book opens with two chapters, one on ship construction, and the other explaining steps taken ashore and at sea to protect the ship from various corrosive effects of cargo and elements. The level of knowledge upon which these two chapters are based, together with Chapter 14 in Part Four concerning the construction of double hulled Supertankers, is that required for various statutory deck officer's examinations. This is supplemented with additional material and expertise provided by two major shipowners and a leading provider of protective coatings to the VLCC industry. Subsequent chapters in this Part discuss much of the gear found on the main deck, including docking and mooring machinery and wires, various pumping and piping arrangements, the non-domestic equipment found in the superstructure/accommodation block, including the radars and other devices necessary in handling and navigating the ship. An elementary insight is given also into the complex principles, which affect specifically this particular class of ship, especially problems of hull and tank stresses for various conditions of ballasting and loading, when the VLCC is both alongside as well as under way.

Part Three examines duties undertaken by deck officers and shows the manner in which the equipment presented during Part Two is used in shipboard routine duties undertaken during the voyage including, sea trials, emergency drills and pre-handing over inspections. The payload, which is indeed the very reason for the existence of a VLCC, is the cargo, and so pre-loading, care and discharging operations are examined, based on actual voyages, as well as treatment of the procedures in tank cleaning, including "Crude Oil Washing" (COW), and Inert Gas System (IGS) practices. The question of how a navigating officer manages his bridge watches will be answered by considering the equipment found in a wheelhouse and demonstrating methods of deep sea and coastal position fixing techniques. This chapter will offer some insights concerning the way in which a VLCC might be handled practically under different everyday circumstances and conditions. It is in this section of the book that I include an insight into the social life on board showing the standards of accommodation and life styles, of officers and crew, with ideas explaining how off-duty time can be spent. Owing to the rigorous measures currently undertaken concerning improved schemes for deck cadet training, together with the considerable repercussions for shipping and allied industries regarding the employment and subsequent availability ashore of ex-seafarers, a few thoughts are included on aspects of both of these important areas of manning. In Part Four, apart from an insight into the construction of double hulled Supertankers, an examination is made into the application of computers to navigation, cargo control, and other areas where computerisation has become involved at sea as these affect the work of a deck officer. The book concludes with some thoughts on aspects of the double hulled construction, and reflections concerning the way in which a navigating officer's role is likely to be affected by future developments in VLCC technology and operational techniques.

Although I genuinely regret the omission from the "anatomy" part of my title, no apology can be offered for omitting all references to the engine room of a VLCC, its equipment and workings. This is a key department aboard any ship, let alone a Supertanker, and I am simply not qualified to understand or even comment on the domain of the engineering officer. Any attempt to do so would lead to misrepresentation. I write merely from the viewpoint of an ex seagoing navigating officer whose wish is to provide a *basic* technical interest book that addresses non-seafaring laity who may well be shore based professionals, and those captains of industry without

*8. ULCC "GLOBTIK TOKYO". Built in 1973 at 476,000 sdwt.
(By kind permission of Globtik Shipping Company.)*

whom there would be no Supertankers. People whose talents and skills include areas such as management, law, finance, design and construction, the various branches of engineering, chartering and broking, insurance, equipping and, indeed, the provisioning of ships, who wish to expand their knowledge by finding out about the more menial tasks of the navigation and deck work practised on the VLCC's with whom they are involved. This, of course, presents me as an author with a serious problem. The potential diversity of readership envisaged is *highly* professional. Each reader is an expert within his respective sphere who possesses a far greater and indeed deeper range of knowledge than I. Qualifications held, for instance, may easily include a sophisticated degree of understanding, including mathematics and science which I, as a mere deck officer, could neither possess nor be expected to possess. In my research, to give a good example, I have found virtually incomprehensible some findings that are reported in technical manuals and reports because they are, by definition, packed with highly complex details and often intricate formulae.

My readers will come to the book with completely different needs and levels of experience involving VLCC's. The requirements of the lawyer could well seek knowledge contained within the entire book, but his needs will certainly differ considerably from those of say the naval architect. On some chapters, the latter will know more than myself about Supertanker construction and stresses, but perhaps very little about cargo operations or bridge watchkeeping in port or under sea going conditions. The oceanographer may well be directly interested in the section on ship stresses whilst under way or alongside, but have only limited knowledge of the procedures involved in actually handling a VLCC in any sort of weather, in ballast or loaded conditions. I do not require my reader to possess a professional maritime qualification in order to understand the problems and life-style of a mate working a very large tanker, or to appreciate the equipment found on board. I would hope that our common bonding is his desire to learn more about the interests of a fellow professional, albeit a seafarer. In most cases therefore, for reasons of continuity, the book is intended to be read progressively, from the Preface to the Conclusion, using the starting point of mutual involvement with the workings of a splendid ship.

Clearly, these factors are bound also to affect both the approach and, indeed, the technical level at which to pitch the material I have used, the extent to which factual information is best omitted, and the detail into which I might probe with the facts that I do include in order to explain adequately my subject. Inevitably, a compromise results. Following extensive thought about the former, I decided that it would be best to regard the world of the Supertanker from three major areas. Initially, I would consider the 'early' generation of VL and ULCC's. This presented an area of concern regarding what I term 'negative regression' or, in the light of the generic nature of the term Supertanker, the extent to which I should look backwards. The most expedient method would be to regard my "first generation" as dating from the initial 160,000 sdwt plus tonnage, built in the 1960's, using the internationally accepted AFRA definitions, and to include within this category all Supertankers constructed up to the 1980's, after the 1970's oil crisis led to a decline in demand for, what had become by then, a considerably large number of VLCC's worldwide. Even this kind of categorisation encompasses an additional difficulty because my loosely applied thirty year time span includes ships known technically as "pre and post MARPOL tankers". In 1973, with major revisions five years later, regulations came into force that were developed by the then International Maritime Consultative Organisation – IMCO – (See Appendix), and accepted world wide, which introduced a considerable range of far reaching safety, operational and design measures that aided prevention of marine pollution. Collectively and colloquially they have become known as 'MARPOL 73/78'.

9. *The ULCC "JAHRE VIKING", at 564,763 sdwt and 460 metres long is the "largest ship ever built".*
(By kind permission of Jahre Wallem Shipping.)

From the technical aspect, I feel that I should very much like to "keep things simple" as much as possible: to draft the information offered in a factual semi-anecdotal way which says what it means and means what it says. To write 'in a seamanlike manner', hopefully without unexplained jargon or blinding the reader with too much science.

Officers serving ashore in supervisory positions, and even in senior ranks aboard tankers of all cargo capacities, may well find interest in that which is written. They may not always agree with my viewpoint, however, because so many practices vary according to personal style and the total uniqueness of individual experience. This leads me to stress that this book is most certainly *not* intended primarily for my colleagues who have served, or remain serving, as navigating officers on VLCC's. Neither is it considered my remit to produce a comprehensive, exhaustive technical manual on the Supertanker for fellow navigating officers, wherever they may be currently serving. There are already a number of admirable works on the market which provide much of the theory necessary to assist a deck officer in the shipboard practice of his craft. Although, in common with most professions, of course, the reading of any amount of books can quite categorically never prove a substitute for practical experience, in this case, actually working a Supertanker at sea.

Non-mariners and other interested people ashore however who have dealings with VLCC's, but have never actually served at sea would not, I hope, be the only level of readership. I have a deep concern with training, both academic and professional, and have taken the opportunity to examine the various vocational schemes available to deck cadets. It is my hope, therefore, that perhaps serving and pre-sea deck/engineering dual certificate cadets in nautical colleges, and even junior officers newly appointed to Supertankers, may find in these pages *some* answers to their possible questions on the mate's side of life on VLCC's. It might even be that merchant navy officers serving aboard dry-cargo vessels may, as a result of their reading, find individual professionalism challenged sufficiently to seek employment on a class of ship which has provided me with a livelihood, and a range of challenging and varied responsibilities, as well as a highly sustained interest.

Raymond Solly
May 2001

10. The ULCC "BURMAH ENDEAVOUR" is the largest Supertanker identified by the Author. (By kind permission of Revd. Brian Coward.)

ACKNOWLEDGEMENTS

I owe a considerable debt to a number of Organisations and also to numerous extremely busy people who have been prepared to offer their time and expertise freely to supply materials and read certain draft chapters of this book. Without their interest and ready assistance my book would quite simply not have been possible. I extend to each of these my **genuine and very grateful** thanks. Worthy of particular mention are the following Companies and senior personnel:

Captain John Enston, the Managing Director of **Mobil Ship Management (UK)**, who allowed me initially to discuss my ideas with him and Captain Peter Webb, hence making available Steve Walker, an engineering superintendent, who kindly hosted me on numerous occasions in the London office and gave free access to plans, drawings and photographs. I am grateful for the friendship extended by many of his colleagues, including: Captain Colin Graham, who cast an eye over much of the draft text in Parts One, Two and Three and hosted me aboard his VLCC. Captain Ian McKenzie, and Ewan Muldane for the loan of personal photographs. Captain "Bas" Appleby, Loading Master, for considerable encouragement and advice. I am indebted to Superintendent Rob Drysdale for his time, patience and professional advice with the chapters on Construction and Ship Stresses in Part Two, and the chapter in Part Five.

Captain D.R. Salmon, Senior Nautical Superintendent (Fleet Operations) and Jim Cusiter, Head of Oil Tanker Construction in **Shell International Trading and Shipping Ltd** who, through the courtesy of their staffs, made available photographs, manuals, plans and diagrams of Single and Double-Hulled VLCC's in the Shell Trading fleet. This Company was my first contact and I retain memories of considerable gratitude for their kind hospitality and encouragement during the crucial initial research stages.

Gerry Vagliano, Director of **Lykiardopulo & Company**, whose offer of ready assistance, interest — and indeed kindness — remain very much appreciated. To David Manning, who was *extremely* positive regarding access and loan from his personal collection of negatives. These were taken whilst he supervised all stages in the construction of the VLCC "AROSA". These gentlemen read much of Chapter Fourteen, offered constructive advice and suggestions regarding amendments.

Captain Mike Graham of **BP Shipping** made available plans and diagrams of ships in their fleet of VLCC's and allowed me free access to their photographic resources library. I am indebted to the various custodians for their patience and assistance (as well as an endless supply of coffee) during the occasions of my visits to the photographic archives.

Roger Green, an ex-supertanker deck cadet and chief officer, now Operations Manager with a London based shipping company, for correcting drafts of the early chapters. David Penny, a senior engineering superintendent with the same company for providing an excellent paper specifically for this book, from which I have been grateful to quote, covering his experience with the construction and running of both single and double-hulled tankers.

I am especially grateful to the following Companies and personnel who readily made available material, offered valuable technical assistance and worked with me either to correct — or approve — draft copy:

John Lewis and Linda Diamond of *Jotun-Henry Clarke (Coatings) Ltd* on Chapter Three.
Alan Marshall, Sales Director of *Victor Pyrate Ltd* on Chapter Eight.
Bill Oswell of *FGI Systems Ltd* also on Chapter Eight.
Jim Gray of *Clyde Marine Training* on Chapter Ten.
Principal Surveyors Messrs. Tony Butt and Alan Gavin of *Lloyd's Register of Shipping* on a number of chapters.

To the following lady and gentlemen for often considerable assistance with proof reading and corrections on various sections within Part Four:

Brian Deer, Managing Director of *Saab Marine (UK) Ltd*, and a number of his colleagues in technical departments in Gothenburg.
Michael Noorlander (and colleagues abroad, including Goran Lindh and Leif Brafelt) in *Consilium Marine (UK)*.
Messrs. David Wilson and Rob Prentice, Senior Marketing Managers with *Litton Marine Systems Ltd*.
Mark Slavin of *Oceanroutes (UK) Ltd*.
Mike Robbins, Managing Director and John Hardcastle, Sales Director of *Transas Marine (UK) Ltd*.
Captain David Croft & colleagues of *Transas Marine (UK) Ltd*.
John Dawson, Manager in the *UK Hydrographic Office*.
Mike Farrow, Senior Marine Sales Engineer, with *Strainstall Engineering Services, Ltd*.
Captain Thomas Grieves, Sales Director, *Dasic Marine Ltd*.
Kevin Taylor, Systems Manager with *Whessoe-Varec Ltd*.
Malcolm Boyd, Managing Director of *Kockum Sonics (UK) Ltd*.
Keith Shaw, Sales Director, and Sara Trimbee from *Sanderson CBT Ltd*.
Captain Geoff Cowap, Director and Marine Consultant of *Energy Marine (International) Company* for providing a technical paper on computerised loading.
Brian Mullan, Maritime and Safety Manager, *INMARSAT plc*, — together with colleagues who kindly made available copies of their professional papers: Gavin Trevitt, Philip van Bergen, George Kinal, Derek Tam, Dale Irish, Chris Wortham and Ian Thomas.
Captain Michael Ridley, the Fleet Director of *P&O Short Sea Management Ltd*, for allowing me passage on the *mv "PRIDE OF BURGUNDY"* and to the Master, Captain Philip Wray, and his officers aboard that ship for their willing help.
Mr. Paul Taylor, an ex deck cadet and second officer with BP Shipping, for providing most of the excellent line drawings that enhance the information offered in a number of chapters.

I acknowledge also assistance, advice or material from:

Lars Sundberg, Sales/Marketing Manager, *AxTrade Inc.* Kenneth Gibbons of *The British Marine Equipment Council*. Professor John Kemp, Captain Paul Young, and the publishers of their Kemp and Young Series of books, *Butterworth-Heinemann*. Revd. Brian Coward. Cliff Tyler, *H. Clarkson and Company*. T. Dargan, Sales/Project Engineer of *Clark-Chapman Marine Ltd*. Inger Stalblad of *Concordia Maritime AB*. Douglas Lang, Director, *Denholm Ship Management (UK) Ltd*. Kari Tofthagen and Terje Staalstrom from *Det Norske Veritas*. Mr. Graham Douglas, serving Chief Officer with *Mobil Ship Management*. *Draeger Company Ltd*. Mike White, Sales Engineer, *Ferguson and Timpson Ltd*. Captain Peter Adams, and John Maddock — Clerk to the Company, *The Honourable Company of Master Mariners*. Captain G.A.B. King, CBE. Roger Kohn, Head of the Information Office, *International Maritime Organization (IMO)*.

Acknowledgements

Simon Bennett, External Relations Adviser, *The International Shipping Federation*, Barbara Viken, Advertising/Sales Manager, *INTERTANKO of Norway*. Jim Smith, Permanent Secretary/Representative to IMO, and David Buck, Data Manager, *International Association of Classification Societies, (IACS)*. Simon Bennett, *International Chamber of Shipping (ICS)*. Barry Cullum of *International Marketing*. Deborah Ansell, Librarian, *The International Tanker Owners Pollution Federation Ltd (ITOPF)*. Captain T.H. Pettersen, *Jahre-Wallem Management AS*. Kardorama of Potter's Bar, Ian Cochran of *Lloyd's List* and London Correspondent of *"Tradewinds"*. Captain Jeremy Howard, Director, *The Marine Society*, The Marine Accident Investigation Board and the Marine Safety Agency of *The Department of Transport Marine Division*. Graham Scales, Marine Sales Manager, *Marlow Ropes Ltd*. Brian Beeston of *MMC (Europe) Ltd*. Paul Sorensen, *Maersk Data AB*. Paul Doughty and Colin Sowman, *The 'Motor Ship' Magazine*. Julian Parker, Secretary, *The Nautical Institute, London*. Captain Graham Hicks of *The National Union of Marine, Aviation and Shipping Transport Officers (NUMAST)*. Dorthe Borke, Marketing Manager, *Odense Lindo Steel Shipyard Ltd*. Dick Oldham and Mr. J.W. Hughes, Directors, *Oil Companies International Marine Forum (OCIMF)*. Alan Pearson, Managing Director of *Ords Ltd*. Dave Wilson, PR Manager, *P&O Ferries Ltd*. Herr. Joachim Pine. Ed. Chandler, Sales and Marketing Manager, *Prosser Electronics Ltd*. August Kjerland Jnr, Manager Marine Division, *Maritime Pusnes AS Norway*. Julie Tranter, Marketing Department, *Racal-Decca Marine*. *The Royal Institute of Navigation, London*. Nils Olsson, Quality Assurance Manager, *Scan Skarpenord AS, Norway*. Eric Hammond, UK Sales Manager and Finn Haugan, Export Manager, *ScanRope of Norway*. Nigel Rose, Technical Services Manager, *Sheen Instruments Ltd*. James L. Shaw of *Milwaukie, USA*. A.S. Lewis, Engineering Assistant, *Stagecoach South Coast Buses*. Dr. G. Patience and Mr. J.A. Clayton of *Stone Manganese Marine Ltd*. A.D. Hewitt, Head of Design, *Swan Hunter Shipbuilding Group*. Gary Welch, Technical Sales Executive, *Unitor Ship Services Ltd*. Captain K. Coyne with *Universe Tankships (Delaware) Inc*. Captain Bernard Gardner and Dr. S.J. Pettit of *University of Wales, Cardiff. The Maritime Centre at Warsash*. Herald Tolpinrud, Project Marine Sales Manager, *Westad Industri AS, Norway*. Mr. Tony Blackler of *World Ship Society's* magazine, "Marine News". Mr. R.A. "Tony" Smith, Custodian of *World Ship Society* Photographic Library. Claire Lloyd-Hickey, Marketing Services Manager, *Zellweger Analytics Ltd for Netronics Limited*.

Publishers Comment:

It could be said that the publication of this book is one of the books, "I always wanted to do" and after over 25 years of Marine Publishing, represents the zenith of my publishing career!? However, publishers never retire and at a publication rate of 20-25 books a year I am sure there are many books still to come.

When Ray Solly walked 'through the door' with his manuscript I immediately said "yes" to its publication. It has taken us both a long time to 'knock it into shape' but here at last it is! It has been a costly and time consuming 'labour of love'. I hope the Marine Industry at large, buys it, it is after all an industry book, for all those throughout the world who know and dare I say, love the business of shipping and the transportation of oil.

It was in 1956 that Captain Gabby King wrote his book *Tanker Practice* remembered to this day as being the authoritative work on the subject. His enthusiastic Foreword speaks volumes for Raymond Solly's work and we are most grateful to him for 'running the rule' over this publication.

Enjoy this absorbing book, it is for you.

Alan Witherby
May 2001

GLOSSARY

ADG	Adaptive Digital Gyropilot
AFRA	Average Freight Rate Assessment Panel representing London freight brokers
After-Peak Tank	Situated below Steering Gear and Deck Store and used for sea-water ballast
ARCS	Admiralty Raster Chart System
ARPA	Automatic Radar Plotting Aid
Back Pressure	Resistance created in oil flow through cargo pipes on VLCC
Ballast Line	Used to carry sea-water for ballast completely separate from cargo lines
Beaufort Scale	Classification system determining sea states and wind speeds
Blind Sectors	Areas not readily observable from Navigating Bridge for lookout purposes
Bosun	Senior deck rating with European crews
Bridge Box	Sandwiches and tea/coffee drinks container placed by stewards in Wheelhouse
Butterworth Store	Situated on Main-deck housing tank cleaning equipment and deck stores
CALM	Computer Aided Level Measurement system used in tanker cargo control
Cathodic Protection	Use of Cathodes/Anodes protecting hull/tanks against corrosion
Catwalk	Safety passage ways from Accommodation to Foc'sle Head
Cement Boxes	Temporary Reinforcement to minor plating cracks. Also Sealing Anchor Spurling Pipe in small ships
Chafing	Longitudinal stress point where rough seas create movement between cargo pipe and cushioning
Chain Locker	Housing for Ship's Anchor cable leading from Windlass to Spurling Pipe
Chiksans	Cargo loading/discharging cranes situated on oil jetty
Cofferdam	Empty spaces separating cargo tanks from engine-room aft and fore-part of tanker
Compressive Stresses	Forces of Sea on Ship's Sides and Bottom leading to hull distortion
Crude Oil Washing. COW	System using crude oil from cargo for tank cleaning
Deck Officer	Navigating officer aboard ship with responsibilities also for deck and cargo work
Derricks	Supports for cargo pipes during loading/discharging
Double Hulled	A VLCC with double bottom and sides
Drag Chain	Used to arrest Momentum of ship during launching
Dynamic Stresses	Forces superimposed on Static Stresses whilst ship is under way
ECDIS	Electronic Chart Display and Information System
Elephant's Foot	Colloquial name for the "Bell Mouthed" shape Stripping Suction Pump
ENC	Electronic Navigation Chart
ETS	Emergency Towing System legally required for all tankers above 20,000 sdwt

Fore-Peak Tank	Situated below Deck Store and Chain locker and used for sea-water ballast
Freeboard	Vertical distance of Hull from Waterline to Main deck
GMDSS	Global Maritime Distress and Safety System
Gross Registered Tonnage. grt	Moulded enclosed capacity of a ship
Handing-over Ceremony	Official passing of ship from builders to owners
Hawse Pipe	Housing place of anchor and leading of cable from Windlass through Spurling Pipe to Chain Locker
HBL	Hydrostatically Balanced Loading crude oil cargo safety system
HCMM	Honourable Company of Master Mariners
Hogging	Longitudinal Stress causing centre area of ship to be comparatively unsupported
Hydrocarbons	Toxic and highly explosive gas remains following crude oil cargo discharge
IAPH	International Association of Ports and Harbours
IACS	International Association of Classification Societies
ICS	International Chamber of Shipping
IHO	International Hydrographic Office of IMO
IMO	International Maritime Organization based in London advocating and protecting all aspects of ship safety
Inert Gas System. IGS	Use of flue uptake from from ship's exhaust gases to provide oxygen-free neutraliser to cargo tanks
INMARSAT	International Maritime Satellite Organisation
INS	Integrated Navigation System
Interaction	Forces causing tendency for a ship to pull towards/away from other ships/objects when passing too closely, and at too high a speed
ISGOTT	International Safety Guide for Oil Tankers produced by ICS, OCIMF and IAPH
ITOPF	International Tanker Owners Pollution Federation
Lightening	The operation of reducing the draft of a VLCC by discharging part-crude oil cargo into a smaller tanker
Load Lines	Hull markings of a ship indicating loading states in various conditions
Luffing Cranes	Replaced Derricks as supports for cargo pipes during loading/discharging
Manifold	Point on a tanker where the cargo is loaded/discharged from/to the jetty Chiksans
MARPOL	IMO Legislation governing the prevention of pollution from ships
Mate, The	Chief Officer in charge of Deck Department and second-in-command of ship
MIRANS	Modular Integrated Radar and Navigation System
MNI	Member of the Nautical Institute
MNTB	Merchant Navy Training Board
Monkey Island	Area above Wheelhouse housing navigating equipment and mainmast with statutory lights

Glossary

MSA	Marine Safety Agency linked with Department of Transport and Coastguard
Net Registered Tonnage. nrt	Volume of cargo space available in a ship
Note of Loading	Owner's Instructions to VLCC notifying cargo and voyage details
Notice of Readiness. NOR	Advice on eta/cargo details sent to next port. Also Advice from deck to engineroom on speed reduction prior to port arrival
NUMAST	National Union of Marine, Aviation and Shipping Transport officers
OCIMF	Oil Companies International Marine Forum
OPA90	The Oil Pollution Act introduced by the United States government in 1990
ORION	Shipboard Weather Routing Service provided by Oceanroutes of WNI (Weathernews) Group
Parallel Indexing. PI	Automatic or Manual Radar Safety Device for Coastal Navigation and Harbour Entrance
Pelorus	Portable, hand operated compass, lined up with ship's gyro-compass, and used to take bearings of objects blind to the repeaters due to obstructions (ie: funnel)
Pitching	Motion of a ship along her fore and aft axis as she heads into a moderate/rough sea
Plan Position Indicator—PPI	Radar picture appearing on Screen of Cathode Ray Tube (CRT)
Pumproom	Space (usually aft on VLCC) housing ship's cargo and ballast pumps
Rathole	Access spaces cut into horizontal tank structures
Safe Working Load. swl	Marking engraved on derricks, cranes and other lifting gear
Sagging	stress in which the ends of a ship are comparatively unsupported
SBM	Single Buoy Mooring with facilities for crude oil loading/discharging
Scantlings	Quality and thickness of steel plating used in tanker construction
Sea Trials	Tests of ship and all equipment prior to handing-over from builders to owners
Segregated Ballast Tanks. SBT	Tanks used for sea-water ballast only with own pumps and piping system
Serang	Senior Indian deck rating
Shadow Areas	Minimal spaces within cargo tanks difficult to clean completely
Shearing Forces	Stresses created on Ship's Frames leading to hull distortion at possible breaking points
Slamming	Force created when fore-part of Ballasted Ship surfaces and pounds into rough sea
Sludge	Semi-solidified residue of Crude Oil cargo deposited on frames etc within tanks
SOLAS	Safety of Life at Sea Legislation of IMO
Sounding	Measurement of water (usually) found in any tank
Springing	Vertical movement of main-deck of VLCC when loaded and in a seaway
SPS	Seakeeping Prediction System
Spurling Pipe	Leading of anchor cable from Chain Locker over Windlass to Hawse Pipe

Squat	Tendency of a VLCC to sit lower in the water when Under Way in shallows
Static Stresses	Forces acting upon Hull of Ship when stationary- loaded, unloaded or in ballast
STCW95	IMO Convention governing the Training and Certification of Watchkeeping Officers
Stripping Pump	Used in final stages of Tank Cleaning to clear residue of cargo from tanks
Swedish Corrosion Institute	International Organisation specifying standards of steel preparation prior to protective painting
Swettenham Fender	One of a series of Fenders fitted to a smaller tanker used when lightening a VLCC
Tankradar	Highly accurate and non-dangerous Tank Gauging system
TCIM	Tank Control Inventory Management system
Topping Lift Gear	Arrangement of blocks and wires for raising derricks
Topping-off	Final stages of tanker loading operation
Tramp Ship	Dry cargo ships and tankers trading irregularly between any of the world's ports
Transversal	Frame extending width (or beam) of ship
Turning Circle	Measurement of maneouvrability when navigating any class of ship
ULCC	Ultra Large Crude Oil Carrier – Tanker in excess of 320,000 sdwt
Ullage	Space between surface of oil cargo and under-deck level at top of tank's ullage point
VDR	Voyage Data Recorder
VLCC	Very Large Crude Oil Carrier-Tanker in excess of 160,000 sdwt
VMS	Voyage Management System
Whipping	Longitudinal Force created on loaded VLCC by heavy sea causing deck movement
'Wind-Chronom'	Aide Memoire to Second Officer reminding him, in olden days, to "wind the chronometer"
Work Station	Total Computerised Cargo Handling system
Zener Barrier	Safety Interface Connector between an Electronic Cargo Handling system and hazardous deck area on tanker

PART 1: Introduction

Chapter One

BACKGROUND

1.1. INTRODUCTION

1.2. THE SHEER SIZE OF A TANKER, LENGTH AND BEAM

1.3. IMO AND THE CLASSIFICATION SOCIETIES. REGULATIONS AND CONVENTIONS: GRT, NRT

1.4. CAPACITY, DWT. LOAD DISPLACEMENT AND LIGHT DISPLACEMENT TONNAGES

1.5. LOADLINES, FREEBOARD

1.6. NAME AND DRAUGHT MARKS

1.7. DECK LINE

1.8. OTHER LOAD MARKINGS

1.9. A LOADING EXAMPLE

1.10 SUPERTANKER MOVEMENTS

Chapter One

BACKGROUND

(Synopsis — Comparisons with Cross-Channel Ferries, Dry Cargo Ships and Familiar Objects — Definitions: International Maritime Organization (IMO) — Classification Societies — Tonnage Rules — Loadline Regulations — Applications of Rules to Specific Examples of a ULCC — Supertanker Trading Areas in Terms of Ports Visited — Typical Voyages of VLCC's)

1.1. INTRODUCTION

It is the ferry, with its constant hustle and bustle, which represents the ship most common both in sea-going experience and the mind's eye of most people. One of the largest ferries operating from any United Kingdom port is the P&O m.v. "PRIDE OF BILBAO" which trades from Portsmouth to France and Spain. (Photo. 1.1). This ship is 37,583 grt and is 176.8m in length overall (i.e. from tip of the stem to the furthest point of the stern, as opposed to the length between perpendiculars bp). Her extreme beam, or width, is 28.4m and she has a maximum draught of 6.2m.

In the realms of deep-sea ships, the m.v. "GOLD VARDA" continued until recently trading as a regular dry-cargo tramping vessel, owned by the Haverton Shipping Company of London, before she was sold for further service. The ship has since been scrapped. I served aboard her as a navigating officer for a voyage from Haifa to Ashdod and Antwerp between trips on a Supertanker

1.1. P&O Line's Ferry m.v. "PRIDE OF BILBAO" probably represents the ship most common to the public eye. (By kind permission of P&O Ferries, Ltd.).

1.2. The mv. "GOLD VARDA" represents a typical dry cargo ship — one of the Country's "shopping baskets".
(By kind permission of World Ship Society Library.)

1.3. VLCC "RANIA CHANDRIS" is a good example of a first generation Supertanker upon which the author served.
(By kind permission of Odense Steel Shipyard.)

during a period of extended leave. The "GOLD VARDA" (or m.v. "VARDA", as I knew her) is 146.6 metres in length overall, around 22.9m in the beam, and has a moulded depth, that is, the distance from the keel to the maindeck, of 13.7 metres. She is a 18,863 gross registered tonnage vessel which draws 8.9 metres, fully laden. (Photo. 1.2).

As an exemplar of a "first-generation" Supertanker, I want to use the Supertanker "RANIA CHANDRIS" which was built in 1973 by the Odense Steel Shipyard in Denmark. (Photo. 1.3). I choose this ship as a typical example because I joined her in the final stages of her construction, worked with the Danish officers during sea trials, witnessed the handing-over ceremony (enjoyed the celebration luncheon) and completed the maiden and second voyages. I returned after leave to continue serving aboard her for many months and came to know most aspects of her with a comforting familiarity as well, indeed, as accumulating a quite considerable collection of builder's plans, diagrams and photographs. Subsequent service was performed aboard other Supertankers.

The VLCC was 347.2 m in overall length which, to all intents and purposes, is not far short of one third of a mile in length. She had a beam of 51.9m and a moulded depth of

28.7 metres. From the bridge to the keel was 50.3m and from the keel to the top of the funnel 64.0m. She was of 286,000 deadweight tonnage, 142,055 gross registered tons and 111,025 net tons. Fully laden, at a working draft, she drew 22.2m. Her largest cargo tank was number two centre which was 63 metres long, 21.7 wide and had a capacity of slightly under 29,000 tons. This one single centre tank held not too far short of the total cargo capacity carried by the "Z" class Shell Tanker which so fired my teenage imagination. It is interesting to note that the height measurement from the keel to the funnel top, and the length of number two centre tank, are virtually the same.

1.2. THE SHEER SIZE OF A TANKER, LENGTH AND BEAM

The following table shows a variety of familiar objects, whose dimensions are, as practically as possible, compared to ships and Supertankers. (Table 1.1). The double decked bus is probably the most familiar method of transport to us all, even if we do not travel by them regularly, and the nine metre length is very much a standard measurement. The implications of these measurements are really quite profound. The VLCC was the length of almost thirty-nine double-decker buses standing bumper to bumper and, from keel to main-deck, would be equivalent to six buses, one on top of the other if this realistically can be visualised. It would require a total of fourteen buses stacked vertically to reach from keel to funnel-top, and six buses end-on to cover the beam. (Photos. 1.4, 1.5 and 1.6).

When looking at a Supertanker, assuming that she is beam-to the sightline of the observer, the eye is drawn invariably from a point at the funnel along the superstructure and then along the overall length of the ship. Comparatively speaking, if a person could imagine him or herself standing at the west entrance of Canterbury Cathedral and looking upwards to the top of the unmistakable central Bell Harry tower, then the funnel of the ship would reach almost to the height of the tower, whilst the ship would be over twice the length of the cathedral. The aerial view captures the length of the cathedral probably more effectively than the height of the tower, but offers a very realistic idea of how the VLCC "RANIA CHANDRIS" would appear were the ship to be superimposed and represented in the cathedral grounds. Alternatively, if the observer stood in Trafalgar Square, and glanced upwards at the statue of Nelson surmounting his column, then an additional twelve metres would have to be added to the height of the column in order to reach the

OBJECT	LENGTH (m)	BEAM/WIDTH (m)	DEPTH/HEIGHT (m)
DOUBLE-DECKER BUS	9.1	2.4	4.6
CANTERBURY CATHEDRAL	157.6	21.3	24.3
(BELL HARRY TOWER)	—	—	71.6
NELSON'S COLUMN	—	—	51.9
m.v. "PRIDE OF BILBAO"	76.8	28.4	6.7 (Draught)
m.v. "GOLD VARDA"	146.6	22.9	13.7
VLCC "RANIA CHANDRIS"	347.2	51.9	28.7
(KEEL TO FUNNEL TOP)	—	—	64.0
(No.2 CENTRE TANK)	63.0	21.7	28.7
ULCC "JAHRE VIKING"	458.5	68.8	29.8

TABLE 1.1. Comparative Table of Familiar Objects.
The Author

Supertankers Anatomy and Operation

1.4. *A 9.1 metre Double Decked Bus is the most familiar form of transport. (By kind permission of Richard Maryan and Stagecoach South Bus Company.)*

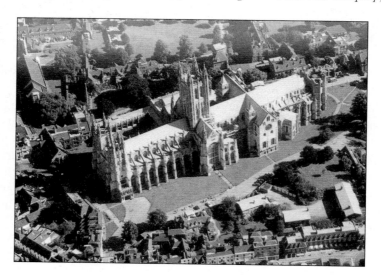

1.5. *Canterbury Cathedral in Kent remains a popular tourist attraction. (By kind permission of Bill Oates and Judge's Cards of Hastings.)*

1.6. *For a Country with a strong sea-faring tradition, Nelson's Column remains a firm land-mark. (By kind permission of Kardorama Company.)*

*1.7. The VLCC "WORLD UNICORN" nearing completion at Swan, Hunter's Shipyard.
(By kind permission of Swan, Hunter Shipyard.)*

funnel-top of the Supertanker. Were the ULCC "JAHRE VIKING" to replace the exemplar Supertanker, then the proportions would be even more spectacular: nearly three times the length of the cathedral, for instance.

Perhaps the accompanying photograph of the VLCC "WORLD UNICORN" whilst she was under construction at Swan Hunter's on the Tyne, now, alas, no longer in the business of very large ship construction, puts the Supertanker into a totally realistic perspective to her shore environment. This VLCC was built also in 1973 and was owned by the Worldwide Shipping Company of Hong Kong. At 252,000 deadweight tons she was 1/6 smaller in capacity than the "RANIA CHANDRIS", but less than half the capacity of the "JAHRE VIKING". The photograph (Photo. 1.7) was taken, of course, from the bridge accommodation looking across the maindeck, so that a considerable part of the stern, or about one third of the total length of the vessel, is excluded from this shot. Even so, the bulk of the deck with its impressive height towers dramatically over the surrounding houses, whilst the forecastle-head, nearing completion, can only just be made out in the background. Even the men working amongst the sprawl of pipes and cables seem insignificant figures against the vast expanse of deck.

The width of the objects selected make interesting comparisons with the beam of a Supertanker, but the solidity of a fully loaded VLCC under way as she approaches her berth, (Photo. 1.8), indicates the impressiveness of this over 50 metre width. The ship can be seen to draw just under 21 metres. When it is considered that the "JAHRE VIKING" exceeds the "WORLD UNICORN" by an additional seventeen metres, in the beam, and over 110 metres in length, then the accolades of the "largest ship ever built" and "the greatest moving entity devised by man" gain enhanced significance. Little more need be said by way of superlatives.

1.8. An impressive shot of a fully laden VLCC as she nears her berth at Finnart in Scotland, shows effectively the beam/width. (By kind permission of British Petroleum Shipping.)

As a matter of historic interest, the VLCC's upon which I served were sold by the company in the mid to late 1970's to a subsidiary within the group. Under various names they continued regular trading for an additional period of ten to twelve years before they were sold for scrap and eventually all three were broken-up in Taiwan.

1.3. IMO AND THE CLASSIFICATION SOCIETIES. REGULATIONS AND CONVENTIONS: GRT, NRT

When any ship is registered, international regulations necessitate that a measurement is made of the craft that is based on its tonnage. The word itself has historical origins and meant originally the number in 'tuns' of wine, each holding about 250 gallons, that could be carried as cargo by a ship. The International Maritime Organization (IMO) is the official body that regulates, from its head office in London, and in direct consultation with the International Association of Classification Societies (IACS), virtually all aspects of shipping concerned with safety. Some details of this organization are offered in the Appendix. Their 1969 International Convention on Tonnage Measurement of Ships controlled directly the classification of commercial craft. The situation regarding tonnage measurement prior to this time was very untidy, to say the least, owing to considerable variations adopted by maritime nations to the British Merchant Shipping Act of 1854 which, with Amendments, had stood previously as the criterion against which world-wide standards had been measured. As they felt it inappropriate, invariably with good reason, nations amended and altered the British Act with the result that, instead of a standard international norm prevailing, a number of widely spread amended versions existed.

The 1969 convention came into force in July 1982, but did not become mandatory for all signatories until July 1994. This apparently staggering twenty-five year delay serves only to emphasise the efforts of the Convention to tidy the tonnage measurement situation. There was an

> **Regulation 4**
>
> *Deck Line*
>
> The deck line is a horizontal line 300 millimetres (12 inches) in length and 25 millimetres (1 inch) in breadth. It shall be marked amidships on each side of the ship, and its upper edge shall normally pass through the point where the continuation outwards of the upper surface of the freeboard deck intersects the outer surface of the shell (as illustrated in Figure 1), provided that the deck line may be placed with reference to another fixed point on the ship on condition that the freeboard is correspondingly corrected. The location of the reference point and the identification of the freeboard deck shall in all cases be indicated on the International Load Line Certificate (1966).
>
> **Regulation 5**
>
> *Load Line Mark*
>
> The Load Line Mark shall consist of a ring 300 millimetres (12 inches) in outside diameter and 25 millimetres (1 inch) wide which is intersected by a horizontal line 450 millimetres (18 inches) in length and 25 millimetres (1 inch) in breadth, the upper edge of which passes through the centre of the ring. The centre of the ring shall be placed amidships and at a distance equal to the assigned summer freeboard measured vertically below the upper edge of the deck line (as illustrated in Figure 2).
>
> **Regulation 6**
>
> *Lines to be used with the Load Line Mark*
>
> (1) The lines which indicate the load line assigend in accordance with these Regulations shall be horizontal lines 230 millimetres (9 inches) in length and 25 millimetres (1 inch) in breadth which extend forward of, unless expressly provided otherwise, and at right angles to, a vertical line 25 millimetres (1 inch) in breadth marked at a distance 540 millimetres (21 inches) forward of the centre of the ring (as illustrated in Figure 2).
>
> (2) The following load lines shall be used:
>   (a) The Summer Load Line indicated by the upper edge of the line which passes through the centre of the ring and also by a line marked S.
>   (b) The Winter Load Line indicated by the upper edge of a line marked W.
>   (c) The Winter North Atlantic Load Line indicated by the upper edge of a line marked WNA.

Extract 1.1 Extract from the International Convention on Load Lines, 1966. (By kind permission of International Maritime Organization.)

additional problem. The Convention introduced new criteria concerning tonnage and the payment of dock, harbour and light dues levied upon all ships as they moved between ports in the world. Countries introduced various systems in an attempt to "compare" their variations with the Convention, but inevitably there was a considerable delay until agreements could be reached. Some nations devised a sliding scale of charges, whilst others adopted a new 1982/94 option of measurement, whereby they would have retained their old tonnage certificate for ships, but were issued with a tonnage statement. New ships are built to the now established criteria, so that there should in future be only one standard measurement for ship's tonnage throughout the world for every ship, thus clarifying an incredibly tortuous system.

There are five tonnage measurements relevant to determining the size of Supertankers. Regulation 3 and 4 of annexe one in the 1969 Convention applies quite complex formulae in order to obtain precise defining values. An extract from these regulations is included for reference: (Extract. 1.1). In its broadest terms, the first factor, the Gross Registered Tonnage (grt) of any ship is expressed in terms of 100 cubic metres and represents the moulded volume of the vessel including nearly all enclosed underdeck spaces. With dry-cargo ships, it is the reference point to size such that mention is made to a coaster as being of 250 grt, or an ocean-going dry-cargo vessel as 15,000 grt. As a measurement it is of only limited interest to the serving VLCC deck officer. Net Registered Tonnage (nrt) is the volume of cargo space available. Thus it represents the revenue earning capacity of a ship, with deductions being made for certain areas that are non-earning such as double-bottom tanks, officers and crew accommodation, engine room spaces etc. The values again are based on quite a complicated formulae. Nrt is again of no particular importance to the seagoing navigating officer other than, of course, as professional interest. As he has to pay the dues on his ship that are based upon nrt values, however, it is very much a concern to the shipowner.

1.4. CAPACITY, DWT. LOAD DISPLACEMENT AND LIGHT DISPLACEMENT TONNAGES

The size of tankers is invariably made by reference to the capacity of the ship based on Deadweight Tonnage (dwt). This is because crude oil (and its products), is a liquid cargo whose properties differ considerably from dry cargoes. Oil is not a constant. It possesses different densities and has also expansion variables, sometimes within each separate tank, due to temperature differences. Also, the cargo never actually fills the complete tank, because allowance has to be made for expansion of the oil due to various temperature changes as the ship experiences different climatic changes over the course of her voyage. This is perhaps more true in the case of product carrying tankers than VLCC's largely because, apart from North Sea crude oil, the cargoes are loaded in ports that are in the hottest parts of the world, namely the Arabian Gulf and Nigeria, hence they tend to decrease in temperature 'en route' to their markets, thus causing a corresponding variation in ullage. Most VLCC's, however, rarely load above 98% capacity, and most loading tables reflect this with calculations for both the latter and a full 100% cargo load. It is more due to the varying specific gravity of crude oil that different tonnage source calculations are used with crude oil carrying tankers, regardless of capacity, in contrast to the more stable volume payload of the dry-cargo ship.

In order to determine deadweight it is necessary to examine two additional tonnage's which provide its constituent parts. The first of these is the 'load displacement tonnage' which is the average weight of the VLCC between being loaded to a loadline with cargo, and consumables such as fuel, stores, lubricating oils, fresh and ballast water etc, when the ship is empty. As a unit it is perhaps of more use to the naval architect than a deck officer. The second factor is the 'light displacement tonnage' which is the actual weight of the ship as built. Calculations have to be made therefore of the remaining consumables and these added to the actual weight of the Supertanker. The payload of the VLCC is based on this tonnage and the shipowner is paid only for each ton which his ship carries. It is implied also that, because the deadweight is dependant upon SG and temperature, that the deadweight of a Supertanker can vary quite considerably on different voyages. It can be argued that Supertankers and all tankers, should be referred to in terms of their summer deadweight tonnage and not by their grt. Many authors and publishers of popular shipping magazines could, perhaps, add enhanced credibility to their often very professional approaches to matters maritime by offering consideration to this factor.

1.5. LOADLINES, FREEBOARD

When discussing tonnage it is impossible to avoid reference to the ship's load-lines and draught marks. Indeed, mention has been made already to loadlines when discussing 'load displacement' tonnage above. Similarly with the tonnage regulations, the IMO recommendations to the Classification Societies concerning loadlines, are very stringent and complex. The 1967 Loadline Act consists of 52 regulations and a number of accepted recommendations. An extract, from Regulation 27 of the act, governing types of ships, gives an idea, for information purposes, of the complexities involved:

Type "A" ships.

(2) *A Type "A" ship is one which is designed to carry only liquid cargoes in bulk, and in which cargo tanks have only small access openings closed by watertight gasketed covers of steel or equivalent material.*

Such a ship necessarily has the following inherent features:

(a) *high integrity of the exposed deck; and*

(b) *high degree of safety against flooding, resulting from the low permeability of loaded cargo spaces and the degree of subdivision usually provided.*

(3) *A Type "A" ship, if of over 150 metres (492 feet) in length, and designed to have empty compartments when loaded to its summer load water-line, shall be able to withstand the flooding of any one of these empty compartments at an assumed permeability of 0.95, and remain afloat in a condition of equilibrium considered to be satisfactory by the Administration. In such a ship, if over 225 metres (738 feet) in length, the machinery space shall be treated as a floodable compartment but with a permeability of 0.85 ...*

Prior to the 1966 Convention, upon which the Act was based, a multitudinous variety of norms had existed world-wide concerning the "freeboard" of ships. Section g, of Regulation 3, defines freeboard as *"the distance measured vertically downwards amidships from the upper edge of the deck line to the upper edge of the related loadline"*. In more basic lay terms, it is the distance between the waterline and the main deck, hence, that part of the hull length which is seen above water supporting the superstructure and deck fittings. Virtually the entire content of the Act stipulates internationally accepted agreements governing all of the very considerable number of facts affecting this distance. A consultative paper, presented to Lloyd's Register Staff Association in 1970, summarises something of the philosophy contained in the articles and regulations constituting the Convention's criteria regarding freeboard:

1. The prevention of entry of water through the exposed parts of the vessel.
2. Probability of deck wetness in relation to bow height.
3. The maintenance of sufficient reserve buoyancy in normal conditions of service.
4. The protection of crew when moving about the vessel.
5. Adequate structural strength of the ship.
6. Stability and compartmentation.

1.6. NAME AND DRAUGHT MARKS

The classification societies stipulate the conditions concerning the painting of names and draught marks on any ship. Generally, the name is engraved on each bow and also, with the port of registry, across the stern. Draught marks are etched on both sides of the bows at the stem and on both sides of the stern so that visible indications port and starboard, fore and aft, show how the ship is sitting in the water. This condition is called the vessel's 'trim'. The draught marks indicate the depth of water against the hull and are usually in metric figures in decimetres, but may also be shown in feet. Supertankers, and many other vessels, have also additional draught marks on both sides amidships.

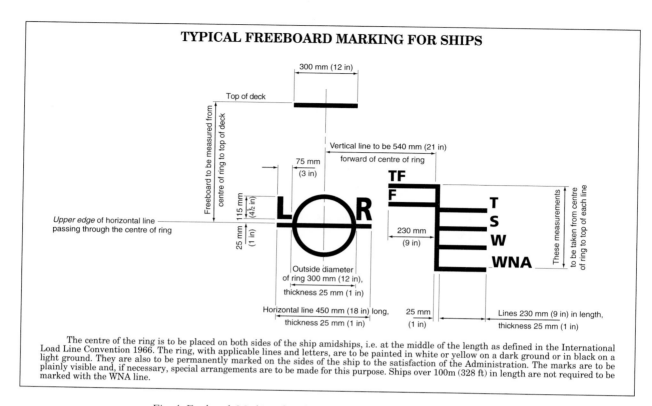

Fig. 1 Freeboard Marking for Ships showing the Plimsoll Load-line logo.
(By kind permission of International Maritime Organization.)

Regulations 5 and 6 of the 1967 Loadline Act made provision for the Plimsoll mark, as it is popularly known, and its attendant markings:

"The Load Line Mark shall consist of a ring 300 millimetres (12 inches) in outside diameter and 25 millimetres (1 inch) wide which is intersected by and horizontal line 450 millimetres (18 inches) in length and 25 millimetres (1 inch) in breadth, the upper edge of which passes through the centre of the ring. The centre of the ring shall be placed amidships and at a distance equal to the assigned summer freeboard measured vertically below the upper edge of the deck line ...."

The Act clearly specifies the actual position on the hull that the loadline should be inscribed and it is generally found just forward of the midships set of draught marks. The distinctive and well-known device, with its centre-line circle and markings, is of vital importance to the stability of the ship. The letters found either side of the circle indicate the name of the Classification Society. (Fig. 1.1). Some of these are:

LR indicating Lloyds Register (the oldest Society founded 1760)
AB " the American Bureau of Shipping
KK " Nippon Kaizi Kyokaie
GL " Germanischer Lloyd
NV " Det Norske Veritas

Regulation 27 of the Act clearly classifies tankers as "Type 'A' ships" and, because they are "designed to carry only liquid cargoes in bulk", and their absence of hatches and possession of "cargo tanks (that) have only small access openings", means that they are allowed by law to sit deeper in the water than a dry-cargo ship.

1.7. DECK LINE

Above the Plimsoll line is the 'deck line'. This indicates where the longitudinal plating of the main deck joins the shell plating of the ship's side and, by provision of Regulations 4 and 5 of the Act, it is situated vertically above the central ring of the loadline. It is this distance which determines the minimum summer freeboard of the VLCC. The construction of some ships, however, may determine an extension to the minimum summer freeboard. The diagram offering examples of applications to Regulation 4 mentions provision for a 'striking plate'. This is indicated by the plate underneath the sloping line of the mark which acts as reinforcement to the longitudinal plating when it is not exactly perpendicular to the shell plating of the ship's side.

1.8. OTHER LOAD MARKINGS

The other markings associated with the loadline are set-down in Regulation 6. Aboard Supertankers, they are represented by etched letters, at precise specifications and distances from the centre-line, and indicate loading limits within the seasonal zones in which the ship is likely to trade.

T   indicates Tropical Zones

S   the by now familiar Summer Zone

W   the Winter Zone.

The zones and trading areas are published in a detailed map, (Map 1.1), produced in the Act, accompanied by specifications in Annexe 2 which determine geographical positions and specific trading dates. The summer zone, for example, is defined as an area in which "not more than 10 per cent of the winds of Force 8 Beaufort (34 knots) or more" are likely to be experienced, with tropical of "not more than 1 per cent". The necessity for taking an all-round draught of the Supertanker, so that she may proceed on her voyage in a totally sea-worthy condition, becomes paramount.

The summer mark is taken as a base-line because it is adjacent to the centre loadline ring and serves as the criterion against which all measurements relevant to draught are assessed. As can be seen from the diagram, it is equidistant to the tropical and winter marks.

The summer freeboard is always a fixed distance for any individual ship, even though the draught of any VLCC would obviously have to be a variable factor. Tropical Fresh, TF, and Fresh, F, water marks are included because, although on initial reflection it may seem improbable that any Supertanker is ever likely to find herself in fresh water, she may well occasionally find herself loading in salt water into which flows fresh waters from rivers. The specific gravity of the water would have to be taken during the final stages and appropriate allowances made.

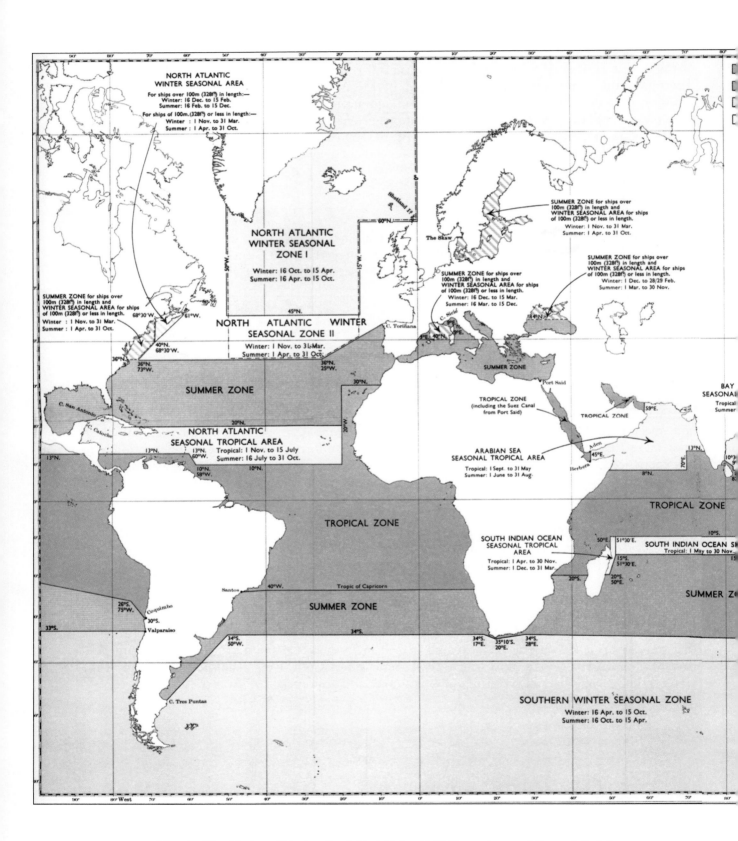

Map 1 The Merchant Shipping (Load Line) Rules 1968-Zones, Areas and Seasonal Periods.
(By kind permission of HMSO and Hydrographic Department.)

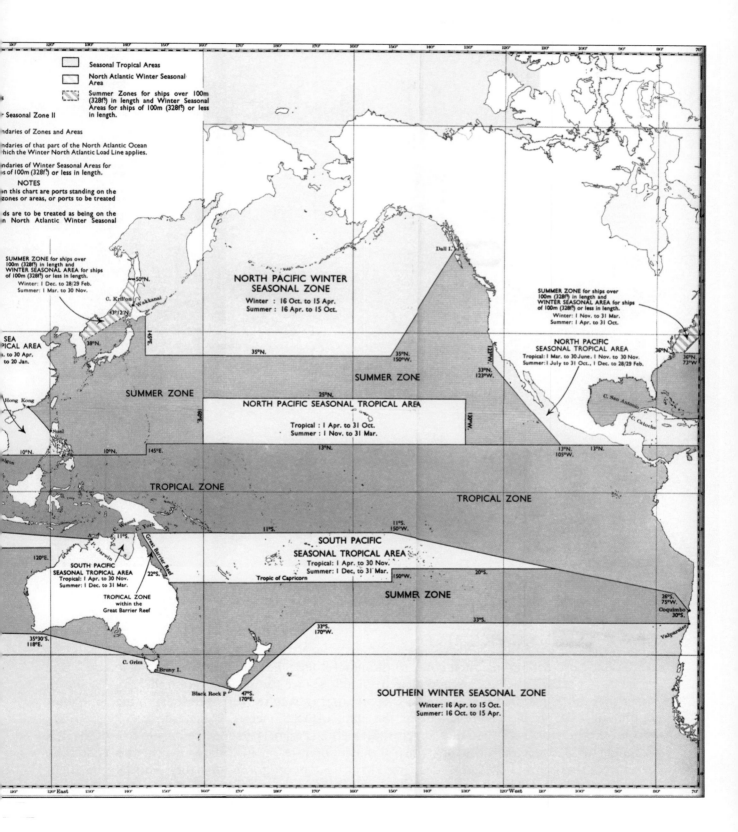

Map 1 continued
(By kind permission of HMSO and Hydrographic Department.)

Supertankers Anatomy and Operation

1.9. An impressive view of the forward draft marks on the World's largest Deep-Sea tanker.
(By kind permission of Jahre, Wallem Ship Management.)

1.9. A LOADING EXAMPLE

Such dry definitions often come to meaningful life when applied to an actual example and the ULCC "JAHRE VIKING" would seem a suitable choice for demonstration. It has been shown earlier that this Supertanker has a moulded depth, that is from the top of the keel to the uppermost continuous deck in the centre-length of the ship, of 29.8 metres. When she is a 'light ship', that is, as she was built without any cargo, fuel, stores or personnel etc., on board she has a draught of 3.5 metres and a light displacement tonnage of 83,192 tonnes. In this condition, therefore, the ship has zero dead-weight. The photograph of this ship in a light condition, (Photo. 1.9), admirably demonstrates how the sixty-five draught marks commence, from the keel line beyond the darker shade of red, until an area is reached that is just below the anchor hawse pipe. As she is loaded with stores and cargo etc., she sinks lower into the water to a level reaching the top edge of the Summer loadline so that her draught is now approx. 24.6 metres. In theory, therefore, the "JAHRE VIKING" would be loaded to a draught of 25.7 metres in tropical fresh water and that would give a deadweight tonnage of 662,313 tonnes. In practice, however, this condition would never occur because the ship would never travel for her entire voyage in tropical fresh water. The realisation that when she is at her Summer line there is something like eighty feet (or about 24.5 metres) of ship below the surface of the water is quite a sobering thought to many people. The following calculations illustrate the appropriate tonnage's at the various marks, and it is interesting to notice that this ULCC has an extended minimum summer freeboard:

From Keel to Centre Summer Line	= 24.6m
From Summer Line to Deck Line	= 4.0m
From Deck Line to Stringer Plate	= 1.2m
Hence Summer Moulded Depth	= 29.8m
The freeboard at the Summer line	= 5.4m

The distance between the loadline marks for the "JAHRE VIKING" has to be calculated using the provision of Regulation 40:

"Tropical Freeboard"

(3). The minimum freeboard in the Tropical Zone shall be the freeboard obtained by a deduction from the summer freeboard of one forty-eighth of the summer draught measured from the top of the keel to the centre of the ring of the loadline mark.

Thus:

Distance from Keel to Summer Line = 24.6m

$24,600 \div 48 = \underline{512\text{mm}.}$

The significance in practical cargo carrying terms, mindful of the specific gravity of the oil, between the loadline differences can be demonstrated:

	S Mark	= 24.6m	=	564,763 dwt
	W Mark	= .512m	=	550,174 dwt
	Differences S to W		**=**	**14,600 tonnes less cargo.**
Similarly:	T mark	= .512	=	579,396 dwt
	S Mark	=		564,763 dwt
	Differences T to S		**=**	**14,600 tonnes more cargo.**

The ULCC when she finally leaves the port would have to be floating in sea water at her Summer mark. When she is thus loaded to her Summer mark this ULCC would have a freeboard of slightly over 5 metres, this being the distance from the S mark to the stringer plate of the deck line, hence the amount of the ship above the waterline.

1.10. SUPERTANKER MOVEMENTS

Legitimately, it could well be imagined that quite rigid trading limits would have to be imposed concerning ports in the world where Supertankers could trade. Surprisingly, this is not completely the case. Certainly, they may well have to off-load a portion of their cargo — in order to sail only partially-laden to their ultimate destination, but their general trading range is almost world-wide. On just one day, in mid 2000, the following movements were recorded of the near 450 Supertankers which comprise the world's fleet: that is, crude oil carriers in excess of 160,000 tonnes dead-weight. The movements include VLCC's in all conditions of steaming, fully loaded, in ballast, 'in transit', awaiting orders as well as the sixteen ships being used for storage purposes. There were no Supertankers laid up on that day.

The ports cited incorporate those of loading and discharging, as well as oil terminals which are often in excess of eight miles off shore. The Louisiana Offshore Oil Port (LOOP), for instance, is some thirty-four miles south-west of Louisiana and is geared to take any draught of Supertanker in its fully loaded condition.

EUROPE:

United Kingdom: Coryton, Fawley, Milford Haven, Shellhaven, Sullom Voe, Tranmere Oil Jetty at Liverpool.

Denmark: Kalundborg.

Norway: Mongstad, Stratfjord, Sture.

Sweden: Brofjorden.

Netherlands: Rotterdam/Europoort.

Portugal: Lisbon/Sines.

France: Le Havre, Port de Bouc, Fos (Mediterranean).

MEDITERRANEAN:

Gibraltar.

Spain: Algeciras, Barcelona, Bilbao, Cadiz, Malaga.

Italy: Brindisi, Milazzo, Taranto, Trieste.

Greece: Megara, Piraeus.

Bulgaria: Varna.

Rumania: Constanza.

Russia: Novorossiysk.

Turkey: Izmir, Iskanderun, Ceyhan.

Israel: Hadera.

Lebanon: Beirut.

Egypt: Alexandria, Ain Surkhana, Sidi Kerrir.

Libya: Bouri Terminal, Marsa el Bregha.

WEST/SOUTH AFRICA:

Islands off: Ponta de Madeira.

Angola: Cabinda.

Nigeria: Bonny, Brass Terminal, Forcados, Qua Ibo Terminal.

River Congo: Djeno Terminal.

South Africa: Capetown, Durban, Saldana Bay.

NEAR/FAR EAST:

Yemen: Ash Shihr Terminal, Ras Isa Terminal.

Trucial Oman: Al Fujairah, Haminyah Terminal, Khor Fakkan, Mena al Fahal, Sharjah.

UAE/Qatar: Abu Dhabi, Arzanah Island, Das Island, Dubai, Halul Island, Jebel Dhanna, Ruwais, Zirku Island.

Kuwait: Mena al Ahmadi, Mina Abdulla.

Saudi Arabia: Juaymah Terminal, Ras Tannurah, Jeddah (Red Sea).

Iran: Kharg Island, Lavan Island.

Pakistan: Karachi.

Bangladesh: Mongla.

India: Madras, Vadinar Terminal, Mormugao.

Vietnam: Ho Chi Min City.

Malaysia: Pasir Gudang, Penang, Pulau Bukom, Republic of Singapore.

Indonesia: Alang, Balikpapan, Widuri Terminal.

Phillipines: Bataan, Batangas.

Thailand: Sriracha.

Hong Kong.

Taiwan: Kaoshiung.

China: Nongbo, Qindao, Shenzhen.

Okinawa Island: Heianza, Mizushima.

Republic of Korea – (South Korea): Busan, Daesan, Kwangyang, Ulsan, Yosu.

JAPAN:

Aioi, Chiba, Kagashima, Kakagowa, Kashima, Kawasaki, Kiiri, Kisarazu, Kure, Mizushima, Nagoya, Oita, Okayama, Onahama, Sakaide, Tokuyama, Tokyo, Tomakomai, Yokkaichi, Yokohama.

AUSTRALIA:

Dampier, Fremantle, Gladstone, Newcastle, Port Hedland, Port Walcott, Sydney.

U.S.A:

Alaska, Baltimore, Barber's Point, Corpus Christi, Freeport (Texas), Galveston Lighterage Area, LOOP Terminal, Los Angeles, Pascagoula Lighterage, Philadelphia, Port Angeles, River Mississippi Lighterage, San Francisco, Seattle, South West Passedena Lighterage area, Valdez.

CANADA:

Halifax, Port Tupper (Nova Scotia), Prince Rupert Sound, St. John, Vancouver.

MEXICO:

Cayo Area Terminal.

PANAMA:

Armuelles.

WEST INDIES:

Aruba, Barbados, Cul de Sac, Freeport (Bahamas), San Nicholas Bay, St. Eustatius, St. Lucia.

SOUTH AMERICA:

Beunos Aires, Covenas, Gebig (Rio de Janeiro), Santos, Sepiteba Terminal, Sao Sebastiao, Tubarao.

It is worth recording that the twenty-five ULCC's in excess of 400,000 dwt traded only in the following areas on the selected day:

Ain Surkhana (Egypt), Durban, Dubai, Europoort,

Fujairah Terminal, Galveston Lighterage area,

Juaymah Terminal, Kharg Island,

LOOP Terminal, Mina al Ahmadi, Piraeus, Port Angeles,

Ras Tannurah, San Nicholas Bay, Sullom Voe.

The ULCC "JAHRE VIKING" was at Juaymah Terminal on that day waiting to be loaded, following a passage in ballast from LOOP.

The 160 or so ports cited do not, of course, show the total limits of trading areas specific to Supertankers because they represent only ports visited on that day. A random sample of some actual passages undertaken is offered showing VLCC's trading between the following ports:

Europoort to: Brazil for orders, Cabinda (Angola), Fawley, Jeddah, Madeira, Mina Abdullah, Mongstad (Norway), Piraeus, Sidi Kerrir (Egypt).

Kharg Island to: Ain Surkhana (Egypt), Brofjorden (Sweden), Durban, Europoort, Oita (Japan), South Korea: 2 x for orders, plus 1 for Yosu (Japan),

Valdez: 1 VLCC to each of San Francisco and Taiwan

Barbados to Nevis

Ain Surkhana to Kharg Island

Fos: 1 VLCC each to Sidi Kerrir, Forcados (Nigeria)

Sullom Voe: 1 to Port de Bouc, and 1 to Saldana Bay (South Africa)

Coryton (River Thames) to Sidi Kerrir

Aioi (Japan) to Singapore

Galveston Lighterage Area to Sidi Kerrir

Ho Chi Minh City to Yanbu (Saudi Arabia)

Cabinda to US Gulf for orders.

PART 2: The 'First Generation' Supertanker

Chapter Two

CONSTRUCTION

2.1. INTRODUCTION

2.2. CONSTRUCTION PRINCIPLES AND DEVELOPMENT

2.3. CLASSIFICATION SOCIETY REGULATIONS AND "SCANTLINGS"

2.4. FURTHER CONSTRUCTION: LOOKING AT THE PLANS

2.5. STERN CONSTRUCTION

2.6. RUDDER

2.7. PROPELLER

2.8. COFFERDAM AND TANKS

2.9. BULKHEADS, TRANSVERSE FRAMES AND STIFFENERS

2.10. PLATES

2.11. MAN AND RATHOLES

2.12. FORE PEAK TANK

2.13. BULBOUS BOW, STEM

2.14. LAUNCHING

Chapter Two

CONSTRUCTION

(Introductory Remark — Construction Principles — Influence of Sir Joseph Isherwood — Two Longitudinal Bulkheads Construction — Remarkable Quickness in the Building Stages of early VLCC's — Classification Society Regulations — "Scantlings" — Uses of Mild and High Tensile Steels — Lloyd's Rules for Oil Tanker Construction Mentioned and Brief Quotation — Stages in ULCC and VLCC Construction — AFTER Part: Cargo Tanks — Engine Room Machinery and Equipment Cited — System of Numbering Frames — Example of a Brief Damage Report to Owners Concerning a Dry-Cargo Ship — Stern Frame — Propeller and Rudder — Types of Rudders on Ships — Styles of Propellers in Use — Accommodation Block fitted — MIDSHIPS Section: Construction and Development of the Hull — Cofferdams Fore and Aft — Slop Tanks — Segregated Ballast Tanks (SBT's) — Yard Practice is to Work on a Number of VLCC's Simultaneously — Oiltight and Swash Transversal Bulkheads — Man and Rat Holes Facilitating Inspections — FORE Part: Fore Peak Tank and Bulbous Bow Construction — Types of Stem — Launching from Slipways and Floating Dry Docks.)

NB: *It may prove useful to read this chapter in conjunction with* **Chapter 14,** *in* **Part Four,** *on double hulled VLCC's.*

2.1. INTRODUCTION

In many respects, a tanker is probably the easiest ship to construct. Reduced to its simplest terms, it is merely a collection of near-identical welded tanks, possessing a pumping and pipeline system, with an engine generating power to turn a propeller and deflect water across a rudder at the rear, and a flared bulbous bow forward that assists progress through the water. A Supertanker becomes, therefore, a larger edition of the smaller version. Like most over-simplifications, this one is near the truth but, of course in practice, the construction of any VLCC is a considerably complex undertaking.

2.2 CONSTRUCTION PRINCIPLES AND DEVELOPMENT

It is the shipowner who decides his requirements by discussing specifications with his superintendents and the Naval Architects department. The decision is made regarding the Classification Society against whose regulations the ship would be built in accordance with IMO Conventions. The port of registry is decided, but this does not necessarily have to possess any connection either with the country or countries involved in building or operating the ship. It is very often a port in the country where the operating company has established a registered office. Then an approach is made to either the company's preferred builder, or other yards could be canvassed for quotations, following which, a choice would be determined that would produce the shipowner's ideas in accordance with precisely drawn plans. Alternatively, it may be decided to purchase a yard specified ship with or without the owner's modifications.

Building for single hull tankers in excess of 150 metres continues to rely on principles of basic construction, comprising two longitudinal bulkheads reinforced by a system of transverse framing, that was a feature first introduced by Sir Joseph Isherwood, Shipwright Surveyor to Lloyd's Register of Shipping between 1896 and 1907. Since the post Second World War period, however, the method of building Supertankers has advanced out of all recognition. VLCC's have become a unique process of mass-production and prefabrication. Even thirty years ago, in the early days, they were built as specific unit blocks, ranging from around 80 to in excess of 100 tons, which were assembled within various sheds or in open areas of the yard, and transported to a building berth or dock. A glance at the drawing outlining the TSU Works of the NKK Corporation of Japan, (Draw. 2.1), offers a typical lay-out of a modern ship building yard. Limitations in the size of construction blocks were imposed largely by restrictions, at that time, in the safe working loads (swl) of the gantry cranes designed to lift them. As the swl increased, so also did the size of the building blocks. One 1974 VLCC, for example, was built in larger sized blocks in under 120 days. Even the earlier Supertankers were completed, prior to the fitting-out stage, in an amazingly impressive period of time, as can be seen from the accompanying series of photographs. The shots show the various construction stages, over a short cycle, of a 206,971 sdwt VLCC belonging to a major oil company, as long ago as 1969, which was built in Japan by the Mitsubishi Heavy Industry shipyard. She was commenced in October and within one month the after end had been brought to the building dock. (Photo. 2.1). In the pumproom, the cargo and stripping pumps have been seated and a start made on the pipework forward and a base for the main engine. After only three months, the ship was working well towards completion. (Photo. 2.2). The men working on various sections of the fore part offer an indication of the dimensions involved. Less than one month later, (Photos. 2.3 and 2.4), showed quite remarkable progress on the main cargo tanks had been made and, in less than another month, just five months from the start of the assembly, this very large tanker was being towed to her fitting-out berth for completion. (Photo. 2.5).

Even today, the construction of VLCC's continues to follow a sequence whereby various parts of the ship are built individually, yet simultaneously, as modules. Often, today, these are constructed under cover in order to prevent corrosion and to save time. The blocks are built onto keel and bilge supports at the bottom of the dock so that the static stresses can be evenly distributed. The height of the supports can be demonstrated by the under-keel, photograph of a 190,000 sdwt supertanker that was built in 1968 by Swan Hunter's Newcastle yard. (Photo. 2.6). The packing, keel and bilge supports stand around two metres and are just sufficient for a man, shown on the left hand side, to stand upright below the hull of the ship. It is a vaguely uneasy emotion to stand under a vessel, of any size, in the knowledge that such vast masses are situated immediately above you. Certainly, that has been my experience when on the couple of occasions I have found myself under the hull of VLCC's, one in a Far Eastern shipyard and the second in a European dry-dock.

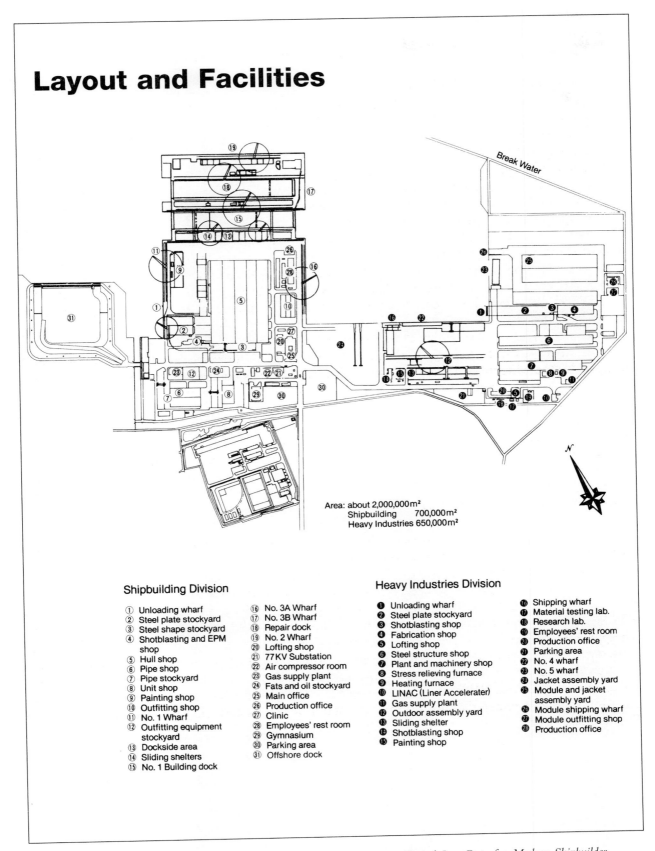

Draw 2.1 The "TSU Works" of the NKK Corporation Represents a Typical Lay-Out of a Modern Shipbuilder.
(By kind permission of NKK Japan.)

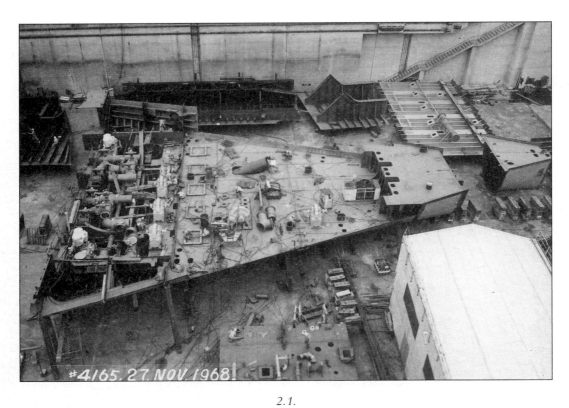

2.1.

Photos 1-5-A series of photographs showing the rapid rates of construction of a 200,000 sdwt VLCC built in Japan in 1969. (By kind permission of Shell International Trading and Shipping.)

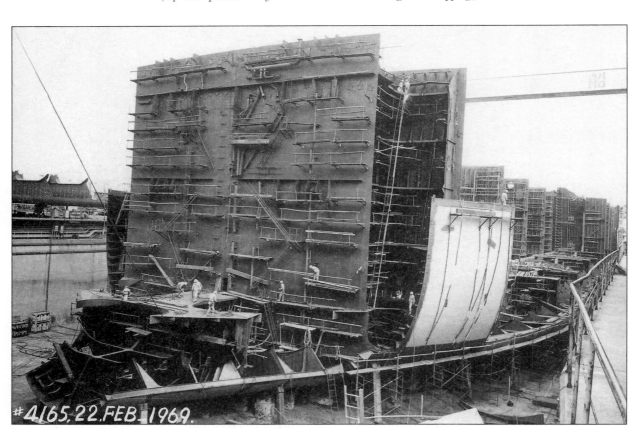

2.2.

Construction

2.3.

2.4.

27

Supertankers Anatomy and Operation

2.5.

2.6. *The underkeel shot of a 190,000 sdwt VLCC showing the packing, keel and bilge supports. (By kind permission of Swan Hunter Shipyard.)*

2.3. CLASSIFICATION SOCIETY REGULATIONS AND "SCANTLINGS"

International Association of Classification Societies regulations stipulate extremely rigid conditions regarding the 'scantlings', or quality and thickness of steel, used in all parts of a tanker's construction. Supertankers of this early generation were generally made from mild steel, with extra thickness at the bottom, and deck areas, where the major regions of stress would be experienced. Mild steels were also used in other areas, such as the accommodation. The single shell plating of the 'early' VLCC's, in the area comprising the ship's bottom, for example, would have been about 25 mm, increasing to around 27 mm at the keel plate. The shell plating on the hull of the ship's side would be about 20 mm thick. High tensile steel was not used extensively until the 1970's and then, generally, only for the areas of decks and hull bottoms. The following table, (Table 2.1), indicates how a combination of HT and mild steels were often used in the hull construction of, in this example, two VLCC's built in (a) 1975, and (b) 1976, from Japanese Yards. It was not until the 1980's and 1990's that HT steel was used more widely in a Supertanker. The advantages of higher tensile steel are largely twofold; it not only possesses greater stress resistance under pressure, but generates also a reduction in the weight of steel used for building so that greater cargo carrying capacity can be achieved.

Hull Materials used on different VLCC's

(a) VLCC of 280,500 sdwt-Built Japanese Yard Alpha in 1975.

Location	Plating	Long/Stiff	Girders
Deck	HT	HT	HT
Bottom	HT	HT	HT
Side	MS	MS	MS
Longitudinal Bulkhead	MS	MS	MS
Transverse Bulkhead	MS	MS	MS

(b) VLCC of 284,400 sdwt-Built Japanese Yard Bravo in 1976.

Location	Plating	Long/Stiff	Girders
Deck	HT	HT	HT
Bottom	HT	HT	HT
Side	HT	HT	HT
Longitudinal Bulkhead	HT/MS	HT/MS	HT/MS
Transverse Bulkhead	MS	MS	MS

Table 2.1 The different use of Mild and High Tensile Steel on two VLCC's built in the mid-1970's.
(By kind permission of Mobil Ship Management.)

The Classification Societies publish stringent construction regulations, with part two of Lloyd's "Rules for the Manufacture, Testing and Certification of Materials", plus parts three and four regarding scantlings of oil tankers. An extract from a Section on Structural Design, from the 'general' heading of this part, is given merely to offer a flavour of the specifications:

*"**4003** The Rules apply to single deck tankers over 90 m (295 ft) in length with machinery aft and having two or more longitudinal bulkheads.*

The bottom and decks are to be framed longitudinally in the cargo tank spaces; the Rules assume longitudinal framing at the sides and longitudinal bulkheads where the length of the ship exceeds 200 m (656 ft), but alternative designs will be considered.

If the proportion of length to depth exceeds sixteen, or the ratio of breadth to depth exceeds 2:1, increased scantlings may be required.

***4004** The length of a tank is not to exceed 0,2L. When the length exceeds 0,1L or 15 m (49 ft), whichever is the greater, a transverse wash bulkhead is to be fitted at about mid-length of the tank.*

***4005** The scantlings of structural items may be determined using direct calculation. Where the length of the ship exceeds 250 m (820 ft), the arrangements will be specially considered and the Society may require certain scantlings to be assessed by direct calculation. In such cases the calculations are to be submitted for approval."*

The following photograph of a 326,000 sdwt VLCC, that was built in 1974 by French shipbuilders at St. Nazaire, illustrates how the construction of a Supertanker proceeds following placing of the initial blocks. (Photo. 2.7). Once the after part centre tank has been positioned, those comprising the base of the wings can be added. Clearly, some investigation and explanation of the maze of steel girders and frames comprising the internal structure of a Supertanker's construction is necessary. It is believed that there is a danger, however, that the non-technical reader, (which might well include some deck officers), could become unnecessarily 'bogged down' by the technical involvement of this task. A glance at (Photos. 2.3 and 2.4), for example, indicates that discussion of each of the vertical and horizontal members used in the cargo tank construction of VLCC's could become quite lengthy and perhaps cumbersome. The difficulty might best be overcome by offering an analytical diagram, (Diag. 2.1), which explains some of the terms used in discussing standard structural nomenclature. From this can be identified, in Photo. 2.7, for instance, the single bottom plate as well as the longitudinal stiffeners that extend the length of each individual tank supporting the main lower transversal frames. A start has also been made to fit the main cargo pipes in the centre tank, as well as towards completion of a vertical after tank transverse oil-tight bulkhead. The figure of a man, climbing up the ladder in the top part of the completed section, gives again an interesting insight into the dimensions involved.

2.7. Once the Centre Tank blocks are positioned, the Wing Tanks can be added and the bulkheads commenced. (By kind permission of Shell International Trading and Shipping).

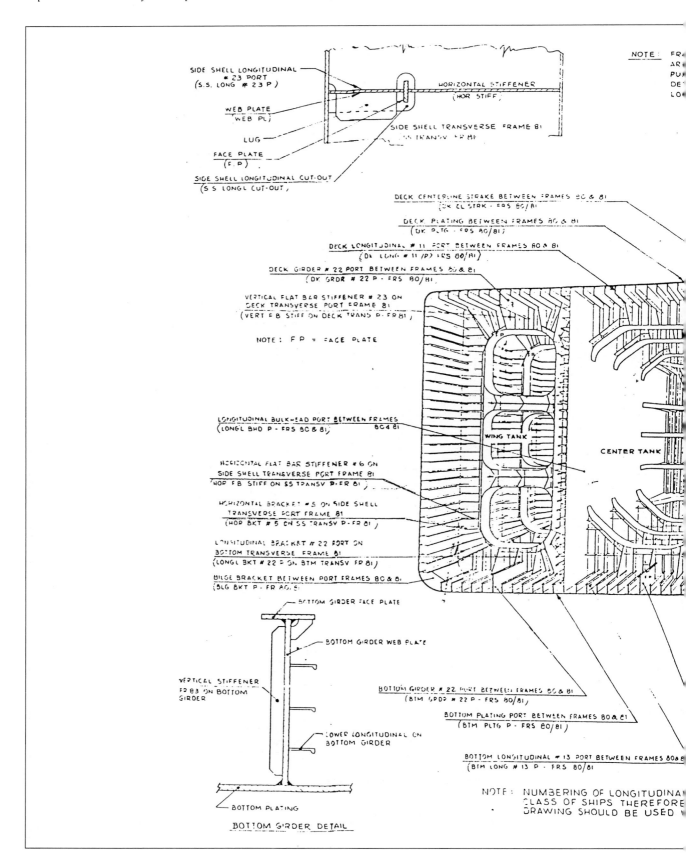

Diagram 2.1 Standard Nomenclature used in referring to Cargo Tank structures.
(By kind permission of Mobil Ship Management.)

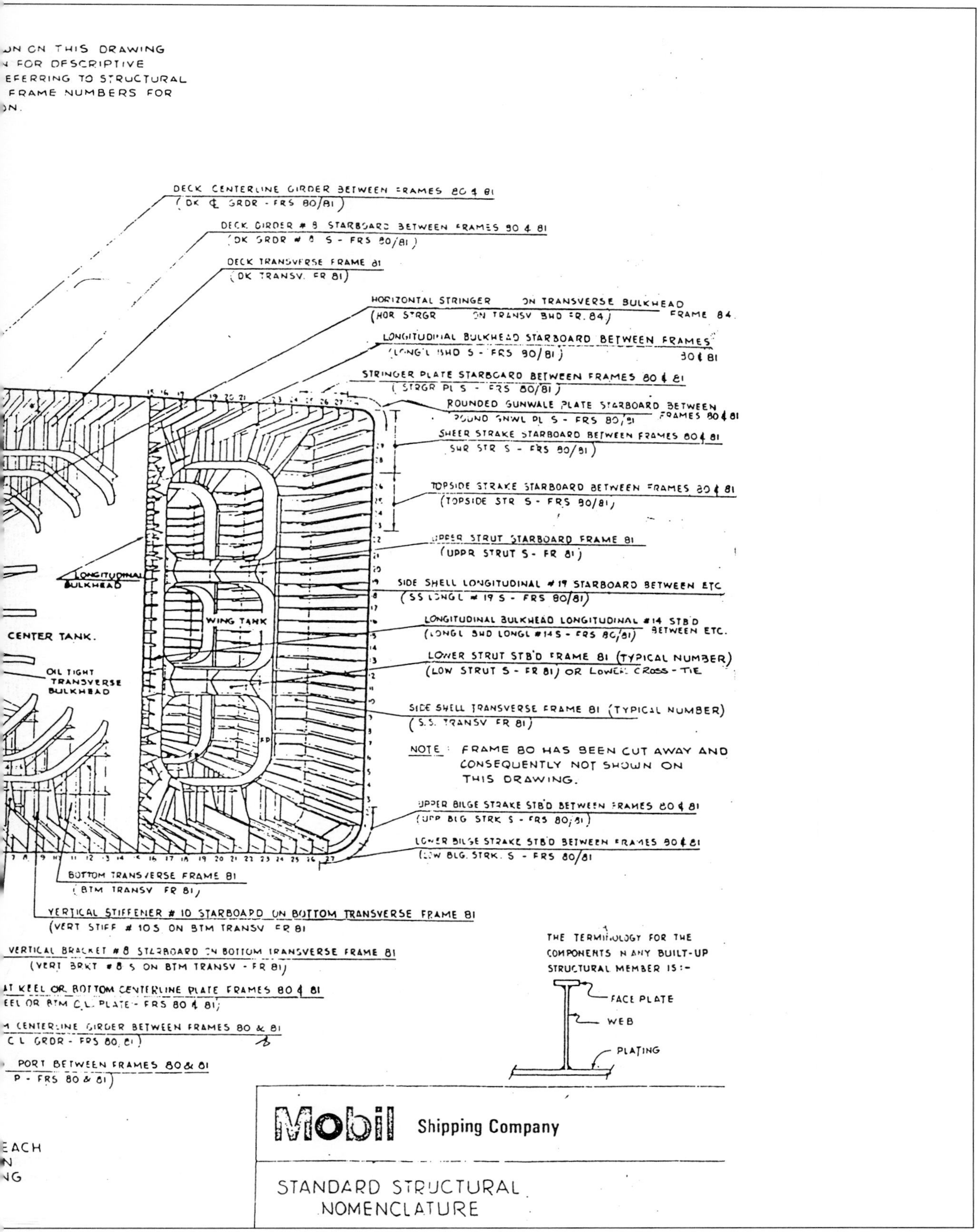

...N ON THIS DRAWING
... FOR DESCRIPTIVE
...EFERRING TO STRUCTURAL
... FRAME NUMBERS FOR
...N.
DECK CENTERLINE GIRDER BETWEEN FRAMES 80 & 81
(DK ℄ GRDR - FRS 80/81)
DECK GIRDER # 8 STARBOARD BETWEEN FRAMES 80 & 81
(DK GRDR # 8 S - FRS 80/81)
DECK TRANSVERSE FRAME 81
(DK TRANSV. FR 81)
HORIZONTAL STRINGER ON TRANSVERSE BULKHEAD FRAME 84
(HOR STRGR ON TRANSV BHD FR. 84)
LONGITUDINAL BULKHEAD STARBOARD BETWEEN FRAMES 80 & 81
(LONG'L BHD S - FRS 80/81)
STRINGER PLATE STARBOARD BETWEEN FRAMES 80 & 81
(STRGR PL S - FRS 80/81)
ROUNDED GUNWALE PLATE STARBOARD BETWEEN FRAMES 80 & 81
ROUND GNWL PL S - FRS 80/81
SHEER STRAKE STARBOARD BETWEEN FRAMES 80 & 81
(SHR STR S - FRS 80/81)
TOPSIDE STRAKE STARBOARD BETWEEN FRAMES 80 & 81
(TOPSIDE STR S - FRS 80/81)
UPPER STRUT STARBOARD FRAME 81
(UPPR STRUT S - FR 81)
SIDE SHELL LONGITUDINAL # 19 STARBOARD BETWEEN ETC
(SS LONGL # 19 S - FRS 80/81)
LONGITUDINAL BULKHEAD LONGITUDINAL # 14 STB'D BETWEEN ETC.
(LONGL BHD LONGL # 14 S - FRS 80/81)
LOWER STRUT STB'D FRAME 81 (TYPICAL NUMBER)
(LOW STRUT S - FR 81) OR LOWER CROSS - TIE
SIDE SHELL TRANSVERSE FRAME 81 (TYPICAL NUMBER)
(S.S. TRANSV FR 81)
NOTE: FRAME 80 HAS BEEN CUT AWAY AND CONSEQUENTLY NOT SHOWN ON THIS DRAWING.
UPPER BILGE STRAKE STB'D BETWEEN FRAMES 80 & 81
(UPP BLG STRK S - FRS 80/81)
LOWER BILGE STRAKE STB'D BETWEEN FRAMES 80 & 81
(LOW BLG. STRK. S - FRS 80/81)
LONGITUDINAL BULKHEAD
WING TANK
CENTER TANK.
OIL TIGHT TRANSVERSE BULKHEAD
BOTTOM TRANSVERSE FRAME 81
(BTM TRANSV FR 81)
VERTICAL STIFFENER # 10 STARBOARD ON BOTTOM TRANSVERSE FRAME 81
(VERT STIFF # 10 S ON BTM TRANSV FR 81
VERTICAL BRACKET # 8 STARBOARD ON BOTTOM TRANSVERSE FRAME 81
(VERT BRKT # 8 S ON BTM TRANSV - FR 81)
...AT KEEL OR BOTTOM CENTERLINE PLATE FRAMES 80 & 81
...EEL OR BTM C.L. PLATE - FRS 80 & 81)
...M CENTERLINE GIRDER BETWEEN FRAMES 80 & 81
... C L GRDR - FRS 80/81)
... PORT BETWEEN FRAMES 80 & 81
... P - FRS 80 & 81)
...EACH
...N
...NG
THE TERMINOLOGY FOR THE COMPONENTS IN ANY BUILT-UP STRUCTURAL MEMBER IS:-
FACE PLATE
WEB
PLATING
Mobil Shipping Company
STANDARD STRUCTURAL NOMENCLATURE

Diagram 2.1

2.4. FURTHER CONSTRUCTION: LOOKING AT THE PLANS

Reference to the after part of the general arrangement plan of an average sized single hulled, 286,000 sdwt Supertanker, (contained in the inside back cover pocket), shows that the curvature towards the stern, in the similar vicinity of the partially built section, (Photo. 2.7), indicates that the wing tanks so far erected are probably intended for conveying fuel oil. The after side elevation on the plan indicates the extent to which this section of the ship is comprised of oil, for the ship's bunkers, and diesel to turn the generators and subsidiary machinery. The engine space is sandwiched, for about ⅕ of its depth from the maindeck, between lubricating oil and distilled water tanks, along with the domestic water (or "dom") tanks for use by officers and ratings. A discussion of the sizes and capacities of these, and the cargo, tanks is made in Chapter 5 on sectional analysis, but comparisons may also be obtained by reference to the capacity plans (also in the pocket).

It is of interest to notice in the plans that the stern of the ship, from the end of the cargo tanks through the engine-room to the stern plate, is roughly the same length as number one tank, thus providing a form of proportional dimension. The main engine itself is bedded on specially reinforced plates, supported by additional transversal webs and girders built onto the existing ship's longitudinals. The following table, from the set of plans issued to one of the VLCC's upon which I served, indicates specifications for the machinery in the engine-room that is concerned with ship propulsion and generators, and pumps for cargo and ballasting purposes. (Table 2.2). Obviously, the equipment provided by a range of numerous international alternative manufacturers is available to shipowners that offers varying permutations regarding the details of main engine output, although this remains similar but, more significantly, concerning the finer details of the supporting machinery.

The length of the hull is divided into transversal frames that are numbered commencing from a point at the centre line of the rudder stock. The frames are spaced at specific distances, in accordance with classification rules, and serve not only the obvious role of hull reinforcement, but act also as a point of reference in identifying parts of the structure. Referring again to the general arrangement plan, the narrower area between the frames in the vicinity of the rudder and propeller, at frames 0-14, distance 10.5 metres, and the engine-room area, frames 14-60, of 32.1 metres, contrasts with the considerably wider spacing between frames 60-70, constituting number five cargo tank at 55.7 metres. Reports to owners of any sustained hull damage invariably require mention of parts of the vessel in terms of her frames. To offer an example showing how this information might be used, I was acting as Mate aboard a dry cargo ship, the m.v. "GOLD VARDA", when we took the mooring dolphins, in way of the after part, on the port side, whilst coming alongside the jetty at Haifa. No bumping was noticed by any of the ship's personnel, but the linesmen reported the incident to the Pilot who, with the Master and myself, sighted the damage. An extract from my report to the company stated:

"... *Two large indents on port quarter traced to engine room in way of engineer's flat. One just abaft of web frame No.30. The second in way of lap weld between Nos 31 and 32 frames just aft port fresh water tank after bulkhead ....*"

Thus the company superintendents could see the extent of the damage, from the positioning of the frames, on their copies of the ship's general arrangement plans. They agreed to proposals, made by Lloyd's local surveyors, that the torn plating at the dents should be welded over, and two Cement Boxes set up in way of the dents as a temporary measure. Once this was done, we were allowed to sail to our next port for more permanent repairs to be effected.

> **MACHINERY PARTICULARS**
>
> MAIN ENGINE: ONE SET REVERSIBLE GEARED STEAM TURBINES. "STAL-LAVAL" TURBINE DRIVING A SINGLE SHAFT THE UNIT CONSIST OF ONE HIGH PRESSURE AND ONE LOW PRESSURE, SINGLE FLOW AHEAD TURBINE WITH AN ASTERN TURBINE INCORPORATED IN THE LOW PRESSURE TURBINE CASING. THE TURBINES ARE CONNECTED TO A REDUCTION GEAR. THE PRIMARY REDUCTIONS ARE OF THE EPICYCLIC TYPE THE LAST REDUCTION OF THE SINGLE PLANE PARALLEL TYPE.
>
> NORM. OUTPUT AHEAD 32440 SHP (METRIC) AT 86 RPM
> MAX. OUTPUT AHEAD 32440 SHP (METRIC) AT 86 RPM
> MAX. OUTPUT AHEAD 15820 SHP (METRIC) AT 56 RPM
> HEIGHT PRESSURE TURBINE NORM. 6010 RPM
> LOW PRESSURE TURBINE NORM. 3652 RPM
>
> AUXILIARY ENGINES: ONE TURBINE DRIVEN GENERATOR "STAL-LAVAL/ASEA" 1280 kW, 450 VOLTS, A.C. 3 PHASES, 60 CYCLES COS = 0.8.
> ONE STAND-BY TURBINE DRIVEN GENERATOR "STAL-LAVAL/ASEA" 1280 kW, 450 VOLTS A.C., 3 PHASES, 60 CYCLES, COS = 0.8.
> ONE DIESEL DRIVEN GENERATOR "HEDEMORA/ASEA" 1250 kW, 450 VOLTS A.C., 3 PHASES, 60 CYCLES, COS = 0.8.
> ONE EMERGENCY DIESEL DRIVEN GENERATOR "GMC/GEC MACHINES" 250 kW, 450 VOLTS A.C., 3 PHASES, 60 CYCLES COS = 0.8.
>
> **BOILERS:**
>
> MAIN BOILER: ONE OIL FIRED V2M 9 MARINE BOILER
> TYPE: COMBUSTION ENGINEERING.
> NORMAL OUTPUT: 95400 KG SUPERHEATED STEAM PER HOUR AT 513°C AND 63.3 ATO.
> MAXIMUM OUTPUT: 120000 SUPERHEATED STEAM PER HOUR AT 513°C AND 63.3 ATO.
>
> AUXILIARY BOILER: ONE OIL FIRED V2M-8 MARINE BOILER
> TYPE: COMBUSTION ENGINEERING.
> OUTPUT: 32000/5000 KG SUPERHEATED STEAM PER HOUR AT 35°C AND 63.3 ATO.
>
> CONDENSER: ONE MAIN CONDENSER "STAL-LAVAL TURBINE" SURFACE 2101
> ONE AUXILIARY CONDENSER "N.D.S.M." SURFACE: 450 m^2.
> ONE AUXILIARY CONDENSER "N.D.S.M." SURFACE: 140 m^2.

Table 2.2. The Machinery Particulars on a 286000 sdwt tanker.
(By kind permission of Odense Steel Shipyard.)

2.5. STERN CONSTRUCTION

Looking forward from directly astern (Photos. 2.8 and 2.9), of a 210,000 sdwt, 1969, French-built VLCC, the stern frame, which acts as reinforcement to take the propeller from the stern-tube, is clearly visible. In both cases, the after peak ballast tank, steering gear, rudder and propeller units have yet to be added. These have doubtless been left towards near-completion in order to gain easy access internally. It can be seen, (Photo. 2.9), that the stern frame has been extended to support the rudder skeg. In many older Supertankers, the entire stem frame construction

Supertankers Anatomy and Operation

2.8. *The After Part of a 210,000 sdwt VLCC in the mid-stages of completion.*
(By kind permission of Shell International Trading and Shipping.)

2.9. *Stages of Construction prior to the fitting of rudder and propeller.*
(By kind permission of Shell International Trading and Shipping.)

would be made from cast steel elements incorporated into the fabricated hull. It was only the later ones that had completely cast stern frames. The strut which could, at first glance, appear to be part of the propeller is actually the extreme edge of the vibration post that contains the propeller boss and surrounds the shaft leading from the main engine. Forward of the vibration plates are the inspection manholes with a series of uprights fitted for additional reinforcement. The extended stern frame of this Supertanker shows the lower, or heel, pintle gudgeon hole. The rudder, once fitted, rotates upon a bronze liner with vitae magnum bearings.

2.6 RUDDER

The function of the rudder is to overcome water pressure and enable the ship to steer, and when necessary to deviate from, her course. Rudders are made often of a higher grade steel plating, with a network of vertical and horizontal web frames that themselves would be stiffened for reinforcement. They vary considerably in construction and appear in a range of different styles, but rudders fitted to Supertankers usually come into one of a narrower range of types. (Fig. 2.1). Roughly half of the smaller deadweight ranges of VLCC's operating today use a "Mariner" semi-balance, or semi-spade, on the horn, which may be fitted with either one or two pintles. A photograph of a 190,000 sdwt German built Supertanker, (Photo. 2.10), shows an 'open' stern frame, ordinary double plate balanced rudder, fitted to roughly one third of the remaining VLCC's. The rudder and pintle on many ships of this class can weigh around 200 tonnes. There are more recent rudder developments, including the Voith Schneider and Swivelling Kort-Nozzle, but these are invariably fitted to considerably smaller craft.

2.10. *The fitting of a propeller and the six-bladed rudder are completed.*
(By kind permission of Shell International Trading and Shipping.)

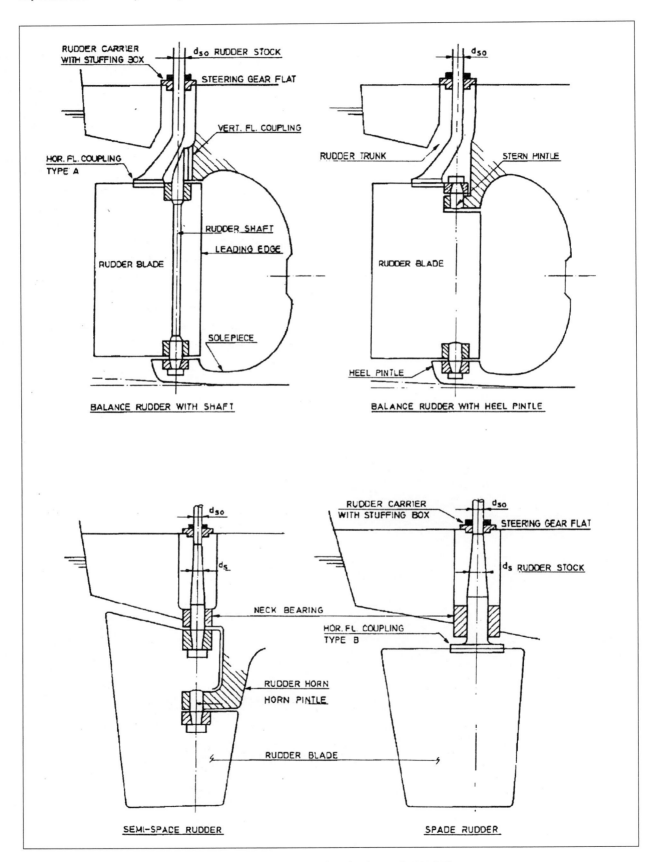

*Fig. 2.1 Rudder Types often fitted to early VLCC's.
(By kind permission of Det Norske Veritas.)*

2.7. PROPELLER

The propeller aboard any ship is essential to transmit power from the engine onto the water and so act upon the rudder. Those fitted to Supertankers are invariably between four and six blades, dependent upon the requirements of the shipowner. Most ships of this class are single propeller, or screw, but this is not always the case. The ULCC "HELOS FOS", for example, of 554,982 sdwt and built in 1979, is currently, after the ULCC "JAHRE VIKING", the second largest Supertanker in the world remaining in ordinary trading and is fitted with twin screws. The propeller of the ship, in Photo 2.10 is of the six bladed propeller construction, whilst examples of five and four bladed propellers are shown in the maker's shop, (Photos. 2.11 and 2.12), prior to being transported by heavy duty lorries and then transhipped for shipyard use world wide. A considerable number of the world's propellers are made from nikalium by Stone Manganese Marine Limited of Birkenhead, and measure approximately 9.1m from end blade to end. They can weigh up to 60 tons and are certainly one of the most expensive single items on the ship. Most are right-hand turning, 'solid', propellers which are used to advantage in practical ship handling and manoeuvring, a few situations of which are examined in Chapter 10. Many propellers are wholly cast in one unit instead of being built with a separate hub and having attached blades. It is important that a critical distance be allowed between the rudder and the propeller. This is essential, not only to permit water generated to flow across the rudder in the required flow, an incredibly complex determination, but also, to help reduce vibration. An 'average' 32,500 shaft horse-powered engine can turn at around 86 rpm to generate a speed, in the case of a VLCC at her load line marks in ballast, of between 16 and 17 knots, and around 15.5 knots fully-laden.

2.11. Examples of a modern five bladed and ...
(By kind permission of Stone Manganese Marine.)

Supertankers Anatomy and Operation

*2.12. ... four bladed propellers supplied to the Shipping Industry and fitted to VLCC's.
(By kind permission of Stone Manganese Marine.)*

2.8. COFFERDAM AND TANKS

Whilst completion work is being undertaken at the stern, the remainder of the cargo tanks constituting the hull are added, the accommodation block is fitted and work commenced on the main cargo deck pipes. (Photo. 2.13). Some ships of this class are built in two or three different sections which are then floated and/or towed to an adjacent part of the yard and welded to complete the ship. (Photo. 2.14).

A glance at the side elevation diagram of the general arrangement plan (inside back cover) indicates that an empty space has been left between the engine-room and the main cargo tanks, extending vertically and horizontally for the depth and width of the hull. Although not immediately clear from a glance at the plan view, the 'cofferdam', as the area is known, actually extends across the width of the ship, including abaft number five centre tank. For it to do so is a legal requirement under the Classification Society rules unless nowadays there are fuel tanks immediately behind. On this particular ship, it has been integrated at the lower end with the pump room. Access is made via a ladder leading to the bottom of the ship and is part of that space. Some, but certainly not all, Supertankers have also an additional cofferdam fitted, forward of the set comprising number one tanks, thus enabling the cargo carrying capacity of the ship to be totally separate from the forward and after ends.

The after part of number five wing tanks is separated into slop tanks into which would be pumped the residue of oily water left after tank cleaning. During a voyage, of course, when the VLCC is in a loaded condition, the slop tanks contain cargo in the usual way. The construction of the remainder of the tanks is very similar to number five, other than the Segregated Ballast Tanks (SBT) at number two port and starboard wings. Under IMO Regulations, these have to be

2.13. *As the Main Deck is completed so the Accommodation Block is added and work commenced on the Deck Pipes.*
(By kind permission of Shell International Trading and Shipping.)

2.14. *The Stern Section of Cargo Tanks nears completion allowing work to be commenced on adding the Main Deck.*
(By kind permission of Shell International Trading and Shipping.)

Supertankers Anatomy and Operation

fitted with their own piping system so that there can be no possibility of their use for cargo. The function of the SBT's is to be partially or nearly full of seawater when the ship is in ballast. This is in order to increase the draught of the ship, covering the bulbous bow and propeller, as well as lowering the centre of gravity which helps reduce longitudinal stress. These tanks would be pumped out when the VLCC is carrying cargo.

2.9. BULKHEADS, TRANSVERSE FRAMES AND STIFFENERS

The next photograph, (Photo. 2.15), is of the same Supertanker seen in a previous shot, (Photo. 2.13), but taken looking from forward to directly aft towards the engine room and showing adjacent cargo tanks. In a further dock, directly astern, a second large tanker is also being constructed. It is not uncommon for a yard to be working on a number of different VLCC's and for these to be in varying stages of completion. In the dock alongside, below the crane, additional cargo tank blocks are in the process of being built and, beside the shed, the sections which will comprise later blocks are being made so that each section moves towards its own varying stages

2.15. The positioning of the Watertight Bulkheads indicates clearly the construction of individual "sets" of cargo tanks. (By kind permission of Shell International Trading and Shipping.)

2.16. The Swash Bulkheads of some VLCC's are of different construction and contain more steel for additional strength. (By kind permission of Shell International Trading and Shipping.)

Supertankers Anatomy and Operation

of completion. The sixth deck of the accommodation block is the wheelhouse. This unit is central to the port and starboard bridge wings, which have yet to be fitted. Looking from the completed sections, just in front of the accommodation and working forward, the construction of the individual 'sets' of cargo tanks may be plainly seen. Particularly clear, is the building of the watertight longitudinal bulkheads that separate the centre tanks from the wings. Immediately down the starboard side, the transversal bulkheads, separating the tanks within each set, and the members constituting the construction of the ship's bottom can also be seen extending towards the shell plating on both port and starboard sides. Above the equidistant bottom longitudinals are built bottom transverse frames together with vertical stiffeners, that reinforce the bilge brackets, and flat stiffeners connecting the shell plating of the bottom with that on the ship's sides. Together, they offer additional strength against plate distortion and bending or buckling in what is often a particularly vulnerable area on a VLCC.

The shot of another very large tanker looking forward, (Photo. 2.16), indicates how work towards completion develops and shows, once the centre and wings have been completed, the manner in which blocks containing the reinforced plating forming the deck are added. All tanks are constructed around a basic binding of upper, midships and lower stringers with reinforcing struts and a series of stiffeners. It is interesting to note the alternative internal construction of the swash bulkheads, or those fitted as transversals within the sets of tanks between the main watertight bulkheads, allowing the cargo to flow within the actual tank. The different internal constructions are produced from different yards. The 'swash bulkheads' are those referred to in Lloyd's Rule 4004, and are so called because they permit the free flow of oil within each tank but, at the same time, cut down considerably the motion when the ship is pitching into even a moderate sea thus reducing resonance of the longitudinal motion. The transverse bulkheads actually separating each set of wing and centre tanks from the next main cargo tank are either corrugated or have attached stiffeners for additional strength, in order to comply with Classification Society Rules. The flow of corrugation, when this is used, generally runs vertically to assist tank cleaning. Far more steel has been used in the construction of this particular VLCC thus providing a far stronger ship.

2.10. PLATES

The deck plates are still in the stages of completion but the flush deck of the forecastle can already be plainly seen. Plating here is invariably made of a high grade steel and, similar to the shell plate at the bottom, is of increased thickness, often to the order of 23 or 24 mm. A deck centre girder is reinforced on each side by transversals and deck longitudinals which is, once again, reinforced with stiffeners in order to alleviate stress. The joining of the deck to the ship's side is made by a specially thickened plate called the 'stringer plate'-whose practical function was discussed when considering the draught/loading relationship with the load line, in Chapter 1.

2.11. MAN AND RATHOLES

The starboard side looking forward, in the VLCC, (Photo. 2.17), offers a quite dramatic view. The near completed fore part of this ship indicates the complete wing tank construction from longitudinals at the ship's bottom to the plating of the main deck. The smaller circular holes that have been cut into the transversal girders, immediately above the longitudinals, are man holes enabling crew and shore staff to walk the actual bottom of the Supertanker, within each individual tank, of course, not the entire length of the ship, inspecting in pre MARPOL classes of tanker the cleanliness of the tanks following discharge of one cargo before loading the next. Post MARPOL's would require such visual inspections only prior to dry docking for safety purposes and periodically between dry docking to check that everything within the tank remains in order. Some tanks, such as the after peak ballast, would also have 'ratholes' inserted into the internal deck plating, thus allowing access to the various levels of deck constituting the tank. The chains

Construction

*2.17. As the Forepart nears completion, so the "drag" chains to assist with launching are prepared.
(By kind permission of Swan Hunter Shipyard.)*

alongside the ship are called 'drag' chains and are used to control the rate of launching. The figures, situated in various places alongside the ship, offer further obvious perspective.

2.12. FORE PEAK TANK

Continuing to work forward, the next photograph, (Photo. 2.18), shows the early stages of construction of number one starboard wing tank, with the transversal bulkhead separating number one tank from the fore peak tank. A further glance at the general arrangement plan shows the extent to which the forepart of the cargo-carrying section of the ship is very much like a large oblong box that curves radically towards the bows. The forecastle space has a comparatively short head-length from the tanks to the stem, and the bows are finely flared and very rounded in order to meet the horizontal thrust upon the Supertanker as she encounters a head sea whilst under weigh.

2.13. BULBOUS BOW, STEM

The downwards and outwards flow of the water is assisted in its deflection by a 'bulbous bow', at the lower part of the stem, submerged to a level that is at optimum depth to the forward

45

2.18. The Forepart of a 200,000 sdwt VLCC nears completion.
(By kind permission of Shell International Trading and Shipping and Ned. Doken Scheepsbouw Mij.)

draught of the ship. A secondary pressure system is then able to occur which is out of phase with the wave motion generated by the forward motion of the ship under weigh. The bulbous bow has therefore quite a significant part to play in the forward propulsion of a supertanker. Clearly, it is subjected to considerable stress and has, therefore, to be very heavily reinforced. There are usually a greater number of transversal stringers, frames and beams fitted in the fore-peak/bulb area than in the complete number one tank construction, over a distance that is about only ⅕th the length of that cargo tank. (Photo. 2.19). In dimension, the bulb can be around 9.1 metres high, and is usually, but not necessarily, of a similar dimension to that of the propeller. A different type of bulb construction is fitted to some VLCC's, as indicated in the following photograph, (Photo. 2.20), of the "WORLD UNICORN", 250,000 sdwt, built by Swan Hunter's in 1974. It was the success of fitting bulbous bows to very large tankers which led to the adoption of a form of bulbous bow to smaller ships not necessarily of the tanker class. Most dry cargo ships, for instance, including the m.v. "GOLD VARDA", have a non-hollow bulb constructed of reinforced steel plates that, on being submerged with the cargo ship under weigh, add up to an extra knot to her speed.

Whilst dealing with the forward part of a Supertanker, it is worth mentioning that some VLCC's have a vertical drop from the bottom of the fo'c'sle head to the top of the bulbous bow, in what is termed a U-shaped stem. (Fig. 2.2). The corresponding height of the bulb can vary considerably. The more streamlined, or V-shaped stem found, for example, in the "WORLD UNICORN", (Photo. 2.21), although possibly more aesthetically pleasing, is not significantly more efficient. The choice is largely a matter of shipowner's taste, acting upon advice of architects and superintendents.

2.19. The Early Stages of Construction of the Fore Peak Tank and Bulbous Bow.
(By kind permission of Shell International Trading and Shipping and Netherlands Dock and Shipbuilding Company.)

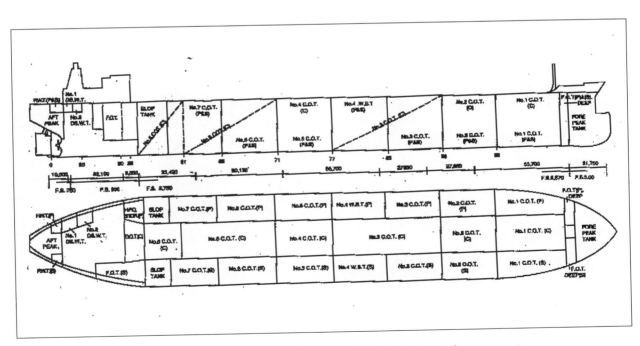

Fig. 2.2 Profile showing a U-Shaped Stem, and Plan indicating tank arrangements.
(By kind permission of Mobil Ship Management.)

2.20. *The Forepart and Bulbous Bow near completion.*
(By kind permission of Swan Hunter Shipyard.)

2.21. The completed Supertanker "WORLD UNICORN" showing the V-Shaped Stem. The "Tracing Wires" assisting launching remain attached to the Hull.
(By kind permission of Turners (Photography) Ltd and Swan Hunter Shipyard.)

2.14. LAUNCHING

In view of their considerable dimensions, and the very real danger of causing stress to the hull, methods of side-launching or using ship's lifts have never been a realistic proposition for Supertankers. The ships were launched in one of two ways. Some of the 'early' generation, such as the World-Wide Shipping Company of Hong Kong's "WORLD UNICORN", were launched stern first from the slipways of Swan Hunter's Yard in Newcastle-upon-Tyne. The VLCC's were built on the traditional moderately angled, greased slipways, apparent in the following construction photograph. (Photo. 2.22). The rate of launching would be slowed to manageable proportions by attaching to the hull the 'drag' chains mentioned earlier, which are supported also by tracing wires. (Photo. 2.23). The final stages of launching as the ship entered the water, proved inevitably to be a fascinating realisation of an original set of plans into a working reality. It is interesting to note the number and disposition of the tugs used (Photos. 2.24 and 2.25) as the VLCC entered the restricted waterway of this particular launch.

The second method, more commonly used as Supertankers increased in size and usual today, is to build the Supertanker totally in dry dock, until completion, and then to launch by a simple expedient of flooding the dock, allowing the massive ship merely to float off the blocks and to be towed into the river or sea. The following photograph, (Photo. 2.26), shows the launch of a 553,000 sdwt ULCC in this way.

Supertankers Anatomy and Operation

2.22. Some "Early" VLCC's were launched from moderately angled Slipways.
(By kind permission of Harland & Wolff and Shell International Trading and Shipping.)

2.23. In the Final Stages of Preparation for Launching, tracing wires are attached to the Hull.
(By kind permission of Harland & Wolff and Shell International Trading and Shipping.)

2.24. *The Final Launching Arrives. A 200,000sdwt VLCC takes to the ...*
(By kind permission of Harland & Wolff and Shell International Trading and Shipping.)

2.25. *... water to be towed to the Fitting Out Berth for final stages of Completion.*
(By kind permission of Harland & Wolff and Shell International Trading and Shipping.)

2.26. Preparations for the Launching, by flooding the Dry Dock, of a 553,000 sdwt ULCC. (By kind permission of Shell International Trading and Shipping.)

Chapter Three

HULL AND TANK PROTECTION

3.1. INTRODUCTION

3.2. RUST AND CORROSION

3.3. INITIAL STEEL CLEANING

3.4. PRIMING

3.5. CAREFUL PREPARATION

3.6. PAINT COMPOSITION AND APPLICATION

3.7. PROTECTION, CORROSION

3.8. PROTECTION, FOULING

3.9. CATHODIC PROTECTION, SUBMERGED HULL

3.10. MARINE AND VEGETABLE GROWTHS

3.11. SIDE EFFECTS OF FOULING

3.12. HYDRO-JETTING

3.13. TANK CORROSION: CARGO, BALLAST

3.14. PROTECTION OF PWB TANKS

3.15. INSPECTION, ELECTROLYTIC DE-SCALING

Chapter Three

HULL AND TANK PROTECTION

(Synopsis — General Comments — Restrictions to Single-hulled VLCC's — Atmospheric Attack on Steel — Swedish Corrosion Institute's Standards — Blast-cleaning — Nature of Rust-Importance of Consultation Between Owners, Builders and Paint Company — Limitations of a Deck Officer's Knowledge in a Highly Specialist Area — Necessity for Adequate Cleaning of all Surfaces — VLCC Painting Area — Types of Painting — Preparation of Plates — Welding Operations — Composition of Paint — Hull Protection: Against Corrosion and Fouling Including Sea-chest, Rudder and Propeller — Protection of Main Deck: Hydro-jetting — Protection of Clean Water Ballast/Cargo Tanks: Brief Quotation from Tanker Structure Co-operative Forum's "Condition Evaluation and Maintenance of Tanker Structures" Manual — Segregated Water Ballast Tank Problems and Cathodic Protection: Incident Regarding a ULCC — General Comments on Metal Corrosion: Stripe Coatings: Electrolytic De-Scaling — Contribution of Jotun-Henry Clark Marine Coatings to this Chapter.)

3.1. INTRODUCTION

A popular political remark coined a few years ago urged everyone to 'go back to basics', an adage that is totally germane to hull and tank protection, particularly of Supertankers. There is something psychologically pleasing on seeing a tidy neatly painted ship and all professional deck officers have not only a justifiable pride, in being associated with the smartness, but also a concern with playing their part in combating the attendant corrosive effects caused by weathering and chemical decay. Chapter 9, in fact, will show how considerable emphasis continues to be placed on the main deck, superstructure and funnel, each area of which falls within the Mate's jurisdiction. There were and remain however parts of the Supertanker, namely the hull, cargo and ballast tanks, in whose protection he would have not only a professional interest, but also a very practical concern for, after all, every serving seagoing officer has more than a passing desire that their ship should remain in one piece beneath them, particularly when she is facing some of the heavy weather conditions commonly found at sea. Certainly, the practices examined in this chapter are relevant to 'early' Supertankers. This applies also to refinements made subsequently to their protection, including all single hulled VLCC's and the 1970-plus ships still in operation, particularly when these have been blast cleaned to the bare steel during routine dry-docking.

3.2. RUST AND CORROSION

The steel from which ships are made comes from iron ore and it is the ore content itself which is under attack from oxygen and water vapour in the air. Owing to this atmospheric oxidisation therefore all ship's plates, after they have been rolled in the steel mill, have a layer of iron oxide on the surface. This is in the form of what is called 'mill scale' which, as long as it is intact, will protect the steel from further corrosion. When this layer has been disturbed however

Supertankers Anatomy and Operation

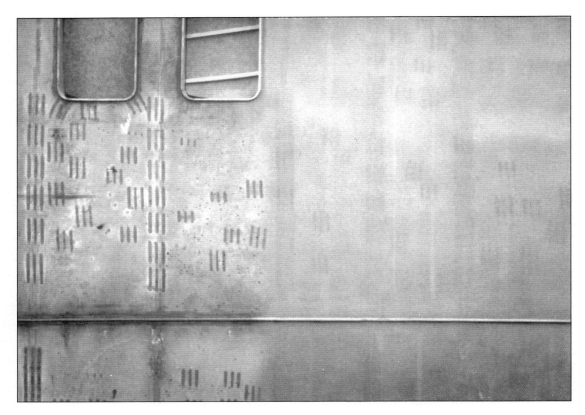

*3.1. Grit Blasting to Swedish Corrosion Institute's ISO Sa 2.5.
(By kind permission of Jotun Marine Coatings.)*

a process known as 'flake rusting' commences, and it is in order to arrest this decay that the plates, immediately on receipt, are shot-blasted by being subjected to a high speed jet of grit or sand. The Swedish Corrosion Institute has produced internationally accepted standards concerning the surface state of steel prior to protective painting. They have produced various scales denoting pictorially the progressive stages of rust as well as determining standards covering grades of subsequent preparation. Paint manufacturers invariably accept the Swedish standard of ISO 805l, at Sa 2.5, as offering probably the best surface preparation for subsequent painting. This standard states:

> *"Blast cleaning to nearly white metal cleanness until at least 95% of any section of surface area is free of all visible residues. Very careful blast cleaning. The blast cleaning is maintained long enough to ensure that mill scale, rust and foreign matter are removed so thoroughly that any residues of these appear only as slight shadows, streaks or discoloration on the surface. Finally, the dust is removed with a vacuum cleaner, or with clean, dry compressed air or with a clean brush"*

These exacting standards are designed to increase the corrosion protection quite considerably, as much perhaps, as 50% in plating used in areas above water, and possibly more over submerged plates. The grit-blasting would be effected also to ISO 8503, which is concerned with standards in the surface roughness of steel before the application of paints. The above photograph indicates a plate whose surface has been prepared by grit-blasting to the ISO Sa 2.5 standard. (Photo. 3.1). The cleanness is indisputable, all foreign particles gained from manufacturing have been removed, whilst the slight streaking and discoloration referred to can be plainly observed.

Hull and Tank Protection

3.2. A rather Aesthetic view of Rust seen under a Microscope.
(By kind permission of Jotun Marine Coatings.)

Rust is iron ore in a very refined state and, although when viewed under a microscope, it may appear rather aesthetic as, undoubtedly, it does, (Photo. 3.2), it is in reality like a cancer eating away at the metal comprising the ship's plates and will ultimately destroy them beyond repair if it is not checked. The darker parts of the picture show the more extensive levels of decay that are approaching the final stages of irreparable damage. Following the process of shot or grit-blasting, the next step would be to give the plates a temporary thin, dry-coat, of shop-primer; one that would be compatible with any subsequent paint treatment, and that offers some months protection from rust. The plates are then ready for transportation to the cutting and construction sheds were they can be shaped into the required sections before being given final coatings.

3.3. INITIAL STEEL CLEANING

As soon as it is possible during planning of the ship's construction, the shipowner, building yard and advisors from the paint company should meet to discuss specifications regarding the numbers and thickness of paint applications. Successful, long-lasting and effective painting cannot be achieved without the correct preparation of the steel surfaces. Most deck officers, by virtue of their own on-the-job training as cadets, understand the problems that are involved in shipboard painting. It is customary for the second mate to be assigned the responsibility of painting supervision/liaison during dry docking, so mates are at least aware of aspects involved in the protective work undertaken by those in the ship-building yard at the various crucial pre and post construction periods. The protection of these areas at the construction stage, however, requires the possession of highly complex and technical knowledge at a level which falls only rarely within the resources of most Masters and Mates. Some of the difficulties which have to be overcome by managers and their artisans within the yard are therefore highlighted. The following photograph, for example, (Photo. 3.3) shows sea salt crystals, in the form of chloride ions, remaining after the paint

3.3. *Salt Crystals in the form of Chloride Ions which remain after paint coating has fallen off.*
(By kind permission of Jotun Marine Coatings.)

3.4. *Effects of applying Paint Coating to an oily surface.*
(By kind permission of Jotun Marine Coatings.)

Hull and Tank Protection

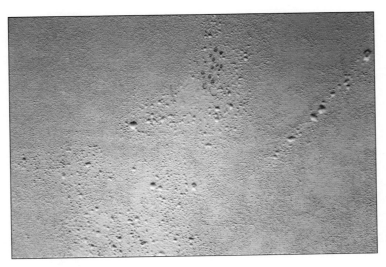

*3.5. The effect of Trapped Solvents below the paint skin causing Blistering.
(By kind permission of Jotun Marine Coatings.)*

coating has come off, and gives clear-evidence of the necessity for ensuring that all surfaces due to be painted are completely clean. To give some idea of perspective on the photo, the imposed grid indicates the dimensions, with a distance of 200 mm being represented horizontally and vertically by the ten-unit 30–40, 40–50 etc. It is interesting to note that the more recent crystals formed are virtually square in shape, loosing their sharpness and diminishing as they are exposed increasingly to the condensation in the atmosphere, until finally disintegrating into the deeper reddish-brown of rust. The only effective way of removing salt crystals is by a high pressured washing down with fresh water. The photograph reinforces how vulnerable is the untreated bare steel to the elements, whilst the next shot, (Photo. 3.4) portrays how even sufficiently thick paint coats will flake-off if they have been applied to an oily surface. Obviously, the oil will have to be removed using, if necessary, the application of detergents in hot water for especially persistent spots. The oil patch has affected the back of the paint layer and penetrated through into the top surface in the form of discoloration. When the coats are being applied, due attention must be paid to the weather so that high humidity and low temperatures do not cause moisture or solvents to become trapped underneath the paint skin and so cause the surface to "blister". (Photo. 3.5). It may even be necessary for artificial heating to be used regardless of the inconvenience, or even potential impracticality, this may cause.

3.4. PRIMING

Ideally, the first coat of paint on new steel surfaces, should be applied by brush because that is the most effective way of covering all of the areas thoroughly and with adequate penetration. A VLCC, however, has a total painting area of around 100 acres, with about 300,000 m^2 of ballast tank painting alone. Clearly, in the light of such enormous surfaces, brush painting would prove far too costly and impracticable in terms of labour and time. Brushing, even though the eventual finish may not look over-smart owing to the inevitable brush-strokes, or be of consistently even application, helps an initial primer to take to the surface more readily. It is the best medium also for ensuring that any moisture content, that might have seeped into the process, may be spread consistently. At best, in practice, a compromise has to be agreed: that certain spaces, such as spots requiring rejuvenation, will be brush-painted whilst others, such as vast extended flat areas, will be either roller or spray painted. Certainly, the latter produces the most pleasing and professionally finished result. With spray painting, however, due attention must be paid to ensuring that the correct settings are made to the equipment. If there is a viscosity, for example, that is

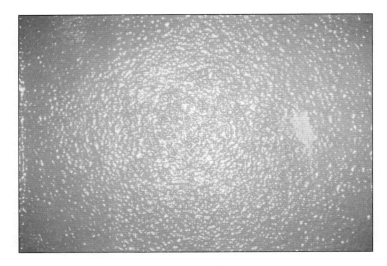

*3.6. "Orange Peel" effect caused by Incorrect Spraying.
(By kind permission of Jotun Marine Coatings.)*

too high, or if the spraying takes place too close to the surface area, then a defect in appearance can be caused which has the appearance of 'orange peel'. (Photo. 3.6). This defect, like all of the potential difficulties examined, can be avoided if sufficient time and care is taken to ensure that the paint is applied evenly across surfaces. Excessively thick coatings should be avoided in order to prevent not only paint wastage, but the eventual build up of successive coats over a period of years into what is termed a 'sandwich' coating that could lead eventually to the paint layers falling from the hull.

3.5. CAREFUL PREPARATION

Much of the preparation in the early stages is conducted under cover in order to keep atmospheric and weather conditions as near as possible to a minimum. The plates are given support just sufficient to enable access to be made, thus reducing initially to a minimum, those untreated spots that would have to be dealt with later. Operations during welding need to be

*3.7. Evidence that rough finish of Welding ...
(By kind permission of Jotun Marine Coatings.)*

Hull and Tank Protection

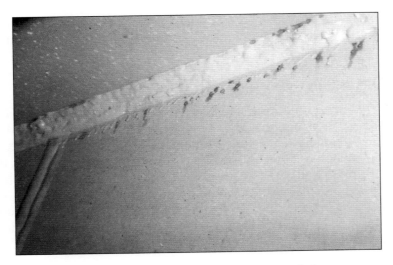

3.8. ... which can lead to Rust, if not smoothed.
(By kind permission of Jotun Marine Coatings.)

3.9. A Small Hole left during Welding will soon lead to Rust.
(By kind permission of Jotun Marine Coatings.)

conducted very carefully and with due attention so eliminating, so far as is practicable, some of the major faults likely to be introduced. For instance, all rough finishes such as 'weld spatter' (Photo. 3.7) have to be detected, and remedied by smoothing, so that rust will not be allowed to permeate (Photo. 3.8). Welding slag and any residues, such as weld flux deposits, also need to be removed. The welding in all areas should be completely finished because no amount of paint can fill any small holes that might have been allowed to pass unnoticed as, again, the enemy rust will soon seek them out. (Photo. 3.9). Seams have to be rounded and rendered non-porous and any burn damage rectified, that might even have gone through to the other side of the plate. It is apparent, even from this elementary and cursory examination, that the professional treatment of steel is regarded as a vitally important procedure. Certainly, any of the cited mistakes, plus others not mentioned, made at this constructional stage will live with the ship until she has reached the end of her useful trading life and is ultimately scrapped. Indeed, it is true to say, that inadequate initial preparation may well reduce that useful period perhaps quite considerably as well as affecting subsequent maintenance costs. On most VLCC's, nearly all of the painting is applied at the block stage before being erected in the building dock. This means that the areas where the sections have been joined are left for subsequent painting, along with the coating of any damaged areas which might occur.

3.6. PAINT COMPOSITION AND APPLICATION

It might be relevant to offer an elementary idea of the quite complex composition of paint in order that subsequent explanations attempted may be better understood. Marine paint is composed initially of a binder, pigments, extenders and solvent. By far the most important ingredient is the resin binder because this is the medium enabling the paint to obtain its film-forming quality and it provides also the adherence by which paint remains on the surface to which it has been applied. It is the binder that also describes the generic type of paint, such as epoxy and alkyd etc., whilst the pigments are able to provide both colour and rust protection as well as the important task of reinforcing the paint film. There may be as many as eight different varieties of paint and coatings available for application to different parts of the ship.

3.7. PROTECTION, CORROSION

The hull is one of the main areas for protection, and any considerations must take into account the combating of two likely problem areas: the first concerns the under-water plating that is attacked by sea-water corrosion, whilst the second confronts the difficulties caused by fouling due to the presence of marine growths. There exists also the potential difficulty of ascertaining that the application of the anti-corrosive and anti-fouling coatings will have compatibility with each other. Obviously, it is desirable also that paint used on the hull and sides of ships, above the water-line, should be water-resistant and durable. In order that this can be achieved, a suitable paint system is applied in a wet film which will dry back to a solid film by the evaporation of solvents and, in the case of epoxy type paints, will cure also a chemical reaction that takes place between the two components. The main paint type used in ordinary practice on the hulls of VLCC's is coal-tar epoxide-resin. This is because epoxy-tar is highly resistant to water and chemicals and has the quality of working well with other protective agencies applied to the hull. On the flat bottom of a Supertanker two coal-tar epoxy coatings, together totalling a minimum thickness of 250 microns, should be applied to which should be added a further coat of anti-fouling. From the bilge-keel to the boot-topping the same anti-protective measures are applied as with the keel, but with an additional two coats of anti-fouling making the latter 150 microns in total. The area from the boot-topping to the topsides of the hull has a slightly thinner coat of 200 microns anti-corrosion, together with 50 microns of black finish.

3.8. PROTECTION, FOULING

Particulars of trading patterns, with specific regard to changes likely to be encountered geographically that will affect temperature changes and salinity levels, are important considerations to hull protection. Operational matters also have some significance, especially the number of days actually spent in port and on passage by the VLCC, as well as the closeness of the ship to her dry-dock survey time. Each of these factors affects the degree of fouling particularly to which the Supertanker is likely to be subjected. For many years, research had been undertaken by leading paint manufacturers, which led to developments in the 1960/70's of a long-term 'five-layered system' of anti-fouling. The difficulty which had to be overcome was that of enabling the anti-fouling agent to be released. With the passing of time, as the agent became used up, it remained as an inactive thick layer which trapped the new agent preventing it from emerging. The solution from one leading paint manufacturer was to invent their "Seamaster" range that used, at the time a revolutionary novel approach, of varying coat thickness applied after the customary blast cleaning to Sa 2.5 and a 15-20 micron application of shop-primer. This was followed by four coats of 100 micron 'Vinyguard' and, on the side bottom underwater, four 75 microns coats of "Seamaster" anti-fouling, with two 75 micron coats of "Seamaster" to the flat keel area. The anti-fouling contained a cuprous oxide coloured pigment which, when exposed to sea-water, gradually faded from deep

*3.10. ULCC "JAHRE VIKING" indicating the effect of Successful "Self-Polishing" Hull Coating.
(By kind permission of Jotun Marine Coatings.)*

red to a lighter red and then finally a brownish grey and, during reactivation, yellow. One year on, following the application of the protective compound, a diver would be sent below who looked for the yellow-coloured layer of the fifth coat and scoured this until the red of the fourth layer appeared. The process proved an ideal protective for VLCC's, especially those which were used as storage vessels and might spend periods in excess of one year at anchor awaiting orders for the cargo. This five-year activation however, although certainly effective in prolonging the anti-fouling life of the ship, was gradually overtaken because, following developments in copolymer technology, it was found that various active ingredients could be released that gave excellent protection and used the actual passage of the ship through the water to clean and smooth the hull. The following photo of the ULCC "JAHRE VIKING" (Photo. 3.10) gives an indication of the effectiveness of this research into what is now known as "self-polishing" coatings. The Supertanker was treated by Jotun Marine Coatings against fouling and corrosion by their combination painting. Their fast-polishing "AF Seamate HB 66", was painted over their long-lasting "AF Seamate HB 33". The photograph shows this impressive ship in ballast passing through the Suez Canal en route for the Gulf, an ideal state and condition in which to examine the protection coatings applied. The patches of dark red, which is their HB 66, is seen against the light red, the HB 33, and the 'time-expired' grey, showing that whilst, the appearance is not, as Jotun's state: *"perhaps as aesthetic as uniform colour, it indicates how the self-polishing mechanism worked to achieve hull smoothness and a fouling-free hull"*.

3.9. CATHODIC PROTECTION, SUBMERGED HULL

The external part of the hull underwater, which has to be protected, includes a considerable extent of boot-topping, or the submerged part of the supertanker when she is down to her marks with cargo, as well also as the propeller shaft, rudder and pintle. It is the process of electrolysis, activated by the immersion of the shell plating into sea-water, which sets up the electrical current that attacks the metal and causes corrosion to the hull. Amongst a number of problems this may cause is the development of alkaline deposits that attack the paint binder thus creating a softening of the paint, a possible blistering effect, and even in extreme instances, the paint falling from the hull to expose the bare metal. It is the combination of a tie-in with the epoxide-based paint that helps to combat this defect. The use of sacrificial anodes has proved effective on smaller tankers, but to use this system on the hull of the average 300,000 sdwt VLCC would necessitate something to the order of twenty tonnes or so of anodes in each of the periods between her

Supertankers Anatomy and Operation

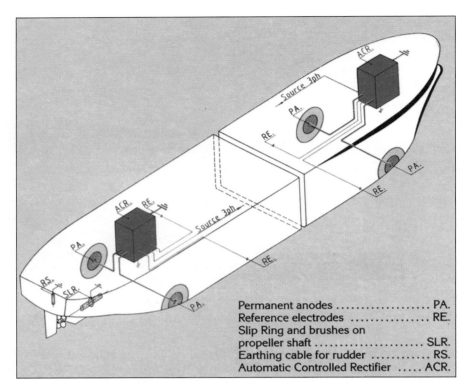

*Diag., 3.1 Diagram showing the layout of a typical ICCP System to a VLCC.
(By kind permission of Jotun Marine Coatings.)*

thirty month intermediate dry-docking periods, midway between the customary five year special survey. This would clearly be quite prohibitive in terms of cost over the life of the ship. In view of this detrimental feature, Supertankers were generally fitted with an "impressed current system" (ICS) of cathodic protection which worked by having a connecting link between the ship's electrical circuits and two rectifier control sources that supplied four power units. Although basically the same, in concept, the techniques in the "early" generation of Supertankers were effective, but quite unsophisticated, compared to the description of the following system currently in practice today, although this modernised technique has been supplied to many of the older generation of VLCC's. The Impressed Current System of Cathodic Protection (ICCP), designed specifically for hull areas, consists of two automatically controlled rectifiers one of which is situated in the engine-room and the other in the fo'c'sle. Each of the units supplies a number of permanent anodes that are fitted to the outside of the hull. The metal to be protected is coupled to the negative pole of a direct current source, whilst the positive pole on the DC source is connected to an auxiliary anode, which dissolves at different rates in the electrolyte depending upon the type of material from which it is made. The diagram above indicates the arrangement offered, with modifications made to suit individual VLCC's (Diag. 3.1). The permanent anodes are made either from platinum or mixed metal oxides, each of which has an extremely low dissolution rate and is provided with an extra thickness of coat painting for additional protection. They are linked by cables that are led through the side of the ship by watertight junction boxes. The reference electrodes are made of high purity cast zinc and their function is to control the current flow. Again, these are recessed in the forepeak and engine-room aft of the permanent anodes. The two rectifiers, which can be plainly seen in the diagram, use the voltage to control an AC supply that is then transformed and rectified in the power unit to provide a protective current which basically opposes the current activated by the sea-water electrolyte. The anodes and electrodes are recessed into the hull and so contribute to the smoothness and hence reduce drag to an absolute minimum, thus not affecting adversely the speed of the ship. Recessing means also that they are protected from potential physical damage and are unaffected by surrounding paint decay. The propeller shaft and propeller are protected by a slip ring and a set of carbon brushes which have the effect of bonding the propeller and its shaft to the hull, so ensuring that these would not be insulated from the remainder of the ship. The rudder is grounded, in a similar manner, by a length of flexible copper cable. The actual current necessary for complete protection can be varied automatically and is governed by a number of factors, including the speed of the ship, her draught, the salinity of the water and, as time progresses nearer to the dry-docking, the condition of her paintwork.

3.10. MARINE AND VEGETABLE GROWTHS

The necessity for providing adequate protection against marine growths and similar agencies has been mentioned in passing. Measures to counter an increasing number of incidents, arising from 'attacks' by something to the order of 4000 animal and vegetable species that comprise the 'fouling community' of marine organisms that are attracted to the hulls of ships, have been the subject of much concern by seafarers. The early Greeks, for instance, used pitch in an effort to counter-attack, whilst the Romans went one better by plating the hulls of their ships with tin, but it was not until advances made in the 13th Century, which remained virtually unchanged until the 19th Century, that a move was made away from the copper-sheathing that had existed adequately over these intervening centuries. Hull-protection against fouling for the early generation of Supertankers was therefore a subject of great concern, and considerable research was undertaken, with a pooling of knowledge between some of the major shipping companies, paint manufacturers (a successful outcome of which was seen regarding the hull of the Jahre-Wallem ULCC), the predominant learned institutes, especially of corrosion, chemical and water engineers, and the Marine Research Agency, to find an effective and long-lasting antidote. The problem was widespread. The most common type of fouling encountered is the barnacle, familiar to visitors at any beach, which can be seen attached to any underwater object, be it man-made such as a jetty or breakwater, or natural, such as rocks and most exposed beach surfaces. Even other marine creatures, such as shellfish and crustaceans, may be observed carrying their unrequited lodgers. Their adaptability is almost a byword for barnacles are to be found in every area of the world and in all extremes of temperatures. One type is the Acorn Barnacle, (Photo. 3.11). The eggs from this species are released into the sea and soon develop into larvae which possess six pairs of cirri for swimming, and two antennules. On meeting an acceptable surface, the larvae emit a very powerful adhesive substance, said to be the strongest glue in the world, from a body gland that helps it to mould permanently to the object and is one that has capabilities to withstand considerable water pressures. On reaching maturity, a new generation of eggs is hatched which develop into larvae and so continue the process. The Gooseneck Barnacle (Photo. 3.12) lives in deeper seas, but around 20 to 30 miles offshore and they act also in a manner similar to the Acorn. The common problem posed by barnacles to VLCC's is derived from the amount of time which this class of ship spends at sea, under weigh, anchored or moored to berths and jetties and the comparatively slow speeds at which they steam. Unlike faster ships, the barnacles can cope easily with the 15 to 16 knots of the Supertanker and both types soon grow and reproduce on the bottom and submerged sides. Other problem growths are Entermorpha which is a slender grass-type weed that is most prolific in the light. Invariably, this grows on the boot-topping areas of the hull and the upper side-plates. Ectocarpus is a short stunted dark-coloured growth that favours the darker surfaces and thus attaches itself to the deeper sections of the hull. The area of the flat-bottom around the keel does not seem to attract much fouling, probably because it is too dark there for even the most tenacious of maritime growths to gain too much of a hold. The only exception perhaps would be the old building-block areas which might, as a result, be slightly more vulnerable. Amongst other major fouling is the commonly – seen Slime (Photo. 3.13). This appears as a fur like micro-organism which is attracted to all under-water surfaces. The great obstacle of the combined influences of these various growths is the increase in hull resistance they provide to the speeds of the ship. Inevitably they

3.11. Marine Growth: Acorn Barnacle.
(By kind permission of Jotun Marine Coatings.)

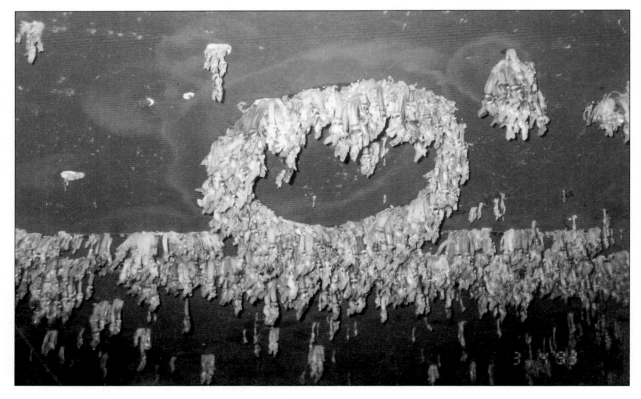

3.12. Marine Growth: Gooseneck Barnacles.
(By kind permission of Jotun Marine Coatings.)

Hull and Tank Protection

3.13. Marine Growth: Slime.
(By kind permission of Jotun Marine Coatings.)

3.14. The Effect of Tin-Free Anti-Fouling on the VLCC "BRITISH RESOLUTION".
(By kind permission of Jotun Marine Coatings.)

slow her down considerably and so increase fuel consumption and reduce the all-important charter speeds, with the very serious implications that this implies.

The remedy against animal and vegetable growth has been to use anti-fouling protection by coating the hull with toxic paints that, basically, use the poisons found in oxides with tin-copper-mercury compounds. The mixture is applied over the anti-corrosive paints, with very considerable care being taken to make certain that no conflict arises between the constituents. Research findings have found that, if the anti-fouling coats are made from an organotin copolymer application, then a very effective, quite long-term, protection can be afforded against most of the marine growths. Certainly, experience proves that the growths feed on the poisons and die, with new growths being deterred from attaching themselves to hulls by the poison released into the water. The concern, of course, is with the protection of the environment and particularly over the use of Tributyl Tin Oxides (TBT), and recent legislation has banned their use in an effort to find the "environmentally-friendly 'green' ship". A plethora of SOLAS (and other) rules and regulations has exploded upon the maritime industry which has certainly had the effect of concentrating thoughts on developing protectives that will be more acceptable. The people involved are all responsible. Most agree that it is the tin in anti-fouling which is the problem: some argue that the tin dissolves in three weeks or so, but remain vague concerning where it goes to from there, others argue that it does not dissolve quite so readily. The debate remains open, but the research is producing some answers. In the early 1990's, for instance, a VLCC of a major shipping company, who remain actively involved in environmental protection, was treated with "Seaguardian" TBT-free self-eroding copolymer anti-fouling and, when it was dry-docked some four years later, was found to be virtually free of marine growths (Photo. 3.14). A later development "Seavictor", then the most advanced tin-free and self-polishing anti-fouling, also represents advances in ridding the oceans of poisons.

3.11. SIDE EFFECTS OF FOULING

A side-effect of marine fouling concerns the points where the intake of sea-water is made into the ship for circulatory purposes. These are the sea-chests in the lower side of the hull into which marine growths enter and whose environmental conditions are conducive to encouraging proliferation. Inevitably, the result is turbulence, with the accompanying factors of cost and time-consuming repairs. A 'CUPROBAN' protective system consists of a rectifier and control panel (Photo. 3.15) that is situated in the engine-room spacing from which can be controlled two anodes, that are made from copper and aluminium, or soft iron, that are fitted into the sea chest and which, by electrolysis, treats the sea water and kills the offending growths, as well as providing anti-corrosive protection. (Photo. 3.16) (Diag. 3.2). Details of a typical Cuproban application found on a range of tankers, not just VLCC's, shows the arrangements between the sea chests and the control lines. Other internal areas that can be safeguarded are the fuel tanks, with epoxy-tar, and the engine-room/machinery spaces with often two 40 micron coats of 'Alkyd Primer'. Areas at the stern where the speed of the ship affects close proximity to rudder, propeller and shaft are particularly vulnerable. This is due to the difference in metals used in construction, compared with that of the hull, creating a potential difference, and also to the speed of the ship. Rudder and pintles require an extra thickness of paint coating, but the propeller shafting is best served by cathodic protection. The diameter of the propeller and the ship's speed will determine the extent of the current, but its flow will occur in the bearings of the shaft. The answer is for a specifically designed steel ring to be prepared for each individual propeller shaft which is bolted around the shaft and joined at the top with a brush holder containing two silver-graphite brushes. (Photo. 3.17). A short length of electrical wire is fitted between the brush-holder, and a bolt is welded to the hull thus ensuring a low resistance path between shaft and hull.

Hull and Tank Protection

3.15. 'CUPROBAN': *Rectifier and Control Panel with a Selection of Anodes.*
(By kind permission of Jotun Marine Coatings.)

3.16. 'CUPROBAN': *Anodes installed in a Sea Chest.*
(By kind permission of Jotun Marine Coatings.)

3.17. *Slip Ring Arrangement to Propeller Shaft.*
(By kind permission of Jotun Marine Coatings.)

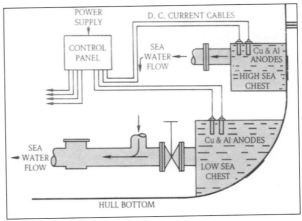

Diag. 3.2 'CUPROBAN': *Typical Installation.*
(By kind permission of Jotun Marine Coatings.)

3.12. HYDRO-JETTING

The safeguard of the main-deck on one particular VLCC due to operational reasons with the crew being simply too busy with port turn around and a continuous heavy spell of bad weather that prevented deck work, reached a state of decay that put it quite outside the Mate's preserve. The remedy was for the deck area to be refurbished whilst the ship was dry-docked for her thirty month survey. The following photographs indicate the state of the maindeck before treatment. (Photos. 3.18 and 3.19). The extent to which the rust had developed can be clearly seen and plainly drastic treatment was necessary in order to preserve the integrity of the deck. The answer was to hydro-jet the area at 20,000 psi and then apply a coat of protective paint. The following shot (Photo. 3.20) indicates the excellent result. Hydrojetting is an alternative method to traditional grit-blasting used in plate cleaning and, in this case, was effective in the removal of all rust scale, soluble salts, loose paint fragments etc. that left all surfaces clean for the painting. Certainly, with reductions in crew size and additional recognition of the extent of the maintenance involved, it has become a current practice for shore-gangs to be put onto supertankers to do whatever blasting and painting might be necessary.

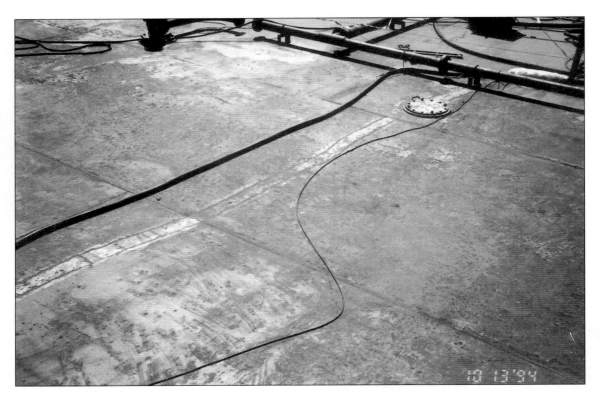

3.18. Rust Problems on the Main Deck ...
(By kind permission of Jotun Marine Caoatings.)

3.19. ... Close-up View.
(By kind permission of Jotun Marine Coatings.)

*3.20. The Same VLCC Main Deck following treatment.
(By kind permission of Jotun Marine Coatings.)*

3.13. TANK CORROSION: CARGO, BALLAST

Other than the fitting of anodes, complete protection of all cargo carrying tanks on a VLCC by coating has never been a viable proposition either economically or practically. Those cargo tanks which are to be used for clean sea water voyage ballast are however a second area of interest to the professional Mate but are again, virtually outside his shipboard resources. The early VLCC's usually had the ballast cargo tanks partially painted, invariably black or brown, with two 125 micron coal-tar epoxy coatings to a height of about three or four metres. Sometimes only the bottom structures were treated but, on some ships, coating was applied also to the top three or four metre area. On virtually all VLCC's, therefore, only the centre tanks used for both cargo and sea water ballast would be partially treated. The dark colour, however, sometimes made it difficult to assess the coatings which, in reality, could be anything between five and nine hundred microns, along with the inevitable 'holidays' caused by the difficulty in seeing that which was being applied. The sighting of corrosive areas, however, was generally easier to undertake, and thus to induce counter-measures, because the tanks were regularly cleaned and inspected, invariably at the end of each discharge of the ballast water and before loading the new cargo. The problem was undoubtedly that of corrosion of the steel which could lead to a reduction in the thickness, hence strength, of the components constituting the tanks. In a 1990 Survey, conducted by the Tanker Structure Co-operative Forum, investigations were made into the wider problem of steel corrosion aboard some 54 tankers, of varying ages, ranging between 150,000 and 300,000 sdwt. Their findings were published in their 'Condition Evaluation and Maintenance of Tanker Structures' and, amongst other observations and recommendations identified on Page 7, with a range of supporting photographs not here reproduced, were the following problems concerning cargo/clean ballast tank spaces:

"1. Except for the steel that is cleaned by water washing (directly impinged areas), structural elements submerged within the tank are protected by the adhering cargo oil. However residual ballast water causes pits, grooves and lake-type patches on the top surfaces of horizontal structural components.

2. Pits are found in the vertical surface of directly water impinged area. These are gradually formed into grooves by the running water."

Coating the tanks offered a low risk of corrosion, but there existed still a risk of pitting.

Supertankers Anatomy and Operation

Water ballast tanks constitute an important structural component in the framework of all VLCC's and by far the biggest cause for concern over recent years, affecting particularly the still operating 'early' generation of ships, concerns the levels of corrosion that have been detected in these tanks. There have been a number of cases in which it has been proved that the framework of Supertankers within the SBT and forepeak has been eaten away by rust to such an extent that, when the ships have been loaded, or have encountered moderate weather, structural weaknesses have resulted leading to complete breakdowns in the frames. A comparatively recent incident concerned a 357,000 sdwt ULCC that experienced large cracks below the waterline in the forepeak-one of which extended across something like 400 square metres of the ship. An enormous hole appeared in the forepeak making it possible to see right through the lower part of this ballast tank. The officers initially were under the impression that they had struck a submerged object and it was only with subsequent investigation that the true cause was revealed. There was no pollution, injury or loss of life, but the effects could have been extremely serious, to put it mildly. The cause of the problem, therefore, needs to be examined.

It will be demonstrated that the introduction of salt-water ballast is essential when a Supertanker is running without cargo in order to maintain stability and prevent uneven stresses on the hull. The problem, of course, is that salt is far more corrosive than fresh-water and so it attacks the steel of the ballast tanks. The other difficulty is that, even in temperate waters, the air inside the ballast tanks is invariably moist and warm which, even if it does not actually cause problems, certainly aggravates them. Ballast tanks themselves are invariably dark and difficult to access and the complex construction of their internal members means that the damage is not uniform throughout the tank, but is selective in that some parts of the tank are more at risk than others. The lack of illumination means also that it is difficult to detect the corrosive effects. The tanks are not always completely filled which implies that there is a differential in the corrosion caused to those parts that are normally completely submerged and those that remain largely above water. The bottom of the tank, therefore, with its various frames and reinforcing members, are attacked at different rates to that of the plating forming the top, under-deck section, of the tank.

3.14 PROTECTION OF PWB TANKS

In accordance with classification rules, governing the individual VLCC, the best form of protection for permanent ballast tanks is by the use of sacrificial anodes that are used in conjunction with protective paint coatings. The latter is clearly preferable, when it is considered appropriate, as the number of anodes required will be reduced quite substantially, a point which is illustrated impressively by the following comparative photographs: (Photos. 3.21 and 3.22).

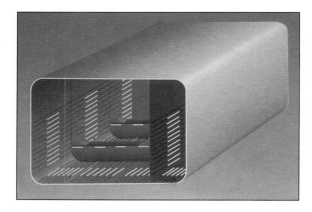
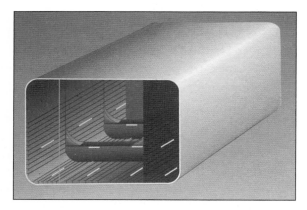

3.21. *A Substantial Number of Anodes are required in an Uncoated Ballast Tank …*
(By kind permission of Jotun Marine Coatings.)

3.22. *… Compared with the Considerable Reduction necessary when a Tank is treated.*
(By kind permission of Jotun Marine Coatings.)

The numbers required in the uncoated tank, for example, along each of the longitudinals at the bottom of the tank and down each of the sides, is considerable in comparison to the few necessary in the coated tank. The correct distribution of the anodes in order to provide maximum cathodic protection, should be noted. Obviously, no anodes would need to be fitted to the top underside as this area would always be left permanently dry and anodes are effective only when they are submerged. No ballast tanks would ever be filled 100% with sea-water. A further advantage is using both anodes and cathodic protection ensures that in the event of the paint coating becoming damaged protection is continued within the tank. On many Supertankers, only those tanks designed to carry ballast which are not fully painted are completely protected with anodes usually, of the 'bar' type that are secured by welding the core rods to lugs or brackets on the tank structure. Those supplied, throughout the fleet with whom I sailed, were zinc anodes from "Seaguard" in the segregated permanent ballast tanks at number two wings with each tank having 683 pieces, and an additional 131 pieces in each of the WB wing tanks aft. Of the designated cargo/ballast tanks, number one centre was fitted with 600 anodes, number three centre with 652 and number four centre with 668. Each anode was designed, in the company's view, for a life of four years which assumed an average ballast time of 25%. Another factor determining sound cathodic protection would include the coating quality where this was appropriate. The Classification Societies laid down no stipulations or restrictions concerning zinc anodes, but did make specifications, according to the individual ship, for the use of any aluminium anodes. The following table (Table 3.1) indicates the coating protection in ballast, slop and cargo tanks aboard

SAUDI SPLENDOUR
LIST OF TANKS
COATING/PROTECTION

TANK NO.	TYPE	COATING EXTENT	OTHER PROTECTION	CORROSION RISK	REMARKS (EG. COAT COND)
Fore Peak	W.B.	C		M	FAIR
1 Centre	C.O.			P	
2 Centre	C.O./CBT	H/L	Anodes	M	GOOD
3 Centre	C.O./CBT	H/L	Anodes	M	GOOD
4 Centre	C.O.		Anodes	P	
5 Centre	C.O./CBT	H/L	Anodes	M	GOOD
6 Centre	C.O.		Anodes	L/P	
1 Wings	C.O.			L/P	
2 Wings	C.O./W.B.			M/P	
3 Wings	C.O./W.B.			M/P	
4 Wings	C.O.			P	
5 Wings	W.B.	C	Anodes	M/P	GOOD
6 Wings	C.O.			P	
7 Wings	C.O./W.B.			M/P	
8 Wings	C.O./W.B.			M/P	
Slop	Slops	C	Anodes	L/P	FAIR
Aft Peak	W.B.	C	Anodes	M	FAIR

Type of Tank:
C.O. = Cargo
W.B. = Water Ballast
CBT = Dedicated Clean Ballast

Coating Extent:
U = Upper
L = Lower
C = Complete Tank
H = Horizontal Surfaces

Risk of Corrosion:
H = High
M = Medium
L = Low
P = Pitting

TABLE 3.1.
(Courtesy of Mobil Ship Management)

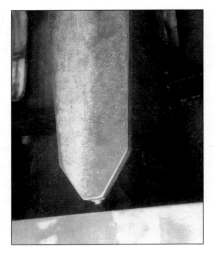

3.23. Unrounded edges can cause later problems of Corrosion.
(By kind permission of Jotun Marine Coatings.)

3.24. Corrosion by Rust Decay in Ballast Tank.
(By kind permission of Jotun Marine Coatings.)

3.25. Rust eating away at Ballast Tank structure.
(By kind permission of Jotun Marine Coatings.)

3.26. Advanced Corrosion in Ballast Tank Bottom Longitudinals.
(By kind permission of Jotun Marine Coatings.)

3.27. Severe Ballast Tank Corrosion.
(By kind permission of Jotun Marine Coatings.)

the VLCC "SAUDI SPLENDOUR", owned by Mobil Ship Management, showing the extent of protection, throughout this particular Supertanker, together with the risk of corrosion and the coat condition within the various constituent tanks. It is interesting to note how the dedicated cargo tanks for clean ballast water differs from that to be found between other VLCC's of a comparative tonnage. (See general arrangement plans in the pocket folder inside the back cover, as well as referring to the various capacity plans elsewhere in the book).

The seams where welding has taken place, the rat, and other access, holes and the reinforcing knees and sharp edges are invariably the first areas to be corroded, as demonstrated by the photographs. If, for example, the sharp edges in the enclosed photograph (Photo. 3.23) had been rounded then far less likelihood of an attack by corrosive agents would have occurred here later in the life of the ship. Clear evidence of the metal being eroded by decay (Photo. 3.24) can be seen in the weakening of the bracket which joins the stiffener at the deck transverse to the longitudinal bulkhead. A similar effect is seen in the top right hand corner (Photo. 3.25) where rust has started to eat away the reinforcing steel. Unless action is taken to stop the process, a more advanced state will be reached (Photos. 3.26 and 3.27) that leads ultimately to a total decay of the plating and a noticeable reduction in the thickness, hence strength, of the surrounding surfaces. It is worth re-emphasising that the prevention of such incipient corrosion must be taken at the initial stages of ship construction when the tanks could be sufficiently prepared and adequately protected. The only remedy, when corrosion has advanced to the state of the last photograph, is for costly replacement of plating and vertical structures. As Jotun state:

"The replacement of steel is a very expensive procedure. It can be estimated (in excess of) $500 per square metre, to which must be added the high cost of several weeks off hire-time."

3.15. INSPECTION, ELECTROLYTIC DE-SCALING

The company emphasise, with considerable justification, that ballast tanks although: "out of sight, hence out of mind" must have frequent inspections so that minor damage can be observed, assessed and treated before it develops into anything more serious. The extent to which it might prove feasible for such inspections to be made by the crew is however arguable. The treatment readily available has to be professionally administered from ashore and conform to strictly regulated procedures. Perhaps one of the most effective measures is 'stripe coating' in which coats of paint of different colours are applied to the most vulnerable areas of edges, welding seams etc. whereby the original coating is as easily identified as the new protection. An effective stripe-coat application is shown in the following photograph (Photo. 3.28). Generally the stripe-coat is applied with a round brush, especially in areas easily accessible, when otherwise other methods such as spray or roller-brush would have to be used. An overlap weld which had not been treated correctly has been photographed after a period of one year. (Photo. 3.29). The rust around the edge is clearly seen, showing indisputable evidence of decay that would need to be cleaned thoroughly and then painted with an appropriate stripe-coating. There is much justification for the application of two coats, especially if the paint contains a binder system that has been formulated specifically for ballast tanks. Balloxy HB Light, for instance, is tar free, light-coloured and has excellent wetting properties when it is painted onto surfaces that have been blast-cleaned, as well also to power-tool prepared steel. The application of two recommended 200 micron coats, in conjunction with anodes, and applied by airless spray, brush or roller affords protection for periods of up to a possible fifteen years. The effectiveness of the applications can be seen in the following photograph (Photo. 3.30) which was taken two years after the provision of the two-coat treatment. The light-beige painted surface is virtually as fresh and crisp as the time that it was applied and certainly any weak spots could be readily identified and repaired.

3.28. An Example of Effective Stripe-Coat Application.
(By kind permission of Jotun Marine Coatings.)

3.29. An Overlap Weld after One Year which was not treated by Stripe coating.
(By kind permission of Jotun Marine Coatings.)

Hull and Tank Protection

3.30. Ballast Tank after Two Years' service following Two Coat Treatment.
(By kind permission of Jotun Marine Coatings.)

3.31. Section of a Ballast Tank showing Advanced Scaling.
(By kind permission of Jotun Marine Coatings.)

Jotun's have designed a system of electrolytic descaling of water ballast tanks, particularly in instances where age or the state of neglect in the tank has allowed deterioration to reach such an advanced state that there has been a break-down of paint to an excess of 70% of the surface area. (Photo. 3.31). It is a potentially valuable process because, following an initial fitting whilst the Supertanker is in port, the actual descaling occurs whilst the VLCC is on passage in ballast to its destination for loading. The process has been the successful outcome of extensive research concerned with finding the appropriate anti-corrosive coating necessary for the task of protecting the ship's plates. Sacrificial anodes are fitted which are in the form of slender alloy magnesium strips that have been designed specifically to deliver a heavy electrical current. The company stresses that there is an important distinction: *"It must be emphasised that descaling with magnesium strips is a method for removing rust not paint"*. The implication being that the protective paint must have been allowed to decay to such an extent that the plates have become rusted. Initially, their surveyors inspect the tank thoroughly and tailor-make a system specifically for each VLCC to remedy the extent of rusting that is discovered. They produce relevant drawings and prepare detailed instructions designed to cover every stage of the process. The material is then delivered in 305 metre bundles, each of which weighs 100 Kilograms, that is cut into the specially active strips (Photo. 3.32) ranging between 20 to 30 metres in length. The magnesium is then removed from both ends so that the iron core is exposed. This is then fastened to contact bolts on the clamps which are welded to the tank at distances of 15 metres apart. (Photo. 3.33).

3.32. Electrolytic Strips received on board and being Prepared.
(By kind permission of Jotun Marine Coatings.)

3.33. Strips being clamped to the steel Tank Sides.
(By kind permission of Jotun Marine Coatings.)

Supertankers Anatomy and Operation

3.34. Magnesium strips being Clamped to the Side.
(By kind permission of Jotun Marine Coatings.)

3.35. Magnesium Strips Fitted to Ballast Tank.
(By kind permission of Jotun Marine Coatings.)

3.36. Tank ready for filling with sea water.
(By kind permission of Jotun Marine Coatings.)

3.37. Calcareous Deposits left following Electrolytic Treatment.
(By kind permission of Jotun Marine Coatings.)

3.38. Following Electrolytic descaling all Residue falls to the bottom of the Tank.
(By kind permission of Jotun Marine Coatings.)

3.39. The Newly Treated Ballast Tank Surfaces reveal the Roughness caused by Corrosion.
(By kind permission of Jotun Marine Coatings.)

A firm electrical contact is thus ensured which is made via the anode, that is the strip, and the cathode provided by the steel in the ship. (Photo. 3.34). The strips are suspended along the tank bulkheads in such a way that as much of the surface plating as possible is covered. (Photos. 3.35 and 3.36). This allows a more powerful and consistent current to be applied to the area being treated.

Once the fitting is considered satisfactory, the tanks are filled with clean sea-water at a minimum temperature of 10°C thus allowing an electrolytic reaction to occur. Thus a breakdown between the oxide of scale and rust is caused so that a soft calcareous layer forms on the surface of the steel. A barrier is created that permits the rust scale to lose its adhesion and drop to the bottom of the tank. The entire process may take between eight and fourteen days, a period that is determined by a number of variable factors including the extent of rusting, the temperature and conductivity of the sea-water, amongst other conditions. The process may be speeded by applying steam through heating coils, where these are fitted. Oxyhydrogen gases are emitted during the descaling operation so it is vital that the tanks are well ventilated for this period in order to eliminate any risk of explosion. The tanks are then emptied and washed down with fresh water in order to remove the greyish-jellylike calcareous deposits, rust scale and salt, (Photo. 3.37). Treatment by fresh water washing has to be applied as soon as possible in order to prevent the calcareous deposits from hardening for, were this to happen, the resulting crust would prove extremely difficult to remove. The calcareous waste and scale then fall to the tank bottom, along with the now expired magnesium strips. (Photo. 3.38). After this cleaning, the tanks are dried with dehumidifying equipment and so prepared for recoating with a combination of Balloxy BH Light anti-corrosive coating, or sacrificial coral anodes, or a combination of both. The newly painted surfaces will undoubtedly show evidence of roughness caused by the corrosion (Photo. 3.39), but this will certainly not affect the renewed life given to such a vitally important area of any Supertanker.

Chapter Four

DECK GEAR

4.1. INTRODUCTION

4.2. BOWS: OBSERVATION PLATFORM
    4.2.1. Towing Brackets
    4.2.2. SBM Mooring Gear

4.3. FORE DECK
    4.3.1. Mooring Bollards
    4.3.2. Windlass/Winch
    4.3.3. Anchors

4.4. PIPE WELL AND CATWALK

4.5. FIRE MONITORS

4.6. MAIN DECK
    4.6.1. Piping System
    4.6.2. Valve Control Boxes
    4.6.3. Gate and Butterfly Valves
    4.6.4. Tank Access

4.7. TANK GAUGES AND ULLAGING

4.8. MID SECTION: MANIFOLD AND DECK PIPES, DERRICKS

4.9. ACCOMMODATION BLOCK

4.10. SERVICE PIPES TO THE MAIN DECK

4.11. LIFEBOATS

4.12. AFTER DECK EQUIPMENT, FUNNEL

Chapter Four

DECK GEAR

(Function and Dimensions (where appropriate) of all Equipment Found on the Main or Upper Deck Including: Winches – Siren – Bollard/Bitt Placing – Windlass – Anchor and Cable: Housings, Dimensions and Weight – Quotation from OCIMF Booklet – High Velocity Ventilation Valves – Fire Fighting Hydrants and Types of Foam Monitors – Tank Access and Inspection/Sighting Ports – Heating Coils – the Older and Modern Types of Whessoe Gauges – Arrangement of Mid-Section Manifold and Deck Pipes – Butterworth Store – Accommodation: External Aspects and Bridge – Lifeboat and Davits – Service Pipes – Engine Room Ventilation – After Deck Gear – Swimming Pool – Funnel – After Emergency Towing System (ETS).)

4.1. INTRODUCTION

An overall view of the forepart of a VLCC, the "RANIA CHANDRIS", (Photo. 4.1), gives an initial indication of the gear that is to be found on the main, or upper, deck. When viewing this area of any such 'early generation' Supertanker for the first time, from the wheelhouse top, or monkey island, (Photo. 4.2), the eye tends to be overwhelmed by the sheer mass of equipment that seems to be cluttered everywhere, including an array of different sized pipe lines; vertical gauges, tank tops and an almost weird assortment of masts, gangways and walkways. Collectively, they present an assortment of shapes that have an almost surrealistic appearance. Re-assuring, however, is the understanding that each apparently haphazard part must be placed precisely where it is for a specific purpose. Starting at the bows, the examination commences with a look at part of the foredeck likely to be found aboard a typical VLCC of this period. (Diag. 4.1).

4.2. BOWS: OBSERVATION PLATFORM

Although not fitted to the VLCC shown in the plan, the stem plating very often has an observation platform, in the 'eyes' of the ship, enabling the duty mate to see directly over the stem in order to have a clear view forward of the bows. This position is useful in enabling him to gain an overall sighting of the run being taken by mooring ropes and wires during berthing operations and when taking on and supervising wires from tugs. It offers also an excellent viewing point from which to have a clear view of the anchor cable whilst this is being raised or lowered and to see how the anchor is housing once it has been weighed. Whilst, on a smaller ship, this position would allow the ship's draught to be read, the distance and angles to the waterline involved on a ship of this size mean that a far more accurate set of readings are generally obtained from draught indicators situated in the wheelhouse aft. (Diag. 4.2). It is normally the custom, however, to obtain visual readings of the marks either from the jetty or, more usually in order to gain an all round set of readings, from one of the boats which have been used to take ashore the ship's lines.

Supertankers Anatomy and Operation

4.1. A Shot taken from an Aircraft looking Aft, offers an overall view of the Main Deck of the VLCC "RANIA CHANDRIS".
(By kind permission of Skyfoto's.)

Deck Gear

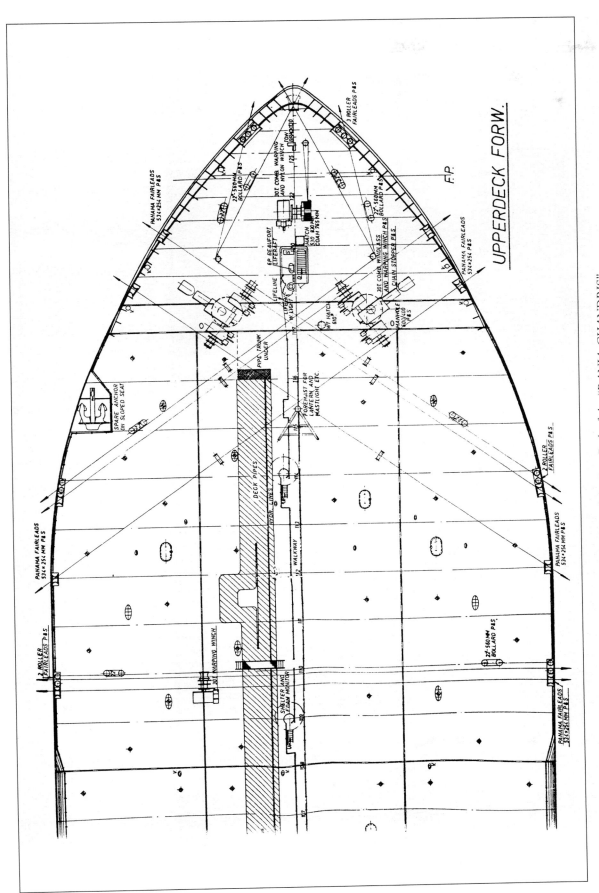

Diag. 4.1 A plan View of the Fore Deck of the "RANIA CHANDRIS".
(By kind permission of Builder's Plans.)

Supertankers Anatomy and Operation

4.2. *Viewed from the Monkey Island, the main deck appears to be cluttered by an apparent maze of different sized pipes. (The Author.)*

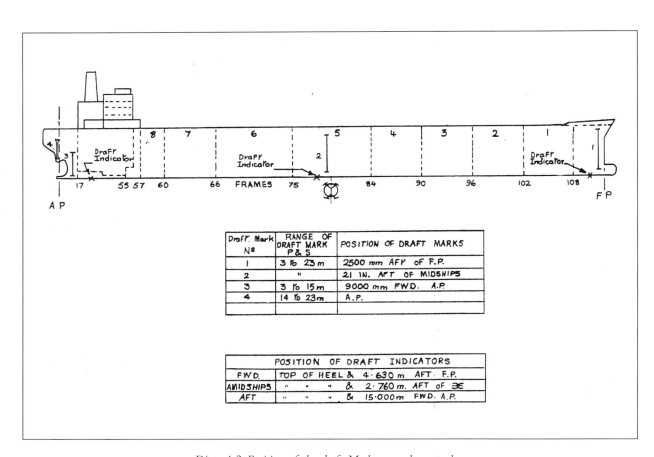

Diag. 4.2 *Position of the draft Marks on a large tanker. (By kind permission of Shell International Trading and Shipping.)*

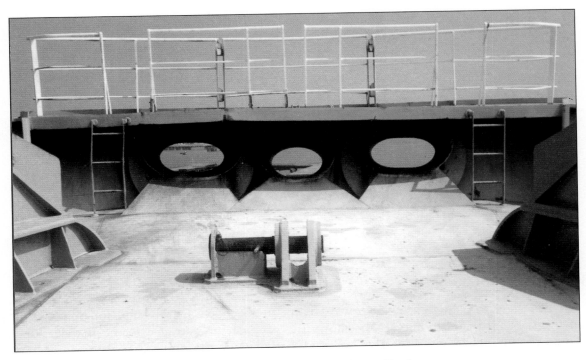

4.3. The Quick Action SMIT style Towing Bracket.
(The Author.)

4.2.1. Towing Bracket

Immediately below the platform lie the mooring fairleads with the centre one leading directly to the quick-action 'Smit' style towing bracket that is situated about three metres from the stem. (Photo. 4.3). The longer left hand side portion of the bracket is a moveable punch or quick-release pin, which slides to the left, thus allowing the towing wire from the tug to be inserted into the smaller RHS section of the bracket. The slide is punched home and secured with a locking pin, thus securing firmly the wire. The construction and safe working load (SWL) of the bracket has to conform strictly to the International Association of Classification Societies regulations, but is generally to the order of 350 tonnes which, bearing in mind that it is easier to tow dead-weights in water, would be sufficient to enable an ocean-going rescue/salvage tug to control any sized Supertanker. On occasions when this is used, it would normally by-pass the towing bracket and be attached, via a sufficiently reinforced thimble, to the anchor cable by a 'D' type shackle. In view of the considerable weight that is involved, this operation would have to be carried out using the forward windlass. It would require patience and be very time consuming.

4.2.2. SBM Mooring Gear

Slightly abaft the towing bracket, on the port side of the fo'c'sle, is a 'Pawl-type' chain stopper which, again conforming to similarly strict rules regarding construction and SWL, would be used to make fast the mooring chain when securing the ship to a single buoy mooring. (SBM). (Photo. 4.4). The reinforced stud-linked chain from the buoy is taken into the tracks through the centre hole. When the chain is correctly straightened, the top lever is lowered vertically into position and secured with a locking pin. There is an alternative bar-type design in use that is fitted to some ships.

Supertankers Anatomy and Operation

4.4. A "Pawl-Type" Securing Bracket for a Mooring Chain to a SBM.
(The Author.)

4.3. FORE DECK

The view from the extreme forward point in the bows of the ship, looking slightly aft down the starboard side, (Photo. 4.5), shows the back of the entrance to the fo'c'sle store just visible on the far right hand side. This houses, (Diag. 4.1) on the port side, a six person Beaufort self-inflatable liferaft for use in the kind of emergency which might leave some of the crew forward and unable to reach the lifeboats in time. This, depending on the length of the ship, could be between one quarter to one third of a mile aft. The lower part of the fore-mast, supports the for'd navigation masthead light, anchor light and the ship's for'd bell. The bell would not be used to indicate the cable laid during anchoring, as would be a function aboard a

4.5. The View from the Extreme Forward Point in the Bows.
(The Author.)

smaller ship, because this information would be conveyed to the bridge by VHF radio due to the distance involved but, under the Collision Regulations, it is sounded to act as a warning to other maritime craft whilst the ship is anchored during fog. The forward siren is also on the foremast and is often a piston horn whistle used either simultaneously with the main whistle on the forward part of the funnel, or independently, to give sound signals, again as required by international Collision Regulations.

4.3.1. Mooring Bollards

Immediately facing the left-hand side of the observer, in the same photograph, can be seen the upper parts of the mooring bollards. The provision of these, similarly to the towing bracket and chain stopper equipment, is designed for safety and operational considerations with exacting rules governing the leadings of mooring wires and ropes stipulated by guidelines from the Classification Societies, following IMO regulations and consultation with OCIMF, the Oil Companies International Marine Forum. (See Appendix). The bollards are just one set of some 30 situated around the main-deck, forward, aft and midships, and are used not only for securing of mooring ropes, but also to take on the towing wires of tugs, and for taking water or stores barges alongside. Two of the pedestal roller fairleads, (referred to in Diag. 4.1), can be seen for'd of the anchor windlass. There were ten of these on this particular ship and their function was to lead wires from the mooring drums of the winches more readily through fairleads and so run lines ashore.

4.3.2. Windlass/Winch

The forward part of one of the two 30-tonne combined windlass and warping winches can be seen immediately abaft the pedestal fairleads. The winch side of the windlass is a twin

4.6. Anchor Windlass and a Mooring Wire Winch Drum.
(The Author.)

drum designed to handle mooring wires, one of which can be seen quite clearly leading over the bows of the ship to the jetty, whilst the drum end is used for nylon hawsers, again during docking operations. The windlass/winch is driven by super heated steam, to a temperature around 210°C, that is fed from the main deck steam line leading from the engine room. It has an exhaust return, which can be seen more clearly from the after part view of the windlass, (Photo. 4.6) along with drainage points that can be used in very cold weather conditions to prevent condensed water from freezing and thus bursting the pipes. A lever can be engaged or disengaged as appropriate, enabling the winches/windlass to be used totally independently of each other. The main function of the windlass is to serve the anchors on the port and starboard sides.

4.3.3. Anchors

The OCIMF booklet, "Anchoring Systems and Procedures for Large Tankers", explains on page 13 the qualities of an effective anchor:

"... the safety of a ship at anchor depends to a very large extent on the degree of efficiency and characteristics of the anchor itself. The philosophy is that the anchor should drag before the cable parts. The design must be such that it must take a grip on any type of holding ground and must be robust enough to take the heavy loads imposed upon it, not only riding to anchor, but also the dynamic loads imposed during the anchoring and when swinging. It must stow easily, neatly, and firmly in the hawse pipe."

There are three anchors carried on most Supertankers and, invariably, each is of the Hall's Patent Stockless type, with swivel ring adapters, and weigh around 21.5 tonnes. Another type of anchor that might be carried is the British Admiralty Cast Anchor Type 14 (AS 14). One of the three is a spare which is invariably housed on the port side of the fo'c'sle (Photo. 4.7) whilst the other two are housed at the end of the port and starboard hawse pipes and settled in reinforced chafing rings. The working load and holding power governing the anchor fitted aboard Supertankers is, once again, subject to strict OCIMF guidelines and photograph (Photo. 4.8) shows the anchor flukes, or arms, tripped so that the anchor head can lie flush to the ship's sides. With either type of anchor, the cable leads from the windlass to the anchor by way of the hawse pipe (Photo. 4.9) inside of which there reside four built-in high pressure water jets, supplied from the main fire pipes, and controlled by valves fitted near the windlass

4.7. Spare 21.5 tonne Hall's Patent Stockless Type Anchor.
(The Author.)

Deck Gear

*4.8. An Admiralty Type Anchor housed against the ship's side on a VLCC.
(The Author).*

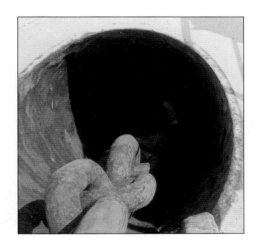

*4.9. Anchor Cable leading from the Windlass through the Hawse
Pipe to the Anchor.
(The Author.)*

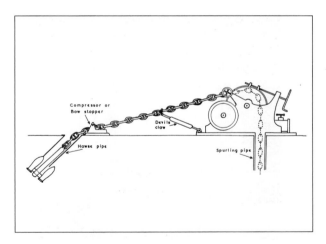

*Diag. 4.3 The Relationship between Anchor Windlass,
Hawse and Spurling Pipes.
(By kind permission of Professor John Kemp &
Captain Paul Young.)*

which wash away mud and debris from the sea bed thus enabling the clean cable to run over the windlass drum through the spurling pipe to the cable locker beneath. (Diag. 4.3). A 'doubling plate' reinforces the deck in the vicinity of the top of the hawse pipe and the cable is led through a bow stopper that runs a locking bar through one of cable links, thus preventing movement of the cable. There are usually 13 shackles of cable to each anchor. One shackle being equal to 15 fathoms or 27.5 metres, which is equivalent to 355.5 metres. Each link of the cable

Supertankers Anatomy and Operation

4.10. Each Link of Cable measures 670 mm and weighs about 3.8 cwt.
(The Author.)

4.11. Cable Meter showing Anchor made fast in Hawse Pipe.
(The Author)

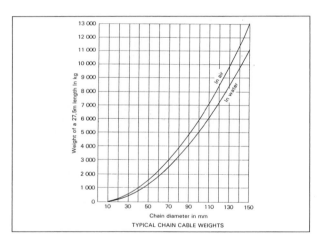

Table 4.1 Typical Chain Cable Weights.
(By kind permission of OCIMF.)

(Photo. 4.10) is 670 mm in length and 114 mm in diameter and made of U2/U3 grade special steel. Referring to Table 11, on Page 17, of the OCIMF Guide to Anchoring Systems for Large Tankers, (Table. 4.1), the weight of the cable can be deduced as being approximately 8,000 Kg (in air) and 7,000 Kg (in water). A length of 27.5 equals 8,000 Kg, so that 305 mm equals approximately 88.7 Kg or about 196 pounds or 1.75 cwt. One link at 670 mm, therefore, weighs not far short of a staggering 3.8 cwt. The inserting and removing of spare links into the Bosun's store, and the fitting of a new anchor or cable clearly has to be undertaken by shore cranes. Needless to say, it is the weight of the cable, not purely the anchor, which holds the ship in an anchored position. The windlass is rated to lift the anchor and 100 metres cable at a speed of between 9/12 metres per minute depending upon local conditions of wind and tide. A cable meter on the port side of the fo'c'sle head indicates the amount of cable laid on deck. (Photo. 4.11). The spare anchor and after-part of the port windlass, with its steam and return control valves, plus the 't-shaped' brake control lever controlling the starboard drum, together with the disengaging lever, can just be seen situated between the drum and the windlass rail, in the Photo. 4.6. Substantial wire cable preventers are always fitted between the windlass and the hawse pipe in order to prevent movement whilst the ship is at sea, and heavy steel stoppers are usually applied which may be used to secure the anchor once the ship has been brought up in position. The use of the stoppers, however, is very much a matter of personal preference to the master of a VLCC. Some captains do not use them in the belief that, by allowing the cable to slip slightly on the brake, an earlier indication is given should the ship start to drag her anchor along the seabed. Other masters would argue that the clamp is, for them, the 'best invention since sliced bread' and that the officer of the watch should be sufficiently alert to notice any movement in the ship's position indicating dragging, from the aids available to him in the wheelhouse, if not visually. The issue is one of those 'personal preferences' mentioned in the Preface.

4.4. PIPE WELL AND CATWALK

Looking forward, from the foremast, (Photo. 4.12) the entrance to the Bosun's Store is very close to the round inspection hatch leading to the pipe well on the starboard side. The tall white ventilator, mounted on a pedestal, is a high velocity ventilation valve standing as a sure sign that this particular Supertanker, at the time when the photograph was taken, was of the early pre-1973 MARPOL and a pre-inert gas system ship. The valve would have been fitted with a gauze spark arrester so that when the ventilator was opened during loading/discharging of cargo, any stray sparks would be prevented from entering the highly volatile gas mixtures in the tank. There was one ventilator to serve each of the centre and wing cargo tanks. The after-part of the port windlass, showing the massive cable links, can be seen on the other side of the railings marking the entrance to the deck-way. This 'catwalk', as it is called, measures about two metres in width and extends the length of the ship from the accommodation. (Photo. 4.13). It was a legal requirement, under the load line regulations previously considered, and provided a safe walkway down the middle of the ship. Many Supertankers are flush deck ships from aft to forward, although with varying practices, some were fitted with a raised fo'c'sle head.

4.12. Entrance to Bosun's Store with Fore Deck Equipment.
(The Author.)

4.13. The Catwalk Offers Safe Passage from the Fo'c'sle to the Accommodation.
(The Author.)

4.5. FIRE MONITORS

There are usually about nine fire monitors fitted along the main deck, each often has a safety shelter built on the after-part of the platform base. (Photo. 4.14). These are an essential part of the ship's external firefighting capacity and, although they can project a powerful jet of water, they carry also foam through uptakes from the deck pipes running along the port side of the catwalk. They can deliver either a foam blanket covering or a spray. The distinctive red monitors added

4.14. Fire Monitor on a raised Platform.
(The Author.)

4.15. Fire Monitor Situated on the Main Deck.
(The Author.)

4.16. Fire Monitor integrated into a Cross Over Walkway.
(The Author.)

a splash of colour to the sombre, but practical, non-reflecting greys and blacks of the maindeck. The artistic eye doubtless finds these aesthetically pleasing but, in an emergency it is good to be able to find them quickly and to know in which direction they are pointing. It is interesting to compare the monitors fitted to other VLCC's, the first of which was the "MOUNTAIN CLOUD", (Photo. 4.15 and 4.16) placed either on the main-deck or integrated into the cross over walkways. Certainly, I believe that those elevated prove far more effective not only in dispersing the foam, but also enabling the officer-in-charge to have a much clearer overall view of the events taking place. It is also possible additionally for high pressure water jets to be directed by hose pipe onto a fire area, or to wash foam coverings over the ship's side in an emergency. The hydrant points for these are taken from the main water lines extending the length of the deck.

4.6. MAIN DECK

The photograph looking directly aft has been taken from the pedestal of the monitor nearest to the foremast. (Photo. 4.17) The sheer beam of the Supertanker can again be appreciated from this shot as well as the distance from the bows to the superstructure. The main cargo manifolds, running thwartship and looking like some kind of barrier, are in the vicinity of the derricks amidships. In the foreground, another style of raised 'bridge-type' constructed walkway permits safe passage over the short set of deck service pipes. A warping winch, of the kind only just visible on the extreme lower right hand side, can best be examined in the next picture, (Photo. 4.18). This winch was welded by way of the longitudinal separating number five centre and port-side wing tanks. The basic printer on the drum, together with that on the panama lead fitted into the ship's side-railing and the Butterworth machine lowering hatch top, indicates part of the maintenance work undertaken by the deck crew in clement weather whilst the ship is under-weigh at sea. The polypropylene rope, which had been used on this particular ship in the berthing operation, can be seen awaiting stowage. This winch, like all mooring winches of a similar construction, is a 'standard' model frequently found on Supertankers and is called a 'split drum' winch. This allows a constant holding power on the brake to be applied as only a single layer of wire is used from the inner, or working drum. The winches on VLCC's are often placed down the port side of the ship only so that, when moored starboard side to a jetty, it is necessary to run the wires across the main deck by way of a pedestal type rolling fairlead.

4.6.1. Piping System

The apparently innocuous square grey box, with the open lid, is a particularly important piece of deck equipment. It is best appreciated in the following photograph, that was taken further amidships on the same ship, (Photo. 4.19), which shows two red painted horizontal

4.17. View From Alongside Pedestal Mounted Foam Monitor Looking Aft.
(The Author.)

4.18. Mooring Winch with Wire Crossing Main Deck to Starboard Side Jetty with Polypropylene Rope on LHS.
(The Author.)

4.19. Main Deck View on Port Side Showing hydraulic Cargo Valve Control Box and Deck Pipes.
(The Author.)

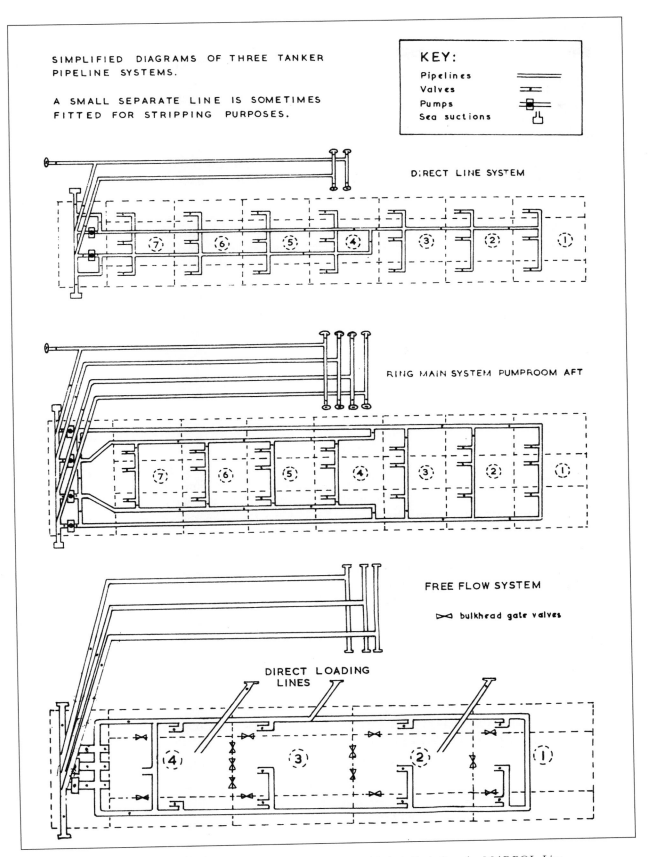

Diag. 4.4 *The Three Major Pipeline Systems Found on Tankers Excluding the MARPOL Line.*
(By kind permission of Professor John Kemp & Captain Paul Young.)

wheels on vertical spindles. Both box and wheels are valve controls and are concerned directly with the flow of the cargo through the ship's piping system. On many VLCC's, the piping system is called the 'Direct Line' type, and it has a totally separate set of stripping lines that are designed to drain the residue oil from the cargo tanks. This enables the ship to carry a number of independent cargoes but, by means of a cross-over arrangement, all tanks could be used when only one homogeneous cargo was carried. The variations in different voyages was quite considerable. Some companies carried only one or two cargoes on any particular voyage, whilst others coped with four or five grades. An alternative pipe system termed 'Free Flow' was usually intended for VLCC's used permanently on only one trade with just the one grade of cargo carried. The longitudinal bulkheads on such ships are fitted with hydraulically controlled sluice valves through two drops into the bottom tank lines which, when opened, allowed the oil to run from the wing tanks into the centres. Thus, with the tanker trimmed aft, gravitational forces assisted discharge of the cargo so that considerable time would be saved due to the reduction in stripping required. A third piping system is called "Ring Loading". In the accompanying drawing, (Diag. 4.4), the differences between the three systems is clearly shown. The ring system is an older type that used two pumprooms to serve the tanks through two different circular loops, usually each working half of the tanks forward and half aft, with a connecting cross-over line. It has been used on some of the older VLCC's but, because it is a rather clumsy arrangement, it has not found favour for widespread use in ships of this particular class, but is more generally used on smaller tankers.

4.6.2. Valve Control Boxes

Returning to direct line pipe systems owing to their considerable popularity on Supertankers. These also have a number of variations according to the requirements of owners and sometimes the provisions of the shipyard. One cargo system with which I worked for a number of years was of the following pipe arrangement. (Diag. 4.5). The lead comes from the two sets of manifolds, only one of which would be in use at any time of loading or discharging, depending on whether the ship was port or starboard side to the jetty. The numbers indicate the valves which control the flow direction and these are found in the square grey boxes on the main deck. A total of 39 such hydraulic valves control the oil flow via ten deck boxes. (Diag. 4.6). The manifolds are fed by valves 117 to 120 (inclusive) for the port side and numbers 113 to 116 on the starboard side. The cross-over lines enable the four cargo pumps with which VLCC's fitted with this system carry, to serve two common lines, such that cargo lines 1 and 4 serve the wing tanks, controlled by valves 121 and 122, whilst Lines 2 and 3, controlled by valves 126 and 127, serve the centre tanks. The important cross over valve 125 enables integration of the system when only one grade of crude oil happens to be carried. The remainder of the valves control cargo as:

   Number One Tank — 103, 104, 102,   for port, centre and starboard tanks.
   "   Two   — 106, 107, 108   do.
   "   Three   — 112, 111, 110   do.
   "   Four   — 129, 130, 131   do.
   "   Five   — 133 and 132 for port and starboard tanks.
         — 135 and 136 for the centre tank.

Valve 134 serves as a slop tank isolator valve.

Valves 137 and 138 are balance line main sanctions for the port and starboard sides.

In accordance with IMO rulings, all Supertankers are now fitted with permanently segregated ballast systems so that the ballast and cargo tanks can be kept completely separate. Each has its own set of pumps, lines and valves so that whatever the pressure exerted on the

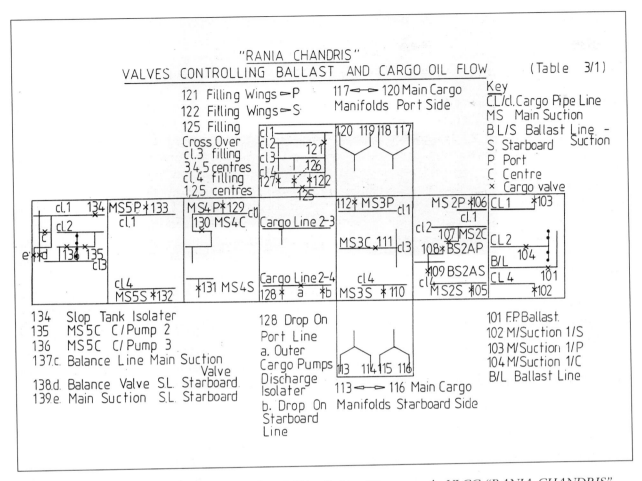

Diag. 4.5 The "Direct Line" Pipe System and Deck Valve Position Diagram on the VLCC "RANIA CHANDRIS".
(By kind permission of Paul Taylor & Builder's Plans.)

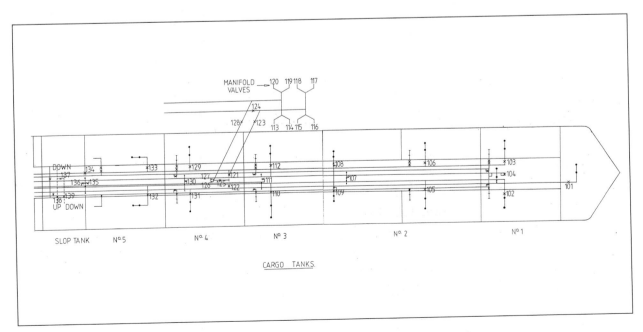

Diag. 4.6 Hydraulic Valve Control Boxes on the Main Deck.
(By kind permission of Paul Taylor & Builder's Plans.)

4.20. Internal controls of a Valve Box on Deck.
(The Author.)

duty mate no cargo can be pumped into these tanks by accident. The ballast pump can be used only to pump ballast water into the after part number two wings, through valves 108 for number two port, and 109 for number two starboard, whilst the forepeak is controlled by valve 101. The afterpeak tank has an entirely separate system operated by and from the engine room.

Considering more closely the cargo valves used on VLCC's. The valve control boxes –to give them their correct name – are situated at convenient points along the main deck. Invariably, these are manually operated using a Kracht hydraulic system. Each valve inside the box has a lever operating handle that works on a volume meter principle so that as the hydraulic oil travels to operate the valve actuator, it in turn drives a volume meter whose rotational movement is transferred by a magnetic coupling. This is geared to a needle on an indicator dial so that as the handle moves to the open or shut position, so the dial turns accordingly. The pressure to which the system works is around the 130 to 150 bar and an alarm sounds in the cargo control room in the event of pressure falling to around 110 bar. The internal controls of a valve box (Photo. 4.20), indicate that this is the second box from forward covering number two tank and showing the condition of the valves serving the ballast suction to 2A Port (number 108) and 2A starboard (109), as well as the main suction for two port (106-on cargo line 1), two centre (107-on cargo line 2) and two starboard (105-on cargo line 4). The main suction on number two tank is open whilst the other valves are closed.

4.6.3. Gate and Butterfly Valves

The red coloured wheels (Photo. 4.19) indicate locally operated main tank cargo valves. They are painted red, indicating port side gate valves, whilst those on the starboard side would be green, and those serving the centre tanks are white. Gate valves are strong and robust, but require a considerable number of turns before becoming effective. They are as a consequence quite slow to operate, taking perhaps one minute from fully closed to fully open, but are totally reliable. As the wheel is turned so a gate is raised or lowered thus opening or sealing the pipe. When the valve is opened the gate is raised into the body of the valve itself so that there is no resistance to the oil flow through the pipe. Gate valves tend to be smaller valves, of less than 250 mm. The third type of valve used on VLCC's is the eccentric butterfly valve that is in excess of 250 mm and is generally power operated as well as being more effective than the

4.21. Eccentric Butterfly Valve.
(By kind permission of Westad Industri AS.)

ordinary 'butterfly' type. Here, the valve is central to its axis, thus permitting a greater control over the oil flow. They are quick and easy to operate, which is a major advantage with the large diameter pipes to be found aboard VLCC's, and have generally been found completely reliable. (Photo. 4.21).

4.6.4. Tank Access

Access to the tanks is by means of a water-tight hatch that is raised from the deck by about a metre. (Photo. 4.22). The dogs have to be first released and then the wheel turned anti-clockwise in order to raise the hatch-lid which then opens across and towards the blue heating coils on its starboard side. The hatch illustrated serves No. 1 port wing tank and the narrow access, just forward of the wheel, called a 'sighting port' would be used for cargo inspection purposes. Again, on this early type of Supertanker, this clipped access allowed either a thermometer to be inserted into the tank, to take temperatures, or a manual ullage tape. At the extreme ends of the photograph in the background can be seen panama leads that are fitted into the bulwark to lead mooring wires/ropes ashore. These are two of 32 situated fore, aft and midships, and have been fitted additional to two of increased thickness, immediately in the bow and at the stern, used to run ashore mooring wires. It is interesting to note that the crew have also been working here on the railings by 'touching-up' rust areas as a prelude to grey painting. The heating coils, shown in the fore-part of the hatch, were rarely either fitted or even used on VLCC's, but were supplied to the cargo tanks of this particular ship, as well as to the slop tanks, enabling her to engage in the trade of conveying very heavy crude oil cargoes, perhaps

4.22. Tank Access and Cargo Tank Heating Coils.
(The Author.)

Supertankers Anatomy and Operation

4.23. Intersection of Tank Top Openings on a Large Tanker.
(By kind permission of Mobil Ship Management.)

from Nigeria, ones that possess a high wax content. Such a cargo could well require heating, prior to discharge, to around 125°F. Many of these very early VLCC's required considerable modernisation, in order that they could comply with MARPOL Regulations, including those rules governing the use of crude oil washing and inert gas systems. As this occurred so the style of tank tops and ventilation valves altered also, as can be seen in the above shot which shows the intersection access tops between numbers one and two tanks. (Photo. 4.23).

4.7. TANK GAUGES AND ULLAGING

Many cargo tanks, on 'early' generation VLCC's, were and continue to be, served by Whessoe Gauges. The older type of marine tank gauge that performed first class service until superseded, (Photo. 4.24), measures, in the words of the manufacturer's brochure:

"accurately and continuously the contents of tank compartments of marine tankers during loading, discharging and ballasting operations ... contained within a sealed drum is the mechanism for maintaining constant tape tension at the float and the means for controlling the rate at which the float will fall freely when the gauge is put into operation."

Basically, the distance is measured between the surface of the cargo oil and the under-deck level at the top of the tank, although this is usually extended to the top of the ullage pipe with appropriate allowances being made by comparison with ullage tables supplied by the equipment manufacturer. It is the opposite to a 'sounding', which is the measurement of the amount of water (usually) that might be found in any tank. Invariably, soundings are references to readings taken in the ballast tanks in the forepeak, afterpeak and midships, whilst the term 'ullages' is restricted to readings from cargo tanks. The same colleague, who had so humorously

Deck Gear

4.24. *Older Type of Whessoe Tank Measuring Gauge Found on Pre-MARPOL VLCC's.*
(The Author.)

4.25. *Vapour Control Valve for Gas Tight Ullage Tape.*
(The Author.)

described the cargo plan for a VLCC as being merely the laconic comment 'oil-oil-oil ...', remarked when being told about ullaging: "I suppose an ullage is really the amount of oil which is not in the tank" – rather droll, but quite accurate. The older type gauges consisted of a steel tape, that reached to the bottom of the tank, which supported a rectangular float which was kept in place by guide-lines. (Diag. 4.7). As the oil entered/left the cargo space so the float rose or lowered. This movement was recorded in feet/inches and centimetres on a clock-dial behind a gas and water proof transparent sealing. Like most mechanical devices, they required a certain routine maintenance in order to prevent the float from sticking to the leading tape. More advanced types, which became fitted extensively to virtually all of the older VLCC's were inserted through the vapour control valves (Photo. 4.25) serving each tank. A portable gas tight ullage tape, that works on a principle similar to the older types, allows additionally for the temperature of the cargo to be displayed via LED. The same instrument is fitted also with a light indicator so that the oil and water interface may be taken simultaneously with the other measurements. The lever on the side of the valve is to allow the pressure of the inert gas system to be maintained whilst the portable equipment is fitted. The following diagrams illustrate the type of triple function gauge that is supplied to the VLCC industry by the international organisation MMC. (Diags. 4.8 and 4.9). There have been some very sophisticated advances in tank measurements aboard Supertankers, that have occurred with the advent of computerisation, some techniques of which are examined in Chapter 18 of Part Four.

Supertankers Anatomy and Operation

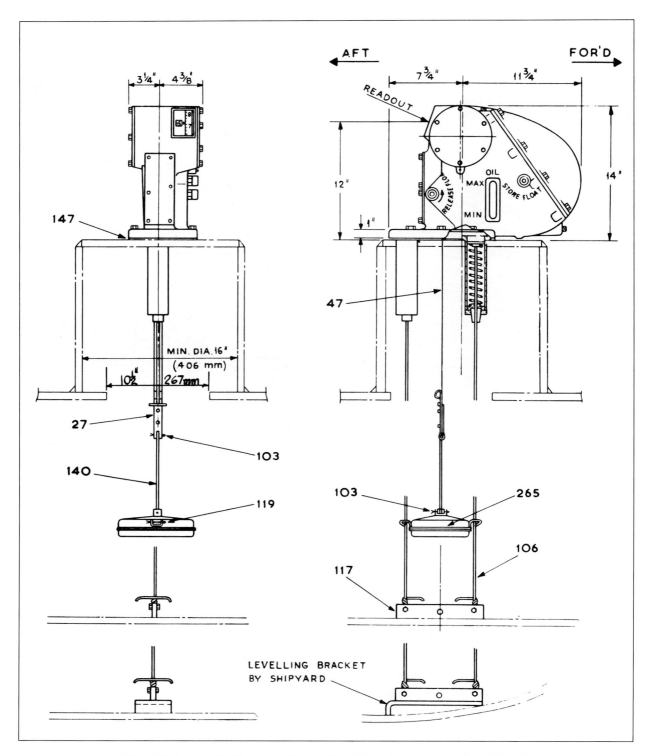

*Diag. 4.7 Internal Workings of a Whessoe Tank Measuring Gauge on Early VLCC's.
(By kind permission of Whessoe Systems and Controls.)*

Deck Gear

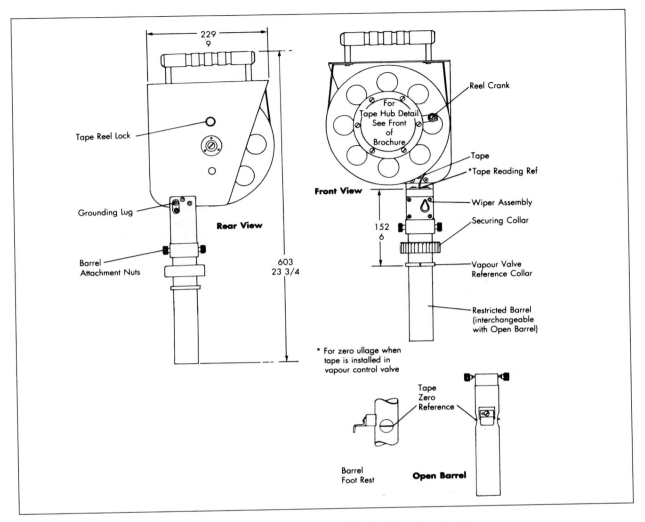

Diag. 4.8 Workings of a Modern Portable Gas Free Gauging Tape ...
(By kind permission of MMC (Europe) Ltd.)

 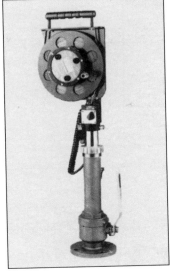

Diag. 4.9 ... Which Replaced the older Pre-MARPOL Gauges.
(By kind permission of MMC (Europe) Ltd.)

105

Supertankers Anatomy and Operation

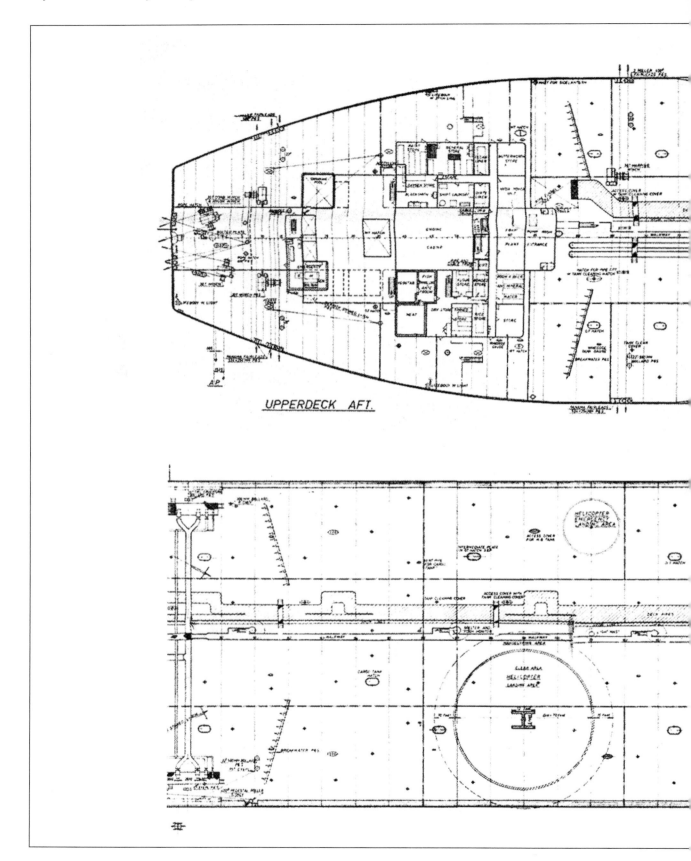

Diag. 4.10 Complete Main Deck Layout of the "RANIA CHANDRIS".
(By kind permission of Builder's Plans.)

Deck Gear

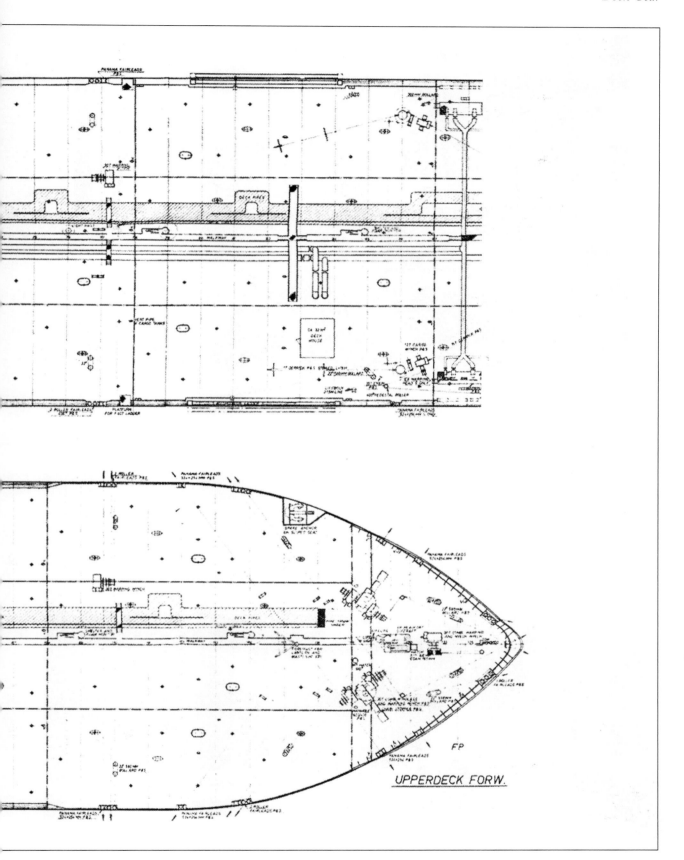

Diag. 4.10

Supertankers Anatomy and Operation

4.26. *Cross-Over Pipe Arrangements Midships Looking From Starboard to Port.*
(The Author.)

4.27. *After-Part of Cargo Manifolds on Port Side Showing Derrick Arm.*
(The Author.)

4.28. Cross-section Midships as an Alternative Arrangement.
(The Author.)

4.8. MID SECTION: MANIFOLD AND DECK PIPES, DERRICKS

From the entire deck layout, (Diag. 4.10), it can be seen that the area about one-third from the stern is the main section of the Supertanker concerned with the loading/discharging of cargo. Immediately abaft the helicopter landing area, with its internationally used, precisely measured letter "H" and surrounding concentric circles, two breakwaters were fitted to some VLCC's. This was intended to deflect the force of any sea-water taken on board during heavy weather when the ship was in a loaded condition, hence with a reduced freeboard, so that the potentially vulnerable sections of pipes and deck level fittings as well, perhaps, as the manifold-ends could be afforded extra protection. On first being appointed to VLCC's, I was confronted with what appeared to me then as an apparent confusion of pipes which surrounded this area. The view (Photo. 4.26) was taken standing between the two main cargo manifolds, on the port side, indicated by red-coloured valve controls, over the after-part number three centre tank, looking athwartships to a point where the manifolds are intersected by cross-over piping, leading to the extreme ends of the shore-connecting flanges. The group of six pipes (leading into the camera) that run parallel with the manifolds run from midships to port and are part of the section of pipes, moulded away from the remainder, which continue to run fore and aft. The long narrow white horizontal 'bar' is, in fact, the arm of the port-side derrick leading toward the topping lift gear fixed at the free end. The lower part of a reduced walkway can be seen at the top of the photograph. The port-side derrick arm can be seen more clearly in the next view (Photo. 4.27) near the gooseneck connection to the mast. The extreme pipes, aft of the manifolds, are for diesel and fuel oil, and the winch drum that serves the derrick can just be

4.29. View of Accommodation Block taken from Entrance to the Butterworth Store.
(The Author.)

4.30. Alternative Accommodation Design from Amidships.
(The Author.)

4.31. Main Deck Looking Aft from Amidships with Accommodation View.
(By kind permission of Jahre Wallem Shipping.)

4.32. The Accommodation End of the Catwalk.
(The Author.)

Supertankers Anatomy and Operation

seen on the left-hand side. On the starboard side abaft of the manifolds is the Butterworth Locker. This was used to store the Butterworth portable tank cleaning equipment in the days prior to the fitting of a variety of fixed cleaning guns. Today, it houses reduction pieces for the manifolds, together with a range of casual deck stores and cleaning materials, including scupper wedges and blocks necessary in the event of a minor spillage whilst loading or discharging that could not be catered for by the drip tanks situated below the manifolds. A similar cross-section beam view, taken aboard a different VLCC, (Photo. 4.28) shows the alternative arrangements that can exist between the near midships piping systems in the vicinity of the manifolds. The figure of the man, standing the other side of the catwalk rails on the lower left hand side, gives an indication of the dimensions involved. A glance at Photo 4.36 helps clarify arrangement of the midships cross-over piping.

Derricks on the port and starboard sides of the early Supertankers, prior to many being retro-fitted with 360° slewing arc luffing cranes, had a Safe Working Load (SWL) of around 15 tonnes. They were used mainly to support the cargo pipes during loading/discharging, as well as for fitting the accommodation ladder and taking on stores. Their use was particularly essential during lightening operations, as will be briefly examined in Chapter 11.

4.9. ACCOMMODATION BLOCK

The view of the accommodation was taken from the entrance to the Butterworth store. (Photo. 4.29) One of the prominent factors about some Supertankers was the absence of bridge-wings. Many a pilot cursed such ships, not so much for what they saw merely as her disfigurement, but because they were unable to see from the extreme bridge wing tip along the side of the ship. Whilst the wings were doubtless aesthetically pleasing, as indicated by the views

4.33. The service Pipes to the Main Deck leading to the Pipe Tunnel and Engine Room.
(The Author.)

4.34. Docking Winch Aft showing Expansion Joint on a Main Deck Pipe.
(By kind permission of Mobil Ship Management.)

of the "SAUDI SPLENDOUR" and the "JAHRE VIKING", (Photos. 4.30 and 4.31) their absence was not as much of a practical disadvantage as some pilots indicated. In the magazine 'Motor Ship', for November 1994, a chief executive of Harland and Wolff's Belfast Yard made the following observation: "Bridge wings are no longer required ... but captains insist that they continue to be added to the bridge deck". Certainly, I preferred to sail aboard VLCC's that were so fitted,- both practically for the wider field of view, and also psychologically (as well as perhaps, romantically). I felt the ship looked more complete and, indeed, felt like a Supertanker. The lamp-post (literally) supporting essential decklights that were necessary when the ship was anchored or working cargo at nights, can be seen along with other items of deck gear so far examined in this chapter. The after breakwater and lifeboat davits, on the left hand side, and the first two walkways over the manifolds are also shown. The large white ventilators, immediately forward of the accommodation block, lead from the pumproom allowing air to circulate within this space. Before leaving the fore-deck, a view looking forward from the accommodation, on the port side may well clarify other items of deck gear. The first photograph (Photo. 4.32) shows the entrance to the catwalk and a portion of the walkway itself, along with relevant deck equipment, much of which will now be readily recognised: the catwalk, number one foam hydrant, lamp-post, roller pedestal, winch, port side Samson post with derrick and valve control box.

4.10. SERVICE PIPES TO THE MAIN DECK

The service-pipes leave the under-tunnel from the engine-room to emerge on deck slightly abaft the ladder from "A" deck port side which leads down to the catwalk. (Photo. 4.33). There are ten pipes which run the length of the ship, and they carry respectively; main fire water, foam, with leadings to fire monitors, heating coil exhaust return, main exhaust return, main steam (discoloured pipe), Butterworth line, fuel oil, diesel oil, air pipe and the electrical conduit. Then, on the other side of the catwalk, follow the stripping and the four main cargo lines

4.35. Open-Style Life-Boats are still very much a Feature on VLCC's.
(The Author.)

leading into the pumproom. It is not unusual for Supertankers to increase in overall length by a surprising half-a-metre or so due to temperature differences found between the Arabian Gulf ports and European waters. In order that the deck pipes will not burst under this additional strain a number of the pipes are moulded away from the remainder, forming what are termed 'Omega' bends, that can be seen clearly from many of the overall deck photographs. On other pipes, a specially constructed expansion joint is fitted with, in this photograph, the dual drum, double mooring winch and fire monitor appearing in the background, along with the wheel of the gate valve operating the centre cargo tank. (Photo. 4.34).

4.11. LIFEBOATS

Just above the main-deck, port and starboard, are carried the two lifeboats, distant views of which have been seen in previous photographs. They are situated there, instead of on the traditionally associated boat-deck, to reduce the waterline drop in an emergency, yet to keep them sufficiently high from the main-deck to be out of the way during deck work and avoid damage during heavy seas. (Photo. 4.35). To conform to SOLAS regulations, each have to be motor-powered and are often equipped with 25 hp, 2-cylinder diesel engines. They range from 7.8 to 10.5 metres in length with a capacity of between 56 and 70 persons – well in excess of the complement of crew – so that all officers and ratings can take to only one boat in an emergency when possibly only one could be launched due perhaps to the list of the ship. They are lowered from gravity davits and recovered using air powered winches. Many lifeboats today are fully enclosed – even though considerable numbers of the open kind remain in service – prior to retro-fitting.

4.36. *Overall Aft Deck and Accommodation View from the Air.*
(By kind permission of Skyfoto's.)

Supertankers Anatomy and Operation

4.37. General View of After Deck.
(The Author.)

4.38. After View of Funnel and Part of the Engine Room Ventilation System.
(The Author.)

4.12. AFTER DECK EQUIPMENT, FUNNEL

The equipment in the first of two after-deck views can perhaps best be appreciated by looking at an overall deck view (Photo. 4.36). This view was taken at the same time as the initial photograph to this chapter whilst the ship was streaming through the Dover Strait en route for the Arabian Gulf for first loading on her initial voyage. The boat covers can be seen plainly, as well as my duty lookout on the starboard bridge well. The derricks serving the engine room were each 5 tonne swl and were used for taking deck, catering and engine room stores as well as landing crates of spares or the occasional broken machinery. The stack of 40 gallon oil drums

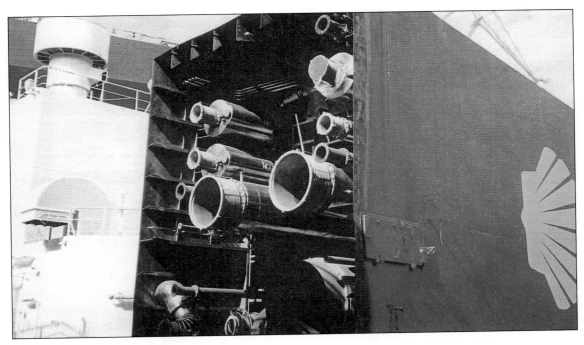

4.39. Internal View of the Funnel of a Modern VLCC.
(By kind permission of Shell International Trading and Shipping.)

4.40. After Deck View.
(The Author.)

on the starboard quarter were of various grades of lubricating oil. The first sectional shot shows the mooring winches with nylon ropes made fast on the bitts. (Photo. 4.37). Referring to the after-deck part of the main-deck plan, (Diag. 4.10), it can be seen that, as with the fore part, the positioning of the winches was very carefully placed to make direct alignment with the bitts and panama fairleads. The lower part of the engine-room venting system can be seen in the background, along with the port after-mast posting. The derrick to the mast, and the swimming pool are each obscured by the winches and after-trunking. The air intake for the engine-room on some VLCC's was quite an impressive affair compared to the more modest systems on board other ships. (Photo. 4.38). It was never made quite clear why the owner's should have desired such an elaborate construction. The ladders leading to the afterpart of the accommodation and serving the back of the bridge-deck show the open-style which helped prevent accumulation of any possible gas pockets. Air ventilation for the engine-room can be seen on the lower right-hand

Supertankers Anatomy and Operation

side, whilst the funnel towers over everything else. When the Supertanker was in ballast, we found that the funnel acted almost as a sail and certainly deflected not only the course of the ship, by catching the wind, but also affected the magnetic compass. When the ship was on course between Sierra Leone and Capetown, for instance, such a considerable error was introduced into the compass that the magnetic course was invariably south-westwards, opposing the actual/true south-easterly course being steered. Internally, most funnels consist largely of pipes including uptakes from the main engine, boiler and alternator, as well as ventilator trunks, thus carrying engine exhaust (and potential sparks) well clear of the deck, indicated in the photograph of a funnel belonging to a new-building VLCC. (Photo. 4.39). Behind the door, at the base of the funnel, there was a ladder leading up to the top below the cowling. A clearer view of a mooring winch aft (Photo. 4.40) indicates the forward aspect and shows also the drum end supplied for handling any ropes to be used as well as the inner working drum on this particular winch.

Whilst an emergency towing bracket similar to that found forward may be found in the centre-line of the ship on the after-deck, this fitting is not compulsory under IMO Regulations and there are a lot of VLCC's without it. From January 1996, however, it was an IMO and OPA90 requirement that a Full Emergency Towing System (ETS) must be fitted on all tankers over 20,000 sdwt. On tankers over 50,000 sdwt, a minimum 90 mm x 77 metre towing wire of galvanised high tensile steel is provided that is stowed on a storage drum situated either on deck or in a locker below. The drum has its own brake and power supply with a separate locker containing the retrieval gear. The latter consists of 50 metres of 20 mm wire, connected to 200 metres of 44 mm rope, a marker buoy and then an additional 50 metre pick up line with a final marker buoy. Both buoys are fitted with lights to assist the operation in restricted visibility or at night. The messenger line, connected to the towing wire, may be passed to a tug either manually or by remote firing from a gun operated by compressed air. The following diagrams of the equipment supplied to the market by Pusnes of Norway, (Diags. 4.11 and 4.12), indicate how the ETS may be fitted on board the VLCC and how the line would be floated to the tug from the disabled ship.

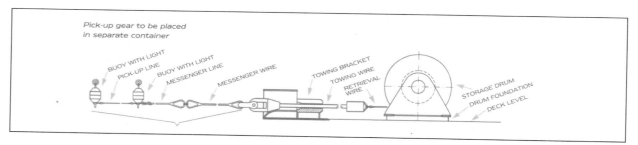

Diag. 4.11 Setting Up the Emergency Towing System (ETS).
(By kind permission of Pusnes AS.)

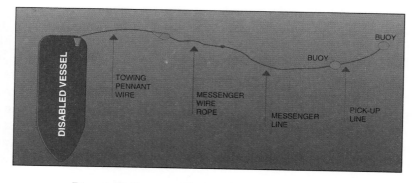

Diag. 4.12 Emergency Towing System (ETS) Procedure.
(By kind permission of Pusnes AS.)

Chapter Five

SECTION ANALYSIS

5.1. FORE-PART SPACE AND STORE AREA
5.2. SUEZ SEARCHLIGHT, ANCHOR CABLES AND LOCKERS FO'C'SLE HEAD, FORE-PEAK TANK
5.3. CENTRE AND WING TANKS
5.4. BUNKERS AND CONSUMPTION OF DIESEL
5.5. CAPACITY OF CARGO TANKS
5.6. PUMPS AND PUMP ROOMS
    5.6.1. Pump Lines
    5.6.2. Butterfly and Gate Valve Operation
5.7. BALLAST SYSTEMS, BALLAST TANKS
5.8. CARGO CONTROL ROOM
5.9. FIRE EXTINGUISHING ROOM AND THE USE OF CO_2
5.10. STEERING GEAR SYSTEM
5.11. GYRO ROOM
5.12. RADIO ROOM, TELEPHONE EXCHANGE
5.13. WHEELHOUSE AND BRIDGE
5.14. CHARTROOM
    5.14.1. Charts and Radars
    5.14.2. Navigation Consoles, Radios, Telephones
5.15. STEERING PEDESTAL
5.16. SIGNAL AND NAVIGATION LIGHTS
5.17. BRIDGE BOX
    5.17.1. Walkie-Talkies
5.18. MANOEUVRING INDICATORS, 'MONKEY ISLAND'

Supertankers Anatomy and Operation

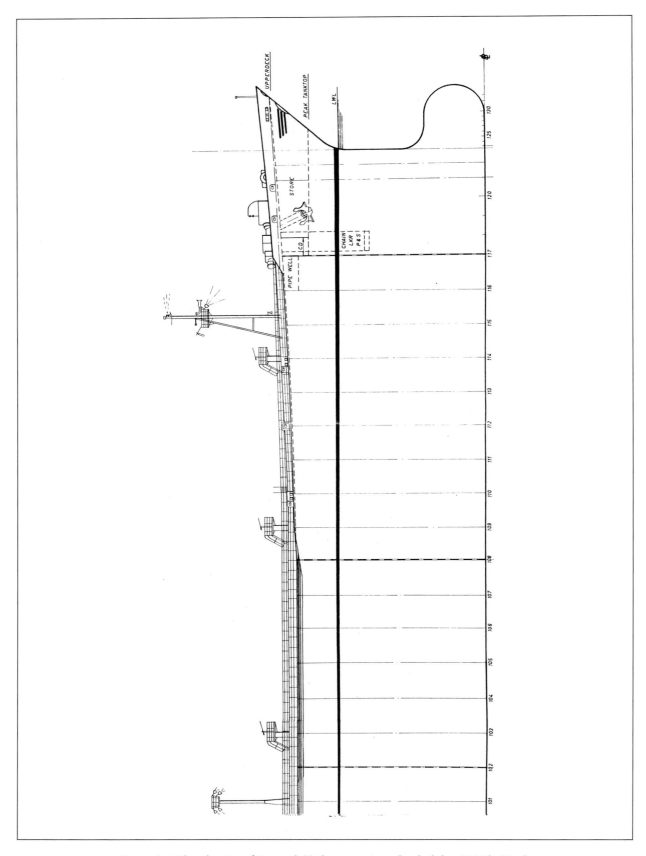

Draw. 5.1 Plan drawing of Forepeak Tanktop area immediately below Fo'c'sle Head. (By kind permission of Builder's Plans.)

Chapter Five

SECTION ANALYSIS

*(The Hull and Equipment Below the Main Deck — Fore and Aft, but Excluding Engine room and Machinery — **FOR'D:** Fo'c'sle Head: Suez Canal Searchlight and Function — Spurling Pipe — Telephone — Spare Anchor Cable Links — **MIDSHIPS:** Variety of Cargo Tank Arrangements — Cargo and Other Tank Capacities — **AFT:** Fuel/Lubricating/Diesel Oil (etc) and Water Tank Arrangements — Pump Room and Equipment — Cargo Tank and Ballast Lines — Cargo Control Room and Equipment — CO_2 Room — Steering Gear System — Gyro Compass Room — Radio Room: Equipment and Updating — Chartroom and Wheelhouse: Navigation Gear — DF/Radar/Depth Sounder/Auto-Pilot — Navigation Control Consoles — Monkey Island Equipment (Including Lights).*

5.1. FORE-PART SPACE AND STORE AREA

There are a variety of internal under deck arrangements aboard Supertankers depending upon the wishes of the Owner and the agreements that have been made with the building yard. Invariably, the area above the fore-peak tank and bulbous bow comprises the Bosun's store. (See Draw. 5.1). The differences, however, although obviously conforming to requirements of the International Association of Classification Societies and International Maritime Organization, could sometimes be quite significant. For example, some Supertankers had a forward cofferdam, or totally empty space separating the forepeak from number one cargo tank, that extended over the total distance from immediately below the maindeck to the ship's bottom plates. Others, the more standard type of VLCC, had only a partial space separating the pipe well and upper part of number one tank from the store area, which terminated below the fore peak tank top. (Draw. 5.2). Whilst this massive fore-part space is virtually empty, significant equipment is sited there. Often, about six or seven spare anchor cable links; the polypropylene mooring ropes on their self-winding reels that would be used forward during berthing operations, and a telephone connection to the main exchange, not that this is used very often as all officers are, as a matter of course, equipped with VHF localised portable radio sets. Apart from routine testing, to ascertain that the gear continued functioning in the event that it might be needed in an emergency, the only call I ever remember was made by the Bosun, before we changed from European to Goanese/Bangladeshi crew. He wanted to contact the Petty Officer's mess on one occasion telling the 'peggy' (or deck boy who looked after the PO's) to keep his luncheon hot as he had been detained over-long by a job forward. There would be lockers containing the lifejackets that are normally kept forward for emergency uses, as well as fuse-boxes covering deck gear and sundry machinery whose function normally came under the jurisdiction of the chief engineer. (Photo. 5.1).

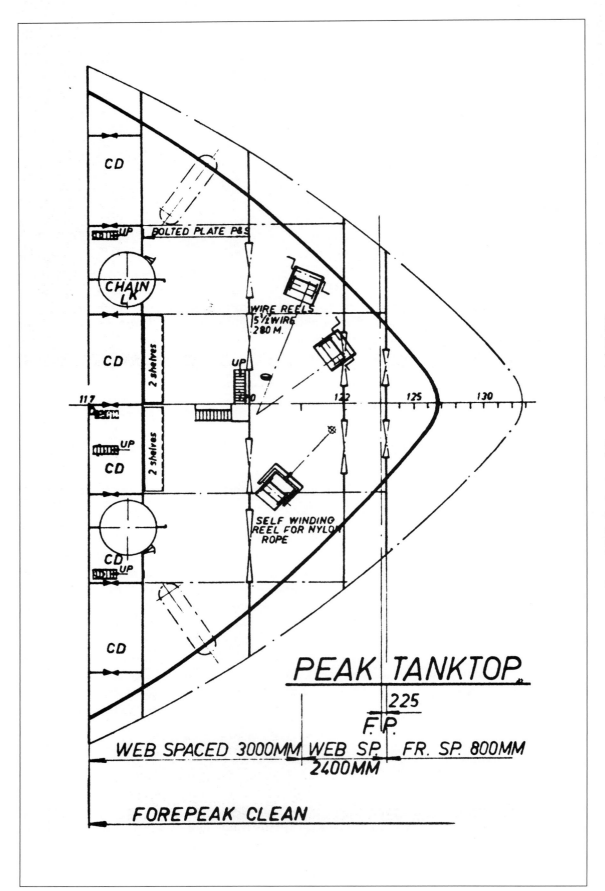

Draw. 5.2 Profile drawing of Forepart of a typical VLCC. (By kind permission of Builder's Plans.)

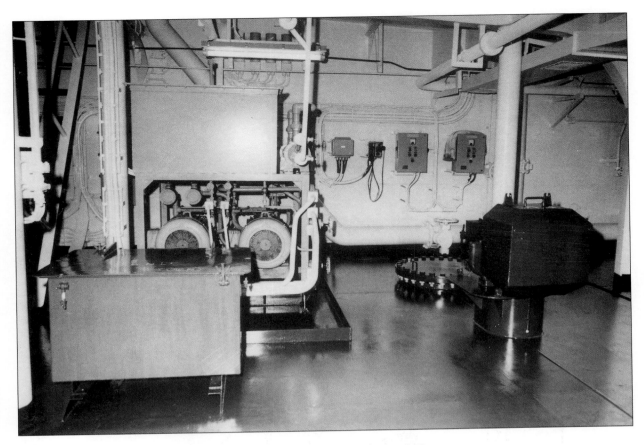

5.1. General view of the Forward Bosun's Store.
(By kind permission of Shell International Trading and Shipping.)

5.2. SUEZ SEARCHLIGHT, ANCHOR CABLES AND LOCKERS, FO'C'SLE HEAD, FORE-PEAK TANK

For the majority of VLCC's, which are not fitted with an overside davit, adjacent to the small mast in the very 'eyes' of the ship on the observation platform where this is fitted, the Bosun's store contains also the Suez Canal searchlight and a circular manhole, with a cover that would be barred and clamped in order to render it completely watertight when not in use. (Photo. 5.2). The searchlight would be rigged prior to entering the Suez Canal and is used to

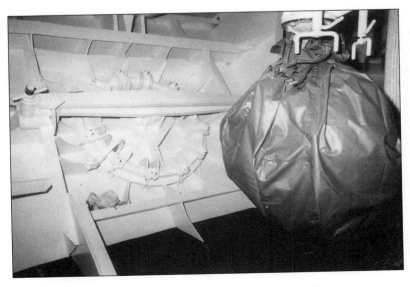

5.2. The Suez Canal Searchlight and Manhole.
(By kind permission of Shell International Trading and Shipping.)

Supertankers Anatomy and Operation

5.3. Starboard Side looking Aft of Fo'c'sle 'Tween Deck area.
(By kind permission of Swan Hunter Shipyard.)

pick out the buoyage indicating the dredged limits in the canal system within which ships of any size would have to navigate. The afterpart of the fo'c'sle contained the housing for the two anchor cables and their chain lockers on each of the port and starboard sides. The cable is led downwards from the windlass through spurling pipes into the lockers which are built with a strong substantial floor, that is equipped with a false bottom for drainage purposes. The sides are reinforced with web frames and the top of the spurling pipe is strengthened with stiffeners in order to alleviate chafing. The end link of the cable is fixed firmly to the bottom by a very large shackle which can be unpinned if the anchor becomes immovably fast on a submerged object and the anchor and all cable let-go, in order to release the ship. I suspect, however, that any Master who allowed this to happen would not be 'flavour of the month' as recovery or total replacement costs would be extremely high. When visiting the chain locker and clambering over the cable I found that, although this area was very roomy, it was also very slippery and damp. I was surprised at the circumference of the spurling pipes themselves, having expected them, for some reason, to be only sufficiently large for the passage of the cable. The photograph of a 250,000 sdwt ship undergoing completion, (Photo. 5.3), shows the construction of the fo'c'sle head, with the fore-peak tank top being completed and the starboard side spurling pipe being built into place. The men standing inside the head next to the spurling pipe and the one on deck emphasise the vastness of this below deck area. The shot was taken looking aft, and the space which is to be part of the fore-peak is clearly shown.

5.3. CENTRE AND WING TANKS

The number and distribution of cargo tanks along the length of the hull varies considerably between Supertankers and there is no fixed determination of the number of sets comprising centre and wing tanks that might constitute the ship's construction. The following series of drawings

Section Analysis

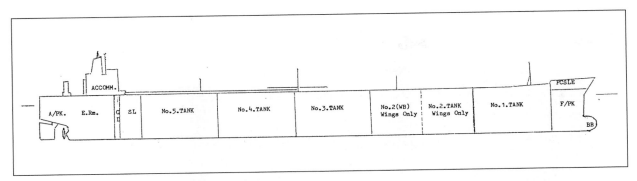

Draw. 5.3 Side Elevation of VLCC "RANIA CHANDRIS" upon which ship the Author served.
(The Author.)

Draw. 5.4 Plan View of the VLCC "RANIA CHANDRIS".
(The Author, based on Ship's Plans.)

indicate merely two of the totally different layouts that might be found. The first are those of one of the VLCC's upon which I served, the "RANIA CHANDRIS" from which the side elevation shows the cargo capacity is divided into five main cargo tanks, (Draw. 5.3), with an after cofferdam only. The plan view, (Draw. 5.4), indicates the arrangements and distribution of ballast against cargo tanks, from which it can be seen that number two wing tanks are separated so that the after parts supply the totally segregated tanks used only for sea ballast. The tank capacities on this particular VLCC are best seen from the following table, extracted from the ship's capacity plan. (Table 5.1). The table shows the cargo oil tanks capacities as well as those for each of the fuel, water, ballast and lubricating oil tanks. Of particular importance, is the siting of the tanks in relation to the ship's frames, a vital element when assessing stresses on the hull, a factor that is examined further (Chapter 6). It should be noted that the summer deadweight capacity of the VLCC "RANIA CHANDRIS" was 286,000 long tons which is, of course, the assessed deadweight at that particular draft. This accounts for the apparent disparity between the sdwt and the 98% capacity shown on the table at 291,483 long tons. For comparative purposes, reference could be made to the two capacity plans, covering each of a single hulled and double hulled Supertanker, contained in the inside back cover folder.

CARGO OIL TANKS

TANK	FRAME NO.	CUBIC METRES 100%	CUBIC FEET 100%	F.W. l.tn. at 35.88 cu.ft./l.tn. 100%	F.W. l.tn. at 35.88 cu.ft./l.tn. 98%	CARGO OIL l.tn. at 42.00 cu.ft./l.tn. 100%	CARGO OIL l.tn. at 42.00 cu.ft./l.tn. 98%
CENTRE TK. NO.1	108 – 117	29409	1038561	28945	28366	24728	24233
CENTRE TK. NO.2	96 – 108	38229	1350044	37627	36874	32144	31501
CENTRE TK. NO.3	87 – 96	28667	1012383	28216	27652	24104	23622
CENTRE TK. NO.4	78 – 87	28667	1012383	28216	27652	24104	23622
CENTRE TK. NO.5	64 – 78	35411	1250527	34853	34156	29774	29179
TOTAL		160383	5663898	157857	154700	134854	132157
WING TK. NO.1 P&S	108 – 117	2×17889	2×631740	2×17607	2×17255	2×15041	2×14740
WING TK. NO.2 P&S	102 – 108	2×13920	2×491569	2×13700	2×13426	2×11704	2×11470
WING TK. NO.3 P&S	87 – 96	2×20906	2×738286	2×20577	2×20165	2×17578	2×17226
WING TK. NO.4 P&S	78 – 87	2×20877	2×737264	2×20548	2×20137	2×17554	2×17203
WING TK. NO.5 P&S	69 – 78	2×19165	2×676812	2×18863	2×18486	2×16115	2×15793
SLOP TK. P&S	64 – 69	2×3921	2×138482	2×3860	2×3783	2×3297	2×3231
TOTAL		193356	6828306	190310	186504	162578	159326
MAIN TOTAL		353739	12492204	348167	341204	297432	291483

FUEL OIL

TANK	FRAME NO.	CUBIC METRES 100%	CUBIC FEET 100%	F.W. l.tn. at 35.88 cu.ft./l.tn. 100%
F.O. WING TANK FORWARD S	46 – 60	2250	79458	2215
F.O. WING TANK FORWARD P	35 – 60	3473	122664	3419
F.O. WING TANK AFT. P&S	A – 35	2×2285	2×80697	2×2249
F.O. SETTLING TANK S	35 – 46	1211	42790	1193
DIESEL OIL TANK IN CL	–1 – 20	418	14756	411
TOTAL		11922	421062	11736

FRESH- AND FEED WATER

TANK	FRAME NO.	CUBIC METRES	CUBIC FEET	F.W. l.tn. at 35.88 cu.ft./l.tn.
FRESH WATER TANK I	26 – 29	61.7	2178	60.7
FRESH WATER TANK II	21 – 26	102.8	3630	101.2
FEED WATER TANK I	32 – 35	61.7	2178	60.7
FEED WATER TANK II	29 – 32	61.7	2178	60.7
TOTAL		287.9	10164	283.3

WATER BALLAST

TANK	FRAME NO.	CUBIC METRES	CUBIC FEET	F.W. l.tn. at 35.88 cu.ft./l.tn.	F.W. l.tn. at 35.00 cu.ft./l.tn.
WING TANK NO.2A P&S	96 – 102	2×13926	2×491777	2×13706	2×14051
W.B. TANK AFT P	60 – 64	940	33199	925	949
W.B. TANK AFT S	60 – 64	951	33580	936	959
FORE PEAK TANK	117 – F	8496	300026	8362	8572
AFT PEAK TANK	A – 20	1130	39908	1112	1140
TOTAL		39369	1390267	38747	39722

LUBRICATING OIL

TANK	FRAME NO.	CUBIC METRES	CUBIC FEET	F.W. l.tn. at 35.88 cu.ft./l.tn.
LUB. OIL STOR. TANK	26 – 29	51.9	1834	51.1
LUB. OIL STOR. TANK	21 – 26	54.2	1913	53.3
LUB. OIL SUMP TANK	39 – 45	30.0	1059	29.5
TOTAL		136.1	4806	133.9

Table 5.1 Tank Capacities on the "RANIA CHANDRIS".
(By kind permission of Builder's Plans.)

5.4. BUNKERS AND CONSUMPTION OF DIESEL

Steaming on passage, for twenty-four hours per day continuously at around fifteen and a half knots, means that a VLCC would cover about 370 nautical miles per day. Consumption of fuel oil between different ships might vary considerably and would be dependant upon the make and type of engine, particularly whether or not she was driven by diesel or steam turbine engines, her loaded condition and shaft horse power etc, and could range between 90 to as much as 170 tonnes per day. Consumption of diesel also can vary between ships and range from four to less than one tonne per day. A round trip between the Arabian Gulf and Europe is about 25,000 nautical miles, taking about two months including waiting time, loading, discharging etc, so that the capacity of bunkers on board would rarely be sufficient for the round trip. We used to bunker to capacity outward bound and take on board additional fuel in the loading port, allowing generally a ten per cent leeway for emergencies. The ship was fitted with a steam turbine engine of 32,500 shp (metric) driving a single propeller shaft at 86 rpm giving a service speed of between fifteen and seventeen and a half knots depending upon the loaded/ballast condition.

5.5. CAPACITY OF CARGO TANKS

It is interesting to make a comparison between this 286,000 sdwt VLCC and the tank arrangements and capacities of the ULCC "JAHRE VIKING". The plan view and side elevations of the latter ship, (Draw. 5.5), indicate the much smaller individual capacity, and therefore greater number of cargo tank arrangements into which the hull of this ULCC has been divided at the preference of the original owners who placed the order for her building. Following construction in 1976, the ship was of 418,610 sdwt, at 376.73 metres length overall. In 1980 she was cut in two and her length extended to around 459 metres, with the deadweight tonnage increased to 564,739. She retained her 50,000 shp engine giving her a speed range between thirteen and sixteen knots. The following table, (Table 5.2), gives the capacities of cargo tanks, using the international American oil tonnage measurement of gross or net barrels. One barrel is equivalent to approximately forty two (42) US gallons which works out to approximately seven barrels to one tonne. The capacities for oil tanks and water tanks are shown in the accompanying tables, (Tables 5.3 and 5.4), which again make interesting comparisons. Reference again to the capacity plans for the 300,000 sdwt VLCC's, enclosed in the folder inside the back cover, indicate further modifications that have been made to later generation, double hulled VLCC's, built in the 1990's. Whilst there are undoubtedly major constructional differences, the tank arrangements, accommodation layouts, and the various tank capacities between different ships, are very similar.

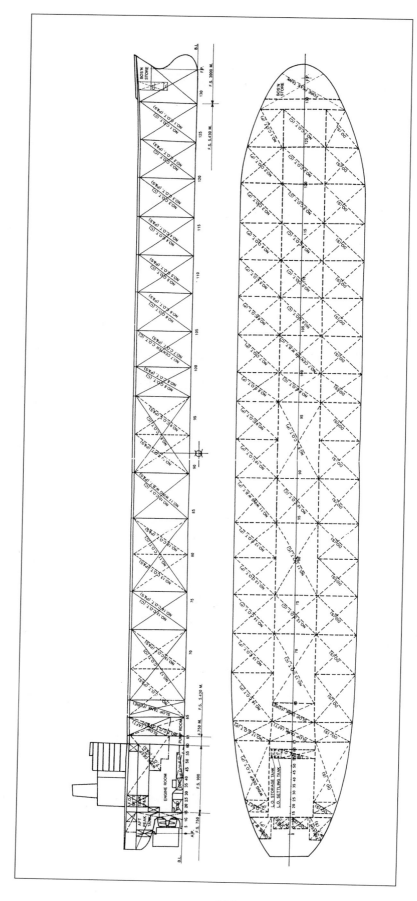

Draw. 5.5 Plan and Side elevation views of the ULCC "JAHRE VIKING".
(By kind permission of Jahre, Wallem Shipping.)

COMPARTMENT			FRAME NO.	CAPACITIES					CENTER OF GRAVITY		MAX. FREE SURFACE MOMENT IN — M⁴
				100 % FULL		98 % FULL			LONGI^L FROM ⊗ IN M.	VERTICAL ABOVE B.L. IN M.	
				CUBIC METERS	CUBIC FEET	CUBIC METERS	CUBIC FEET	BARRELS			
NO. 1 CENTER C.O.T.			124-128	15,965	563,799	15,646	552,523	98,409	-188.76	16.03	22,311
NO. 2 " "			120-124	15,474	546,459	15,165	535,530	95,382	-166.96	15.53	22,311
NO. 3 " "			116-120	15,456	545,824	15,147	534,908	95,271	-145.24	15.51	22,311
NO. 4 " "			112-116	15,456	545,824	15,147	534,908	95,271	-123.52	15.51	22,311
NO. 5 " "			108-112	15,456	545,824	15,147	534,908	95,271	-101.80	15.51	22,311
NO. 6 " "			104-108	15,456	545,824	15,147	534,908	95,271	-80.08	15.51	22,311
NO. 8 " "			97-100	11,592	409,368	11,360	401,180	71,453	-39.35	15.51	16,733
NO. 9 " "			89-97	30,911	1,091,612	30,295	1,069,780	190,536	-9.49	15.51	44,622
NO. 10 " "			84-89	19,320	682,279	18,934	668,634	119,089	25.81	15.51	27,889
NO. 11 " "			76-84	30,911	104,612	30,295	1,069,780	190,536	61.11	15.51	44,622
NO. 12 " "			72-76	15,456	545,824	15,147	534,908	95,271	93.69	15.51	22,311
NO.13 " "			65-72	27,048	955,191	26.507	936,087	166,724	123.56	15.51	39,044
SLOP TANK (FORE)			63-65	7,728	272,912	7,573	267,454	47,636	147.99	15.51	11,155
" " (AFT)			61-63	5,876	207,509	5,758	203,359	36,220	158.85	19.72	12,772
NO. 1 WING C.O.T.		(P)	124-128	10,034	354,347	9,833	347,260	61,850	-187.63	16.49	8,242
NO. 1 " "		(S)	"	10,034	354,347	9,833	347,260	61,850	-187.63	16.49	8,242
NO. 2 " "		(P)	120-124	13,712	484,235	13,438	474,550	84,521	-166.68	15.46	17,904
NO. 2 " "		(S)	"	13,712	484,235	13,438	474,550	84,521	-166.68	15.46	17,904
NO. 3 " "		(P)	116-120	14,800	522,657	14,504	512,204	91,227	-145.21	15.23	21,445
NO. 3 " "		(S)	"	14,800	522,657	14,504	512,204	91,227	-145.21	15.23	21,445
NO. 4 " "		(P)	112-116	14,876	525,341	14,578	514,834	91,696	-123.52	15.19	21,594
NO. 4 " "		(S)	"	14,876	525,341	14,578	514,834	91,696	-123.52	15.19	21,594
NO. 5 " "		(P)	108-112	14,876	525,341	14,578	514,834	91,696	-101.80	15.19	21,594
NO. 5 " "		(S)	"	14,876	525,341	14,578	514,834	91,696	-101.80	15.19	21,594
NO. 6 " "		(P)	104-108	14,876	525,341	14,578	514,834	91,696	-80.08	15.19	21,594
NO. 6 " "		(S)	"	14,876	525,341	14,578	514,834	91,696	-80.08	15.19	21,594
NO. 7 " "		(P)	100-104	14,876	525,341	14,578	514,834	91,696	-58.36	15.19	21,594
NO. 7 " "		(S)	"	14,876	525,341	14,578	514,834	91,696	-58.36	15.19	21,594
NO. 8 " "		(P)	97-100	11,157	394,006	10,934	386,126	68,772	-39.35	15.19	16,196
NO. 8 " "		(S)	"	11,157	394,006	10,934	386,126	68,772	-39.35	15.19	16,196
NO. 9 " "		(P)	93-97	14,876	525,341	14,578	514,834	91,696	-20.35	15.19	21,594
NO. 9 " "		(S)	"	14,876	525,341	14,578	514,834	91,696	-20.35	15.19	21,594
NO. 10 " "		(P)	89-93	14,876	525,341	14,578	514,834	91,696	1.38	15.19	21,594
NO. 10 " "		(S)	"	14,876	525,341	14,578	514,834	91,696	1.38	15.19	21,594
NO. 12 " "		(P)	80-84	14,875	525,306	14,578	514,834	91,690	50.25	15.19	21,594
NO. 12 " "		(S)	"	14,875	525,306	14,578	514,834	91,690	50.25	15.19	21,594
NO. 13 " "		(P)	76-80	14,817	523,257	14,521	512,792	91,332	71.95	15.24	21,594
NO. 13 " "		(S)	"	14,817	523,257	14,521	512,792	91,332	71.95	15.24	21,594
NO. 14 " "		(P)	72-76	14,476	511,215	14,186	500,991	89,230	93.62	15.52	21,594
NO. 14 " "		(S)	"	14,476	511,215	14,186	500,991	89,230	93.62	15.52	21,594
NO. 15 " "		(P)	68-72	13,527	477,702	13,256	468,148	83,380	115.23	16.16	21,635
NO. 15 " "		(S)	"	13,527	477,702	13,256	469,148	83,380	115.23	16.16	21,635
NO. 16 " "		(P)	63-68	14,079	497,195	13,797	487,251	86,783	139.18	17.38	25,296
NO. 16 " "		(S)	"	14,079	497,195	13,797	487,251	86,783	139.18	17.38	25,296
NO. 17 " "		(P)	61-63	4,116	145,355	4,034	142,448	25,371	158.58	19.20	7,705
NO. 17 " "		(S)	"	4,116	145,355	4,034	142,448	25,371	158.58	19.20	7,705
TOTAL				671,803	23,724,503	658,362	23,250,083	4,141,004			

Table 5.2 Cargo Tank Capacities of the "JAHRE VIKING".
(By kind permission of Jahre Wallem Shipping.)

OIL TANKS

SPECIFIC GRAVITY: BUNKER OIL = 0.950   DIESEL OIL = 0.880   LUB OIL = 0.920

COMPARTMENT		FRAME NO.	CAPACITIES								CENTER OF GRAVITY		MAX. FREE SURFACE MOMENT IN M4
			100 % FULL		98 % FULL						LONGI.L FROM ⊠ IN M.	VERTICAL ABOVE BL IN M.	
			CUBIC METERS	CUBIC FEET	CUBIC METERS	CUBIC FEET	BARRELS	BUNKER OIL IN M.T.	DIESEL OIL IN M.T.	LUB. OIL IN M.T.			
WING DEEP F.O.T.	(P)	27-61	6,593	232,830	6,461	228,173	40,638	6,138			178.11	22.93	13,467
" " "	(S)	27-61	6,616	233,642	6,484	228,969	40,783	6,160			178.10	22.90	13,468
FUEL OIL SET. TANK	(P)	20-27	554	19,564	543	19,173	3,415	516			201.73	25.89	843
" " "	(S)	20-27	554	19,564	543	19,173	3,415	516			201.73	25.89	843
HEAVY F.O.T.	(C)	39-60	907	32,030	889	31,389	5,592	845			178.33	1.79	10,166
DIESEL OIL TANK	(C)	55-60	526	18,576	515	18,204	3,239		453		171.28	27.92	2,532
LUB OIL SUMP T.	(C)	34-38	37	1,307	36	1,281	226			33	190.63	4.51	48
LUB OIL SET TANK	(INN)	55-60	45	1,589	44	1,557	277			41	171.28	27.22	3
LUB OIL	(OUT)	55-60	45	1,589	44	1,557	277			41	171.28	27.22	3
BUNKER OIL TOTAL			15,224	537,630	14,929	526,877	93,843	14,175					
DIESEL OIL TOTAL			526	18,576	515	18,204	3,239		453				
LUB OIL TOTAL			127	4,485	124	4,395	780			115			
BILGE TANK	(C)	25-29	129	4,556							198.65	3.12	212
DIRTY BILGE TANK	(C)	29-31	78	2,755							196.01	3.14	186
VOID SPACE (IN ENG. RM.)		16-25	162	5,721							203.70	2.83	
" " "		31-39	214	7,557							191.15	1.93	

Table 5.3 Fuel and Service Oil Tank Capacities "JAHRE VIKING".
(By kind permission of Jahre Wallem Shipping.)

WATER TANKS

COMPARTMENT		FRAME NO.	CAPACITIES				CENTER OF GRAVITY		MAX. FREE SURFACE MOMENT IN M⁴
			100 % FULL		W. B.	FR. W.	LONGITUDE FROM ⊠ IN M.T.	VERTICAL ABOVE B.L. IN. M.	
			CUBIC METERS	CUBIC FEET	S.G.=1.025 IN M.T.	S.G.=1.000 IN M.T.			
FORE PEAK TANK		128-F.E.	15,942	562,986	16,341		-207.78	14.81	59,292
NO. 7 CENTER W. B. T.		100-104	15,456	545,824	15,842		-58.36	15.51	22,311
NO. 11 WING W. B. T.	(P)	84-89	18,596	656,712	19,061		25.81	15.19	26,993
NO 11 " "	(S)	84-89	18,596	656,712	19,061		25.81	15.19	26,993
AFT PEAK TANK		A.E. -16	2,790	98,528	2,860		215.83	21.42	38,816
DRINKING WATER TANK	(P)	A.P. -16	524	18,505		524	213.34	27.98	947
FRESH WATER TANK	(S)	A.P. -16	524	18,505		524	213.34	27.98	947
DISTILLED WATER TANK	(P)	8-16	312	11,018		312	211.03	27.93	296
"	(S)	"	312	11,018		312	211.03	27.93	296
BALLAST WATER TOTAL			71,380	2,520,762	73,165				
FRESH WATER TOTAL			1,048	37,010		1,048			
DISTILLED WATER TOTAL			624	22,036		624			
VOID SPACE (BELOW A.P.T.)		12-16	58	2,048			209.16	6.97	

Table 5.4 Water Tank Capacities "JAHRE VIKING".
(By kind permission of Jahre Wallem Shipping.)

5.6. PUMPS AND PUMP ROOMS

Although practices vary considerably, many VLCC's have pump rooms that are isolated from all other spaces on the ship, but in some designs, they are integrated into the after cofferdam. Whatever the designed layout, the pump room is always situated at the bottom of the ship, adjacent to the engine room, and contains housing for the four main cargo pumps, the pump operating the stripping lines, and those that are required for ballasting. General views of pump-room arrangements can be seen in the following selection of photographs taken aboard 'first-generation' VLCC's. (Photo. 5.4 and 5.5.).

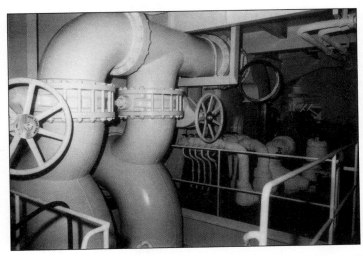

5.4. Piping Arrangement in Pump Room Entrance.
(By kind permission of Shell International Trading and Shipping.)

5.5. Piping Arrangement in Pump Room.
(By kind permission of Shell International Trading and Shipping.)

5.6.1. Pump Lines

In the following arrangement, (Diag. 5.1), the four main cargo pumps are linked by a series of valves through a centi-strip system, a feature of this particular 326,000 sdwt VLCC, to the cargo and slop tanks which allows a separator tank in the pump suction line to provide automatic control of the pump discharge valve, thus reducing the need for a more conventional stripping pump system. Following the convention seen in chapter four, the two inner lines serve the centre tanks and the two outer, the wings. Cross connections between number one and four lines and numbers two and three lines are made at the forward end and between numbers one and two, and numbers three and four lines at midships. The cargo pump discharges commonly into two lines which pass to the manifolds. Two lines are run from the manifold down to the two outboard lines for loading. These pumps are each turbine driven, with an output of 2750 hp, at 60 bar steam pressure, and a capacity of 3500 m³ S.W. per hour. In practical terms, each

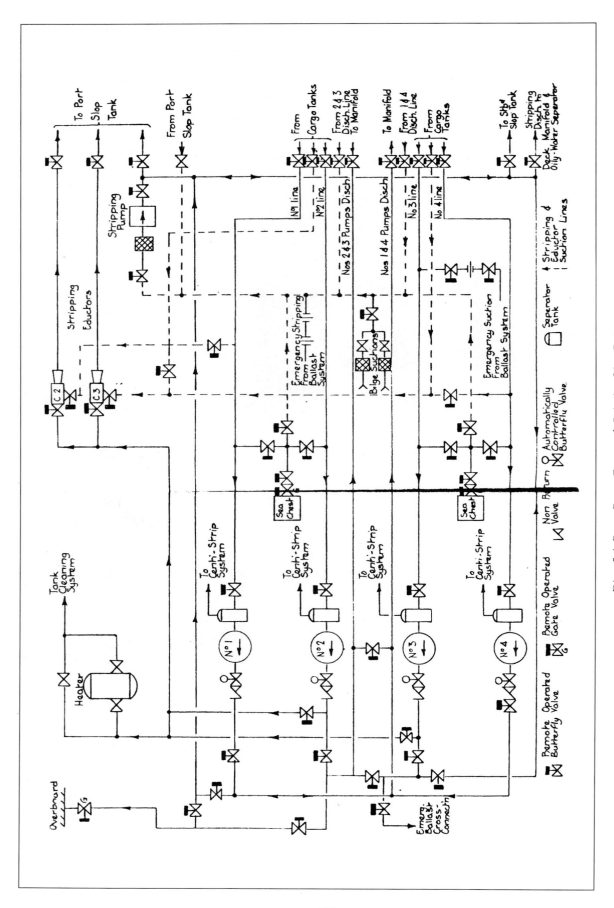

Diag. 5.1 *Pump Room Cargo and Stripping Line System.*
(*By kind permission of Shell International Trading and Shipping.*)

Section Analysis

*5.6. Typical Example of a Main Cargo Pump
aboard a VLCC.
(By kind permission of British Petroleum Shipping.)*

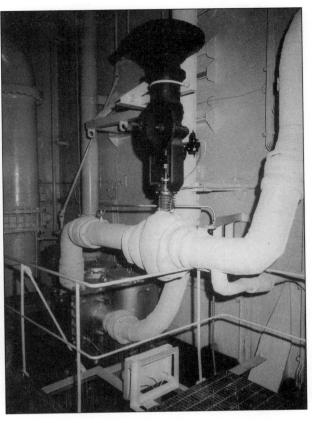

*5.7. A Single Stripping Pump
serves all Cargo Tanks on VLCC's.
(By kind permission of Shell International Trading and Shipping.)*

operates at a maximum discharge rate of around 2,500 tonnes per hour, and is fitted with a Governor that has a shaft speed ranging between 490-1200 rpm. Fresh water is often used as a cooling medium. A typical cargo pump, (Photo. 5.6), gives some idea of its appearance. Whilst systems aboard different ships may vary quite considerably, a single stripping pump serves all cargo tanks on this particular VLCC. The stripping pump is often steam driven and this one has a capacity of about 350 tonnes per hour. The much smaller size can be seen by a similar pump found in (Photo. 5.7). In the layout of the other VLCC, which depicts a slightly smaller capacity of double hulled construction, (Diag. 5.2), only three conventional main cargo pumps handle the cargo and there is no centi-strip system in use. The stripping pump is labelled "TCP", or tank cleaning pump.

5.6.2. Butterfly and Gate Valve Operation

The next diagram, (Diag. 5.3), indicates how the cargo and stripping lines run through into the tanks. The deck operated extended spindle butterfly and gate valves would be controlled from either the main deck red, white and green wheels, or from the hydraulic valve controls, both of which were examined in Chapter 4 when considering the deck gear on VLCC's.

Supertankers Anatomy and Operation

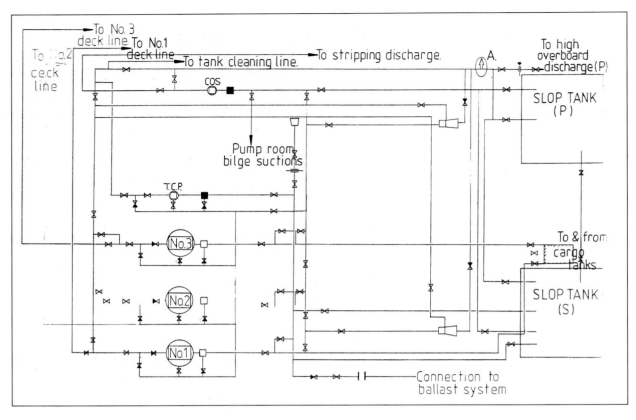

Diag. 5.2 Cargo Pumps, Lines and Valves in the Pumproom.
(By kind permission of Shell International Trading and Shipping, and Paul Taylor.)

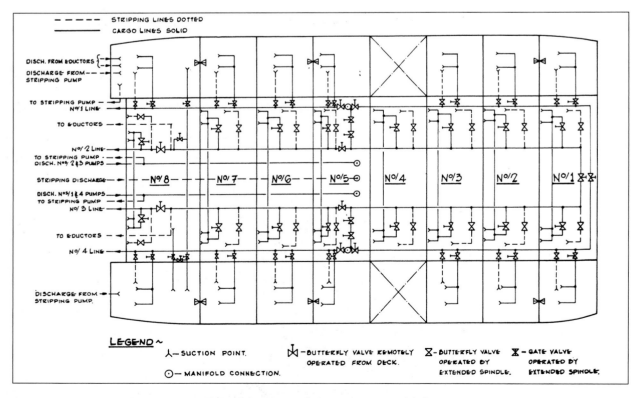

Diag. 5.3 Cargo and Stripping Lines in Tanks.
(By kind permission of Shell International Trading and Shipping.)

5.7. BALLAST SYSTEMS, BALLAST TANKS

The ballast system on all tankers, regardless of size, is kept totally separate from the cargo system and, in this particular ship, (Diag. 5.4), the ballast pump may be used only to fill or empty the permanent ballast tanks, in this case number four wings, or to fill numbers two, three and six centres through the ballast discharge main on deck and then via the appropriate purge pipe to the tank. The single ballast pump, (Photo. 5.8), serving all clean sea-water tanks is fitted with a similar Governor. It is again turbine driven, at a speed of 6500 rpm and has an output of 1250 hp. On a double hulled VLCC, the ballast system in this instance, (Diag. 5.5), consists of two pumps which can serve either line and are used for both the port and starboard wing ballast tanks. This diagram indicates the total segregation of the system from the cargo lines so that no cargo can ever enter the ballast tanks.

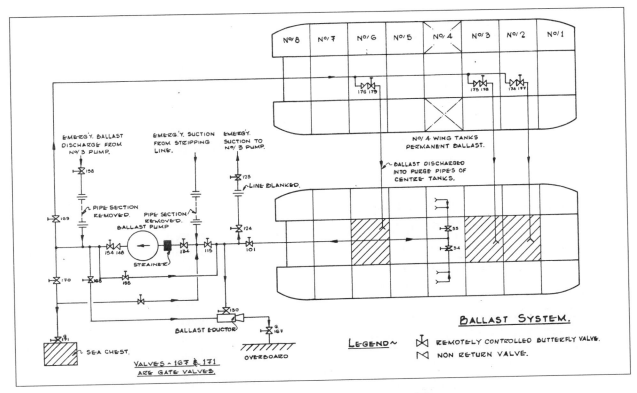

Diag. 5.4 Ballast Line System on a VLCC.
(By kind permission of Shell International Trading and Shipping.)

5.8. Main Ballast Pump in Cargo Room
On Board a Large Tanker.
(By kind permission of British Petroleum
Shipping.)

Supertankers Anatomy and Operation

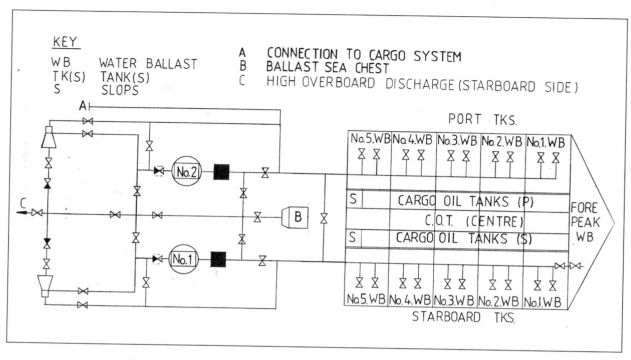

*Diag. 5.5 Ballast Pumps, Lines and Valves in Tanks and in the Pumproom.
(By kind permission of Shell International Trading and Shipping, and Paul Taylor.)*

5.8. CARGO CONTROL ROOM

An essential office, usually on 'A' or the boat deck, is the cargo control room from which is run the entire operation of controlling the flow of cargo during loading/discharging and when engaged on ballasting. A 'Mimic' diagram (Photo. 5.9) is an outline of the hull projected onto the bulkhead, which shows the relative positions of every cargo and ballast valve in the ship. The board is a main feature confronting the eye as the room is entered. Each valve is numbered for reference, a facility which certainly helps avoid error when the mind is concentrating on organising the flow of oil throughout the ship. Light indicators show the condition of operation with, on some displays, green lights showing open valves and red the ones which are shut. There are various layouts available depending upon the requirements of the ship (Photo. 5.10 and 5.11), but they serve a similar purpose. There is also the facility not only to operate, but to monitor the progress of cargo/ballast pumps. Again, on some VLCC's, another Mimic, 'Graphic', board indicates the layout and operating state of the pumps (Photo. 5.12) whilst other displays offer merely the pump name and controls. (Photo. 5.13). Access is available to each pump that regulates speed and checks the suction and discharge pressures. The readings of suction and discharge pressures, as well as oil flow are recorded at regular intervals in specially tabulated books that are retained for official purposes. Additionally, for each pump, displays enable temperatures to be checked, especially of seals and bearings, and pump casing and turbine lubricating oil returns would also be displayed. On ships where this was fitted, the controls would be operated from the cargo control room for the centi-strip system separator gauge, pump trip/reset switch and, finally, the 'ready to start' switch and its indicator. Needless to say, fire alarm switches are available, as well as accommodation and pumproom ventilation fan switches incorporated. Communication between watchkeeping officers on deck and in the engine room are generally excellent and maintained by VHF personal radio's as well as telephone links to all parts of the ship. The control panel for the inert gas system, that is to be examined later, would also be contained in the cargo room. (Photo. 5.14).

Section Analysis

5.9. The "Mimic" Diagram shows the Relative Position of all Cargo and Ballast Valves.
(By kind permission of Mobil Ship Management.)

5.10. Cargo Control Panel and Pump Room Control (LHS).
(By kind permission of Shell International Trading and Shipping.)

Supertankers Anatomy and Operation

*5.11. Valve Control Panel in the Cargo Control Room.
(By kind permission of British Petroleum Shipping.)*

*5.12. Graphic Board of Pump Room in Cargo Control Office.
(The Author)*

*5.13. Pump Control Panel in Cargo Control Room.
(By kind permission of British Petroleum Shipping.)*

5.14. The Inert Gas Control Panel is on the LHS of the Photograph.
(By kind permission of Mobil Ship Management.)

5.9. FIRE EXTINGUISHING ROOM AND THE USE OF CO_2

A glance at the aft side elevation (Draw. 5.6) indicates that a ladder, which serves as an emergency escape route in the event of a serious incident in the engine-room, leads into the CO_2 fire extinguishing room. This is a total flooding system that has been fitted to protect the engine-room and pump room so that, in the event of fire, a blanket of gas can be led into the affected area quickly and efficiently. A typical system may consist of 289 steel carbon dioxide cylinders, each of which contains 45 kgs of CO_2. A piping arrangement (Diag. 5.6) enables 231 cylinders to expel in the machinery space, whilst the remainder would discharge into the pump room. A separate operating compartment, inside the CO_2 room, allows the system to be activated. The valves on the top of the cylinders have first to be collectively opened, then the pilot cylinders have to be filled, which act as feeder tanks, before the main gas cylinders can be released. An essential action in the emergency space would be for the ventilation fans in engine/boiler/pump rooms to be tripped, otherwise hot gases expelled by the fire could cause convection currents that, in turn, would draw air into the space and thus aggravate the problem by feeding the fire. Generally, the fans are capable of remote control outside the spaces for which the gas is intended. For safety purposes, warning klaxons are invariably sounded before the gas is emitted thus allowing personnel to evacuate. Considerable care has to be exercised in the servicing of CO_2 systems and this task is generally carried out by shore engineers provided by the manufacturers or contractors. Even so, incidents have been recorded of an undesirable accidental release of gas into the engine-room which then had to be extracted by use of ventilation fans.

Supertankers Anatomy and Operation

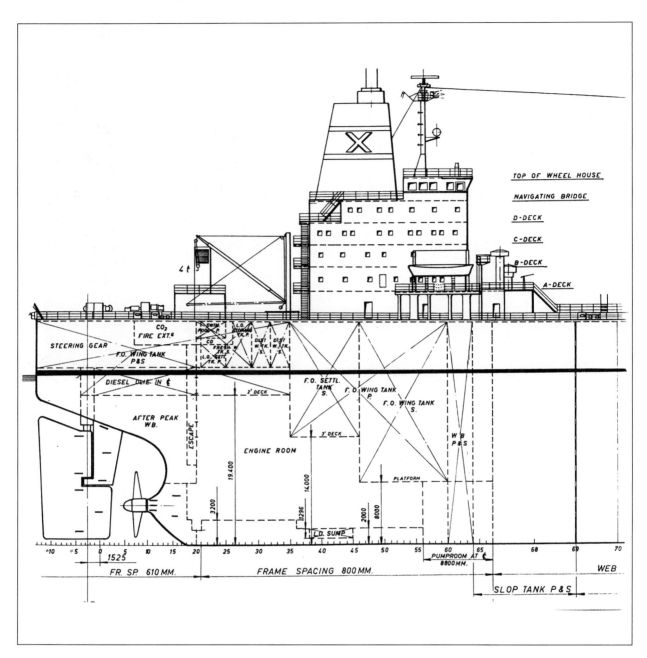

Draw. 5.6 Side Elevation of the After Part of the VLCC "RANIA CHANDRIS".
(By kind permission of Builder's Plans.)

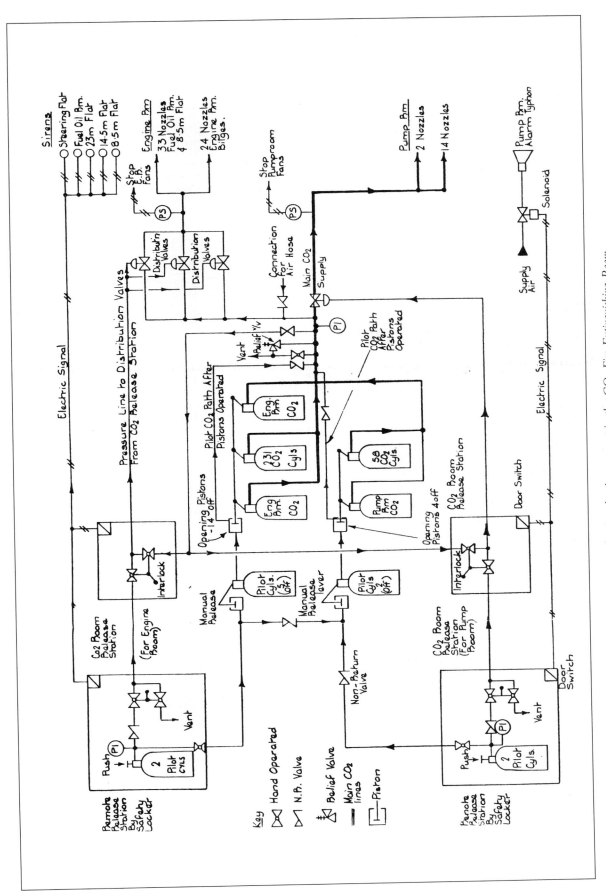

Diag. 5.6 Diagram showing the Arrangement in the CO_2 Fire Extinguishing Room.
(By kind permission of Shell International Trading and Shipping.)

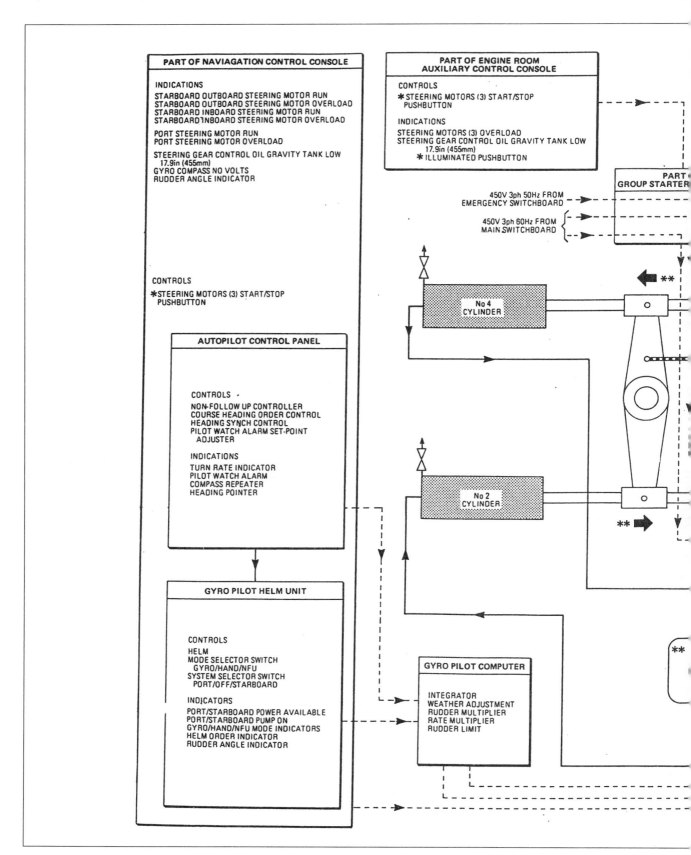

Diag. 5.7 Schematic Layout of Steering Gear.
(By kind permission of British Petroleum Shipping.)

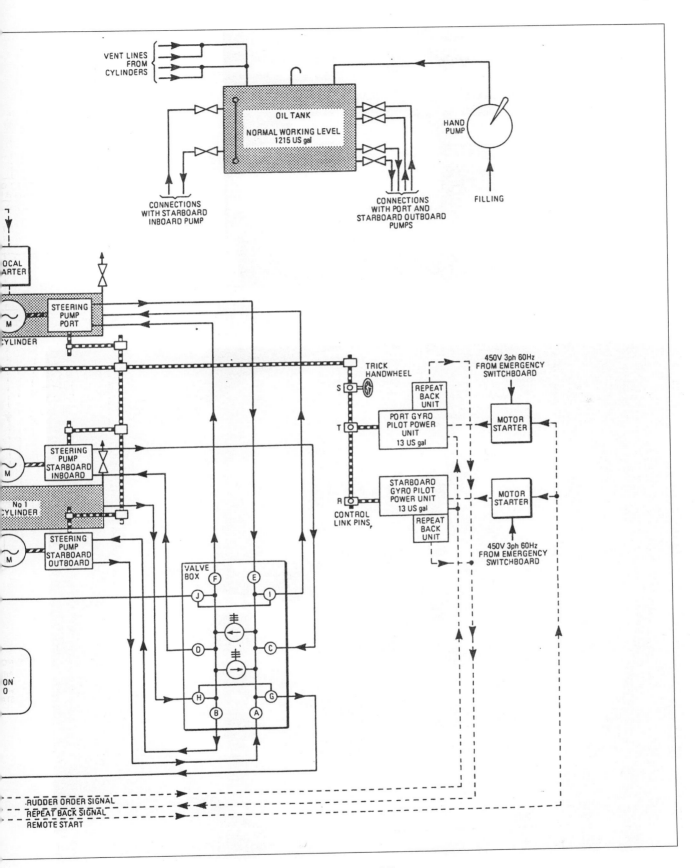

Diag. 5.7

Supertankers Anatomy and Operation

5.15. *Testing Hand Steering in the Emergency Steering Flat.*
(By kind permission of British Petroleum Shipping.)

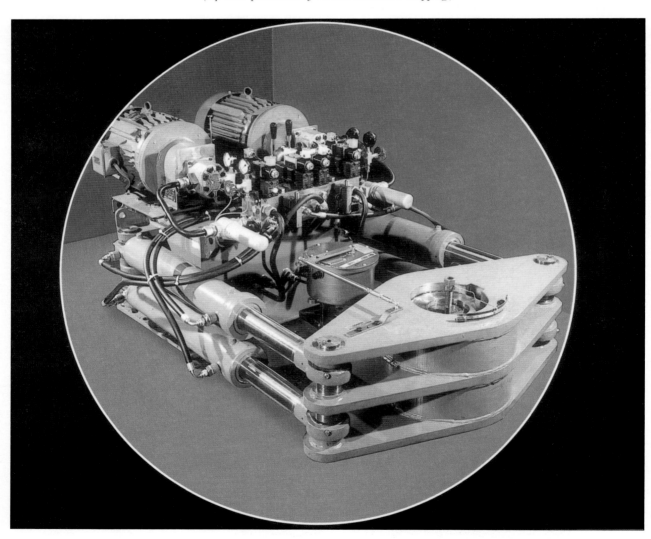

5.16. *4-Ram Steering Gear.*
(By kind permission of Clarke-Chapman Company.)

5.10. STEERING GEAR SYSTEM

A VLCC's main steering compartment is often found abaft the CO_2 Room. The steering gear system is generally electro-hydraulic and its function is to turn the ship's rudder from 35° in the helm hard a port position to 35° hard a starboard in around thirty seconds, with the ship fully loaded and at her maximum speed. This is to conform with Classification Society regulations. The size of the equipment is determined by comparing the rudder area with the square of the speed of the ship, so that the calculation tends to vary marginally with different sized Supertankers, bearing in mind that the speeds of all VLCC's are roughly the same. Often the steering gear fitted is a four-ram hydraulic system that is powered by electric motors which drive variable delivery pumps, and run at around 500 rpm. It is controlled remotely from the wheelhouse hydraulic steering wheel, as part of the ship's navigational control console, but it can be started and worked in the steering flat by pushbuttons fitted to the system. The local control so provided is particularly useful for testing purposes. (Photo. 5.15). The following plan indicates a typical steering gear system. Two pairs of hydraulic rams are operated by two of the three available variable delivery pumps (Diag. 5.7) to control directly the movement of the rudder situated immediately below. Although Clarke Chapman Marine of Tyne and Wear did not supply steering gears to many Supertankers, their 'Donkin' range of equipment has universal application and offers an excellent illustration of their four-ram gear. (Photo. 5.16).

5.11. GYRO ROOM

On the port side, forward of the engine casing or, in some ships, situated immediately above the engine control room, is a separately air-conditioned gyro-room in which is housed the master gyro compass. The makes of these vary considerably. On some ships, it could well be a 'Sperry', which proved totally reliable, and accurate to within one quarter of a degree. Whichever system is fitted, the gyro always points to true north and remains the main direction indicating instrument on the ship. Until superseded by microcomputer or even perhaps laser technology, the gyro compass remains the most important single piece of equipment aboard. A number of repeaters lead to the chartroom, steering compass in the wheelhouse, the compass bearing binnacles on the port and starboard bridge wings or spaces, the two radars and the direction-finder. The older-type was very large indeed, as can be seen from the photograph, (Photo. 5.17), required regular maintenance and took a long time to settle once it had been started: often in excess of twenty-four hours. Here, the second officer is shown instructing the cadet in the servicing of the equipment, perhaps by lightly oiling the cosine groove.

5.17. Second Officer teaching deck cadet to service the Master Gyro Compass. (By kind permission of British Petroleum Shipping.)

Supertankers Anatomy and Operation

5.18. *Radio Officer using Morse code on a First Generation VLCC to Send Message to Company Head Office via Portishead Radio.*
(By kind permission of British Petroleum Shipping.)

5.12. RADIO ROOM, TELEPHONE EXCHANGE

Today, marine communications are in a state of transition that is not entirely completed. This means that the radio room found on 'early generation' Supertankers, where this continues to be used today, contains now an amalgam of communication transmitters and receivers. There is also associated gear such as automatic alarm monitors to alert the radio officer if a distress message is received whilst he is off-watch, as well as advanced computerised technology. Much of this will be examined in Chapter 16 of Part Four when considering a wider application of computers in navigation generally and communications in particular. Radio officers continue to be carried on some VLCC's. In fact, I spoke to a serving R/O aboard a Supertanker in Fawley, early in 1999, who was due to be 'phased out', a dryly ironic term that appealed to both our senses of humour, when next his ship was in for survey and re-fit. Much of the traditional equipment, and indeed radio officers aboard ships, used to be supplied by Marconi International Marine Communications Company, or ITT (Marine), but often major shipping companies employed their own radio officers. The Type ST-1400 transmitter, an emergency transmitter and the main radio receiver Type R408 were typical of the gear found on board the average VLCC. The auto alarm system was also ITT Marine equipment, Type STR-60T. Some representative examples of Marconi's receivers/transmitters and sundry gear can be seen in the accompanying photograph (Photo. 5.18). The lay-person is sometimes surprised to learn that Supertankers, and most ships, for that matter, are fitted with their own automatic telephone exchange. Very often, this is situated on the Captain's deck and contains a comprehensive system of some 100 lines connecting areas of the ship in which it might be necessary to make contact quickly with the occupant, ranging from the cadet's quarters to the senior petty officers' cabins. I used the telephone most to speak with the engine room, asking for change over of the steering motors, requesting power or water on deck, and generally keeping them informed of situations which might involve us co-operating in any manner of situations. It was also useful to keep them informed by courtesy of any large, unexpected alterations of course, when the ship might suddenly swing to her helm or run into heavy weather. Other calls frequently used by deck officers would be to the 'Old Man', or Captain keeping him informed, where appropriate, or giving a relieving officer his "preliminary fifteen minute shake", enabling him to rouse from a deep sleep in the early hours of the morning in sufficient time to wash and "put away his liver", so to speak!

5.13. WHEELHOUSE AND BRIDGE

The lay-out of individual chartrooms and wheelhouses comprising the ship's navigating bridge varies considerably and there is no one standard design that can be considered the norm. Nearly all follow a similar pattern, however, determined only by the need of bringing together in one place similar equipment necessary for navigating the ship safely. A single plan is shown to act as a representative diagram. (Draw. 5.7), but the general arrangement plans (inside back cover) indicate other versions of bridge layouts. Much talk continues to be bandied about in certain marine circles regarding the need of the 21st Century for a 'one-man bridge', but, certainly during day-light hours deep-sea, such individual operation has been both the requirement and the practice now for over thirty years aboard all VLCC's and, indeed, most merchant ships, for that matter. Some Supertankers have separate chartrooms, but the majority combine chartroom and wheelhouse into one, a layout which many officers consider the most efficient method of keeping a navigating watch at sea. It means easy access to both chart and radars whilst, of course, maintaining more readily the essential visual lookout.

5.14. CHARTROOM

The chartroom space is taken up by the large table, which provides a working surface for charts of the area to be spread out, as well as the instruments necessary to run course lines and the navigation equipment essential to electronic position fixing. (Photo. 5.19). The secondary radar set is also situated there. The black set, elevated by hinged bracket over the chart table, is an older type Decca Navigator, a type still in use, but replaced by a smaller model showing a light emitting diode display nowadays. The Decca and LORAN C systems have worked in conjunction with each other, but the Decca part has been phased out owing to the loss of the

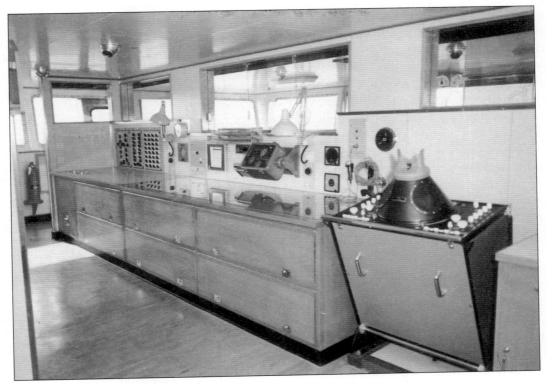

5.19. Radar, Chart Table and some Navigating Equipment.
(By kind permission of Shell International Trading and Shipping.)

Supertankers Anatomy and Operation

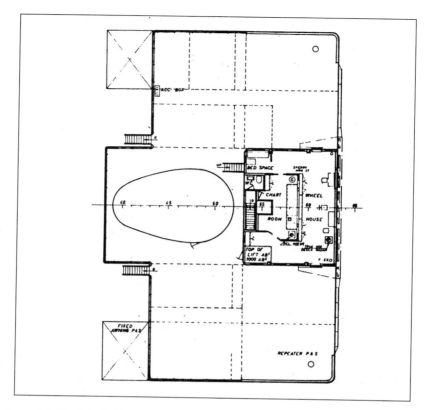

Draw. 5.7 Combined Chart Room and Wheelhouse on the "RANIA CHANDRIS".
(By kind permission of Builder's Plans.)

Decca transmitting stations, due as much to political reasons as any other, (including GPS coverage) which leaves operating the LORAN C long range navigation system. From a purely personal point of view, I regret the passing of Decca which has invariably proved an accurate and reliable system and, with recent equipment, the readings in terms of a very useful latitude and longitude enable an immediate position fix. My own experience of LORAN C has been somewhat mixed: much reception has been subject to what I term "atmospheric wow and flutter". Admittedly it is some years prior to writing since I have used the system and believe that reception has improved considerably. Referring to the plan of the Supertanker, portrayed in Drawing 5.7, on the port side of the of the chartroom table is a gyro-repeater whilst, on the starboard side, under glass covers, resides the two quartz chronometers essential for the timing of accurate sextant work in fixing the ship's position by celestial means. They are checked daily by internationally broadcast time signals from Greenwich. Modern quartz chronometers first introduced at sea in the early seventies, proved generally reliable for prolonged periods of use and served as an excellent check on each other, although I recall that the first quartz we carried gained about one second per day, a regular little 'galloping Gertie', until we exchanged it for a more stable model. They prove a very useful substitute to the old 'windey-windey' type, which preceded the quartz, and required regular winding at set periods every twenty-four hours: all too often the bane of the second officer who was entrusted with the responsibility of nursing the things. A never fading image retained in my mind, and doubtless that of many mates, is of the reminder, written in the girl friend's waterproof lipstick on the mirror above the sink in my cabin as a second mate, reminding me to "Wind Chron". On the fore-part rack, immediately behind the half-wooden cover/perspex curtained partition, allowing the duty mate to see visually whilst working on his charts in daytime, yet eliminating light at night, is very often to be found the direction finder that is a legal requirement under IMO regulations. The Marconi

"LODESTONE" Direction Finder (D/F) is the model that has become virtually a standard piece of equipment found aboard all merchant ships that I have encountered.

5.14.1. Charts and Radars

The charts carried in the drawers under the chart table working space are not world-wide, but cover only those areas into which the Supertanker would be likely to steam under her charter or normal trading areas. These include the entire route between the UK and Continental Europe, completely rounding Africa, with all port entrances to major oil terminals, to and including the Arabian Gulf, plus areas extending to the Far East and out to the American Gulf ports. Additional charts for other waters could, of course, always be provided by the ship's agents in the event of a re-routing. Chartroom pencils, erasers and other navigating impedimenta are lodged on a rack fixed to the bulkhead, each in its allocated place, along with perspex parallel rules and other devices essential to transfer position lines across the chart. The Decca Navigator, Mark 12 type, is to port of the D.F, in the plan whilst, on the extreme starboard side of the table, is the Racal-Decca True Motion Radar, Type 829. This is a set of extreme discrimination and accuracy. I recall in the Kattegat, on low range during a very calm passage, picking up the distinct "V"-shaped pattern of a seabird, frightened by the approaching ship, taking off from the water. The Relative Motion Radar is also a Racal-Decca, the standard RM1226. The positions of these sets were actually reversed to that which appeared on the ship's plans providing a very useful modification, at least so far as I was concerned. Modern radars, and indeed much additional computerised equipment have been retro fitted to virtually all of the 'early' generation of Supertankers invariably as this has become commercially available. All of this gear is examined briefly in Part Four of this book. Amongst other equipment found in the chartroom would be included echo sounders to give an indication of water depth below the keel, electronic doppler log to indicate the ship's speed over the ground, and the draught indicators.

5.14.2. Navigation Consoles, Radios, Telephones

The consoles in the fore part of the wheelhouse also vary considerably in design. The following plan, (Diag. 5.8), showing the layout of a bridge navigation Console where the overall system, for this particular VLCC, works on three separate sections, "X", "Y" and "Z". Console X contains controls for internal communication and the fire detection and alarm systems covering the accommodation, turbine, pump/engine room areas. Compartment flooding alarms are often situated on this console. The alarms emit quite loud piercing shrieks, especially when they go off suddenly in the early hours of the morning, whenever a failure in the appropriate system is detected. There are also direct link telephones to the Captain's cabin, radio or engine room, and to the automatic telephone exchange. An external address system is also on this console. This could well be used to contact men working on deck, if a potentially dangerous action is observed by the officer of the watch which might lie outside the range of vision of those working below. The inert gas pressure monitoring alarm is also here. The Phillip's personnel address system is an inter communication loudspeaker that is active throughout the accommodation and is used for general announcements such as docking stand-by, emergency drills, etc. Console "Y" gives the rudder indicator and propeller shaft rpm. indicators, both of which are essential when manoeuvring the ship. The engine room telegraph indicator and a warning button are housed, so that the duty engineering officer may be contacted immediately in the event of unexpected decrease in revolutions or any equipment failures below. Console "Z" contains the echo sounder readings indicating the depth of water below the keel, together with the VHF radio telephony set used to contact other ships and, depending upon atmospherics, offering a range of up to thirty miles. The off-course alarms, usually set for ten degrees deviation from the course port or starboard, and the gyro alarms as well as the switches for the 'Kent' circular clear-view screens on the port, midships and starboard wheelhouse windows are placed on this console. The 'Kent' screens when activated revolve at enormous speeds so deflecting the very heavy rain often

Supertankers Anatomy and Operation

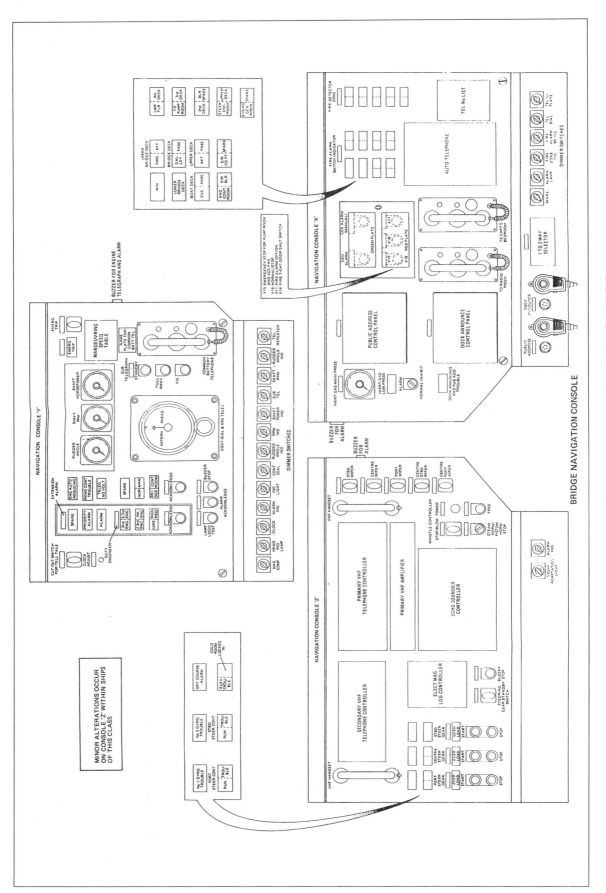

Diag. 5.8 An Example of a Bridge Navigation Console aboard a VLCC. (By kind permission of British Petroleum Shipping.)

Section Analysis

encountered during squalls, enabling an effective visual lookout to be maintained. The engine room telegraphs are sometimes separate from the consoles and are situated near the steering compass. Different designs can be seen in the following photographs. (Photo. 5.21 and 5.22 and 5.23).

5.21. (Top) These three photographs offer examples ...
5.22. (Centre) ... of other styles and Bridge/Wheelhouse layouts ...
5.23. (Bottom) ... found on "First Generation" VLCC's.
(The Author and by kind permission of Shell International Trading and Shipping.)

Supertankers Anatomy and Operation

5.15. STEERING PEDESTAL

The main steering pedestal, which appears prominently in the latter two photographs, operates both manually, for close quarter work at sea or during berthing operations, as well as automatically, when the ship's course is set whilst on passage. The following drawing indicates the plan of a popular model used at sea, the Decca ARKAS 750. (Draw. 5.8). The small inner wheel is spring loaded and over-rides the automatic function in an extreme emergency, although I can never recall allowing circumstances to develop to the extent that this was ever used, other than for testing purposes, whilst the larger wheel is for manual steering. It was quite an interesting conjecture to realise that 300,000 sdwt of ship, that is nearly one third-of-a-mile long, can be steered automatically by the very small button placed in the centre of a control panel. The view of the top plate features (Draw. 5.9) indicates the controls and their function, and

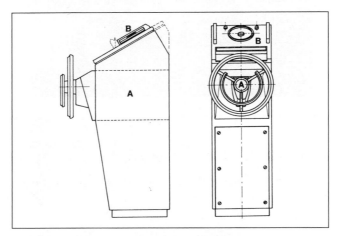

Draw. 5.8 Decca 'Arkas' 750 Steering Pedestal.
(By kind permission of Decca Marine.)

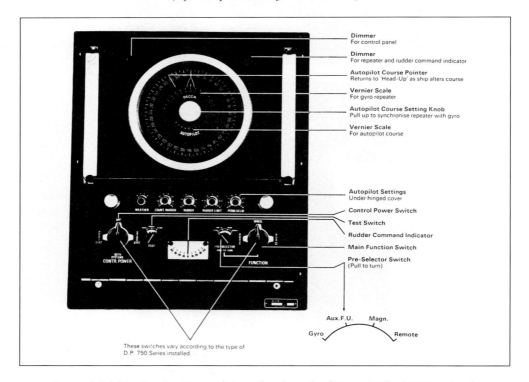

Draw. 5.9 Main Steering Face and Autopilot Controls of Decca "Arkas" 750 Pedestal.
(By kind permission of Decca Marine.)

5.24. Radio Officer Servicing the Steering Console.
(By kind permission of British Petroleum Shipping.)

shows how compensation may be allowed for adverse weather conditions etc. Most radio officers are qualified additionally to make repairs to radar sets and other instruments and this shot (Photo. 5.24) shows 'Sparks' maintaining the 'Arkas' auto-pilot fitted aboard a VLCC.

5.16. SIGNAL AND NAVIGATION LIGHTS

The signal lights switches for the lights on the mainmast are also activated from the wheelhouse, and plans displayed indicating the 'blind sectors', situated forward from the bridge, along with IMO pilot boarding notices. Running along the forepart of the partition is often a battery of telephones leading to the Captain's quarters, engine-room, radio room, steering gear compartment, and chief steward, where these are not fitted on the forward console. The controls for the ship's navigation lights could well be panelled there, as well as a locker containing the International Code of Signal flags. The navigation lights, whose precise definitions are clearly stated in the Collision Regulations, consist of white Foremast and Stern lights, as well as sidelights. The latter shows a red light on the left hand, or port side, of the ship looking forward, and a green light to starboard.

5.17. BRIDGE BOX

The navigating bridge area is very much self-contained, with toilet/washing facilities and the ever-necessary 'Bridge box' containing sandwiches and tea/coffee/cocoa making facilities. It should, perhaps, be pointed out that the bed provided is not intended for the officer-of-the-watch to have a periodic doze, but is more of a settee intended for use by the captain in circumstances which require his presence close-to the bridge for prolonged periods either to proffer advice or to be on hand and assume direct command. Certainly, it is rarely used for this purpose and is generally regarded as a useful place to 'dump' coats and any temporarily superfluous piece of gear. The only time I can recall the Master being on the bridge in such circumstances was aboard a dry-cargo ship, the m.v. 'MATRA' owned by the Brocklebank Line, when she ran into the tail-end of a hurricane off the American Coast. The captain was on call to us navigating officers in the wheelhouse for just over three days.

5.17.1. Walkie-Talkies

Above the settee, the battery charger is found which houses the ship's six portable 'walkie-talkie' stornophone localised VHF sets. These are invaluable for communicating with the bridge during docking operations fore and aft and, indeed during tank ballasting, as well as many other occasions when close contact between a control and officer becomes

Supertankers Anatomy and Operation

necessary. Owing to the mass of steel involved, their use can be rather limited between the bridge, cargo tanks and engine-room. Care has to be exercised when using them as sometimes signals from officers on other ships are intercepted, and interrupted, but so long as station call-signs are used sensibly, mistaken identification rarely occurs.

5.18. MANOEUVRING INDICATORS, 'MONKEY ISLAND'

Just outside of and above the wheelhouse door, on both sides, are to be found two indicators essential when manoeuvring the ship in confined waters. (Photo. 5.25). These are, on the right hand side, the helm indicator enabling the position of the wheel and rudder to be seen at a glance and, next to it, the revolution counter, showing the condition of the engines. Some VLCC's are fitted also with an additional indicator showing the position of the engine room telegraph, (Photo. 5.26). The letters would light up to notify, at a glance, the speed that has been called from the main engine. DS would indicate dead slow (white being ahead, and red astern) with slow, half speed, full speed, and ST indicating stop engines. The extreme top of the wheelhouse is known as the 'monkey island' and houses the ship's main mast with both radar scanners, the after mast head light showing within a specified arc across the stern, so constructed that it does not allow the light to be seen forward of the beam. The radio direction finding loops are also carried on the forward part of the ship's mainmast. (Photo. 5.27). The ship's magnetic compass, in its binnacle is situated on the centre line of the ship. This compass is fitted with a periscope reflector that leads to the steering position and is visible above the gyro steering gear. The white 'Domed' aerial in, (Photo. 5.28), is for INMARSAT, the Satellite Communications Receiver, a system that is examined briefly in Chapter 17, when considering

5.25. The Helm Indicator on the RHS and Engine Revolution Counter above Wheelhouse Entrance ...

5.26. ... along with Indicator showing position of engine room telegraph.
(The Author)

Section Analysis

5.27. Mainmast showing the Radar and RD/F Aerials with GMDSS on Lower RHS.
(The Author)

5.28. The INMARSET aerial and Compulsory Fitted Magnetic Compass on the Monkey Island
(By kind permission of Mobil Ship Management.)

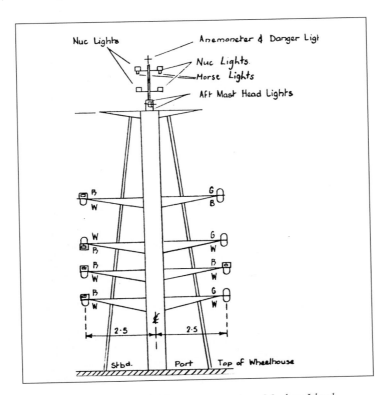

Draw. 5.10 Signal Mast arrangement on Monkey Island.
(By kind permission of Shell International Trading and Shipping.)

aspects of bridge computerisation. A range of emergency lights are incorporated on this particular arrangement with an all-round white signalling light operated from a morse code key often on the port side of the wheelhouse. A typical 'Christmas Tree' (as it is known colloquially) signal mast light arrangement is shown (Draw. 5.10) with the lights being used for the following purposes: the vertical red lights have international legal significance, under the Collision Regulations, where Not Under Command (NUC) lights would be two red lights in a vertical line one above the other visible all round the horizon at a distance of at least two miles, hence being situated immediately at the top of the mast head. These lights would be shown if the ship was unable to manoeuvre within the Collision Regulations due to such incidents arising from engine failure or a steering defect. One all-round red light is the internationally recognised 'danger' light indicating a ship operating with hazardous/explosive cargoes. This light would be displayed whilst loading/discharging. Three red lights in a vertical line indicate a ship, of any size, that is restrained to a dredged channel by virtue of her draft. The remainder of the assorted coloured lights would only very rarely be used and are intended for occasions when local port authorities have their own specific signals for ships of any size manoeuvring in tidal rivers and estuaries. On passage, for example, an outward bound ship might be required to show two or three vertical green lights, with other combinations for inward bound that might include a sequence of green and blue lights. I can recall few incidents, however, during my service aboard VLCC's in which we had occasion to use the local lights.

Chapter Six

SHIP STRESSES AND STABILITY

6.1. ANALOGY
6.2. STRESS CONDITIONS
6.3. STATIC STRESSES – THE EMPTY VLCC
    6.3.1. Longitudinal, Archimedes' Principle
    6.3.2. Shearing Forces
    6.3.3. Transversal
    6.3.4. Hogging and Sagging
    6.3.5. Trim Calculations, Bending Moments
    6.3.6. VLCC in Ballast, Compressive Stresses
6.4. LOADING/DISCHARGING CONDITION, POSSIBLE DANGER
    6.4.1. Righting Moment
    6.4.2. Effects of Listing
    6.4.3. Tipping Centre Calculations
    6.4.4. Cargo Distribution Whilst Loading
    6.4.5. Kockums A.B. 'Loadmaster'
    6.4.6. Ballasting Operation at Sea
    6.4.7. Importance of Correct Ullaging
    6.4.8. Examples of VLCC Properly Ballasted/Loaded
6.5. DYNAMIC STRESSES – METEOROLOGY
    6.5.1. Sea Forces, Swell, Corrugation
    6.5.2. Reducing Hull Stress
    6.5.3. 'Tender' and 'Stiff' Conditions
6.6. INSTANCES OF VLCC'S IN VARIOUS SEA STATES
6.7. THE GREATEST THREATS – 'WHIPPING' AND 'SPRINGING'
6.8. CRACKS AND CHAFING
6.9. FRACTURES AND VIBRATION
6.10. COMPUTERISED SOLUTIONS

Chapter Six

SHIP STRESSES AND STABILITY

*(Practical Analogy – **STATIC STRESSES**: Alongside Empty – Longitudinal Stresses: Tipping Centre – Shearing Forces – Bending Moment – Archimedes' Principle-Centres of Gravity and Buoyancy – Transversal Stresses – Metacentre – Conditions of Hogging and Sagging – Application of Shearing Forces and Bending Moment to a Specific Supertanker – Ballasting – Compressive Stresses – Alongside Loading/Discharging – 215,000 sdwt VLCC Incorrectly Discharged – Righting Moment – Inclinometer – Moment to Change Trim (MCT) Theory – Brief Comments on Early Tanker Loading Instruments – Kockum AB Digital "Loadmaster": Basic Description and Simplified Explanation of the System – Practical Ship Board Ballast Shifting Operation – Ullage Restrictions – Shearing Force and Bending Moment Diagrams for Ballast and Fully Laden Conditions. **DYNAMIC STRESSES**: Fundamental Sea and Swell Theory – Pitching – Rolling – Yawing – Bending Moments Common to all Ships – VLCC in Ballast – Semi-Laden Conditions in Severe Seas – Hogging and Sagging Stresses Due to Wave Motion – 'Corrugation' effect – Loaded Supertankers: Metacentric Height – Free Surface Movement of cargo – Force 10/12 off Port Elizabeth – Reducing Speed – Altering Course – Tender and Stiff Conditions – VLCC's in Various Sea States – Transversal and Longitudinal Stresses – Whipping and Springing Stresses – After Part Stresses: Vibration and Effects.)*

6.1. ANALOGY

A collection of empty shoe boxes, plus a suitably curved section representing the bow, has often been found useful during the course of group talks to illustrate the hull of a VLCC. The analogy thus drawn is not perfect, but four boxes suitably trimmed, totalling roughly 1067 x 178 x 89 mm, demonstrate fairly adequately an effective visual representation of the approximate 1:6.6 and 1:1.8 or, more evenly for practical purposes, 1:6 and 1:2 ratio approximation of the 1200" x 200" x 100" length, beam and depth dimensions of an average 300,000 sdwt Supertanker. This impressive, if somewhat unwieldy package, placed on a side table and reinforced by a scale drawing, can then be used to offer an elementary demonstration of the stresses to which the ship is subjected as these affect in practical terms the work of a deck officer.

6.2. STRESS CONDITIONS

There are two major sets of conditions concerning forces acting on the hull of a ship. First are the packet of static stresses, applicable when the ship is stationary alongside a jetty either in loaded or a totally unladen condition following cargo discharge, or when she is ballasted prior to departure for the next voyage. The second set of major stresses are those dynamic forces which are actually superimposed on the static stresses at work when the ship is under way whilst on passage, either carrying a full or partial cargo, or in ballast. Most of the moment under these circumstances works through the corners of the tanks.

6.3. STATIC STRESSES-THE EMPTY VLCC

The static stresses on a VLCC act through her centres of gravity and buoyancy and affect the way that she will sit in the water both longitudinally and transversally. A comparison is frequently made between any ship and a girder, but the VLCC once she has totally discharged her cargo does not, unlike the unladen girder, have all of her weight distributed evenly along the length of the hull. It is only the light displacement tonnage of the ship itself, in terms of total steel, engine, rudder, propeller and accommodation, together with any fuel, lubricating and sump oils, fresh and feed water, plus stores, personnel and their effects, which apply sufficient weight to keep the ship submerged. In this state, she sits almost literally on top of the water, with the bottom of the bulbous bow and entire fore length exposed towards the mid part of number one cargo tank. There is then a very gradual submerging along the bottom of the hull until an after draught of just a few metres is reached such that both propeller and rudder would not be fully submerged.

6.3.1. Longitudinal, Archimedes' Principle

This uneven distribution of weight in an unladen ship alongside acts as downwards pressure at unequal amounts which creates a series of longitudinal stresses causing a tendency for an imbalance or distortion to occur at various points along the hull. Clearly, most of this weight acts at the stern, with that of the cargo tanks exerting a disproportionate downwards force on their respective transversal water tight frames. As the tanks are each of a slightly different construction and dimension, a series of unequal forces is produced. (Draw. 6.1). The application of Archimedes' Principle has to be taken into consideration because the upthrust caused by water pressure provides the resultant positive buoyancy necessary for the ship to float. The longitudinal moments relating to a VLCC affect the trim fore and aft so that the ship goes down more or less by the head or the stern and is of unequal draught, as shown in the following drawing. (Draw. 6.2).

GB (In Draw. 6.3) is the fulcrum around which the length of the Supertanker swings for small angles of trim, during static loading, and is called the 'tipping centre'. It works at the point along the hull where the forces interact to keep the ship on an even keel. Generally this point is at the area of the midships, or 'mean' draught marks, within the vicinity of the manifolds and, in most fully laden VLCC's at rest, of around the 300,000 sdwt capacity, this point is slightly under one-half the length of the hull from the stern. It cannot be assumed, however,

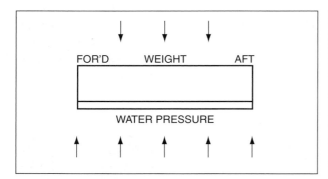

Draw. 6.1 A Series of Uneven Forces is Produced along the length of the Hull.
(The Author and Paul Taylor.)

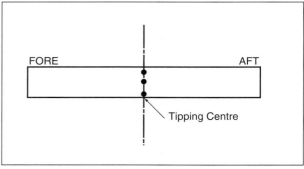

Draw. 6.2 The Tipping Centre is the Point along the Hull at which Longitudinal Forces Interact.
(The Author and Paul Taylor.)

TABLE 1   CALCULATIONS OF SHEARING FORCE AND BENDING MOMENT IN STILL WATER   VOYAGE:

LINE NO.	ITEMS	WEIGHT K. TONS 100 (a)	TRIM FACTOR (b)	TRIM MOMENT (c)	BHD NO.	BUOYANCY (KT) (d)	FRACTIONAL WEIGHT (KT) (e)	LOADED WEIGHT (KT) (f)	SHEARING FORCE (KT) (g)	FACTOR FOR BEND. MT. (M) (h)	BENDING MOMENT (KT-M) (i)	TOTAL BEND. MT. (KT-M) (j)
1	LIGHT WEIGHT	438.07		133.76				17		57.99	986	
2	STORE (AFT)	0.17	6.516	1.11						46.06		
3	AFT PEAK TANK		6.054					102		38.46	3,923	
4	No.2 DISTILLED WATER TANK	1.02	5.761	5.88				118		33.89	3,999	
5	No.1 DISTILLED WATER TANK	1.18	5.584	6.59						29.30		
6	No.1 DRINKING WATER TANK		5.407							29.21		
7	No.2 DRINKING WATER TANK		5.403					87		20.36	1,771	
8	STORE (ENG. RM.)	0.87	5.061	4.40				62		19.72	1,223	
9	WATER & OIL IN ENGINE ROOM	0.62	5.036	3.12						15.53		
10	PROVISIONS		4.867					30		13.38	401	
11	FEED WATER TANK (P)	0.30	4.791	1.44				30		13.00	390	
12	FEED WATER TANK (S)	0.30	4.777	1.43				11		10.27	113	
13	CREW & EFFECTS	0.11	4.671	0.51						8.53		
14	FUEL OIL WING TANK (P & S)		4.604							2.80		
15	FUEL OIL SETT. TANK (C)		4.382									
16	SUM	—	—	—	57	—	7,954			—	197,881	
17					58	—	502			2.250		
18	No.4 WING TANKS (P & S) (3.2%)		4.184									
19	SLOP TANK (S-OUT SIDE)		4.007									
20	SLOP TANK (F)		4.004									
21	SLOP TANK (S-IN SIDE)		4.002									
22	No.4 WING TANK (P & S) (6.0%)		3.688									
23	No.4 CENTER TANK (0.8%)		3.995		59	—	602			2.555		
24	No.4 D.B. TANKS (P & S) (7.2%)		3.969									
25	No.4 WING TANKS (P & S) (90.8%)		2.835									
26	No.4 CENTER TANK (95.2%)		2.818									
27	No.4 D.B. TANKS (P & S) (92.8%)		2.671									
28	SPARE PROPELLER	0.61	1.730	1.06	70	—	7,820			28.105		
29	No.3 D.B. TANK (P & S)		0.573									
30	No.3 CENTER TANK		0.546									
31	No.3 WING TANKS (P & S)		0.545		82	—	8,708			30.660		
32	TOTAL (+) (Lines 2 ~ 31)	0.10	- 1.035	- 0.10								
33	STORE (MID.)		- 1.207									
34	No.2 D.B. TANK (S) (50.0%)		- 1.233									
35	No.2B CENTER TANK		- 1.233									
36	No.2 WING TANKS (P & S) (50.0%)		- 1.233		88	—	4,831			15.330		
37	No.2 D.B. TANK (P) (51.6%)		- 1.233									
38	No.2 D.B. TANK (S) (50.0%)		- 2.392									
39	No.2 D.B. TANK (P) (48.4%)		- 2.392									
40	No.2A CENTER TANK		- 2.416		94	—	4,332			15.330		
41	No.2 WING TANK (P & S) (50.0%)		- 2.418									
42	No.1 WING TANKS (P & S)		- 4.128									
43	No.1 D.B. TANK (P & S)		- 4.131		106	—	6,974			30.660		
44	No.1 CENTER TANK		- 4.195									
45	FUEL OIL DEEP TANK (P)		- 5.497									
46	FUEL OIL DEEP TANK (S)		- 5.498									
47	STORE (FORE)	0.27	- 5.812	- 1.57								
48	FORE PEAK TANK		- 5.902									
49	TOTAL (-) (Lines 33 ~ 48)	—	—	—								
50	DIFFERANCE (Lines 1 + 32 + 49)	—	—	—								
51	SUBTOTAL (Lines 2 ~ 48)	—	—	—								
52	T. MOMENT CORR. Line 50 (c) × CT	—	—	—								
53	DISPLACEMENT (Lines 1 + 51) × 100	—										

1) CORRESPONDING DRAFT (d)

$$\frac{DISP'T}{1,000} \times Ck = d$$

2) TRIM (T)

$$\frac{}{1,000} \times = \text{ m}$$

LINE 52(c) + CM = cm

3) DRAFT FOR'D & AFT

FOR'D (df) = d ± T × CF

" " = "

APT (da) = df ± (T−T×CF)

" " = "

NOTE ; FACTORS OF CM, CK, CT & CF SEE TO TABLE 3.

Table 6.1 The Components on a VLCC which have to be considered in order to determine the Shearing Forces and Bending Moments.
(By kind permission of Mobil Ship Management.)

that the fulcrum point will always remain in this position. Although it will continue to work at the waterline, it will move slightly forward as the ship commences loading and changes draught. In a light ship, it will be further aft of the manifolds. Thus, the centres of flotation and gravity do not always act in the same vertical position along the hull.

6.3.2. Shearing Forces

Ship's frames on a VLCC have, therefore, a considerable stress placed upon them and act as the resultant points of a range of forces. Thus a series of 'Shearing Forces' are created which tend to exert pressure on the frames, and thus distort various parts of the hull to a possible breaking, or shearing, point. The resultant creates a 'load', which acts through either side of a transverse frame that serves as the division between cargo tanks, such that the shearing

Supertankers Anatomy and Operation

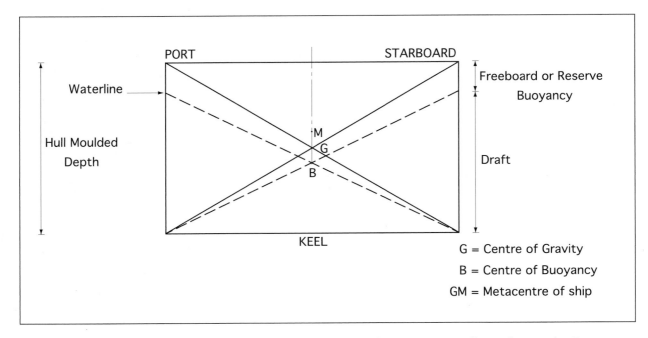

Draw. 6.3 The Metacentre of the Ship, M, is the point around which any Moment will occur between the Centres of Gravity, G, and Buoyancy, B.
(The Author and Paul Taylor.)

force about any specified transverse frame along the length of the hull is determined by the bending moment. Elementary physics, shows us that this is the force through the frame at which the weight is applied, multiplied by the horizontal distance from the point along the hull aft to the tank bulkhead from which the moment is taken. The sum of the moments is expressed as weight in kilo tonnes, multiplied by the distance in metres, and is usually taken about amidships.

6.3.3. Transversal

Concurrent with the longitudinal stresses acting on the hull that affect the trim are transversal stresses which determine any possible port or starboard list which the VLCC may develop. In working out the stress loads on the ship, therefore, there are two centres which have to be considered; the weight of the ship, acting through its centre of gravity, and the upthrust of the water, acting through her centre of buoyancy. The upward and downwards forces must obviously be balanced overall for otherwise the ship would either sink, fly or break up. The forces act in a vertical direction, equal and opposite through each of the water tight transversal bulkheads so, at various points along the hull, there will exist differences between the weight and the buoyancy.

The above diagram illustrates the comparative positions of two centres showing how they act through the hull (Draw. 6.3). The box like shape of the hull helps determine the distances of each above the keel. The Metacentre (M) of the ship is that point around which any change of moment will occur in both longitudinal and transversal directions.

6.3.4. Hogging and Sagging

Except in the case of an empty double bottomed ship, the constant weight at the ends of an unloaded VLCC with machinery towards the stern, and heavily reinforced ballast tanks in

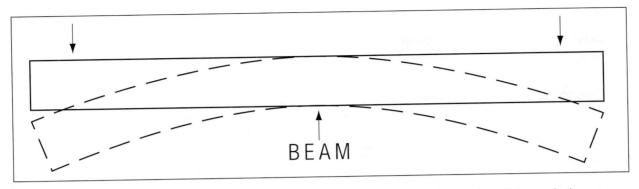

Draw. 6.4 A VLCC is said to be 'Hogged' when the ends of the ship have a tendency to be pulled towards the water with the midships section left unsupported ...
(The Author and Paul Taylor.)

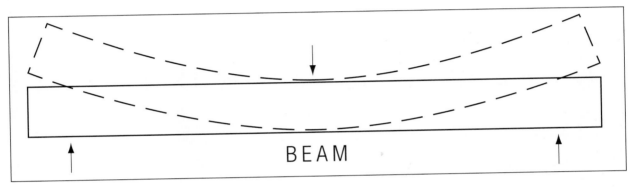

Draw. 6.5 ... the opposite condition is called 'Sagging'.
(The Author and Paul Taylor.)

the fore part, leaves the centre comparatively unsupported. The ends therefore have a tendency to be pulled downwards towards the water with the midships section being very slightly raised. This condition of longitudinal stress is called 'hogging' and is one to which all VLCC's, by virtue of their construction, are particularly prone. (Draw. 6.4). The opposite condition occurs when the overall weight of the ship is unsupported along her length, at the centre, whilst the ends are supported by the upthrust of the water. This is known as 'sagging'. (Draw. 6.5). The condition may be aggravated by incorrect loading.

6.3.5. Trim Calculations, Bending Moments

The theory examined so far might perhaps be better understood by considering the data of a practical example which, in this case, is a 'typical' standard 280,000 sdwt Supertanker belonging to a leading shipping company which was, incidentally and unusually, one of the few VLCC's to be fitted with a double bottom. Referring to a simplified table, (Table. 6.1), the items next to the line number, indicate the components which have to be considered in order to determine the 'Shearing Forces' and hence 'Bending Moments'. The data is used as an example whose principle, based on a range of quite complex calculations, can be considered common to that used for all VLCC's. The specific ship is in still water and unladen with cargo, fuel oil, or water for ballast or domestic uses. The result can be seen in a diagram format. (Diag. 6.1). The trim calculation has been obtained by comparing the weight of any particular loaded space (in Column C) with a 'constant' (in Column B) that is unique to that Supertanker, based on a

Supertankers Anatomy and Operation

Part 1

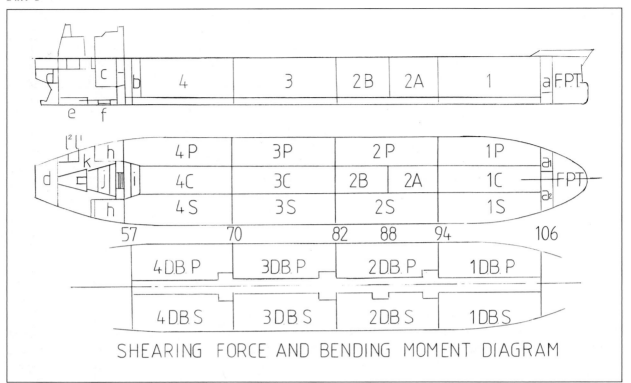

Part 2

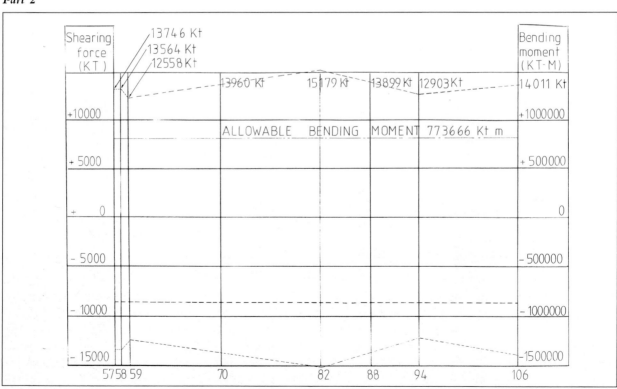

Diag. 6.1 The Shearing Force and Bending Moment Diagram relating to the Condition in Table 6.1, with the VLCC in Still Water and Unladen.
(By kind permission of Mobil Ship Management and Paul Taylor.)
(Refer to page 178 for Key to Diagram.)

formula concerning the distribution of particular spaces about midships, and represents the longitudinal centre of gravity around frame 82. The conventional formula multiplying force by distance is then used so that the 'moment' (Column C) is obtained as a product. The shearing force is the difference between a first integration of the weight and the buoyancy from the after end of the ship, abaft frame no 57, and each transverse bulkhead in the cargo tanks. The loaded weight of the aft tanks exclude, in this example, the fuel oil bunkers, fresh water, provisions, crew effects and water ballast. The light weight of this VLCC, at 43,807 Kt, has been obtained as a fractional weight at each bulkhead from the aftermost part of the forward bulkhead in each cargo tank, and these values entered on each of the bulkhead lines (in column E). The moment of fractional light weight (to the aft part of bulkhead 57), is tonne metres, and the moment of loaded weight in tanks aft of bulkhead 57 has been obtained by multiplying respectively these weights and each level about bulkhead 57. The bending moment has been obtained by integrating the shearing forces and the calculation has been separated into two operations derived from the difference moment of the total fractional weight and the buoyancy from the ship's after end to the aftermost bulkhead in the cargo tanks. At the cargo tanks, the bending moments have been derived by multiplying the shear force at the aft and forward bulkhead, by half length of a selected cargo tank, and then the product added to the total bending moment at the after bulkhead of that tank. A 'half length' is taken owing to the weight of the tanks being evenly distributed between the transversals constituting the particular tank and the next one. The positive algebraic sign indicates a hogging moment and the negative sign a sagging moment. It can be seen that maximum point is reached at the intersection of numbers two and three tanks, within the vicinity of frame 82. Along the total length of the hull, therefore as shown, there would exist a series of uneven forces acting vertically over the frames and plating such that in places where the bending moments are maximum then the shearing forces will be minimum.

6.1. The "RANIA CHANDRIS" at Coryton with Cargo Discharged and Correctly Ballasted ready for sea.
(By kind permission of Photo Reportage Ltd and Port of London Authority.)

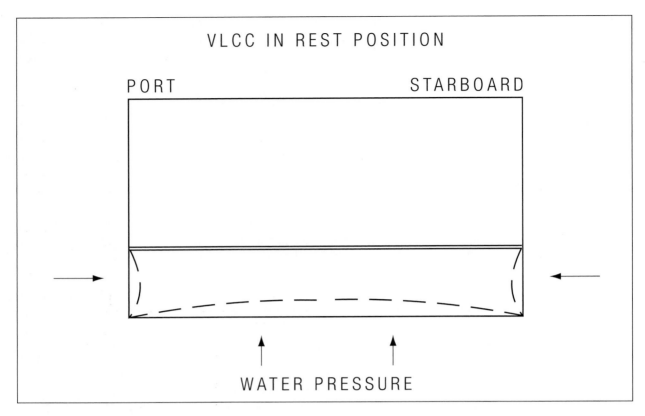

Draw. 6.6 Compressive Stresses occur on the Ship's Sides and Bottom tending the Shell Plating to be forced inwards and the Keel area upwards.
(The Author and Paul Taylor.)

6.3.6. VLCC in Ballast, Compressive Stresses

It would clearly be unsafe for an unladen and unballasted ship to proceed to sea in what is really a very unstable longitudinal condition. Not only would it be quite impossible for her to be handled, with both bulbous bow and propeller partially clear of the water, but also because of the strain caused by the load stress. Consequently, in the example of a non double-bottomed ship, additional to the fore and aft peak and the permanently segregated ballast tanks, certain cargo tanks on most Supertankers would be designated to contain proportions of sea-water ballast during the non-cargo carrying passage. Centre tanks are usually chosen according to the peculiar characteristics of the ship. On some Supertankers, for example, the designated cargo/ballast tanks are numbers one, three and four centres: whilst on other ships, the selected centre tanks could be numbers two, three and six. The determining factor would be the number of complete centre and wing tanks which constitute the cargo carrying capacity of the hull. In the former instance, that VLCC would probably have only five sets of tanks each with a larger capacity whilst, in the latter, there might be up to eight smaller capacity sets.

The unloaded berthed Supertanker, in calm water and correctly ballasted by the chief officer, has attained a balance between two major stress areas introduced by the weight of the ship displacing water, and the force of the water itself acting upon the hull. (Photo. 6.1). Compressive stresses are created on the ship's bottom and sides transversally and along the extended ship's length. (Draw. 6.6). The tendency is for the side plates to be forced inwards, exaggerated in the diagram for demonstration purposes, and the keel to be pressured upwards. These forces are counteracted by the reinforced plating of transversal bulkheads and girders, as well as the longitudinal frames and brackets of the ship's construction, examined in Chapter 2.

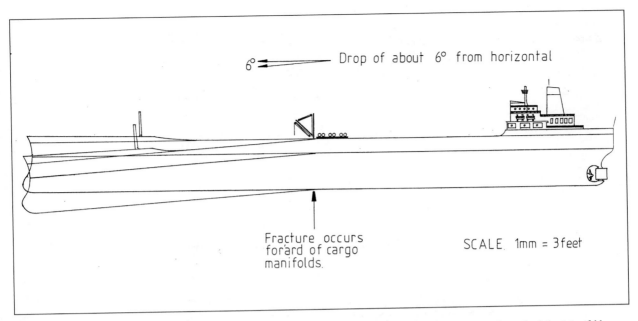

Draw. 6.7 The Effect of unchecked stresses during discharging caused this VLCC to split virtually in two forward of the Manifold area. (The Author and Paul Taylor.)

6.4. LOADING/DISCHARGING CONDITION, POSSIBLE DANGER

It is during the loading and discharging of cargo that considerable care has to be exercised. As the crude oil is pumped into the ship, or ashore, it has been shown that the shearing forces and bending moments introduced act unevenly upwards and downwards over the length of the cargo carrying capacity. A longitudinal bending moment is created, therefore, which can cause hull distortion. If the stresses are permitted to develop unchecked then the hull would split, probably about its mid-point area which is the weakest point where the forces on the hull of a Supertanker loading or discharging can be generally assessed as being resolved. This point, depending on the tank construction, could be either slightly forward of the manifolds or aft of the permanent ballast tanks.

Such an incident occurred on a 215,000 sdwt VLCC a few years ago whilst she was discharging a cargo berthed at the Rotterdam oil harbour in Europoort, (Draw. 6.7). In this accident, the cargo was being taken out of the centre tanks so leaving the weight at both ends of the ship. The aggravated hogging stress increased until the critical momentum was exceeded. The angle of depression was to the order of 6° from the horizontal. In this instance, the spillage, of around ten tonnes, was extremely small compared to what it might have been and the crew of 43 escaped with little more than shock. One suspects that had the inert gas system not been operating then a potentially explosive situation could have occurred. The Supertanker virtually broke in two, was towed away and eventually scrapped. The incident serves, however, as more than a mere salutary lesson of what can happen if careless cargo handling practices exist.

6.4.1. 'Righting Moment'

The metacentre is affected as the ship loads or discharges so that transversal and longitudinal moments are created which cause the ship to list and heel from her norm such that the centres of gravity and buoyancy no longer remain in the vertical position of B being directly

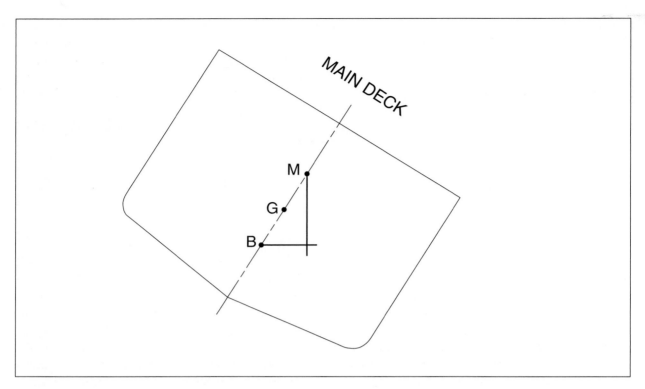

Draw. 6.8 As cargo is loaded into the VLCC, she has a tendency to list from the norm so that the Centres of Gravity and Buoyancy are no longer vertical and ...
(The Author and Paul Taylor.)

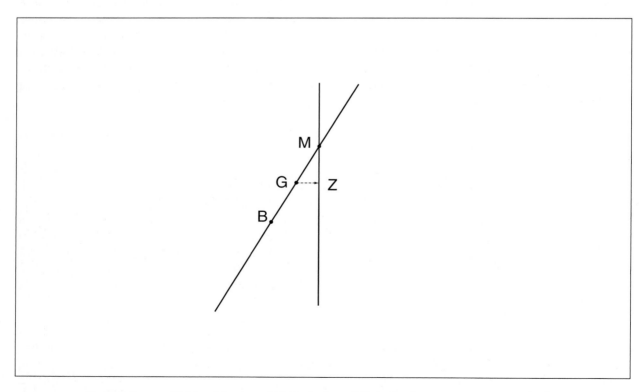

Draw. 6.9 ... a Righting Moment, which is the lever GZ, is therefore necessary to correct the list.
(The Author and Paul Taylor.)

below G. If the VLCC takes a list to port, for example, by a few degrees, then the following situation is introduced. (Draw. 6.8). G and B are moved from the vertical and are no longer equal, thus a list develops and a couple is formed which can be resolved only by righting the ship to bring her back onto an even keel. The oil has to be distributed to the other side of the ship and into her other set of wing tanks, in order to correct or 'right' the heeling and return her to the initial stable condition but, of course, that much lower in the water. The amount of cargo necessary to do this forms the 'Righting Moment' and is the lever of the couple GZ. (Draw. 6.9).

6.4.2. Effects of Listing

A number of tried and tested mathematical formulae resolve the situation, such as, for the mathematically inclined reader, Simpson's Waterplane coefficients but, in practical terms, the duty mate identifies a list developing by glancing at the Inclinometer in the cargo control room. It also becomes more visually observable as he moves around the main-deck, particularly by glancing at the angle made to the chiksans. It should, perhaps, be emphasised that a 1° list on a Supertanker is very pronounced far more than would even be appreciated aboard a dry cargo ship, and all on board will most certainly know that it has happened. In fact, if it increases by much more than 1°, and the duty officer does not notice it, then it is highly probable that the Master may well 'direct his attention towards it', as the latter is perhaps thrown out of his bunk. The Inclinometer in any case helps him determine the extent of the list so that he can then open or shut appropriate valves in order to apply corrections before it reaches serious or critical stages.

The obvious danger is that unequal strain may be put onto either the ship/shore connections at the manifolds causing these perhaps to part or, more probably, stress could be created between the internal structural members within the hull. Far more important and potentially dangerous is an excess in trim. If some VLCC's are allowed to sit too much by the stern this could have an extremely adverse effect on the ship's boiler levels and bring perhaps more than a mere caustic comment from the chief engineering officer. Generally, the terminals themselves will not allow any VLCC to sit by much more than 6° or 8° by the stern and will simply stop the cargo until an adjustment to the trim is made to bring her back a few degrees on a more even keel. Clearly, therefore, constant attention must be paid whilst loading or discharging to the condition in which the ship sits in the water.

6.4.3. Tipping Centre Calculations

The capacity plans of all VLCC's, provided by the builders, (See Plans in end pocket) gives an accurate indication of the moment around the tipping centre that would change the trim. Clearly, the amount of cargo necessary to effect this will be determined by the draught of the ship, with the amount decreasing as the Supertanker sits lower or higher in the water depending on whether she is loading or discharging. To cite, as a practical example, the deadweight scale of the ULCC "JAHRE VIKING". (Extract. 6.1). Column one indicates the MCT, or, the moment to change the trim of the ship by 1 cm. It is measured in metric tonnes per metre and represents the moment necessary to change the trim by 1 cm distance. At a draught of eleven metres, for example, the amount of moment necessary to change the trim by one centimetre is around 7,210 tonnes/metre. At the same draft, to lower the ship by one centimetre, (The TCP in Column Two), would require an increase in cargo of approximately 262.3 tonnes, in either case a considerable amount of crude oil to produce such comparatively short moments. It is interesting, but as expected, to notice the quite rapidly increasing MCT as the deadweight of this VLCC increases as she submerges with loading beyond 300,000 tonnes.

Supertankers Anatomy and Operation

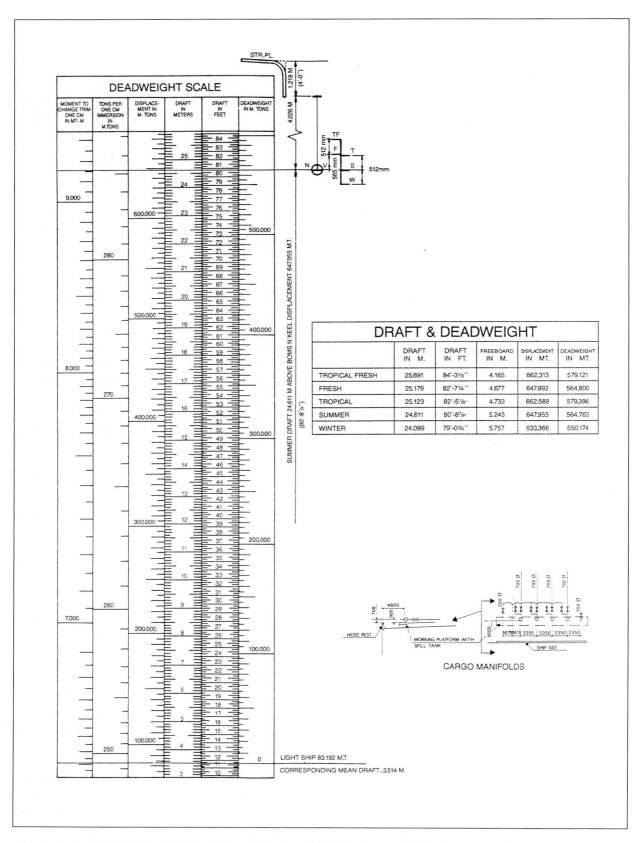

Extract 6.1 The deadweight Scale gives an indication of the Moment around the Tipping Centre necessary to change the trim of a ULCC.
(By kind permission of Jahre, Wallem Shipping Company.)

Ship Stresses and Stability

6.4.4. Cargo Distribution Whilst Loading

Prior to the advent of the very large crude oil carrier, tanker officers were assisted in the task of determining the distribution of oil whilst loading their ships by standard stability/calibration tables, such as 'Tunnard's Tanker Tables,' 'Petroleum Tables' by William Davies, or those of the American Petroleum Institute (API), all of which are 'classics' known to all serving tanker deck officers of an older, and perhaps not so old, generation. In the 1960's, as ships increased in size, simplified longitudinal stress tables were supplied to these large tankers of the day, along with primitive machines, that were used whilst loading/discharging crude oil cargoes, as well as during ballast transfer operations, to give a single bending moment amidships. Captain Geoff Cowap, an Extra Master, marine consultant and Director of Energy Marine International Limited, Leighton Buzzard, in a paper specially prepared for this book, made the following observation concerning the development of loading instruments:

"By this time, technical advances in electronics made it possible to provide instruments that would show shear forces and bending moments at the positions of each bulkhead .... The method of calculating shear forces and bending moments was not satisfactory and at best was a good estimate because of the many assumptions that were made in preparing the pre calculated data.

In the early 1970's, the French Classification Society-Bureau Veritas-took the lead and required the larger tankers to be fitted with multipoint loading instruments as a class requirement. Other classification societies soon followed and, during the extensive ship building programme of the 1970's, all tankers were fitted with loading instruments .... Most of the ship building at this time was being undertaken by Japanese shipyards and many of these yards developed their own loading instruments. Technology was changing and digital computing power was becoming available and so the technology was towards dedicated electronic digital processing."

6.4.5. Kockums A.B. 'Loadmaster'

Whilst I have often used a 'Gotverken' lodicator, an equally popular model is the 'Loadmaster' that was made by Kockums AB of Malmo, Sweden, and served numerous 'early generation' Supertankers with considerable success. The early digital 'Loadmaster' was a fore runner to Kockum's more sophisticated 'Loadmaster' Stress Computer. To quote from the introduction of the digital Manual:

"It is, however, impossible to foresee all possible cargo distributions. It is thus necessary to have a small easy to handle computer on board which can calculate all the appropriate stresses for every load distribution case. The Loadmaster computer is such a small easy to handle computer."

The theory of each instrument is similar, and a figure of a Loadmaster, serving a 250,000 sdwt VLCC, illustrates the major components. (Diag. 6.2). The potentiometer, at 4, gives an indication of the dead weight tonnage and has to be adjusted until the null instrument, 5, indicates a zero reading. For loads in excess of 100,000 tonnes, the push button to the left, at 6, adds extra an 100,000 tonne units. A fine adjustment control, 7, enables a closer reading to be obtained as the zero reading is approached. The draft scale, 10, shows the mean draught at the flotation centre of the ship, reading in this instance about twenty metres. The dial at number 11 is the trim instrument that indicates the load which has to be fed in to obtain a required trim. The fifteen potentiometers, examples at 1 and 2 and stationed on the instrument at a number of points along the length of the hull, are for inserting the proposed tonnage's into the tank spaces as loading proceeds. Turning this to the left or right indicates the load points on the adjacent meter in terms of tonnes. It is at these points that are indicated the maximum allowable

Supertankers Anatomy and Operation

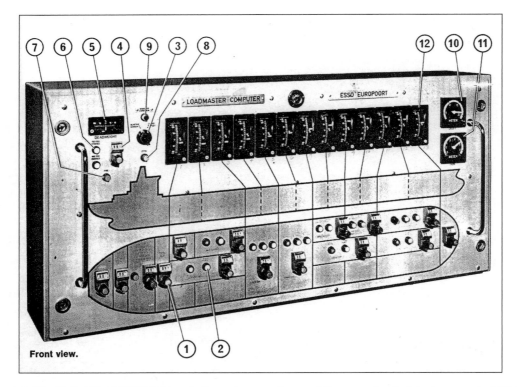

Diag. 6.2 The "LOADMASTER" Stress Indicator has served successfully many hundred 'early generation' VLCC's. (By kind permission of Kockum AB. Malmo. Swden.)

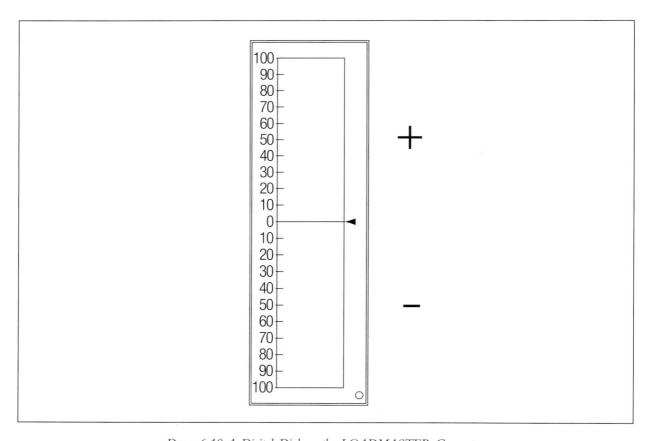

Draw. 6.10 A Digital Dial on the LOADMASTER Computer. (By kind permission of Kockum AB. Malmo. Swden and Paul Taylor.)

Ship Stresses and Stability

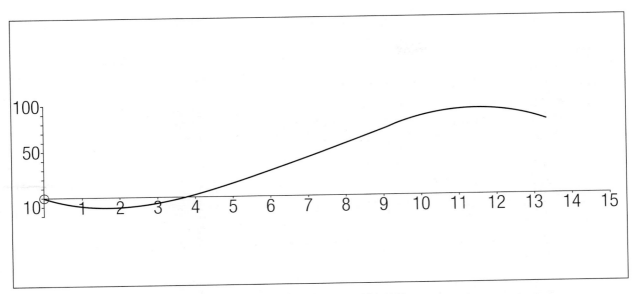

Graph 6.1 A graphical illustration of the loading condition of the ULCC in Diagram 6.2. An unusual curve distribution on first appearances, but one within the safety limits. (By kind permission of Kockum AB. Malmo. Swden and Paul Taylor.)

shearing forces and bending moments, stresses which clearly cannot be exceeded whilst working cargo or ballast. The buttons for the wing tanks are situated on the starboard side of the machine only as wings are always loaded or discharged with equal amounts in these tanks. Pointer 3 indicates the choice between shear forces or bending moments. In the diagram it is set to record the bending moments across each of the selected frames. The dials along the top of the instrument, at 12, are the digital computer read out points. A close up of the indicated dial is offered in the accompanying drawing. (Draw. 6.10). The reading is for the frame separating the fore peak from number one cargo tank and, as would be expected for a ship in this particular condition of partial loading or discharging, the reading is slightly above zero per cent. The dials enable the percentage value of the shearing forces or bending moments on each frame to be read off.

The bending moments for this specific VLCC can be plotted into the conventional curve indicated in the following graph. (Graph. 6.1). In the bending moment condition, a positive reading indicates a tendency towards hogging, with a negative reading showing sagging. A similar set of curves can be produced for shearing forces, thus providing the basic diagram and, in this instance, a positive sign would indicate excess buoyancy aft of the indicated frame, with the negative sign offering a Classification Society based norm where the readings are below zero.

The flow of oil can thus be regulated by judicious use of the valves so that cargo can be pumped within the tank system thus relieving pressure on any particular set of frames. Each loading device has to be designed for each specific VLCC and a comprehensive manual provided for reference which gives precise limits within which the ship can operate consistent with the Classification Society's stress regulations. At pointer 8 there is a special zero button which controls all of the readings and reduces the draught reading to that of the light weight tonnage of the ship. On some models, this button indicates the shear force correction. The harbour, or still water condition of the VLCC, is indicated by a spring loaded button, 9, used, as the manual suggests, when *"stresses may exceed the seagoing limits, but still be approved in port by the Classification Society."*

Supertankers Anatomy and Operation

The following photograph (Photo. 6.2) shows the Chief Officer of a Supertanker, belonging to a leading shipping company, using that vessel's 'Loadmaster'. In the photo, the Mate is assessing stress areas within the vicinity of number one and two tanks using pre-determined test calculations. The 'yellow on black' dials, on the extreme right, indicate the trim and draught of the ship and this shot may well have been taken prior to discharge, looking at the position of the top, draught, indicator. On this ship, the permissible trim error was 10" when she was loaded, and 21" when in ballast. This particular VLCC was not fitted with the 'Harbour Condition Button', (Number 9 in Diagram 6.2).

6.4.6. Ballasting Operation at Sea

An example of a practical operation aboard a Supertanker offers an indication of the procedures involved. The process might perhaps be easier to follow if reference were made to Figures 5 and 6 (Chapter 4) together with the accompanying explanations. On one occasion,

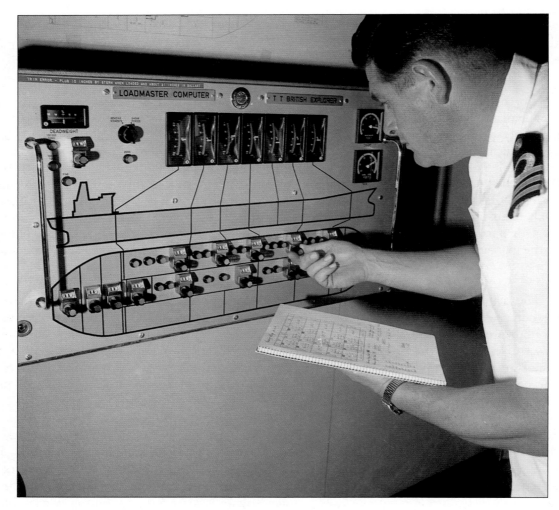

6.2. The Chief Officer aboard a VLCC assesses stress areas likely to result within the vicinity of Number's One and Two cargo tanks.
(By kind permission of British Petroleum Shipping.)

we needed to adjust the ballast in order to ease stress on frames 87-96 by way of number four centre tank. Our plan was to put more ballast into four centre whilst removing the same amount from five centre. We operated with a starting ullage in number four of 11m which we intended to increase to around 6.9m whilst, simultaneously, decreasing the ullage in number five from 14.6m to about 16.5m, thus leaving about six metres in the tank. We worked purely from the cargo control room using the valves and controls situated there, via the mimic board, and not the hydraulic manual valves in the control boxes on the main deck. At 13.30, my records show, we liaised with the duty engineering officer, started number two cargo pump and opened number four centre suction valve, as well as valves 34, 40, 41 and 47, in order to clear the lines. (See Diagram 5 in Chapter 4). At about 14.30, we had an incident in the engine-room which put out of action the boiler serving the pump, for about half-an-hour, thus halting temporarily the operation. (Yes, this sort of "gremlin" does happen quite often in real life!).

By 15.00 hours, however, we were able to slow the now operational pump from 5,500 tonnes per hour to 4,000 and opened valves 62 and 63. We closed also number four suction and five suction on number two line, valve 135, and then cracked opened valve 28 thus opening the drop-over line. Shutting numbers 34 and 41, along with the drop-valve 128, meant that we had lined-up the lines sufficiently for the ullage in five centre to reach 14.8m which we considered adequate to remove pressure from the frames. Thus, the operation was concluded by 16.00, with the opening of valves 34 and 41 and shutting of number 28. Careful use of the lodicator was essential in order to monitor all stages of what was really a common place transfer.

6.4.7. Importance of Correct Ullaging

In ballast, as discussed, Supertanker's have a tendency towards the condition known as 'hogging' so that the opposite condition is a tendency by loaded Supertankers, termed 'sagging'. This is because the cargo weight is concentrated away from the two extreme ends of the ship overall, not just the cargo carrying area, notwithstanding the permanently empty and segregated ballast tank. On the first VLCC in which I served the SBT's were the after parts of number two wing tanks. (See Plan View Drawing 4 in Chapter 5). This meant that ullages had to be kept to within three metres between tanks on all occasions when handling cargo, otherwise the hogging and sagging tendencies would have been severely aggravated. Other causes of hull stress could arise from weakening in parts of the internal structure within the hull. This seems to be particularly true of the 'early' generation of Supertankers most of which are now well approaching many years of service, with some having already taken their fifth five-year survey.

6.4.8. Examples of VLCC Properly Ballasted/Loaded

The situation concerning a VLCC filled to 98% capacity, and with the double bottom, fore and aft peak tanks empty, produces the following force and moment diagram. (Draws. 6.11 and 6.12). Obviously, when carrying cargoes of different densities, an alternative set of values would be depicted. The tendency for the loaded VLCC to sag is demonstrated quite clearly in the bending moment bar of the graph between frames 65/66 and 94. The same Supertanker arriving at a loading port in light ballast can be seen in the following diagram. (Draws. 6.13 and 6.14). The double bottom ballast tanks and number three centre tanks are filled with sea water and there is now a distinct shift in the moments creating the shearing forces and bending moments. The resultant continues to act around frame 82. The different ballast and loaded conditions make an interesting comparison with the still water unladen condition found in, (Diagram 6.1).

Supertankers Anatomy and Operation

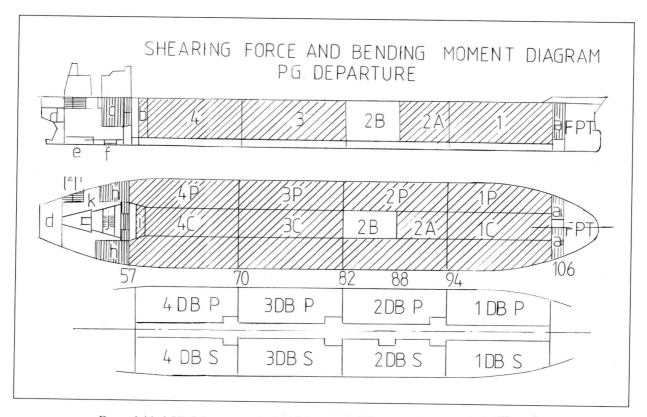

Draw. 6.11 *A VLCC in a Loaded Condition prior to Departing the Arabian/Persian Gulf.*
(By kind permission of Mobil Ship Management and Paul Taylor.)

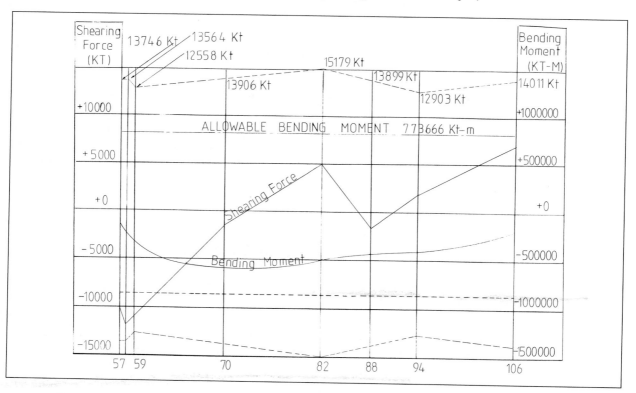

Draw. 6.12 *The Shearing Force and Bending Moment Condition for a Loaded VLCC prior to Departing the Arabian/Persian Gulf.*
(By kind permission of Mobil Ship Management and Paul Taylor.)

Ship Stresses and Stability

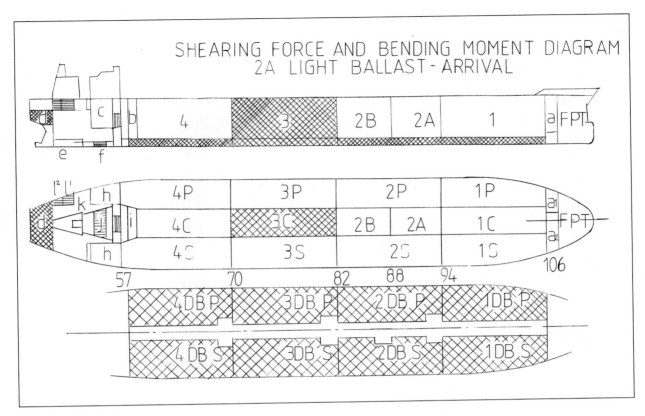

Draw. 6.13 A VLCC in Light Ballast Arrival Condition.
(By kind permission of Mobil Ship Management and Paul Taylor.)

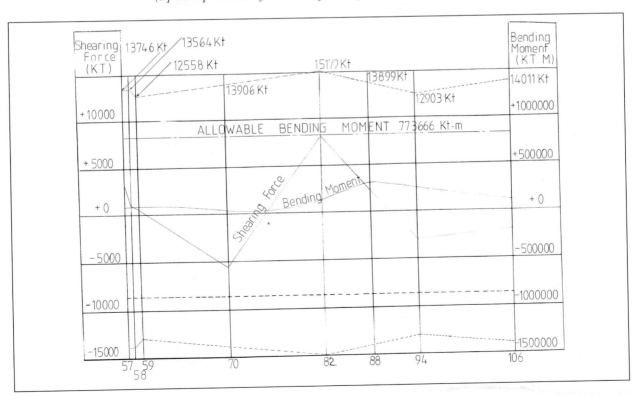

Draw. 6.14 The Shearing force and Bending moment Condition for the VLCC in Light Arrival Ballast.
(By kind permission of Mobil Ship Management and Paul Taylor.)

Supertankers Anatomy and Operation

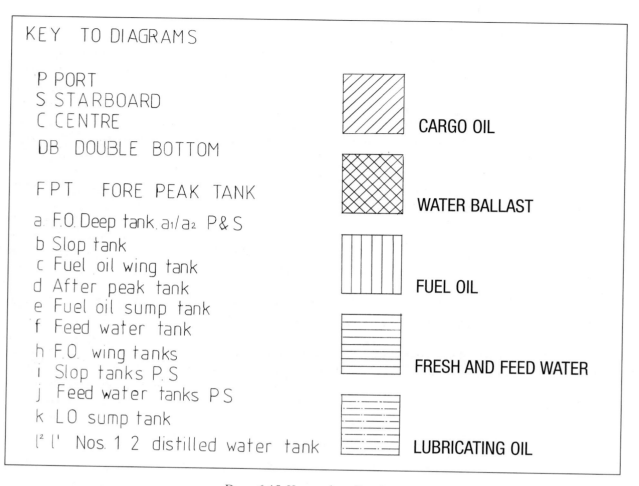

Draw. 6.15 Key to above Drawings.
(By kind permission of Mobil Ship Management and Paul Taylor.)

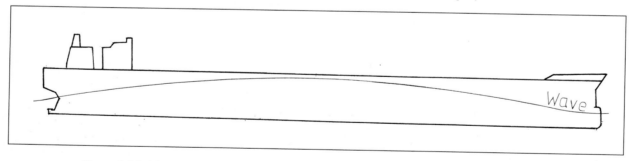

Draw. 6.16 Adverse Weather Conditions can cause a tendency towards Hogging on a VLCC ...

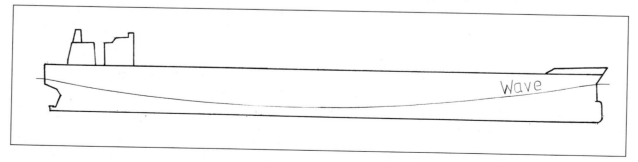

Draw. 6.17 ... and Sagging.
(The Author and Paul Taylor.)

6.5. DYNAMIC STRESSES-METEOROLOGY

The earth is subjected not only to atmospheric or air pressure, but also to its own movements, both elliptical around the sun and a daily rotation about is own axis. The latter motion leads to Coriolis forces that cause wind currents by affecting directly the dissipation of air pressure from high to low areas. The wind currents have minimum potential at the Equator, thus causing that area beloved by sailing ships known as the 'Doldrums', with alternating north easterly and south easterly 'Trade Winds' and then the basic wind patterns of variables, 'Westerlies and Polar Easterlies,' in both northern and southern hemispheres respectively. The wind movements, in turn, cause waves to appear across the surface of the water or seas, hence an elementary understanding becomes necessary of a highly complex range of sciences, including oceanography, meteorology and probability mathematics. Before computerisation sea forces were virtually impossible to calculate because there exist far too many variables but they follow, in very general terms, variations on the basic sinusoidal curve. The wave motion that seafarers refer to as 'swell' also has an influence on ships. Nautical Publication, NP100, 'The Mariner's Handbook', is an authoritative guide which is carried in the professional libraries aboard all military and merchant ships. It tells a navigator virtually everything he needs to know about meteorology, and ice in particular, but has an excellent chapter on wave formation and irregularities. On the subject of swell, it informs us:

"Swell is the wave motion caused by a meteorological disturbance which persists after the disturbance has died down or moved away. Swell often travels for considerable distances out of its generating area maintaining a constant direction as long as it keeps in deep water. As the swell travels away from its generating area its height decreases though its length and speed remain constant giving rise to the long low regular undulation so characteristic of swell.

The measurement of swell is no easy task. Two or even three swells from different generating areas are often present and these may be partially obscured by sea waves also present. For this reason a confused swell is often reported."

This admirable publication continues by defining the various length and height for swell waves and considers, amongst other things, the syndrome of abnormal waves.

6.5.1. Sea Forces, Swell, Corrugation

All ships tend to pitch, which is a longitudinal movement, into any kind of seaway, and it is a condition that is aggravated by increasingly butting and slamming into heavy head seas, the force of which causes a pounding motion that sets up considerable stresses upon the forepart of the vessel. Simultaneously, unless in a flat calm and no swell, of the kind noted in the Log Book as 'light airs', or to Force One or Two on the Beaufort Scale, there is a transversal motion which sets any ship rolling to varying degrees according to the severity experienced from, in this instance, a beam sea. When the sea is encountered from either bows or quarters a tendency is created for the ship to yaw. This is a 'corkscrew' type motion resulting from a combination of pitching and rolling.

The problem faced by the seafarer aboard any ship is how to determine, as accurately as possible, the bending moment of the craft when she encounters various sea states in terms of length, height and frequency, and very often pressure, of the waves compared with her own length, shape of hull, draft and speed. Calm seas offer a more regular wave length and height into which has to be taken consideration of any existent swell, this normally being short, average or long in length, and low, moderate or heavy regarding height. A disturbed or stormy sea often leads to waves that are uneven in height, length and frequency, but the general direction of such waves is often more or less the same. Such seas are frequently accompanied by a short low swell that runs within 20/30° of the mean wind direction. For many classes of ships the conditions are more severe when they are in ballast than when fully loaded. Far more dangerous than a rough sea, however, is the situation that is created when any ship encounters a long heavy swell where the wave length and height is compatible with her own length and draft. Extreme combinations of storm force seas and a low heavy swell could mean that waves may reach as much as fifteen metres in height and anything up to and even exceeding, two hundred metres in length. Unusual wave motion, caused by such severe meteorological disturbances, can lead also to both hogging and sagging stresses to occur, even on a ship that is as pronounced in length as a Supertanker. (Draws. 6.16 and 6.17). The Bending Moments created along the hull can, therefore, be really quite considerable and may often exceed three or four times those experienced in calm waters.

The VLCC under way, in ballast or semi/fully laden, is similarly subjected to these considerable ranges of dynamic stresses. The situation on Supertankers is not quite so pronounced when in ballast because a greater measure of control over draught can be exercised by the regulation of additional sea water dependant upon local conditions experienced. Although it is far from desirable, due to the uneven stresses that could be created upon the hull, additional ballast may be taken, with direct permission from the Master, in order to alleviate stresses caused by the meeting of a series of exceptionally heavy seas.

There exists also a danger, when running in a very light ballast condition, for an effect known as 'corrugation' to occur if the bulbous bow is allowed to pound directly into heavy head seas. The pressure of continual crashing could lead to the fore part plating to take the familiar appearance of corrugated cardboard, with the possible splitting of the peaks or troughs at their maximum points, which would allow minor cracks to appear that might lead to a seepage of the sea into the fore peak tank. It is quite impossible, however, for this condition to occur in a fully or even partially loaded VLCC.

6.5.2. Reducing Hull Stress

Another method of reducing stresses is to alter the course of the ship. This latter action, although a preferred alternative on the more manoeuvrable cargo and smaller ship, is not generally regarded as an option on VLCC's, unless the safety of the ship is directly involved. It is generally considered easier to reduce speed rather than to take the more cumbersome action of manoeuvring from the course line. Whether or not the option taken is to alter course or speed, or even both, the underlying principle is to break the synchronisation of the wave period compared with the ship's movement.

6.3. The "RANIA CHANDRIS" heads into heavy seas off Port Elizabeth.
(The Author)

Such a situation occurred on an occasion approaching Port Elizabeth when, on board a fully loaded VLCC, we found ourselves heading into a full Force 10 storm that was gusting Force 12 at times. The photograph, (Photo. 6.3), indicates the extent of the head sea encountered on this occasion. Considering the length of the Supertanker, the seas can be seen breaking right over the Fo'c'sle and extending to the lamp post just forward of the manifolds. We were 'shipping it green', as the saying goes, and I could feel the biting effect of the spray when venturing onto the Monkey Island above the wheelhouse to take the snap shot. From a distance of not far short of a third of a mile from the bows the change in motion experienced in the wheelhouse was really quite minimal. We thought that we had better slow the ship to four knots, however, in order to maintain steerage and reduce concussion on the fore part of the ship otherwise, as we generally concurred, *"we stand a good chance of pounding her to pieces"*.

6.5.3. Tender and Stiff Conditions

Most loaded VLCC's have a small metacentre and are really quite stable. They tend to roll quite slowly in a seaway and are said to be 'tender'. This is caused by the 2% cargo expansion allowance that permits free surface movement of the oil in each tank. It is for this reason that they are constructed with their specific beam and depth/draft proportions which enables them to sit lower and more easily in the water than other classes of ship and gives them a reduction in metacentric height. For the average 300,000 sdwt VLCC, this could lie between 1.4 to 1.8 metres. In a ballast condition, the metacentric height is obviously much larger and the tendency then is for the Supertanker to roll more rapidly. The ship is then said to be in a 'stiff' condition.

6.6. INSTANCES OF VLCC'S IN VARIOUS SEA STATES

In the following series of photographs, a number of different loaded VLCC's are seen under weigh in a variety of sea states. The elements shown are very much representative norms of the conditions likely to be met in the course of ordinary everyday passages. The first, (Photo. 6.4), shows a Supertanker sailing before a diffused following sea, about Force 5/6, off the port of Nagasaki in Japan. The regular pattern of wave motion alongside the hull can be seen quite plainly and the ship is shown on an even keel moving easily and well. A calm sea, about Force 3/4, with an average length and moderate height beam swell is seen being encountered by the VLCC (Photo. 6.5). Again, there is no particular problem experienced as the ship is well supported along the length of her hull by the period of the crests. A Supertanker making heavy weather, again through a completely calm sea, but a longer and much heavier beam swell is seen acting on the VLCC in the next two photographs. (Photos. 6.6 and 6.7). This ship is rolling considerably in the Atlantic rollers that are invariably met off Capetown, and southern African waters. The rollers, according to The Mariner's Handbook:

"... are swell waves emanating from distant storms, which continue their progress across the oceans till they reach shallow water when they abruptly steepen, increase in height and sweep to the shore as rollers."

Undoubtedly, they result from cold frontal systems acting upon the south westerly movement of water masses that round up from the southern polar regions. The transversal stresses observed on this fully laden VLCC are quite considerable causing heavy green seas to be taken on board, notwithstanding her possible four to five metre freeboard.

Although correctly loaded and properly handled supertankers are quite safe, the considerable forces to which they are subjected must not be dismissed too casually. The transversal stresses mentioned are countered by the reinforced frames, beams, knees and bulkheads examined in Chapter 2. More serious problems are those caused by the series of longitudinal stresses to

Supertankers Anatomy and Operation

6.4. *VLCC in a following Force 5/6 sea.*
(By kind permission of British Petroleum Shipping.)

6.5. *VLCC moving easily in beam sea with moderate swell.*
(By kind permission of Shell International Trading and Shipping.)

Ship Stresses and Stability

6.6. *VLCC rolling to Port in the Cape Rollers off South Africa and ...*

6.7. *... rolling heavily to Starboard.*
(By kind permission of British Petroleum Shipping.)

which these very large ships are subjected. Certainly, the strongly supported bulbous bow, stem and fo'c'sle, as well as the thicker bow and side shell plating, offer resistance to the fore part of the ships' structure, but these are of little effect in countering the stresses induced along the hull as the powerful engines thrust a deadweight of a few hundred thousand tonnes through the water, aggravated as it so often is by the sea state. Ships of this class have been known to break into two in even moderate seas owing to the shearing stresses through their frames exceeding the allowable critical momentum. An incident occurred a few years ago when the fo'c'sle of a VLCC dropped away completely leaving the fore part of number one cargo tank exposed to the seas. In this incident, however, the condition was believed to have been considerably aided by severe corrosion problems. The ship had to be towed stern first into a port for cargo discharge and eventual scrapping. If the VLCC is, as another example, supported by buoyancy at the ends of the hull length then the bending moment created is likely to cause the condition of sagging that was examined previously when loading/discharging was considered. When the VLCC is in reverse wave conditions then hogging is likely to occur.

Supertankers Anatomy and Operation

6.7. THE GREATEST THREAT – "WHIPPING" AND "SPRINGING"

The greatest threat to Supertanker safety, however, is from the longitudinal stresses of whipping and springing. It is quite a common occurrence to see and feel the maindeck of a fully laden VLCC actually move underneath the feet when standing or walking whilst the ship is under weigh. This is particularly pronounced when heading into even a quite moderate sea or swell. Although initially a little disconcerting, it is perfectly natural for this to happen. In fact, there would be serious dangers of the ship breaking up if it did not. The movement results from the structure of the hull giving way to the enormous strains placed upon the ship. These conditions can cause a springing stress of a few centimetres, rising to considerably more when storm force seas are encountered, leading to potential problems.

6.8. CRACKS AND CHAFING

First, small cracks might be introduced which could cause internal cracking of the frames and possibly total disintegration of an area of internal reinforcement of the hull. The difficulty with these small cracks is their early identification because they are virtually impossible to spot during the course of ordinary tank inspections. The problem, simply, is that the tanks themselves are too vast and dark for a sufficiently thorough examination to be possible. Secondly, high stress points can be introduced by these longitudinal forces into the pipework along the maindeck. (Photo. 6.8). The horizontal movement of the pipes might disturb the protecting rubbing cushion so that they end up lying flush to their supporting brackets. The chafing stresses so caused, if unattended, would lead the pipes either to wear thin, or even completely through, with the obvious possibility of leakage. This might be particularly difficult if the pipe concerned carries crude oil. I have experienced the results of this cause to a steam line on the main deck that required quite urgent attention from the ship's engineering officers.

6.8. Gradual moving of a Main Cargo Pipe from the Rubbing Strip due to hull stresses when heavy laden.
(The Author)

Ship Stresses and Stability

*6.9. Vibration Analysis of Machinery in the Engine Room on a VLCC.
(By kind permission of British Petroleum Shipping.)*

6.9. FRACTURES AND VIBRATION

Stresses on the after part of a supertanker are twofold. First there exists the possibility of fracture damage around the aft peak tank bulkhead, although this has been largely eradicated by the additional stiffening advocated over the years by the International Association of Classification Societies. Vibration has by far the greater potential for damage in the after part of a VLCC. It is induced mainly by propeller forces, but also by wave motion from disturbed seas. Local stiffening in the vicinity of the propeller (as seen in Chapter 2) most certainly reduces vibration considerably, but it still exists as a considerable problem particularly when the VLCC is in ballast. In the accompanying photograph, (Photo. 6.9), engineer officers aboard a very large crude carrier record the vibration of machinery in order to create an analysis of the problem. In practical terms, I recall having to wedge the steering console in the wheelhouse with copies of the Admiralty Pilots, selected from areas in which the ship would be highly unlikely to sail, in order to reduce shaking effects that tended to throw the steering: an extremely practical if rather primitive measure on what was then a very modern supertanker. Our sextants had to be seated on thick wedges of sorbo type material, provided by benevolent engineering officers who had a vested interest in the ship arriving somewhere near its destination, in order to reduce errors that we found were directly attributable to vibration. The handwriting in the log book was decidedly uneven and looked as if all of the navigating officers suffered from a peculiar form of ague. The situation lasted until we could adjust our engine revolutions without impeding too much our estimated time of arrival, usually from when we commenced discharge of ballast water as we approached our first loading port in the Gulf, or wherever, until we sailed with a cargo on board. Manoeuvring the ship astern, even at very slow speeds, also creates severe vibration regardless of the loading condition. This is probably because the forward motion is the norm, for which the ship was designed, so that reverse operations present out of mode phases.

6.10. COMPUTERISED SOLUTIONS

This elementary enquiry into the sea and loading stresses which affect specifically the stability of Supertankers indicates something of the complexity in assessing these factors and the essential necessity to safety of "getting things right". Chapter 17 examines recent developments in computerised sensors that both measure, and predict, forces affecting the hull in even moderate sea conditions, whilst Chapter 18 shows briefly how computers are used in management software packages to monitor hull stress during loading. Virtually all of the 'early' generation fleet of VLCC's have been retro-fitted with these systems.

PART 3: *Operational Techniques*

Chapter Seven

CARGO VOYAGES

7.1. LOGBOOK ENTRIES, "NOTE OF LOADING", SPECIFIC GRAVITIES

7.2. PRE-LOADING PROCEDURES AND CALCULATIONS
- 7.2.1. Loading Example
- 7.2.2. De-Ballasting
- 7.2.3. Inert Gas
- 7.2.4. Pre-COW Tank Cleaning
- 7.2.5. De-Ballasting Duties
- 7.2.6. Manifold Connections
- 7.2.7. Extra Deck Work, Commence Loading

7.3. LOADING: ULLAGING
- 7.3.1. Duty Officer Responsibilities
- 7.3.2. "Topping-Off", Loading to Capacity
- 7.3.3. "Two Parcel" Cargo

7.4. CARE DURING TRANSIT, SBM DISCHARGE

7.5. SECOND DISCHARGING, BACK PRESSURE
- 7.5.1. Heating Heavy Crudes

7.6. CARGO DISCHARGE PROGRAMME

7.7. BALLASTING FOR DEEP-SEA

Chapter Seven

CARGO VOYAGES

(Logbook Entries – Loading: Advisory Radio Message – 'Parcel' – Chief Officer's Discretion – Loading Planning – Specific Voyage Loading – Various Deck Duties – De-Ballasting – Precautions – Commencing – OCIMF Recommendations for Manifolds and Chiksans – Ullaging – Topping off – Specific Gravity – Sampling – Preparation Duties Prior to Departure – Care of Cargo on Voyage – Discharging: Plan – Operation – Back Pressure – Ballasting).

7.1. LOGBOOK ENTRIES, "NOTE OF LOADING", SPECIFIC GRAVITIES

Assuming that the Supertanker is outward bound from a British Port and in ballast, heading for the Iranian/Persian Gulf, the entry in the ship's log book would read "Vessel bound from Coryton towards AGFO" (the standard abbreviation indicating 'Arabian Gulf for orders'). The ship would be heading for an unspecified port that would be decided by the owners whilst she was on passage. Sometimes, the VLCC would call in at Las Palmas for bunkers because, as a free port, fuel oil there is often cheaper. The decision would often be made, however, after the company had 'tested the bunker market'. Fuel oil is the largest single operating cost to the shipowner so that, naturally enough, he would investigate bunkering costs extremely thoroughly. This could mean that some European ports where, perhaps, she may have discharged her previous cargo might be chosen, or the ship may even go over to a Continental port, such as Flushing for example, purely to bunker.

The radio message indicating the prescribed loading port would be received anywhere between the Comoro Islands (off the east coast of Africa) and the island of Socotra, about two days prior to entering the Arabian Gulf. The log book entry would then be changed to specify which port, and the preparations that have been made for the loading. The ship would be advised also of the nature of the cargo, in terms of whether or not it is heavy or light crude, together with indications giving the estimated temperature of the oil and its specific gravity. Initially, only approximate totals would be given because a number of variable factors could occur which would mean that minor allowances might have to be made to the loading programme. The specific gravities cited, for example, are not always completely accurate because they cannot be assessed until commencement of the loading. There are also variations within the specific gravities of apparently homogeneous parcels of cargo themselves that are discovered only during loading. The variables could easily be between two and four per cent of the total cargo but, on occasions, might even be higher. Generally speaking, specific gravities within crude oil cargoes depend on the port of loading and the loading temperature. In the Arabian Gulf ports, for example, SG can vary quite considerably. Arabian, Heavy and Medium crudes, can have specific gravities between 0.8877 and 0.8745 respectively, whilst Arabian Light and Extra Light range around 0.8581 and 0.8398. Iranian Heavy, on the other hand, is about 0.8708. Iranian Light has a SG 0.8560, but Kuwait Export crude is 0.8686, and that from Oman around 0.8514.

Draw. 7.1 A Four 'parcel' cargo loading plan for a VLCC of Seven Centre and Wing tank construction.
(The Author and Paul Taylor)

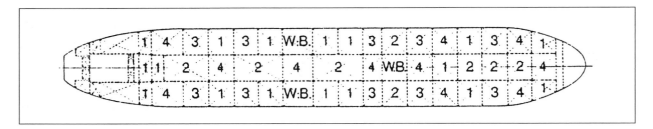

Draw. 7.2 A Four 'parcel' loading plan distributed over a 16 Set of tanks, using Slop Tanks.
(By kind permission of Jahre Wallem Shipping Company.)

Up to five 'parcels' comprising different grades of crude oil can be carried, although it would be extremely unusual for a Supertanker to lift that many over one voyage. Three or four is not quite so usual, but not uncommon, normally a maximum of two different grades is the norm. The drawings indicate the tank arrangements necessary in order to accommodate various sized parcels. Two differently constructed ships have been chosen: the first, (Draw. 7.1), shows a four parcel load spread over a range of seven centre and wing tank combinations. The permanent segregated ballast tanks are at number three wings and the grades would be distributed proportional to weight, and discharge port(s). The second, (Draw. 7.2), indicates the distributions, again for a four parcel grading, but this time arranged over sets of seventeen tanks. I am indebted again to Jahre Wallem's for permission to copy their segregation diagram for the ULCC "JAHRE VIKING".

The company mangers at head office, or the ship's charterers, often offer a loading plan to the ship, although some continue to follow the traditionally established pattern of allowing the chief officer to work out his own loading plan for ultimate discussion with the master. Whichever method is used, it is this officer who has to accept 'on the spot' responsibility for loading his ship and he should therefore be given some necessary latitude to deviate from any plan that has been externally prescribed. As discussed in the Preface, the loading of a Supertanker is a very complex business. I no longer possess copies of the detailed plans of voyages for which I was accountable, even though deck note books have been retained that provide the general indications of cargo loading and discharging.

7.2. PRE-LOADING PROCEDURES AND CALCULATIONS

In all cases, a number of procedures and calculations would have to be undertaken that have been and remain common to all Supertanker cargo operations necessary for the voyage. As a first step in calculating the amount of cargo to be lifted, the trading areas and zones between loading and discharging ports have to be determined in order to assess load line limits. The steaming time, in days and hours, is a straightforward arithmetical calculation comparing the charter speed of that particular voyage against the total distance involved using the known daily consumption of fuel. From this, the amount of bunkers required, together with a safety margin of about ten per cent, can be readily deduced. Other considerations necessary include nominal estimates of the weight of officers and ratings, personal effects and steward's stores, as well any remains on board of the previous cargo as slops, and voyage consumption likely in terms of diesel and other oils, domestic and engine water requirements, ballast and draining and any tonnage that may be retained in the fore and aft peak tanks. There is also a constant factor which is unique to each individual ship that, on a VLCC, could be in the region of 150 to 250 tonnes. These calculations are deducted from the summer deadweight, to give an initial indication of the cargo which can be taken on board. The difference between nominated cargo tonnage and that which the ship will actually load and ultimately sail with can, in some ports, have quite profound implications for the sailing captain and his chief officer. It is not unusual for loading instructions to be amended almost up to the last minute that could affect, with very serious ramifications, the final totals to be taken. An incident occurred a few years ago when a VLCC loaded more than the declared nominated quantity, following amended changes to instructions by the charterers, which led to these two senior deck officers being arrested and ultimately jailed by the port authorities on the charge of stealing the country's oil. Notwithstanding the legal consequences, the incident led to considerable political and commercial intervention before their release could be secured.

In keeping with virtually all maritime organisations internationally, accurate record keeping is as important at sea as it has always been ashore so that all companies require the completion of cargo loading charts. These present, initially, the empty condition of the ship, with additions for remains, from which can be deduced the amount of cargo that can be lifted. The following table, (Table 7.1), indicates the loading details of a very large crude carrier in excess of the 300,000-plus sdwt range, owned and managed by a major shipping company, that was preparing to load a full cargo of light crude oil on a voyage from the Bonny SBM in Nigeria bound for the LOOP terminal in the United States Gulf. It shows clearly the information

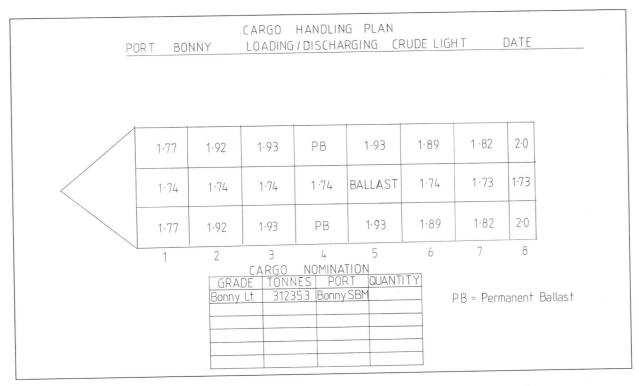

Diag. 7.1 Cargo Handling Plan for a ULCC, showing completion Ullages prior to sailing.
(By kind permission of Shell International Trading and Shipping, and Paul Taylor.)

1	2 QUANTITY ON BOARD PRIOR LOADING				3	4	5 ULLAGE			6	7 = 6 − 2d	8	9	10 Table 53	11 Table 54 A	12 = 7 × 11	13 Table 56	14 = 12 × 13	15
Tank No.	a) Free Water M³	b) Sludge M³	c) Free Oil M³	d) Total M³	Grade Loaded	Code	Observed M cm. Ft. in.	Trim Correction M cm. Ft. in.	True Ullage Ft. in.	Gross M³ in Tank @ Observed Temperature	Gross M³ Loaded @ Observed Temperature	Obs. Temp. °C	Observed Density	Density @ 15°C	Volume Correction Factor	Gross M³ Loaded @ 15°C	Density @ 15°C in Air	Weight in Air Loaded (Tonnes)	Berth
1C	-	-	-	-	BONNY LT		2.03	+0.035	2.065	20555.8	20555.8	31.6		.8431	.9857	20261.8	.8420	17060	SBM 2
2C	-	-	-	-			10.60	+0.035	10.635	14408.8	14408.8	31.8			.9855	14200.0		11956	
3C	6.3	-	-	6.3			13.91	+0.035	13.945	12031.0	12024.7	31.7			.9855	11850.3		9928	
4C	7.6	-	-	7.6			2.05	+0.035	2.085	20540.9	20533.3	31.6			.9857	20239.7		17042	
5C	-	-	-	-			2.00	+0.06	2.06	30897.7	30897.7	31.6			.9857	30455.9		25644	
6C	-	-	-	-			10.88	+0.06	10.94	21322.6	21322.6	31.7			.9855	21013.4		17693	
7C	-	-	-	-			2.00	+0.035	2.035	20566.6	20566.6	31.6			.9857	20272.5		17069	
8C	24.7	-	-	24.7			1.77	+0.01	1.78	10301.4	10276.7	31.7			.9855	10127.7		8524	
1P	-	9.1	-	9.1			1.85	+0.015	1.865	11559.3	11550.2	31.1			.9862	11390.8		9591	
1S	-	-	-	-			1.81	+0.015	1.825	11578.1	11578.1	31.0			.9862	11418.3		9614	
2P	12.6	-	-	12.6			1.95	+0.02	1.97	14528.0	14515.4	31.1			.9862	14315.1		12053	
2S	-	12.6	-	12.6			2.00	+0.02	2.02	14502.7	14490.1	31.1			.9862	14290.1		12032	
3P	12.6	-	-	12.6			1.94	+0.02	1.96	14637.1	14624.5	30.9			.9862	14442.7		12144	
3S	-	16.5	-	16.5			1.90	+0.02	1.92	14657.1	14640.6	30.8			.9864	14441.5		12160	
5P	5.0	-	-	5.0			1.93	+0.02	1.95	21909.0	21903.0	31.1			.9862	21606.6		18193	
6S	10.8	-	-	10.8			1.94	+0.045	1.985	21944.2	21933.4	31.1			.9862	21630.7		18213	
6P	2.1	-	-	2.1			1.93	+0.045	1.975	21659.9	21637.8	31.0			.9862	21339.2		17968	
6S	24.0	-	-	26.1			1.90	+0.045	1.945	21662.6	21638.5	30.5			.9862	21337.9		17967	
7P	-	6.8	-	6.8			1.62	+0.02	1.64	12728.6	12721.8	31.0			.9862	12546.2		10564	
7S	-	30.1	-	30.1			1.65	+0.02	1.68	12756.0	12725.9	31.0			.9862	12550.3		10567	
8P	-	-	-	-			13.76	+0.02	13.78	2549.4	2549.4	31.0			.9862	2543.8		2142	
8P	1933.2	-	364.4	2337.6	SLOPS	SEGREGATED AND EXCLUDED FROM TOTALS	40.02												
						SLIGHTLY IN SIMPLY DURING UNAGING													
	83.8	-	99.1	182.9						347311.8	347128.3					342254.5		288178	

PIPELINES EMPTY/FULL BEFORE/AFTER LOADING V/L ROLLING + PITCHING

*) OBSERVED DENSITY SLOPTANK: USE AVERAGE DENSITY OF SAME GRADE LOADED INTO OTHER COMPARTMENTS (LOADING ON TOP ONLY)

CARGO CALCULATION BASED ON:
SHIP : ASTM Edition 19   54A
SHORE : ASTM Edition 19 6A

FREE WATER AFTER LOADING:
GRADE
M³   207.8

MASTER:
CHIEF OFFICER:

ORIGINAL

Table 7.1 Factors to be taken into consideration when considering the loading of a VLCC.
(By kind permission of Shell International Trading and Shipping.)

necessary for loading. All cargo tanks are considered and ship board remains of water, sludge and oil residues are tabulated, together with details of the proposed loading, and final estimated ullages, for each tank. It is interesting to note the extent to which ullages vary within this apparently homogeneous cargo, illustrating the difficulty mentioned earlier in offering to the ship precise figures. This information is then used to construct a cargo handling plan which consists of a profile view of the ship, showing the individual tanks, and giving the final ullages for each one. (Diag. 7.1). The procedure for loading would follow below the profile and indicate the order for filling the tanks, following the necessary de-ballasting. The plan informs the various duty officers of the chief officer's instructions concerning the routines and when they are required to be on deck and what specific part he wishes them to undertake during their deck watch. At all stages, strict adherence is made to MARPOL/SOLAS anti-pollution regulations to protect the environment.

7.2.1. Loading Example

During one particular voyage, our loading was ordered for approximately 120,000 tonnes of Arabian heavy into all wing tanks and about 160,00 tonnes of Iranian light crude oil to the centres. The ports were to be Ras Tannurah for the former and Lavan Island for the light. Like most VLCC's, we also often loaded at other Gulf ports, such as Kharg Island at the northern end of the Arabian Gulf, about eighteen miles from the Iranian coast which caters for both light and heavy Iranian crude. Details of the discharging ports were given to the vessel and, for this trip, we were to proceed to Sete (opposite Marseilles in the Mediterranean Gulf of Lyons) for discharging 50,000 tonnes of heavy and 50,000 tonnes of light oil. From thence we were to proceed to Le Havre and there discharge 80,000 tonnes of light, with the balance of 30,000 light and 70,000 heavy to be discharged at Coryton in the River Thames.

7.1. CHIKSAN's, or loading arms, on an oil jetty.
(The Author)

Supertankers Anatomy and Operation

7.2.2. De-Ballasting

Once alongside the berth, a decision would be made concerning the method to be adopted for the discharge of ship's ballast water. Very often this would be in consultation with the shore authorities, but sometimes the decision was made on board and passed ashore. In many instances, all of the remaining ballast would be completely discharged in one operation. In this case, it would generally take between five and six hours for this to take place before commencing loading. Another method would be for the VLCC to partially load, then to de-ballast, and then continue with loading. The choice would be made purely from a safety point of view and to reduce any possible risk of pollution. Either way, the ballast is normally clean sea-water from the permanent ballast tanks at the after-part of number two wings. Additional sea water from numbers one, three and four centre cargo tanks, that were also our designated water ballast tanks, (on this particular VLCC), may also have to be discharged in port. This could total as much as fifty thousand tonnes and would be pumped over board until any minute traces showed on the oil/water interface detector. Once observed, the oily-water solution would then be pumped into the slop tanks, abaft number five tank, by the stripping line pumps. The contents of this tank could, at a maximum, total around 2,000 tonnes which was often discharged ashore at the loading port where it was taken into separator-processing tanks in order that the oil could be re-cycled. On occasions, however, the limited residue, often a couple of hundred tonnes, was left on board in the slop tanks and the new cargo 'loaded on top' and mingled with the old. The possible differences in grades for such small quantities compared to those loaded could be safely ignored. The water would be discharged into the Gulf and the tanks used for cargo. Generally, we would pay close attention to ballasting so that the chiksans, or loading arms on the jetty-(Photo. 7.1), and the manifold connections would not be strained by uneven distances.

7.2.3. Inert Gas

For similar reasons, the weather would also have to be watched, particularly with respect to storms or tidal movements if the jetty was an exposed oil island some miles off-shore. If it happened to be a calm day, close monitoring to detect possible gas pockets, particularly in sheltered areas such as abaft the accommodation, would be necessary. It is a legal requirement

7.2. Deck Inert Gas Tank Valves opened and locked into position during loading of crude oil cargo.
(The Author)

and ordinary practice for the inert gas system to the tanks to be used if loading at the same time in order to sustain tank pressure, particularly if the rate of loading happens to be lower than the rate of discharging ballast, a condition which could well occur when commencing the taking of cargo. As will be examined in Chapter 8, when considering in greater detail the tank cleaning procedures necessary after discharge of crude oil, considerable care has to be exercised if loading cargo and discharging ballast simultaneously. Otherwise, during loading, the system was shut down completely, with the master and isolation valves, and the deck tank inlet vents being not only opened, but chained and locked into position. (Photo. 7.2).

7.2.4. Pre-COW Tank Cleaning

Before the advent of Crude Oil Washing (COW) procedures, it was a common practice to tank clean completely throughout the ship on every voyage which meant, in those days, that any hydrocarbon residues were dissipated long before arrival at the loading berth. The total accumulation of this mixture could well have been as much as 420,000 m^3 that had to be released through the vents simultaneously as the cleaning programme was undertaken. Also, in pre COW days, the tanks had to be thoroughly inspected to make certain that they were completely clear of previous oil. This time-consuming task was undertaken by the deck officers accompanied by surveyors representing the shore authorities and shipper's agents. Practices varied on different VLCC's, and some ships commence loading the wing tanks concurrently with ballast discharge but, on these early generation of very large tankers, prior to fitting with an inertial gas system, we waited until about one hour or so after completing the ballast discharge before we commenced loading.

7.2.5. De-Ballasting Duties

The officers, during the short period of de-ballasting would not be idle. A considerable number of jobs have to be done and certain precautionary measures put under way. The gangway would have to be rigged and then manned at all times, with a log commenced recording all crew members proceeding ashore and their times of returning as well as any visitors to the ship, with their times of arrival and departure. This was necessary so that, in any kind of emergency, the Master had a clear idea who was on board or ashore. The time and date of sailing board and 'No smoking' reminders would be rigged at the gangway, notwithstanding the rigidly enforced regulations preventing smoking anywhere within the vicinity of the refinery. The pump room extraction fans would have to be activated and all of those in the accommodation shut down in order to prevent sparking. All outer watertight doors would be closed and secured by their dogs. Fire wires are rigged fore and aft, being led over the side on the foc's'le and poop so that, in an emergency, a tug would be able to immediately take this wire on to her towing hook and move the ship away from the jetty. These, and the mooring wires, would have to be monitored carefully by the crew and shortened as the draught of the ship increased with loading. The scuppers would have to be blocked with wooden bungs, kept in the Butterworth locker, in case of an accidental overflow of oil and the drain valves of the manifolds closed, with any lines that were not going to be used blanked off and bolted fast, as well as those lines serving the offshore side manifolds. Great care had to be taken with the tools used. Non-spark torches are issued to all officers, and non-ferrous metal wheel spanners, and the like, used in order to prevent sparks.

7.2.6. Manifold Connections

Other jobs undertaken by officers, not engaged in the above duties, would be to supervise the crew and shore workers in the rigging of the manifolds. The following sequence of photographs, (Photo's: 7.3, 7.4, 7.5 and 7.6), indicate the manifolds and the process of connecting these to the chiksans ashore. OCIMF recommendations supplement IMO regulations for safety and design standards of marine vapour recovery systems. They insist also on the number of reduction pieces which have to be carried in order that the sizes of the manifolds

7.3. The Chiksan connections to the cargo manifolds ...

7.4. ... are very often joined by a number ...

7.5. ... of reducing pieces, so making an oil-tight connection.
(The Author)

Cargo Voyages

7.6. *Main Cargo Manifold blocked off prior to cargo operation, and afterwards.*
(The Author)

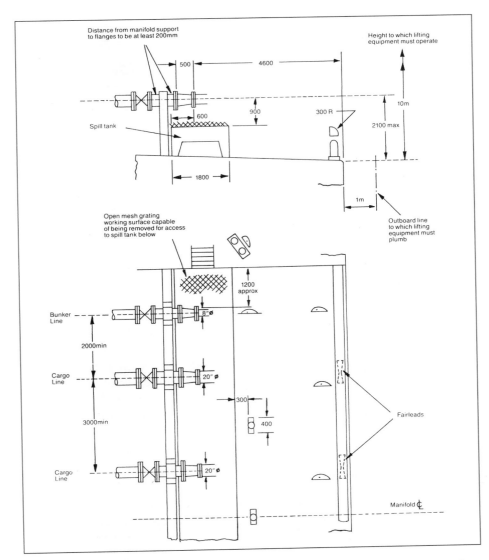

Diag. 7.2 *"Category D" — Standard Manifold Arrangement for Tankers over 160,000 sdwt.*
(By kind permission of OCIMF.)

Supertankers Anatomy and Operation

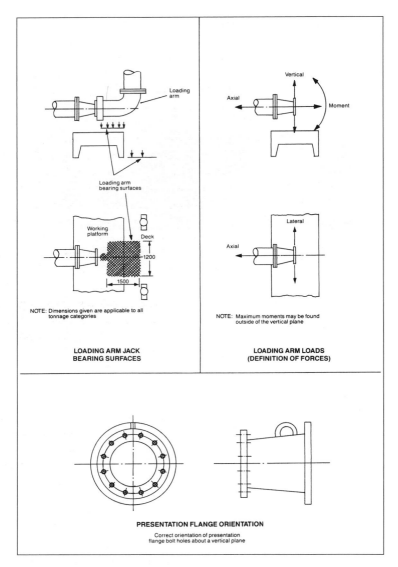

Diag. 7.3 Loading Arm Bearing Surfaces, Forces and Flange Orientation.
(By kind permission of OCIMF.)

and chiksans may be compatible in oil tight connections. For category "D" ships, which include all VLCC's in excess of AFRA's 160,000 sdwt, these include for cargo handling, 8 x 508mm plus 4 x 305mm and 4 x 406mm pieces, including reserve supplies. Bunker reducers are generally required to include 4 x 203mm plus 1 x 305mm – 1 x 254mm and 1 x 152mm. Diesel oil, lubricating oil and fresh water connections are also fitted, as well as the Marpol line for the final stages of clearing the stripping pipes. The following diagrams indicate the criteria necessary for safe management of manifold design and chiksan rigging. (Diags. 7.2 and 7.3).

7.2.7. Extra Deck Work, Commence Loading

Amongst other duties carried out by the deck officers, include reminding the engineering officers (very tactfully) about possible sparks from funnel smoke, and arranging for International Code flags to be flown from the halyards on the main mast above the monkey-island. These would be invariably the flag letter "B", of red bunting, indicating a vessel loading/discharging explosives or inflammable or volatile oils: together with one of the combination flags "TE", or "RY", which inform other vessels that they should pass the station flying the signal at a slow speed. This was in order to prevent the ship surging at her moorings with possible damage to the pipe connections, or at least, that is the theory. Generally, on Supertankers, it would have to be a pretty big surge to cause any damage. An additional duty would be to keep a close eye overboard checking that no oil was being discharged accidentally from the ship. This could occur from a number of sources, including possible seepage from the stern tube gland, or some contamination from ordinary ship board overflows.

Upon arrival at the berth, an officer would have to take the draught of the ship, fore, aft and midships on either side in order to gain an indication of the trim of the vessel, together with a sample of the density of the dock water. Generally, the draught is again taken following the de-ballasting operation immediately prior to the commencement of loading. Whilst these duties were being undertaken by the deck officers, the Mate would discuss fully the loading plan with the shore superintendents and agree the rate at which this should be commenced. I noticed, from the records kept in my deck work book that, on the two parcel loading mentioned earlier, we started initially with numbers one, two and three wing tanks at a slow rate of about 1000 tonnes per hour. This was gradually increased once we were satisfied that the cargo was going where we wanted it to and that the correct valves were open and functioning efficiently. It gave also the shore staff in the jetty control house a check that their pumps too were operating satisfactorily. After about twenty minutes, once it was observed that the loading was proceeding to our plan, we increased to 3000 tonnes/hour and opened numbers four and five wings in order to maintain equilibrium over the length of the ship.

7.3. LOADING: ULLAGING

Our next task would be to ensure that the Whessoe Gauges were working properly and recording accurate ullages. Prior to the fitting of inert gas, ullaging was very often done by hand, in a process called 'dipping', using a steel or cloth tape that was graduated in inches/centimetres with a wooden float on the end. The accompanying photograph, (Photo. 7.7), shows deck cadets aboard a very large tanker manually taking ullages from one of the after cargo tanks. The method serves as a double-check, ensuring that the floats serving the gauges have not become stuck at any level or that the tapes themselves have not become jammed in the indicator housing. In an emergency, on one occasion, we made our own ullage tape by using a block of wood as a floater to which we attached a coil of light rope and a measuring tape. It was quite effective, if rather cumbersome to take around the deck. We discovered that the only way to ullage

7.7. Deck Cadets manually taking Ullages from an After cargo tank aboard a VLCC.
(By kind permission of British Petroleum Shipping.)

effectively was by taking continuous readings as the loading gained momentum. Within an hour the rate of loading had been increased to 15,000 tonnes per hour, distributed over the entire length of the ship in the wing tanks. We opened also both slop tanks so that cargo could be taken there. It was noticed, on one loading that, after about two and a half hours, the ship started to develop a list to port approaching one degree so, in order to correct this, I partially shut numbers one, two and three port, by the hydraulic deck control valves (in the grey square boxes) in order to reduce the flow of oil into that side and bring the ship back onto an even keel. It took a couple of hours before the wings could be fully re-opened and the overall flow established.

7.3.1. Duty Officer Responsibilities

Other duties relevant to the deck officer, during his tour of cargo watch duty would be, initially, to observe and record on a hourly basis gas levels in the pump room, and the temperatures and working pressures of the cargo pumps, sea chest readings and the bilge's. Also hourly, he will check the IG main and all tank pressures updating and recording all stresses on the lodicator, briefly examined in Chapter 6. On a product tanker, if the readings approached 90%, it might be prudent to stop the cargo for a while in order to review overall progress and make any adjustments to the loading programme that might be considered necessary but, on a VLCC, stresses in excess of 90% would be recorded by every dial. As Captain Colin Graham, a serving Master with Mobil Ship Management pointed out: *"A loaded VLCC is very over stressed. 94% to 96% is quite normal in all cases. It is a fact of life and it has to be lived with"*.

The duty officer should visit also the tanks being worked, continuing to take the ullages and temperatures of the oil from each tank, in order that any variations arising from discrepancies between the radio information and the actual could be rectified in order to prevent over-loading. So far as the specific gravity of the cargo is concerned there is really very little that the officers on board can do. We have to accept the 'final' figures supplied by the terminal which form the basis of the Bill of Lading figures against which we compare the ship's estimated figures. My notebook recalls that, on this particular loading, the 'discrepancy' of the heavy crude was minimal which enabled us to take on board the full anticipated 120,000 tonnes. Being of the 'old-school', I suppose, I also looked regularly whilst passing at the Whessoes serving the centre tanks, just to be on the safe side and ascertain that no oil was entering into these. I took also a written note of the state regarding all deck valves in the boxes so that I knew, wherever I happened to be on this vast expanse of deck, the condition of every valve on the ship. It was essential 'for the records' and helped when passing over the watch to my relief as it gave him a complete picture of what was happening to the loading of the ship. Keeping alert whilst touring the main deck means precisely that. The eyes and ears of the duty mate have to be everywhere looking and listening for anything which seems out of the norm. It is only by this way that leaks in steam pipes and hydraulic lines, unexplained disparities in tank levels etc. may noticed and accidents averted. All loading, and any cargo operation for that matter, is essentially a matter of total co-operation and team work. It becomes a matter of course to maintain regular communication with the shore and engine room duty officer. A cargo watch aboard a VLCC entails a considerable amount of responsibility on the part of the duty mate. There is plenty of scope for things to go wrong and for accidents to occur and it is essential for him or her to keep alert for the entire duration of duty. It was during this watch that I was handed a number of sample bottles of the oil loaded, inscribed with all relevant details, for safe-keeping in the cargo office so that it could be handed to the inspectorate at the discharging ports.

7.3.2. "Topping-Off", Loading to Capacity

The Chief Officer invariably appears on deck in time to supervise the final stages of loading. This operation is called 'topping off' and the process is begun about two hours before completion. Logically enough, as loading commences with the forward tanks, so it is with these that we complete. The idea is to stagger the process so that only one tank is dealt with at a time in order to reduce the workload to manageable proportions. The sun is very hot in the Gulf for most of the year and the atmosphere is often very humid so that moving around the deck can be extremely tiring. Hurrying is wherever possible best avoided. More importantly, the chaos which would result from completing all tanks simultaneously can be left to the imagination, and the ever-present danger of an accidental spill could well be brought to fruition. Our rate of loading was slowed to around 6,500 tonnes/hour and number one wing tanks were brought to an ullage of about two metres. About one hour later, as numbers two and three wings were completed, the loading rate was reduced further to around 4,000 tonnes/hour and, shortly before close-down, was slowed to 2,500 tonnes/hour. The rate was gradually stopped as the slop tanks were filled to the same overall ullage and the operation was completed. This particular loading was commenced at 1100 hours on one day and completed, with 120,000 tonnes loaded, by 2300 hours the next, a total time of 36 hours, which was rather lengthy, but one that made allowances compensating for some minor delays. Normally, a VLCC could be loaded in about 24 hours at an average loading rate of 12,500 tonnes per hour, but this would entail a completely problem free operation which, although something of a rarity, does happen.

Having put the vessel back into order by dismantling and reducing the manifolds, closing ventilation shafts, striking flags and generally reversing the opening process, the final act before sailing was to take the draught. It is surprising to many people that a considerable proportion of a navigating officer's duties are taken up so much with deck work. Normally, because the time spent in port is so short, VLCC officers do not adopt the twelve-hour cargo watch that would be the practice on dry-cargo vessels, thus giving a clear twenty-four hours for a run ashore, but retain their usual four-hour sea-watches.

7.3.3. "Two Parcel" Cargo

On this specific voyage, the ship left her berth at 1620 hours and proceeded to an anchorage in order to load stores, and engine-room equipment that had been flown out from Finland and arrived too late to be taken on board whilst alongside the jetty. We stayed at anchor until 2000 hours when we entered the buoyed channel, piloting under the Master's supervision, for the twenty minutes or so until we could ring on passage and proceed to Ras Tannurah. Arrival there, I noted, was some twelve hours later at a distance of around two hundred miles, giving a speed of about 15.25 knots, roughly 'par for the course'. The loading of the 160,000 tonnes of Arabian Light crude was one of those virtually straight-forward operations and took about twenty-six hours. We loaded to just under two metres ullage in each centre tank and, due to variations in specific gravity, were able to take an extra 6,000 tonnes of cargo, at a temperature of 104 degrees. When we left Ras Tanura our draught was 21.97m mean and the ship sat comfortably between 22.10m aft and 21.84m forward and with a cargo of 286,000 tonnes of crude oil, the ship was at her maximum cargo (hence payload) capacity.

7.4. CARE DURING TRANSIT, SBM DISCHARGE

There is not a great deal of work to be done with the cargo whilst on passage. The duties which have to be done are concerned mainly with the daily taking of ullages and temperatures to serve as a check that conditions in the tanks remain as they should. It would be essential, of course, for the inert gas pressures to be monitored. About three days' prior to arrival at the first discharge port, all ship's valves are checked and, inevitably, a discharge plan formulated. In the port of Sete, the cargo, on this particular voyage, was discharged via a single point buoy mooring, or SBM. A pipe was taken on board from a fixed buoy six or seven miles off-shore, made fast and, using the ship's pumps, a steady discharge was made initially of the heavy crude, building-up to a rate of about 7,000 tonnes per hour. Once the heavy had been discharged, there was an almost immediate start made on the light. The discharging programme for 50,000 tonnes heavy/50,000 tonnes light took about twenty-four hours, after which, we sailed for Le Havre to discharge 80,000 tonnes light with the remainder for Coryton on the Essex Coast of the River Thames. After we had cleared Cape St. Vincent we received a change of orders notifying that we would not be calling at Le Havre, but were to proceed directly to Europoort in Holland to discharge the 80,000 tons there, after which we were bound for Coryton, as previously arranged. Such changes in orders are not at all unusual and result from demand changes in the oil markets.

7.5. SECOND DISCHARGING, BACK PRESSURE

Once alongside our berth in Europoort, the usual procedures of draught-lifting, ullages, manifold preparation etc. began. We took also samples from each of the seventeen cargo tanks, which included the two slop tanks, for shore analysis and, before fitting with IGS, opened the ventilators. Whilst discharging, the reverse procedure regarding the cargo applied, although it was not so much the risk of explosion that caused the problem, but the probability that the deck plating would be sucked into the tank owing to the pressure of the oil rushing out. It was also necessary, particularly during discharge, to consider any potential back pressure that might be set-up in the pipelines. This is caused by two factors: the reserve of oil, which already exists in the shore tank receiving the cargo, that brings pressure to bear upon the main valve connecting the shore to the ship, and by the extensive pipework system aboard the ship itself. This latter, with its twists and bends, creates resistance within the pipes that helps pressure build-up in the oil flow. Back pressure is not so much of a problem during loading because its maximum is at the shore end of the system and very little exists at the manifold points themselves. VLCC pipework is designed to withstand levels of back pressure and there exists a gauge in the cargo control room indicating the amount experienced at any moment in each of the main cargo lines.

Although I have intentionally over-simplified the difficulty, the coefficient of linear expansion in pipes could prove potentially a complex problem. Owing to the importance of this factor, the shore-side installation also monitors and records the rate of back pressure and, as there are such excellent communications between ship and shore, the cargo could be stopped literally within seconds. Generally speaking, there are few practical problems involved. There is another aspect of pressure which needs attention. Great care has to be taken when opening the ullage covers in case a build up of pressure has been created in the tank during passage. If there happens to be a fault in the pressure valve or the vent, then there could be quite a considerable blow if pressure was suddenly released. It is the practice to commence discharging very gradually from only one tank at a time in order to give the pumps and pipelines time to warm and to monitor progress. Pumping warm oil through cold pumps and pipelines could contribute to the back pressure.

7.5.1. Heating Heavy Crudes

Although it is not necessary with the majority of crude oil cargoes, some grades are of such a high specific gravity, hence so heavy, that they require heating during the voyage. In these instances, the system has to be turned off before discharge commences in order to prevent damage to the coils. The heavy crude oil would be heated very lightly by running steam through the tanks. This could be due to the comparative coldness in temperature of the sea compared to that in the port of loading. The heating would have to be undertaken with considerable care in order not to over-heat and thus impair the specific gravity which, in turn, could have caused uneven stresses and strains on the hull. Currently, there are only one or two VLCC's which are fitted throughout with coils that handle such cargoes.

DISCHARGE CORYTON

VOYAGE 145

Arrival drafts:
F. 12.70m
A. 12.70m
Deadweight: 142 857mt

Cargo to discharge:
Arab Light: 348 000nbbls
Arab Extra Light: 646 000nbbls

Connect chiksans to No1, 2, 3, 4 manifolds.
Maximum allowed BP is 10kg.
IG to run between 500 - 700mmWg.

THIS WILL BE AN EMISSION FREE DISCHARGE, THEREFORE THE MAST RISERS WILL NOT BE OPENED WITHOUT SPECIAL PERMISSION FROM THE CHIEF OFFICER.

---

1).
DISCHARGE ARAB LIGHT.
Open manifold x'overs 303, 306, 309. 312 Block valves 323, 324, 326. ALL OTHER BLOCKS, DROPS TO BE SHUT. OPEN BOTTOM X'OVERS 123, 99, 98, 106, 107. BLOCK 124 to remain closed.

IG Plant on. Ensure **ALL** mast risers are shut.

Cargo to be discharged from 2c, 1ws, 6ws. Finally Slop tanks.

Start No 2 MCP direct suction No2 bottom line and discharge ashore via manifold x'overs from 2c, 1ws, 6ws. Start No1 MCP suction down No1 line and discharge via manifold x'overs. Then start No 4 MCP suction via v/v's 106, 107 and discharge via manifold x'overs. Open Slop tanks and discharge until 6 000mt ROB (9.05m ullage).

At ullage of 29m or pump's cavitating, stop No1 and 4 pump's and open prima vac valve 283, slow down No2 pump and drain tanks. When tanks are empty stop No2MCP and start No 3 MCP suction from St'd Slop tank through black line driving St'd eductor and return to St'd Slop tank. Educt all tanks then strip the ballast lines, No2 and No1, and No 1, 2 MCP's. When line strip completed No 3 MCP will discharge Slops down No4 line and via sea x'overs and ashore via manifold valve. Before discharge open Block valve 326, 325 and manifold valve 314, 311. Start No4 MCP direct suction and discharge via manifold x'overs.

Have stripping pump warmed through ready for use.
On completion of Slop tanks, line up for next grade.

Permanent ballast: Pump out Fore Peak to empty. Gravitate then pump in 10 000mt to 5ws (10.62m ullage).

PTO▷

Extract 7/1. Example of a partial two parcel crude oil discharge from a VLCC. (By kind permission of Graham Douglas.)

2

SF -72 @ 66
BM 87 @ 75
Mn draft 9.9m   Trim 2.9m

2).
DISCHARGE ARAB EXTRA LIGHT.
Open block valves 325, 326. ALL OTHER BLOCKS, DROPS TO BE SHUT. OPEN BOTTOM X'OVER 124.

Start No4 MCP suction down No4 bottom line from 1c, 4c and discharge direct to manifold No4. Commence transferring from 6c to St'd Slop via v/v's 189, 190 (St'd Slop 16.80m ullage) then discharge 6c ashore. When pump at max speed, prepare for taking dirty ballast.

3).
SIMULTANEOUS BALLASTING OF 2W 7W
NO.2 manifold shut. All manifold x-overs shut.
Mast risers SHUT. Valve 98 open. Block v/v 324 open. No.2 Drop valve 321 open.

BALLAST OPERATIONS AS PER FOM CHAPTER 3.3

PUT STRIPPING PUMP ON TO THE ST'D SEA CHEST AND NO.2 PUMP SUCTION LINES AND ENSURE THERE IS A VACUUM BEFORE STARTING THE PUMP.

Bleed valves on the sea chest to be opened to check for vacuum, as well as check on pressure and vacuum gauges.

No.2 pump will start rolling slowly, still keep the vacuum on the sea suction. THIS MUST BE VERIFIED BY THE BLEED LINE ADJACENT TO THE SEA VALVE.
With the pump running at slow speed, slowly open the sea valve.
Suction can be confirmed by;
DISCHARGE OF SEA WATER FROM THE PUMP BLEED VALVE
DISCHARGE PRESSURE AT PUMP
CHANGE IN SOUND PITCH OF PUMP

SEA VALVES TO BE CLOSED IMMEDIATELY IF SUCTION IS NOT ACHIEVED, OR IF THE PUMP STOPS

Personnel requirements
Officer and pumpman at bottom of pumproom, with radio
Officer at pump controls
Person watching overside for traces of leakage from sea chest.

No.2 pump will take suction from the Std sea chest and discharge up no.2 top line and down no.2 drop to the bottom line and into 2w and 7w.

BALLAST
2W   20 000mt 7.64m ullage.
7W   23 000mt 4.72m ullage.

Maximum allowed rate 8 200 m3/hr

PTO ➣

3

Note ; With ballast in 2w and 7w , No.2, 1, lines will remain isolated.

The vessel will be discharging with no.3 and no.4 MCP at the same time as ballasting . IG fan will also be running. The pressure in the tanks MUST be constantly monitored. If the pressure drops to 400 mm wg , decrease the speed of no.3 & no.4 MCP and stop if necessary.

The tank pressure **MUST NEVER EXCEED** 1000mmwg

On completion of ballast shut sea valve and all valves to No2 pump and lines.

----

3).
When ballasting operations are settled down Start No3 MCP suction down No3 bottom line from 3ws, 4ws 8ws and direct discharge via No3 manifold.
At ullage of 29m or pump cavitating, stop No4 pump and open prima vac valve 284, slow down No3 pump and drain tanks.
When tanks are empty, change No 3MCP to eductor drive and educt out all tanks. Take suction from black line and drive St'd eductor. St'd slop to St'd slop.

COW
No discharge ashore. Manifolds shut. IG off. top blocks shut. All drops shut.

THAMES NAVIGATION TO BE INFORMED ON VHF CH. 12 ONE HOUR PRIOR TO STARTING COW.

Ensure O2 in tanks is less than 8%.

Wash 1c, 6c for one hour each tank.

Strip out other tanks as well as No.4 MCP and lines - top and bottom.

Ensure IG pressure is not less than 500mmWG for COW.

Check round that all tanks are empty.

When all tanks drained and No4 line and pump stripped then discharge Slop tank ashore, using No 3 MCP.
Final line and No3 pump strip using stripping pump and discharge via Marpol line.

Permanent ballast: Start pumping in 5ws immediately to a total of 22 500mt (4.28m ullage), Fore Peak to 1 250mt (27.40m ullage).

SF 87 @ 70
BM 60 @ 63
Sailing draft: F 5.50m   A 10.50m
Deadweight 72 000mt

Sign below when read.

G.M. DOUGLAS
CHIEF OFFICER

Supertankers Anatomy and Operation

7.6. CARGO DISCHARGE PROGRAMME

An example of a partial two parcel discharge, also at Coryton, of Arabian light and very light crude, by a VLCC with similar dimensions and summer deadweight, follows. (Extract. 7.1). The cargo is nominated in net barrels which, using the approximate standard measurement of seven barrels to the tonne, converts approximately to 50,000 tonnes of light and 92,000 tonnes of extra light oil. The back pressure (BP) is set to a standard 7 kg, whilst the IG pressure is at the discharge norm between 500 to 700 mmWg. This particular discharge followed a partial off loading at Antifer, and was undertaken by Mr. Graham Douglas, a chief officer with a major tanker company. I am indebted to him not only for supplying his detailed plans of this recent operation, but also for giving his permission for their publication.

7.7. BALLASTING FOR DEEP-SEA

Our next task was to arrange ballasting of the VLCC in order to bring her down on an even keel to her sailing draught. This was done by opening sea valves and taking water into the segregated ballast tanks and, spread proportionally over numbers one, three and four centre tanks that were designated for dirty ballast, Generally, we would ballast until the ship reached her required marks, at about ten or eleven metres, which was generally sufficient to cover the bulb and propeller, and take into consideration consumption of bunkers, fresh water and stores etc. whilst on passage. Care nevertheless had to be exercised on this operation because slight oil residues would almost certainly remain in the pumps and valves. We therefore used the stripping pump to build up a vacuum in the main cargo lines and pumped internally for awhile until we could check that the pumps were, if I might put it so, sucking instead of blowing, before opening the sea valves so that sea water could ingress.

We eventually sailed from Coryton outward-bound AGFO. Before crude oil washing, which revolutionised the process, there was very little cargo work that could be done until the Supertanker reached the warmer waters south of Las Palmas, when the tasks of tank cleaning could commence. This was, in those days, a comparatively complex and very long job, but one that was essential before the loading of the next cargo could commence.

Chapter Eight

TANK CLEANING PROCEDURES

8.1. CRUDE OIL RESIDUES, SLUDGE

8.2. HYDROCARBON GASES

8.3. GAS EFFECTS

8.4. PRE-MARPOL TANK CLEANING

8.5. GAS POISONING

8.6. GAS MONITOR

8.7. MARPOL 73/78, CRUDE OIL WASHING (C.O.W.)

8.8. FIXED TANK CLEANING MACHINES

8.9. SHADOW AREAS IN TANKS

8.10. TANK CLEANING MACHINES – HOW MANY?

8.11. LINES

8.12. INERT GAS SYSTEMS (I.G.S.), INERTING

8.13. PORTABLE FANS

8.14. EFFECTIVENESS OF C.O.W. AND I.G.S.

Chapter Eight

TANK CLEANING PROCEDURES

(Crude Oil residue: Physical Nature – "Sludge" Necessity of Removal – Chemical Composition – Flammability Limits – Gas Pressure – Fire Triangle – Static Electricity – Loss/Damage to VLCC's – Pre-MARPOL Tank Cleaning – Gas Detection – Explosimeter Principles – Gas Monitors – MARPOL 73/78 – Crude Oil Washing (COW) – Fixed Tank Cleaning Machines: Equipment Description and Operation – Shadow Areas in Tanks – COW Programme – Lines. Inert Gas System (IGS): Initial Research – MARPOL Regulations – IGS Plant Equipment – Description – Distribution Lines – Pre-Inspection/Survey – Gas Freeing – Safe Operation Procedures and Incidents.)

8.1. CRUDE OIL RESIDUES, SLUDGE

Crude oil, as it settles after loading and during passage, exudes sand and waxy asphalt-like substances. Although, even in the 'early' generation of Supertankers, virtually all the cargo would have been quite effectively removed by the stripping lines there is a residue which remains as sediment on the sides and bottom of the tanks. It has been seen, from previous chapters, that the cargo tank construction of the single-skinned VLCC, necessitated the fitting of considerable longitudinal and transversal intercostals and girders in order to provide strengthening and thus reduce hull stress. This intricate system of horizontals and verticals combined with the physical and chemical properties of the oil presented a number of problems when cleaning sediment from the tanks. The accompanying shots (Photos. 8.1 and 8.2 and 8.3) show the physical nature of

8.1. The Nature of 'Sludge'-A Sediment left in the Tank following Cargo Discharge.
(By kind permission of Victor Pyrate Ltd.)

8.2. "Sludge" – Some Problems of Removal ... (1) ... and ...
(By kind permission of Victor Pyrate Ltd.)

8.3. ... (2).
(By kind permission of Victor Pyrate Ltd.)

the 'sludge', as it is called and something of the difficulties caused in its removal as it becomes sandwiched between the oil-cleared areas around the pipe, on the left-hand side of the picture, and the bottom frames on the right. The sludge has clung also to the horizontal surfaces of stringers, frames and areas abaft tank ladders, as well as the tops and undersides of stiffeners and cross-ties and, to a lesser extent, to the vertical sides of web frames and bulkheads. It has remained in large quantities in places difficult to access.

Sludge has to be taken from the tank because, were the sediment allowed to accumulate, then ultimately a number of problems would result. Due to the viscous nature of the oil a gradual build-up of inherent wax could eventually cause a blockage of pipes and valves. It would prove also an unnecessary weight for the ship to carry that would quite considerably reduce the payload. On an average 300,000 sdwt Supertanker, a build-up of around 15 cms across all tanks could significantly alter up to 6% of the cargo Bill of Lading capacity in sludge, thus representing a possible 2,000 tonnes or so in lost cargo. Although 15 cm may sound an exaggerated figure a glance at the first series of photographs indicates, by the thickness and volume of the residue, that such a build-up most certainly would not be beyond the grounds of feasibility. The non-removal of sludge would result also in serious corrosion within the tank and, with the carriage of different grades of light and heavy crude oil in subsequent voyages, there might exist some risk of contamination.

8.2. HYDROCARBON GASES

There is not only the physical problem of clearing away the substances. The tanks, with the oil residue, contain a mixture of different levels of hydrocarbon gases, dependant upon the grade of cargo previously carried and, of course, air which has a 21% oxygen content. For any hydrocarbon in air, there is one particular mixture that could be prone to ignition which is called the 'ideal mixture' of air and hydrocarbons. The following diagram helps illustrate the relationship between oxygen and hydrocarbons in any tank and shows the conditions that may exist to cause an explosive mixture to occur. (Diag. 8.1). The Table is reproduced, for demonstration purposes only, with permission of ICS, OCIMF and IAPH, from their ISGOTT Manual, page 151. The guide offers the following explanation:

"Every point on the diagram represents a hydrocarbon gas/air/inert gas mixture, specified in terms of its hydrocarbon and oxygen contents. Hydrocarbon gas/air mixtures without inert gas line on the line AB, the slope of which reflects the reduction in oxygen content as the hydrocarbon contents increases. Points to the left of AB represent mixtures with their oxygen content further reduced by the addition of inert gas.

The lower and upper flammability limit mixtures for hydrocarbon gas in air are represented by the points C and D. As the inert gas content increases, the flammable limit mixtures change as indicated by the lines CE and DE,

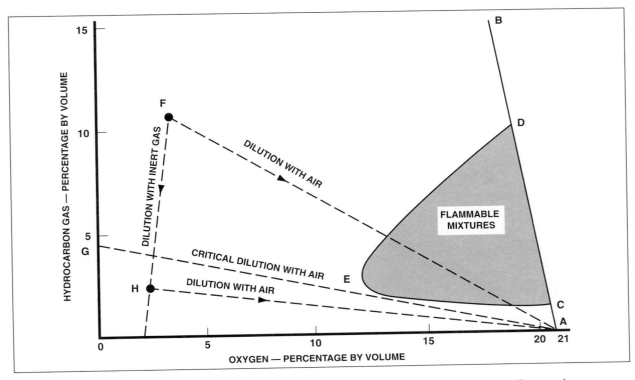

Diag. 8.1 The Relationship between Oxygen and Hydrocarbons as Dangerous Mixtures in Cargo tanks.
(By kind permission of ISGOTT Manual)

which finally converge at the point E. Only those mixtures represented by points in the shaded area within the loop CED are capable of burning."

If more air is added then a reduced hydrocarbon concentration is reached at which ignition could not occur. This specific concentration is called the 'lower limit of flammability' (LLF) C on diagriam, in which the mixture is deficient in hydrocarbon and is thus known as "too lean to be explosive". If, on the other hand, additional hydrocarbon gas is permitted to enter into the ideal mixture so that there is less air then, again, a concentration is reached in which ignition would be impossible. This is called the 'Upper Limit of Flammability' (ULF), D on diagram, where the mixture is over rich in hydrocarbon and thus above which ignition could happen. Thus, if air is added to an over rich mixture, the diluted concentration will first become flammable and finally non-flammable when the amount of air that has been added brings the hydrocarbons below the lower limit. These limits vary for each hydrocarbon and the limits, for the mixture of hydrocarbon vapours emitted by crude oil, depends on the composition of the mixture. The lower limit for crude oil is generally assessed at about 1% and the upper limit about 9.5%. The shaded area representing the flammable limits varies quite considerably with refined products, which is why Diagram 8.1 serves only for demonstration purposes. Benzine, for example, could lie between 1.5 and 8%, whilst methane gas can have a lower limt of around 5% and an upper limit as much as 15%. The safety within a tank is determined therefore by the amount of air permitted to enter, such that the relative proportion of flammable gas and air mixed in the VLCC cargo tank, must lie between the flammability range of 1% and 9.5% hydrocarbon. Put another way, the volume of the oxygen limits in a cargo tank must be in excess of 9.5%, and the hydrocarbon gas up to 9.5% for the ideal mixture and conditions for explosion to occur.

Liquids have a tendency to turn into vapour and, if the transformation occurs in an enclosed space, the molecules pass from the liquid into the area above the surface and so form gas. As more gas is added, increased pressure builds until a state of equilibrium is reached between the remaining liquid in the tank and the gas. The vapour pressure so formed is dependant upon a number of variables, of which the composition of the liquid and its temperature are supreme, such that higher temperatures produce enhanced vapour pressure and, in turn, a greater amount of gas. The constitution of crude oil consists of many compounds that have this tendency to vaporise so that, as the ship is loaded, vapour is produced which has a very high gas concentration. As the liquid fills the tank so the resulting gas layer completely fills the ullage space, although this remains in equilibrium with the oil. The total pressure within the tank is made up of the vapour pressure of the liquid and the remaining air such that the mixture is atmospheric. An example would be: if a particular cargo of crude oil has a vapour pressure of 7 psi., and the total pressure of the atmosphere is 14 psi., then the gas concentration will have a volume of around 50%. Assuming 9.5% as the UFL, and that the vapour pressure is over 1.4 psi., then an over-rich mixture will exist in the ullage space. This therefore becomes the dangerous area and, because gas/air mixture fills the entire space it occupies, it is immaterial whether the tank is loaded to 98% capacity or is

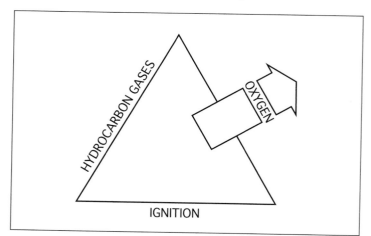

Diag. 8.2 The Traditional 'Fire Triangle': the Emphasis in Cargo tanks is on the Removal of Oxygen.
(The Author and Paul Taylor)

half-full or even less. As the cargo is discharged so more air is permitted to enter the tank and the vapour becomes diluted. The equilibrium is disturbed because additional gas is generated from the vaporisation of the oil, and the sludge adhering to the internal parts, once the tank has been emptied. It is impossible to predict the extent of this concentration of gas because the inconsistencies within the competing processes of dilution and vaporisation. The sludge generates its own gas as it lies dormant within the tank especially if the concentration of gas in the layer immediately above remains below the equilibrium value.

This layer prevents vaporisation. The cavernous cargo tanks of a VLCC can withstand about one third of an atmosphere over pressure, or about 4% of the possible maximum pressure build-up. The bottom longitudinals of most Supertankers are around one metre deep, in a tank depth frequently exceeding 25 metres, so that the build-up of pressure within the enclosed empty tanks would soon reach the 4% or so limit which the tanks would be capable of withstanding before rupturing. This means, in practical terms, that if 4% of the space contains a flammable mixture, and a source of ignition is introduced, then all of the conditions would be fulfilled for the classic 'fire triangle', (Diag. 8.2), and inevitably an explosion would occur. The resultant flame and explosion could well lead to total structural failure within, and possible total loss of, the ship. The greatest danger lies within the wing tanks. Unlike the centres, with the concentration of sludge and rich vapour preventing rapid vaporisation at the bottom of the tank, the sludge in the wings is more widely congregated around the tank and the layer above this is low in mixture so that more rapid vaporisation occurs. The sludge is also at its thickest in the wings so that it possesses an additional gas content. The fundamental object of cleaning and gas freeing of tanks, therefore, is to ensure that the operation is conducted in complete safety, as well as thoroughly, by eliminating flammable mixtures.

8.3. GAS EFFECTS

There was also another very real danger during tank-cleaning operations aboard the early VLCC's. This arose from the spray that was generated in the very powerful rotating operational jets of either portable (or, more rarely in those days, the fixed) machines used for directing the water around the tanks. It was found that a spark, again caused by a build-up of static electricity, could easily be generated within the oily-watery mist and hence an explosive situation created within the tank. Some evidence indicates that such conditions contributed to the explosions which led to the now well known severe damage and, in one case the total destruction,

*8.4. A 200,000 sdwt VLCC burns furiously following an explosion in a Non-Inerted cargo tank.
(By kind permission of Shell International Trading and Shipping.)*

Supertankers Anatomy and Operation

8.5. *A 200,000 sdwt VLCC which has experienced an explosion in her Cargo tanks during Non-Inerted Tank Cleaning operations. (By kind permission of Shell International Trading and Shipping.)*

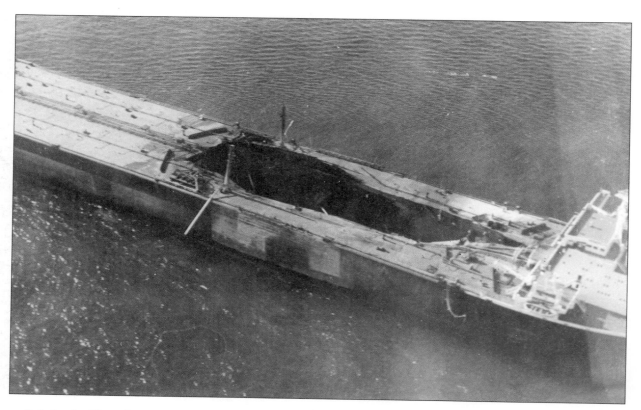

8.6. *Another View of a 200,000 sdwt VLCC showing the considerable Main Deck damage following a Tank Explosion. (By kind permission of Shell International Trading and Shipping.)*

8.7. Deck damage to a VLCC following a Tank Explosion. Collapse of the Main Deck including the Manifolds.
(By kind permission of Shell International Trading and Shipping.)

8.8. Further view of Main Deck Collapse ... (2)
(By kind permission of Shell International Trading and Shipping.)

Supertankers Anatomy and Operation

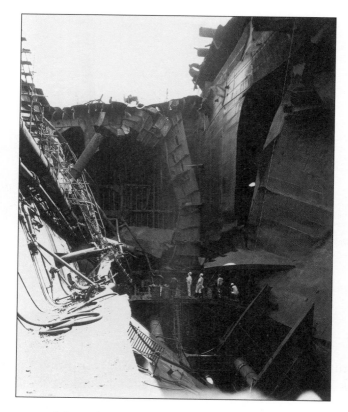

8.9. *Further view of Main Deck Collapse ... (3).*
(By kind permission of Shell International Trading and Shipping.)

8.10. *Internal Tank Damage.*
(By kind permission of Shell International Trading and Shipping.)

of three Supertankers within the space of two weeks in December 1969. The preceding series of photographs shows damage to two of these VLCC's which were sister-ships (Photos. 8.4 and 8.5 and 8.6) indicating far more graphically than words the need for provision of a comprehensive neutralising agency during tank cleaning operations. It is as well for us to be reminded of the quite frightening damage to these two magnificent 200,000-plus sdwt ships. On one, the entire deck abaft the manifold breakwater, including the manifold area, was ripped open to a length of over 122 metres due to the force of the explosion. (Photos. 8.7 and 8.8). This measurement is roughly one third the overall length of the Supertanker and, when it is considered that overall length of the mv. "GOLD VARDA" was about 146 metres the full extent of the damage may be more realistically visualised. Entire sections of pipes and plating were buckled, warped and deposited on the bottom of the tank due to the force of the explosion. (Photos. 8.9 and 8.10). Sadly, two of the crew-members on one ship, the third officer and a seaman, were killed and in the subsequent fire a number were burned and injured. The hull of the other VLCC was damaged aft so severely that two of her crew also were killed and the ship eventually sank. She had, at that time, the tragic distinction of being the biggest tanker lost by any cause.

8.4. PRE-MARPOL TANK CLEANING

In the 'early' pre-MARPOL, generation of VLCC's total tank cleaning and gas-freeing, by washing with sea water as described, was a regular occurrence every voyage and a very complex operation it proved to be. It is essential today to work to this level of thoroughness only prior to the statutory dry docking for surveys along with any repair work that might be necessary to pipework or valve rods. It is worthwhile, I believe, examining briefly. When some of the very heavy grades of crude oil (i.e. a few West African and Indonesian cargoes) had been carried, the water had to be heated to around 150° F, or, within the tank coating limits, 65°C with a pressure of about 10,500 kgf/cm3 (i.e.: 10.5 Bar) in order to prevent wax from forming on the sides and at the bottom of the tanks. Depending on the SG of the oil, however, most of the lighter grades of crude more normally carried, from the Iranian Gulf for instance, would have been washed with cold sea water that provided a sufficiently thorough cleaning to enable the residue to flow easily down the sides of the tanks. The equipment used consisted of a rotating single-or double-nozzled cleaning machine inserted into the tank through a water-tight aperture in the main deck known universally as the 'Butterworth hatch' (after the brand name of a leading tank cleaning machine manufacturer). Long before the fixture of permanently fitted machines, the entire operation was conducted using a considerable number of portable machines. The latter, along with 50m or so of rubber hosing, for each of the sprinklers used per tank, were lowered through a series of about five 'drops', the number being determined generally by the need to clean above and below the stringers in each tank. The drops would be between 30 and 45 minutes duration and the time taken with each set of wings and centre tanks would have been about 2 days duration per set for the total operation, thus taking between ten and twelve days for the entire cargo carrying capacity to be cleaned. The portable machines were stowed in the Butterworth locker on the starboard side abaft the manifolds usually along with the remainder of the equipment to complete the tank cleaning. This consisted of portable ventilation fans, mucking winches and portable gantries, cleaning agents for casual drainage and drippings, rubber buckets and shovels. Once the tank had been ventilated, the crew could descend to the bottom of the tank and remove any sludge, which had not been dispelled by the water, along with the inevitable rust scale. There existed, in the early days, an additional problem of potential absorption of hydrocarbon poisons through the clothes and skin, with a risk of skin cancers developing, (particularly testicular), so that full protective clothing had to be worn as well, occasionally, self-contained breathing apparatus. The discomfort involved in wearing this gear in hot weather, whilst working manually at the bottom of a tank in order to carry out routine inspections of the tanks for cleanliness, valve testing, inspection of pipes and lines and ordinary maintenance duties, may well be imagined. The scale had then to be winched by hydraulically operated hoists through the 25 metres or so of tank to the crew member on the main-deck. A considerable quantity of water had to be used and the entire operation was long, tedious and, in the days before Inert Gas systems, was potentially very hazardous. The machines had to be earthed in order to reduce the risk of static electricity that could build-up on the metal nozzles. Removal of sludge in the "early" generation of VLCC's had to be carried out, therefore, with due care and a very close regard for safe practices.

It is worth recording, as a comment contrary to some remarks heard or occasionally read sometimes from quite authoritative sources, that the tank cleaning system using portable machines aboard VLCC's was really very efficient. During my service as a mate aboard Supertankers in the 300,000 sdwt class, prior to the fitting of more sophisticated techniques, this method of tank cleaning had been used over a number of voyages. The inspection of all tanks with port officials, necessary before receiving clearance to load the next cargo, was extremely thorough and, in virtually all instances the tanks were certainly acceptably clean. Having also, in earlier days, lugged all of this paraphernalia around the main deck of a Supertanker, and worked exhaustively for about ten days or so supervising the crew on a highly labour intensive operation, I had no regrets in sharing the considerable relief expressed by all mates on the implementation of the MARPOL 73/78 Regulations. It is worth noting, however, that a range of portable tank cleaning machines continues to be found on many tankers, but their use is restricted to those well below the VLCC range.

8.5. GAS POISONING

It was clearly essential in the early days of 'tank diving', as it was known, to exercise considerable caution before entering recently cleaned tanks. Even now, in the ordinary interests of self-preservation, it is prudent when going down for survey purposes, to be aware of the ever-real possibility that hydrocarbon gas might be existent in the tanks and not only residing within any parcels of sludge. The gas was, and remains, hard to detect because it is not visible and, unless it was heavily sulphurous, has no distinctive aroma. It is also toxic such that inhalation would soon cause brain damage or death and, sadly, there are far too many recorded incidents where such casualties have occurred. Whilst there is often experienced, however, a slight prickling sensation in the eyes, by the time this has been experienced and assimilated, there is probably insufficient time to climb the 25 metres or so necessary to reach fresh air. A greater danger arises from the effects of oxygen deficiency and most recorded deaths have arisen from this cause than the direct effect of hydrocarbon poisoning.

8.6. GAS MONITOR

The only effective way of detecting the presence of the gas was, and remains, by frequent use of meters. Explosimeters are excellent for detecting explosive gases, but will not reveal the hydrocarbon content of a tank. The tanker industry, however, has always been well served by a range of reliable and easily portable instruments for the measuring of the various gases likely to be encountered aboard the largest of ships including hydrocarbons. The older models were effective, if cumbersome, but modern instruments are totally reliable and easily portable, although some users of models carried around the waist have found on occasions that the machines, compact as they are, have caused obstruction problems when climbing down, in and around, tank space. As a brief explanation, many models operate by detecting the rise in temperature which occurs when an electrically heated platinum filament is exposed to a combustible air/gas mixture. The filament is one limb of a Wheatstone Bridge circuit and, as the gas/air mixture is passed over it, combustion takes place, heat is generated and the resistance increases. This increase creates an inequality across the bridge, which is directly proportional to the combustible vapour present, and is measured by the instrument and recorded as a current flow in the millimeter. Many of the models emit an audible warning once pre-set gas levels are reached thus allowing sufficient time to take quick and evasive action. In view of the fact that the

Tank Cleaning Procedures

*8.11. Chef Officer using an Explosimeter to check Pump Room Bilge for Explosive/Toxic Gas.
(By kind permission of British Petroleum Shipping.)*

*8.12. A "NEOTRONICS" Mini-Gas Monitor.
(By kind permission of Neotronics Ltd.)*

combustible energy for all hydrocarbon gases regardless of the crude oil previously carried, is similar at their LFL, the meter registers gas concentrations within this limit. It is as much for this reason than any other that the 'too-lean' approach has found favour with many major oil companies. In the above photograph, (Photo. 8.11), a chief officer, in the customary "officer's uniform for deck work, the ubiquitous boiler-suit", still very much in use on many ships today, is seen using a portable meter to test the bilge in the pump room for any evidence of gases.

Supertankers Anatomy and Operation

8.13. *A Personal Explosive and Toxic Gas Monitoring Machine.*
(By kind permission of Draeger Ltd.)

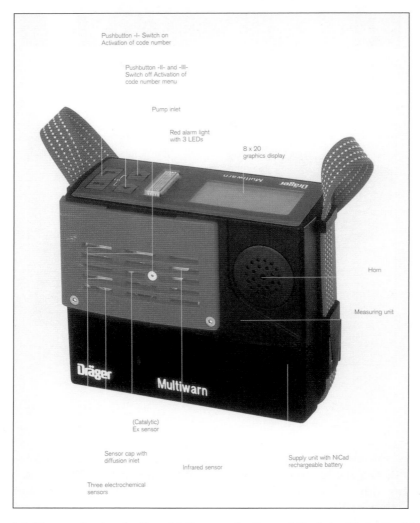

Diag. 8.3 *Diagram showing the Essential Features of a Personal Explosive/Toxic Gas Monitor.*
(By kind permission of Draeger Ltd.)

Nearly all of the current meters in use have also a mode for measuring the oxygen content of tanks, obviously a very necessary survival factor before entering any tank or previously confined space including pump room, bilge's, etc, let alone cargo tanks.

To offer an example of some modern instruments, Neotronics Limited have produced their 'MiniGas Monitor' (Photo. 8.12) a portable, highly compact and extremely rugged portable instrument. It has been designed specifically for shipboard use with channels capable of monitoring from one to four gases: oxygen, flammable, carbon monoxide and hydrogen sulphide, likely to be found on VLCC's. Personal monitors are also available to officers which fit readily into the pocket or can be worn on the belt, such as the 'Multiwarn 11' model, marketed by Draeger's. (Photo. 8.13). This versatile instrument allows for the measurement of up to thirty-five gases via a number of independent sensors. An infra-red sensor offers a detection choice from O_2, CO_2, CH_4 or LFL. A catalytic ex-sensor is also provided for the measuring of LEL across a wide range of explosive gases, whilst three electrochemical sensors, fitted with fifteen interchangeable pugs in "intelligent" sensors, allow for the detection of up to thirty-five gases. The following drawing (Diag. 8.3) shows a displacement of the various components within a very compact case.

8.7. MARPOL 73/78, CRUDE OIL WASHING (C.O.W.)

MARPOL 73/78 had an immediate and far-reaching effect on the tanker industry world-wide, particularly covering ships in the VLCC class, by making mandatory a series of anti-pollution, and later SOLAS regulations. Specifications of the various regulations and their all-important annexes were further revised by resolutions passed in 1981 with a number of subsequent amendments, so enabling the latest ideas and technological advances, along with ideas proven by the light of experience, to be incorporated into existing rules. Amongst the early initial measures introduced, was the idea of using crude oil from the cargo as an agent to clean tanks, instead of an all-water wash, following the complete discharge of a cargo at the terminal and before sailing for a new loading port.

The system of Crude Oil Washing (COW), as it became known, was tailor-made for each specific ship, and the details contained in an appropriate Crude Oil Washing Operation and Equipment Manual produced by the shipping company and approved by the appropriate Classification Society. Each design was based upon similar basic principles, in accordance with Regulation 13B of ANNEX 1 of MARPOL 73/78. It was intended, and I quote from the manual provided for use aboard a 300,000 sdwt VLCC, operated by a leading company:

"The purpose of the manual is to meet the requirements for Crude Oil Washing in accordance with the revised Specifications for the Design, Operation and Control of Crude Oil Washing Systems (IMO Assembly Resolution A 446 (xi)). It provides standard operational guidance in planning and performing a Crude Oil Washing programme and describes a safe procedure to be followed."

The manual made provision for individual drawings of the system to be operated by the specific ship, provided a description of the system and the equipment parameters, and covered, amongst other things, various precautions to be undertaken with regard to leakage and electrostatic hazards, use and control of the inert gas system, programmes for conducting the wash, checklists and programme. The qualifications necessary for the personnel were also stipulated as well as criteria assessing shadow diagrams, stripping line efficiency and drainage of all pumps and lines used in operations and subsequent ballasting and precautions prior to tank entry.

Basically, crude oil washing occurs during the latter stages of cargo discharge and uses oil that has come from the cargo itself, a proportion of which is re-circulated around the tanks.

Supertankers Anatomy and Operation

As a system, therefore, it has to be used totally whilst the ship is in port and can be commenced at any time after one of the tanks has been discharged. The crude is taken from a special COW line into the tanks where the oil is directed by means of a high pressure jet through the fixed tank cleaning machines. MARPOL insisted that both the older 'early' types of Supertanker and all subsequent new buildings had to be fitted with tank washing machines that conformed to strictly defined criteria.

> *"4.2.1. The tank washing machines for crude oil washing shall be permanently mounted and shall be of a design acceptable to the Administration.*
>
> *4.2.2. The performance characteristic of a tank washing machine is governed by nozzle diameter, working pressure and the movement pattern and timing. Each tank cleaning machine fitted shall have a characteristic such that the sections of the cargo tank covered by that machine will be effectively cleaned within the time specified in the Operations and Equipment Manual."*

8.8. FIXED TANK CLEANING MACHINES

The provision was for permanently fixed machines that entered the cargo tanks through the upper deck, that were piped from the cleaning main, as well as other machines which had to be fixed at the bottom of the tank. The number and location of the machines were also clearly directed. Owners have a choice between a number of fixed types of tank cleaning machines, dependent upon the requirements dictated by the number and construction of the tanks with the proviso, of course, that the machines conform to IMO Regulations. Although the products of different manufacturers vary, there are basically two types of machine in use: the single-nozzle and the twin-nozzle. Each type has its own cleaning pattern and is adaptable for use with either hot or cold water or crude oil. Generally, the procedure followed is to commence washing at the top level of the tank and to work progressively along the walls downwards, completing with a thorough vertical bottom wash.

The range of single-nozzle tank cleaning washing machines varies considerably, but each type has fundamentally a very similar construction. A stand pipe of between 4.2. and 5.2 metres penetrates into the tank through a 'drop pipe' and has at the lower end a 360° rotating nozzle.

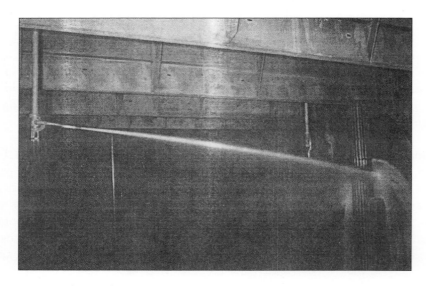

8.14. Under MARPOL Regulations, the Single Nozzle Tank Washing Machine is Always Fitted to the Underside of the Tank Top and possesses a very powerful jet.
(By kind permission of Consilium Marine.)

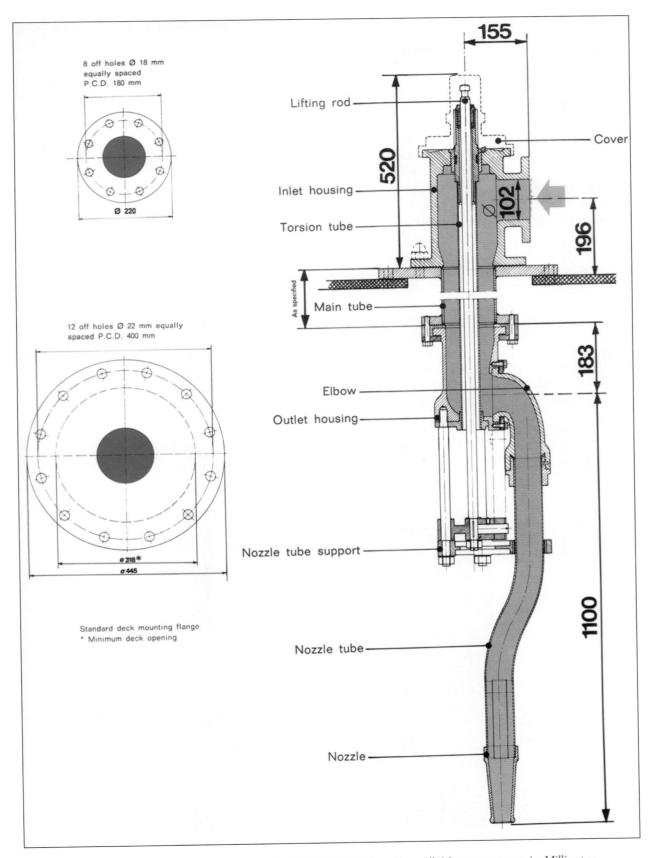

Diag. 8.4 Dimensions of the Single Nozzle "GUNCLEAN" machine. All Measurements are in Millimetres. (By kind permission of Salen and Wicander.)

Supertankers Anatomy and Operation

By its very construction the tank washing machine is always fitted to the underside of the tank top, (Photo. 8.14), and possesses a powerful thrust necessary for a thorough wash. The following diagram (Diag. 8.4) shows the dimensions and specifications for the fitting of a 'Gunclean' machine indicating how it is built through the main deck into the tank. Whilst variations will obviously occur with different models, many are powered by turbines or compressed air that offers a capacity for each machine to the order of 120 tonnes, at 9 bar pressure, which corresponds to a pump discharge pressure of 12.5 bar. The gun is thus capable of producing an extremely powerful jet of either water or crude oil which is directed in a very concentrated stream over the majority of the tank's surface. The action of the nozzle is really very slow horizontally, around two/three revolutions per minute, with an equally slow vertical rotation, thus ensuring a thorough wash. The sediment and minute solid rust pieces are broken down into particles which makes their removal easier to flush, thus reducing to a virtual minimum, the necessity for manual 'mucking out'.

Other popular models for the VLCC trade are the range produced by Victor Pyrate in Essex a company who have supplied well over 1000 tankers with tank cleaning machines, since their formation in 1952, and who stand as one of the first companies to develop fixed automatic washing machines. Their 'VP Monomatic 2' series, (Photo. 8.15), with its accompanying literature, illustrates and describes the equipment and gives additional information regarding the specification of the machine. The photograph on the left shows the upper deck fitting and the drop pipe leading down to the single nozzle, whilst the smaller shots, on the right hand side, indicate a closer view of the part of the pipe above deck that is connected to the appropriate water or COW line, and the fitting of the nozzle with its rotating arm. The centre drawing indicates that the 'VP Monomatic 2', like most types of deck mounted machines, lends itself

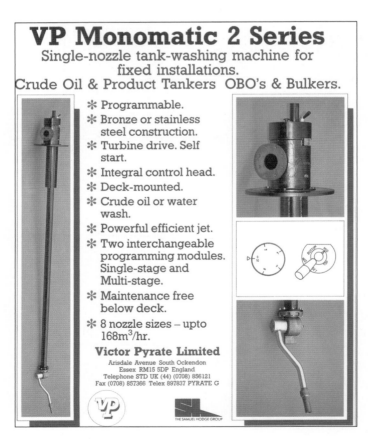

8.15. The 'VP MONOMATIC 2' Deck Mounted Single Nozzle Fixed Tank Cleaning Machine.
(By kind permission of Victor Pyrate.)

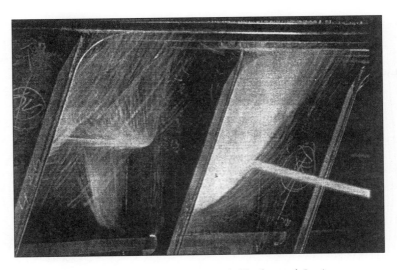

8.16. *The Jet from the Rotating Nozzle Touches and Impinges ...*
(By kind permission of Consilium Marine.)

8.17. *... on Tank Bottom and Sides ...*
(By kind permission of Consilium Marine.)

readily to being programmed thus offering a choice of washes. This enables the cleaning pattern within the tank to be varied so catering for the thoroughness of washes demanded by different grades of cargo.

The chief officer can set the machine to whatever he considers to be the desired pre-determined wash and leave it to run for the varying periods necessary to complete the cleaning cycle. A tank, for example, that has been used to transport a heavy grade of crude would clearly require a different, perhaps more intensified, wash programme, than that necessary if the previous cargo had been of a lighter grade. The wash pattern of the single nozzle machine, (Photo. 8.16), indicates the tank side and bottom pattern that is described by the rotating nozzle within the tank, showing how the jet strikes the sides walls and forces the residue oil downwards into the tank floor. (Photo. 8.17) and (Diag. 8.5). The intention is to drive the oil towards the pump suction so facilitating discharge. This is particularly relevant on the bottom wash programme when the machines are simply removing final residue. The wash options available on the programme are indicated in the following Diagram (Diag. 8.6). The top

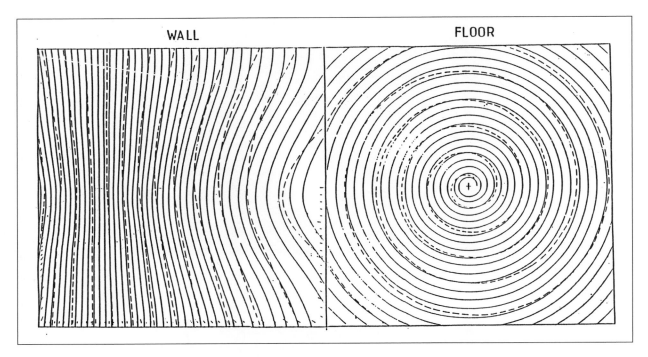

Diag. 8.5 ... in a Distinctive Wash Pattern.
(By kind permission of Victor Pyrate.)

left hand side drawing describes the action of the single stage wash option, and indicates how the nozzle is raised and lowered to give a top-bottom-top wash, through the arcs of 160°–0–160° respectively in order to cover the tank walls and corners. The lower drawing describes the function of the multi-stage wash. It is interesting to study the jet lengths for the relevant nozzle aperture sizes, in the table at the bottom of the leaflet, and see how the impact of the jet can vary quite considerably when used with water as opposed to crude oil, and gives a thorough bottom wash. The following photograph shows the deck hands on a large VLCC setting the controls of the programme drive unit of a tank cleaning machine. (Photo. 8.18). Although I have never encountered problems personally, it has been heard that some users find a disadvantage of the single-nozzle machine is the frequent re-settings of the equipment necessary when a multi-stage washing has to be undertaken. The problem encountered however seems to be not so much with the setting of the actual machine wash arc, because this is an extremely simple and quick operation to perform, but perhaps with the setting of the wash *programme* that is such a vital part of multi-stage tank washing.

The alternative to the single nozzle permanently fitted tank cleaning machine is that fitted with twin nozzles. (Photo. 8.19). The fitting to the underside tank-top is similar to that of the single, as the following diagram of the Victor Pyrate's 'VP Magna' indicates (Diag. 8.7). The jet lengths and wash times are each comparatively shorter to those of the single nozzle machine, but the wash pattern is considerably finer. Having two jets means that the machine is better balanced than the single nozzle but, for some purposes, the main disadvantage is that the twin is non-programmable and has to be monitored manually. The helical washing pattern is frequently compared to the old fashioned ball of string such that each successive strand overlaps that of the previous winding, thus offering a more comprehensive wash to the walls of the tanks. The cycle of the progressive wash can be seen in the following diagram (Diag. 8.8),

VP Monomatic 2 series
Single-nozzle tank-washing machines

Technical Specification

Material: Bronze cleaning head. Selected stainless steel parts. Mild steel downpipe and flanges (epoxy coated).

Housing: Cast iron and copper (2%).

Inlet Flange: 65mm/80mm DIN/BS/ASA/JIS – Junior – smaller vessels. 100mm – Senior – larger vessels.
Filter screen included.

Deck Flange: To suit standard 318mm opening or built to requirments. (10 or 12 studs.)

Wash Pattern: Helical (see separate sheet).

Design: Single 'maximum energy' nozzle with smooth jet flow for efficiency. Self start turbine drive with integral programmable control head.

Cleaning Head: Maintenance free. Self draining. Wash media lubricated.

Operation: Single-nozzle helical wash. Rotation speed 1 rev per 50 secs. Nozzle pitch on each rotation – *variable on single stage (set by operator).*
On multi-stage pre set arcs 2% DOWN – 6% UPWARD PASS nozzle pitch.

Wash Cycle: Variable time.
(Example 160°-0°-160° at (2°/6° pitch (106 revs = 320° = 90 mins)
This cycle time is typical for 38mm nozzle. Faster times apply for smaller nozzles.

Downpipe: Fabricated to suit vessel tank dimensions and vibration frequency. Mild steel shotblasted and epoxy coated.

Machine: MONOMATIC 2J (JUNIOR)
for smaller vessels (18 – 30mm nozzle)
MONOMATIC 2S (SENIOR)
for larger vessels (32 – 38mm nozzle)

APPROVALS:
VP Monomatic accepted & approved by class societies worldwide.

Full design service. Shadow drawings and operating manuals. VP will calculate machine quantity and type.

WASH PROGRAMME OPTIONS
These two programming modules are interchangeable should vessel enter different cargo trades needing different wash requirements.
* © * Patent applied for.

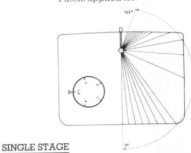

SINGLE STAGE

Washes from TOP-BOTTOM-TOP (160°-0°-160°) in one programme. Can be stopped and re-started at any point. Cam driven unit. Position of nozzle shown on indicator. Nozzle angle pitch can be set between ½°-10° in ½° variations to give fine or coarse wash, *and are adjustable while machine is operating.*

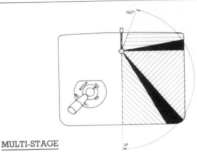

MULTI-STAGE

Washes in pre-determined arcs that are factory set according to vessel tank design or to maker's standard. 4 overlapping wash arcs available using a selector spool in the drive system.
Typical
standard = FULL 160° – 0° – 160° MIDDLE 100° – 35°
TOP 160° – 90°
BOTTOM 40° – 0°
Nozzle will wash between the pre-selected vertical limits and automatically reverse. Position of nozzle shown on indicator. Nozzle pitch is pre-set at factory to give 'fine' wash on down pass and 'coarse' wash on upward pass.

MACHINE PERFORMANCE DATA

⌀ NOZZLE SIZES	18mm			20mm			22mm			24mm			28mm			30mm			32mm			38mm		
Pressure kg/cm²	8	10	12	8	10	12	8	10	12	8	10	12	8	10	12	8	10	12	8	10	12	8	10	12
Flow m³/hr	30	35	38	36	42	46	44	48	51	58	63	67	75	84	92	88	100	108	106	118	126	133	154	168
*Jet length-m (C.O.W.)	19	21	23	22	24	26	23	26	28	25	28	30	28	30	33	30	32	35	31	34	37	35	38	42
*Jet length-m (water)	32	35	38	34	37	40	37	40	42	38	41	43	39	42	45	40	43	46	41	44	47	42	45	49

*Jet – (C.O.W.) = Classification Society approved length of wash jet with maximum solid effect.
*Jet – (water) = Maximum length of wash jet – horizontal throw.

Victor Pyrate Limited

Diag. 8.6 Technical Specifications of "VP MONOMATIC 2" Single Nozzle Machine.
(By kind permission of Victor Pyrate.)

Supertankers Anatomy and Operation

8.18. *Deck Hands on a VLCC setting the Controls of a Deck Mounted Permanently Fixed Tank Washing Machine. (By kind permission of British Petroleum Shipping.)*

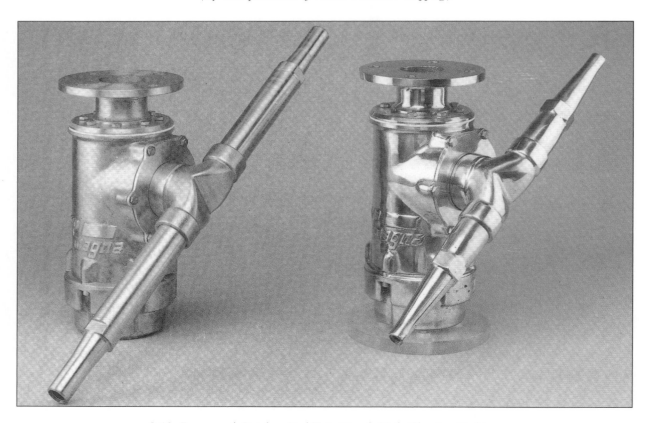

8.19. *Bronze and Stainless Steel Twin-Nozzle Tank Cleaning Machines. (By kind permission of Victor Pyrate.)*

Tank Cleaning Procedures

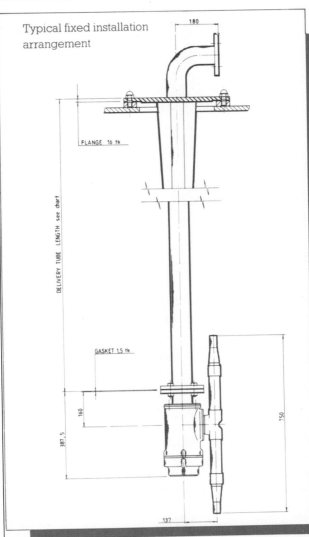

VP Magna
Twin-nozzle tank washing machine

Technical Specification

Material: Bronze or stainless steel 316L.

Inlet flange: DIN 65 NP 16, 65 JIS 16K, or 2½" ASA 150lbs.

Machine weight: 27kgs.

Wash pattern: Spherical.

Design: Twin orbital nozzles with smooth jet flow. Integral turbine drive. Enters standard 318mm deck opening down to 280mm opening.

Gearbox: Grease packed and sealed against leakage – or –
Self-lubricating and self-draining, no grease.

Downpipe fabrication: Any length to suit vessel. Mild steel – shot blasted and coated – or stainless steel.
Deck flange – to suit requirements of vessel. Above deck inlet bend to meet new or retrofit installations.
Composite steel/stainless steel fabrications available.

APPROVALS:
VP Magna accepted and approved by class societies worldwide.

*Full design service. Shadow drawings and operating manuals.
VP will calculate machine type and quantity.*

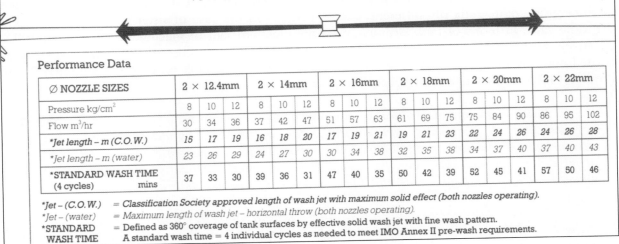

Performance Data

⌀ NOZZLE SIZES	2 × 12.4mm			2 × 14mm			2 × 16mm			2 × 18mm			2 × 20mm			2 × 22mm		
Pressure kg/cm²	8	10	12	8	10	12	8	10	12	8	10	12	8	10	12	8	10	12
Flow m³/hr	30	34	36	37	42	47	51	57	63	61	69	75	75	84	90	86	95	102
*Jet length – m (C.O.W.)	15	17	19	16	18	20	17	19	21	19	21	23	22	24	26	24	26	28
*Jet length – m (water)	23	26	29	24	27	30	30	34	38	32	35	38	34	37	40	37	40	43
*STANDARD WASH TIME (4 cycles) mins	37	33	30	39	36	31	47	40	35	50	42	39	52	45	41	57	50	46

*Jet – (C.O.W.) = Classification Society approved length of wash jet with maximum solid effect (both nozzles operating).
*Jet – (water) = Maximum length of wash jet – horizontal throw (both nozzles operating).
*STANDARD WASH TIME = Defined as 360° coverage of tank surfaces by effective solid wash jet with fine wash pattern.
A standard wash time = 4 individual cycles as needed to meet IMO Annex II pre-wash requirements.

*Diag. 8.7 Installation and Specification of the "VP MAGNA" Twin-Nozzle Machines.
(By kind permission of Victor Pyrate.)*

Supertankers Anatomy and Operation

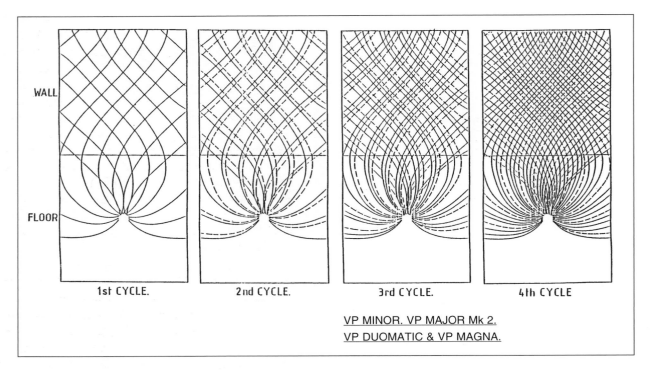

*Diag. 8.8 The Progressive Wash Pattern of the Twin-Nozzle Machine.
(By kind permission of Victor Pyrate.)*

where the 'unravelling' of the four-cycle phased machine builds up its distinctive pattern. Although, in essence, both types of machines could be used in submerged positions in the bottom of a tank, access difficulties to alter the programme on single nozzle machines usually negates their use in tank bottoms. The function of the submerged machine in the lower levels of the tank is to wash a precise area of shadow and the two nozzle machine is able to do this very effectively without the need to be programmed.

8.9. SHADOW AREAS IN TANKS

The construction of each 'early' generation VLCC has been shown to be marginally, and in some cases, quite radically different, except perhaps in the case of sister ships produced from the same yard. In terms of tank cleaning, even the most efficient of machines cannot clean 100% effectively all areas of the tank. This implies that there exist a number of different shadow areas created not only for each VLCC, but also within each individual tank on the same ship. MARPOL Protocol 4.2.8 makes quite unequivocal stipulations:

> "The number and location of the tank washing machines in each cargo tank shall be such that all horizontal and vertical areas are washed by direct Impingement or effectively by deflection or splashing of the impinging jet. In assessing an acceptable degree of jet deflection and splashing, particular attention shall be paid to the washing of the upward facing horizontal areas and the following parameters shall be used:
>
> (a) For horizontal areas of a tank bottom and the upper surfaces a tank's stringers and other large primary structural members, the total area shielded from direct impingement by deck or bottom transverses, main girders, stringers or similar large primary structural members shall not exceed 10 per cent of the total horizontal area of tank bottom, the upper surface of stringers, and other large primary structural members."

Tank Cleaning Procedures

The rules stipulate that up to only 10% of horizontal tank members, and, in sub-section (b) of the same regulation, not exceeding 15% of the vertical faces, should exist within a tank as shadow areas. The manufacturers' of tank cleaning machines have therefore to treat each ship individually and prepare a tailor-made plan for each set of tanks showing the areas actually impinged, or cleaned, by the jets and indicating specific shadow areas which the jets are unable to reach. It is a MARPOL requirement that the diagrams covering VLCC's in excess of 200,000 sdwt should be drawn to a scale of 1:200 and must give plan, profile and elevation views indicating clearly the shadow areas. The following specifications (Plans 8.1 and 8.2) give examples of proposals successfully presented by Victor Pyrate's in the 1980's for the supply of machines to three American owned VLCC's each of around 264,000 sdwt. This particular contract offered a quite unique challenge to the company owing to the considerable volume of steelwork in the lower levels of each tank that required the fitting of multiple quantities of the company's machines in order to provide effective cleaning. The specifications were for C.O.W. and, from the total of nine drawings necessary to indicate calculations showing the shadow areas for the trio of ships, two plans have been chosen showing the intricacy of the work involved. In number two centre tank, (Plan 8.1), for instance, the disposition of the deck and submerged machines within the tank are clearly shown in relation to the stringers so that maximum impingement or cleaning by deflection may be obtained. A horizontal and vertical cleaning of 97.5% and 93% respectively indicates the efficiency of the fourteen submerged double nozzled and four deck fitted single nozzle machines. Wing tanks always create potential cleaning problems due to the additional structural members that are necessary for strengthening purposes. The shadow areas for each of numbers three and six wings, on the same class of VLCC's, can be seen clearly in (Plan 8.2.) Each tank is fitted with four deck and one submerged machine offering an excellent horizontal and vertical impingement that is well in accord with IMO Regulations. Again, the disposition of the machines can be plainly seen.

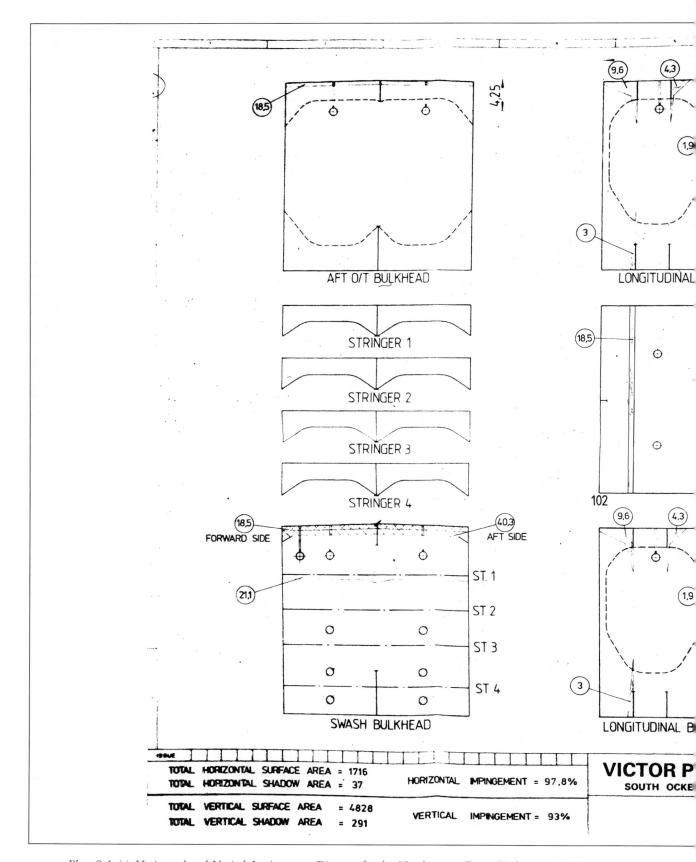

Plan 8.1 (a) Horizontal and Vertical Impingement Diagram for the Number two Centre Tank on a VLCC ...
(By kind permission of Victor Pyrate.)

Tank Cleaning Procedures

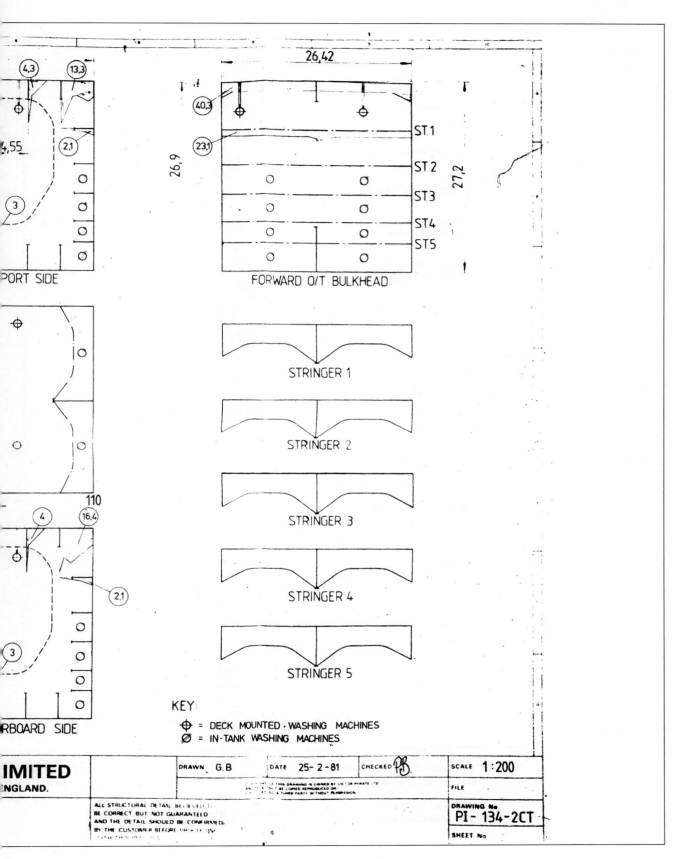

Plan 8.1 (b)

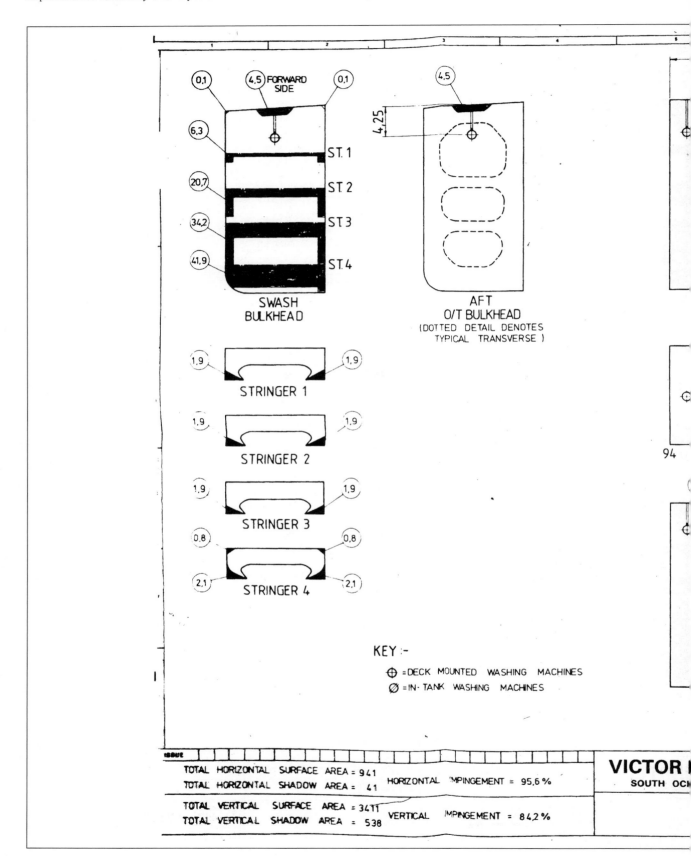

Plan 8.2 (a) ... and in Numbers Two and Six Wing tanks on the Same Supertanker.
(By kind permission of Victor Pyrate.)

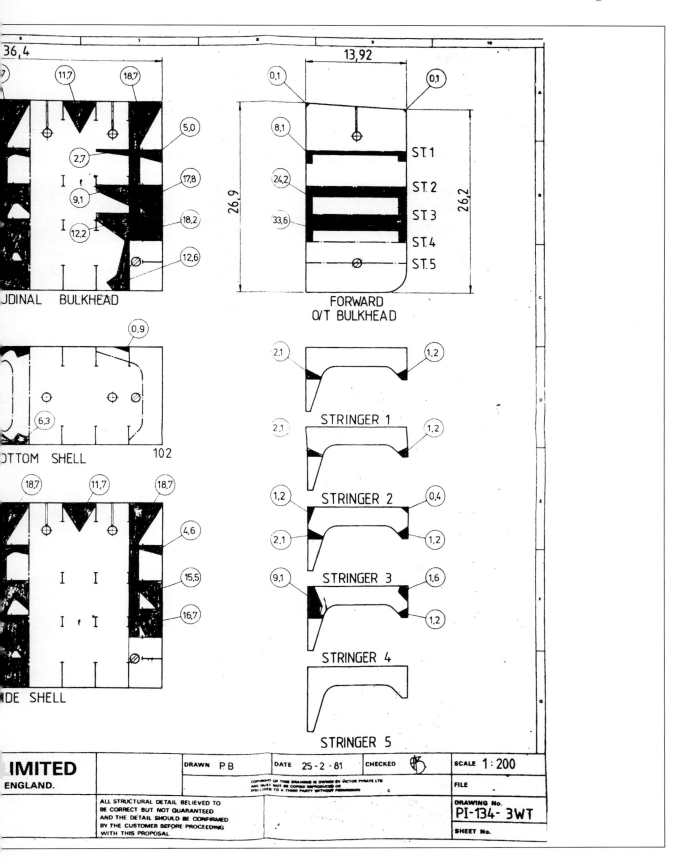

Plan 8.2 (b)

8.10. TANK CLEANING MACHINES – HOW MANY?

It is interesting to compare the numbers of deck and submerged machines in this 'series' of rather complex fittings with other VLCC's of a similar size which, for clarification, might be conveniently termed 'standard' class. Invariably, it is the needs of individual ships in the light of their unique construction, particularly regarding wing tanks (See Chapter 2) that determines the greater or lesser number of machines that it is necessary to fit. The distribution plan, (Table 8.1), of machines in the 'standard' VLCC of similar deadweight tonnage shows a combination of Victor Pyrate's machines, deck mounted, with also submerged machines. The numbers of top mounted machines serving the centre tanks compares quite favourably, there are 24 fitted to the 'standard' ships as opposed to 28 on each of the ships in the series. (Table 8.2 and Table 8.3). The numbers of submerged machines in the centres, however, differs considerably with 13 machines against 98 necessary for each of the 'series'. An even larger difference is to be found in the wing tanks on a 'standard', with 32 deck fitted and 20 submerged, compared to 51 top and only 14 submerged. Thus a total of 56 top and 41 submerged machines might be required to clean efficiently a VLCC of a 'standard' construction, compared to the 79 and 112 necessary for the extra steel distribution in the lower parts of the special constructions undertaken. It should be stressed again that the 'standard' VLCC, used to demonstrate this comparison, would itself be of a unique construction. It could be pointed out also that number four wing tanks on each of the ships considered has been reserved for permanent ballast and would, therefore, not require to be fitted with cleaning machines.

NUMBERS OF DECK AND SUBMERGED CLEANING MACHINES ON A 270,000 sdwt VLCC		
Tank Number	Top Machines	Submerged Machines
1 Centre	3	1
2 Centre	4	5
3 Centre	6	1
4 Centre	4	5
5 Centre	5	7
6 Centre	2	2
1 P/S Wings	8	4
2 P/S Wings	6	4
3 P/S Wings	6	4
5 P/S Wings	6	4
6 P/S Wings	6	4
TOTALS:	56	41

Table 8.1 Distribution of Top and Submerged Machines on a 270,000 sdwt VLCC.
(By kind permission of Mobil Ship Management.)

C.O.W. SUMMARY
SHEET 1

TANK	IMPINGED SURFACE		No. OF MACHINES	
	VERTICAL %	HORIZONTAL %	DECK	IN-TANK
CENTRE TANK				
No. 1	96.9	95.7	3 *	–
No. 2	93	97.8	4	14
No. 3	93.9	97.8	4	14
No. 4	93.9	97.8	4	14
No. 5	93.9	97.8	4	14
No. 6	93.9	97.8	4	14
No. 7	94.4	98.4	6	28

*(OFF MONOMATIC (SEE SHEET 2)

MACHINE SPECIFICATIONS:—	TOTAL Nos.	
VP Matic 32mm Nozzle 4m Del. Tube Jet Length		
34m Thro'put 118M.TON/HR @ 10Kg/cm^2 = 4.82 Hz	28	
VP Major 11 Jet Length 10m Thro'put		
28M.TON/HR @ 10 Kg/cm^2		98.

Table 8.2 *Percentile Horizontal and Vertical Impingement of Deck and Submerged Tank Cleaning Machines in Centre Tanks of a 264,000 sdwt VLCC ...*
(By kind permission of Victor Pyrate.)

C.O.W. SUMMARY
SHEET 2

TANK	IMPINGED SURFACE		No. OF MACHINES	
	VERTICAL %	HORIZONTAL %	DECK	IN-TANK
WING TANK				
No. 1 P & S	95.6	97.8	3	–
No. 2 P & S	84.1	95.6	4	1
No. 3 P & S	84.2	95.6	4	1
No. 5 P & S	80.1	92.9	4	1
No. 6 P & S	84.2	95.6	4	1
No. 7 P & S	81.7	92	6	3
AVERAGE TANK	88.4	95.9		
AVERAGE AREA	88.6	96.2		

* INCLUDES 1 OFF MONOMATIC IN NO 1 CT

MACHINE SPECIFICATIONS:—	TOTAL Nos.	
VP Monomatic 32mm Nozzle 4m Del. Tube		14
Jet Length 34m Thro'put 118M.TONS/HR		
@ 10 Kg/cm^2 = 4.82 Hz.	51*	
	79	112

Table 8.3 ... and for the Sets of Wing Tanks on the same Ship.
(By kind permission of Victor Pyrate.)

Tank Cleaning Procedures

8.11. LINES

The following diagram indicates the crude oil washing lines aboard a large VLCC belonging to a major operating shipping company (Draw. 8.1) and indicates also the position of the valves and tank washing machines. The drawing is quite self explanatory. Each of the centre tanks is served by a set of four points with each wing tank having two. The slop tanks on each of the port and starboard sides are served by one point. A companion diagram, (Draw. 8.2), shows the positions of the hand dipping holes and tank gauges. Again, the distribution is quite clearly indicated. The above diagrams and plans, together with previous explanations, assists understanding of the arrangement of pipes and lines on the main deck. In the following photographs, (Photo. 8.20 and 8.21), for example, the foam monitors and cargo manifolds are now readily recognisable, together with the vent to the pump room and the hand valves for the port wing tanks situated on the cross over walkway. In the foreground, are the tank tops, tank cleaning and COW/IGS lines, as well as the vapour sampling points and hand dipping holes.

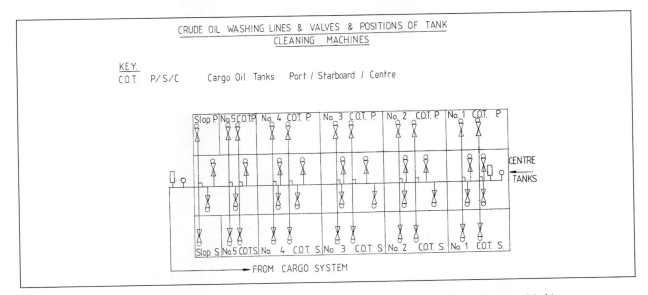

Draw. 8.1 The Crude Oil Washing Lines and Valves and the Positions of Tank Cleaning Machines.
(By kind permission of Shell International Trading and Shipping and Paul Taylor.)

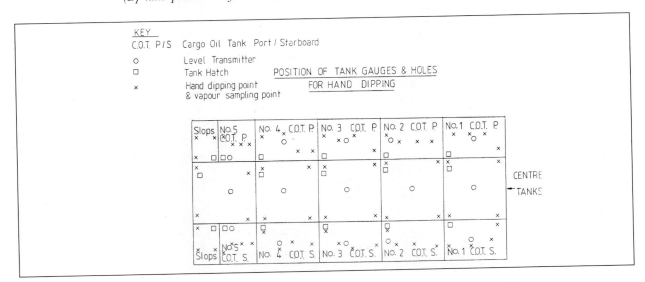

Draw. 8.2 Positions of Tank Gauges and Holes for Hand Dipping on a VLCC.
(By kind permission of Shell International Trading and Shipping and Paul Taylor.)

*8.20. Distribution of Pipes and Lines on Main Deck (1) ...
(By kind permission of Mobil Ship Management.)*

*8.21 ... and Distribution of Vapour sampling pipes, Tank Cleaning and Vent Pipes (2).
(By kind permission of Mobil Ship Management.)*

A typical crude oil washing programme operated by a leading tanker company, (Prog. 8.1), indicates the crude oil sequence in relation to the various stages of cargo discharge and represents a basic procedure that is readily adapted to virtually all conditions of cargoes and ports. The sequence number indicates the operation undertaken with the crude oil wash, as the cargo discharge operation proceeds, and indicates also the various pump sequences utilised. Sequence number 34 refers to a "special small diameter stripping line" as the final stage in discharge. This is called the 'MARPOL' line which is used to eject the last residue of cargo from the tanks. It is operated by opening a vacuum breaker on the main lines, thus letting in more air to assist with the draining.

Quite clearly, safety at sea has always had to be of paramount importance and hard won lessons were learnt from the series of explosions that had occurred in the late-1960s, during tank cleaning operations, which had resulted in loss of human life and severe damage to ships. For some years prior to 1925, and in the years prior to the Second World War, various kinds of tests had taken place to see how a neutralising agent could be introduced into tanks which would enhance safety. British and American shipping companies had experimented with the use of flue gases from a tanker's engine exhaust to create an inert gas by pumping carbon monoxide or nitrogen mixtures into partially and fully empty oil tanks, so reducing oxygen contents in the air to acceptable levels, and thus virtually eliminating risk of explosion. These experiments were resurrected in the 1960's by major oil companies associated within the International Chamber of Shipping (ICS) and the Oil Companies International Marine Forum (OCIMF).

8.12. INERT GAS SYSTEMS (I.G.S.), INERTING

The Classification Societies – especially the Technical Association of Lloyd's Register, as early as the 1950's, had also launched extensive research into tank cleaning practices. Collected findings of such interested working groups were ultimately presented as Protocols by IMO, directed initially at new-tonnage tankers, in excess of 100,000 dwt, a figure based on the assumption that only ships in this class would use the more powerful tank cleaning equipment that worked in excess of 60 m^3/hr, which seemed to be a cause of the problem. Far more universal adoptions were made by IMO in Regulation 62 of the 1974 SOLAS recommendations and, in 1981, it had become compulsory for all 'new and existing tankers' carrying crude oil, regardless of size, to be fitted with what were then the new 'Inert Gas Systems' (IGS). The Introductory paragraph to Chapter 11 – 2 of Regulation 62 defined the function of an inert gas system. This ruling stated that:

"The inert gas system referred to in para (a) of Regulation 60 of this chapter, shall be capable of providing on demand a gas or mixture of gases to the cargo tanks so deficient in oxygen that the atmosphere within a tank may be rendered inert, i.e. incapable of propagating a flame."

In practical terms, any gases used had to reduce oxygen levels to around 5%, or less, and so neutralise hydrocarbons so that they would be incapable of combustion. This is of course not only essential for safety, but also a legal requirement that under no circumstances should a crude oil wash be undertaken without the inert gas system being operated. The gas pressure should be raised initially to around 1000mm W.G. and this should be checked every fifteen minutes during the course of the wash. Paragraph 6.6. of IMO Regulations stipulated the requirements for close monitoring of oxygen levels within the inert gas:

"Before each tank is washed, the oxygen content should be checked with the portable oxygen analyser to ensure that it does not exceed 8% by volume. The sample should be taken at a point one metre below the deck and at the middle region of the ullage space well clear of the inert gas inlet point."

Supertankers Anatomy and Operation

TYPICAL CRUDE OIL WASHING PROGRAMME

This section contains some typical washing programmes under various conditions of discharge such as single or multiport discharge and single and multigrade cargoes.

Crude Oil Washing Sequence In Relation To Cargo Discharge

The following sequence is the basic programme and is easily adapted to all conditions of cargoes and ports.

SEQUENCE NUMBER	CARGO OPERATION	PUMP	COW OPERATION	PUMP
1	Commence discharge of Centre Slop Tank	No. 1		
2	Commence discharge of Nos. 1, 2, 3 & 4 Wing Tanks	Nos. 1, 2, 3 & 4		
3			Wash Centre Slop Tank	No. 1
4			8 Metre ullage. Wash 1 & 2 wings 1.5 cycles 120°-70°	GP on 4 wings.
5	Refill centre slop tank with dry crude using direct load line			
6			Wash 3 & 4 wings 1.5 cycles 120°-70°	GP on 4 wings
7			17 metre ullage Wash 1 & 2 wings 1.5 cycles 70°-40°	GP on 4 wings
8			Wash 3 & 4 wings 1.5 cycles 70°-40°	GP on 4 wings

Programme 8.1 (a) A Typical Crude Oil Washing Programme for a VLCC.
(By kind permission of British Petroleum Shipping.)

9			26 metres ullage. Wash 1 & 2 wings 1.5 cycles 40°-0°	GP on 4 wings
10	Transfer No. 1 pump to discharge wing slop tanks.	No. 1 Pump		
11			Wash 3 wings 1.5 cycles 40°-0°	GP on 4 wings
12	Transfer No. 2, 3 & 4 pumps to centre tanks. Discharge 1, 2, 3 & 4	Nos. 2, 3 & 4		
13			Wash requisite wing slop tank 1.5 cycles 120°-0° and wash 4 wings in rotation 1.5 cycles 40°-0°	GP on centre tank
14	Transfer No. 1 pump to discharge centre tanks.	No. 1		
15	Strip 4 wings and wing slop tanks using educators.	GP on centre Slop tank		
16	Stop 1 & 4 cargo pumps. Drain Starboard deck line into wing slop tank. Drain drop lines into wing tanks.			
17	Strip all wing tanks to centre slop tank using ring main suctions and direct suctions. Check dips in after end of wing tanks less than 25mm	Stripping pump		

Programme 8.1 (b)

18	Strip 1 & 4 pumps plus their suctions and discharge lines to centre slop tank	Stripping pump		
19	Commence ballasting wing tanks using Nos. 1 & 4	Nos. 1 & 4		
20			9 metre ullage. Wash 1, 2 & 3 cente tanks 1.5 cycles 120°-70°	GP on 4 centre
21			Wash 4 centre tank 1.5 cycles 120°-70°	GP on 4 centre
22			17 metre ullage. Wash 1, 2 & 3 centre tanks. 1.5 cycles 70°-40°	GP on 4 centre
23			Wash 4 centre tank 1.5 cycles 70°-40°	GP on 4 centre
24	Complete ballasting into wing tanks. Nos. 1 & 4 pumps stopped and their valves shut		24 metres ullage. Wash 1, 2 & 3 centre tanks. 1.5 cycles 40°-0°	GP on centre tank
25	Stop discharge of cargo for bottom washing of 4 centre tank.			
26			Wash 4 centre tank. 1.5 cycles. 40°-0°.	GP on centre slop tank
27	Strip 4 centre to centre slop tank using educators.	GP on Centre Slop tank.		

Programme 8.1 (c)

28	Drop port deck line to centre slop tank and drain drop lines into centre tanks.			
29	Strip all centre tanks to centre slop tank using ring main suctions or direct suctions.	Stripping Pump		
30	Check dips in after end of centre tanks less than 25mm.			
31	Strip deck lines, pumproom lines (except GP pump lines) and Nos. 2 & 3 pumps to Centre Slop Tank.	Stripping Pump		
32	Discharge Centre Slop Tank to shore.	GP Pump		
33	Strip GP Pump, lines and manifold to centre Slop Tank	Stripping Pump		
34	Discharge slop tank to shore using the special small diameter stripping line. "MARPOL" Line	Stripping Pump		

Programme 8.1 (d)

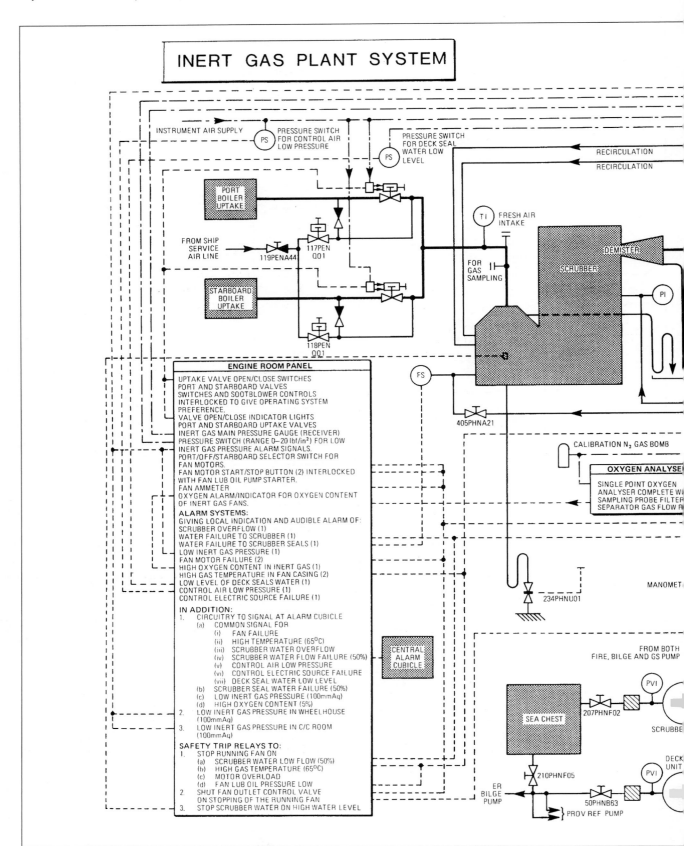

Diag., 8.9 (a) *Typical Inert Gas Plant System on board a 250,000 sdwt VLCC.*
(By kind permission of British Petroleum Shipping.)

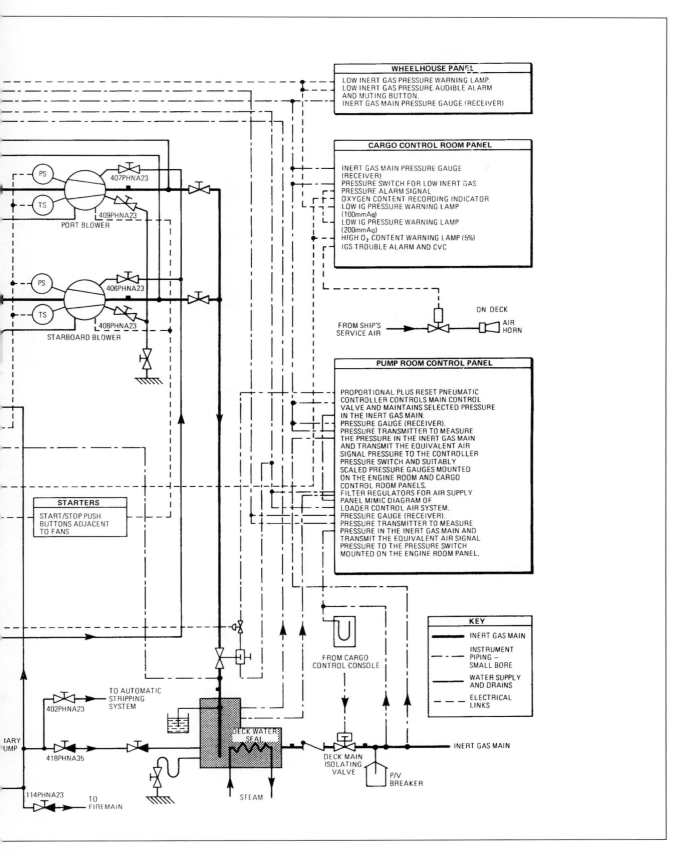

Diag., 8.9 (b)

If the oxygen content is noted to rise to 5%, or the IG pressure falls to around 100mm W.G after which, IMO instructions command that: *"the washing operation must be stopped immediately until the IG pressure is restored."* The following table (Table 8.4) indicates the operating pressures, under varying circumstances in which the IG system is used, recommended by a leading VLCC shipowner to its officers, based on their experiences and regulated within IMO/SOLAS specifications.

Generally, an inert gas system aboard an average VLCC is derived from a flue uptake from the ship's boilers, usually the main, which has the following properties. There is an element of around 80% nitrogen (N_2) content. Of the remaining approximately 20%, between 13 and 14% is carbon dioxide (CO_2). The oxygen (O_2) content is 2–4%, with 0.2–0.3% sulphur dioxide (SO_2), and traces only of carbon monoxide (CO) and oxides of nitrogen (NOX). There are, of course, elements of water vapour and some minute solid particles. Inert Gas generators, found commonly aboard chemical and liquid gas tankers, are often fitted to Supertankers. Rarely, (except in the case of motor-driven VLCC's) as an independent supply but, more usually, as a topping-up unit in conjunction with flue systems from boilers, to save the main system on occasions when only a short "booster-burst" is required in order to maintain pressure within the tanks. This proves especially useful during passage when the cargo pumps are inoperative, hence the boiler is working only partially, and there would be a need to fire an additional boiler.

IMPORTANT OPERATING PRESSURES IN INERT GAS SYSTEMS

IG Deck Pressure mm.w.g.	Remarks
1600	P/V breaker blows; Slop tank P/V valves lift
1500	High level alarm
1400	P/V valves lift
1000	Recommended for cargo discharge and tank washing
500	Maximum for opening purge pipe and high velocity vent covers
300	Maximum for opening horizontally swinging hatch covers
175	High velocity vents open
100	Low level alarm
50	Maximum for opening vertically swinging hatch covers
−250	P/V valves break vacuum
−400	P/V breaker blows

Table 8.4
(By kind permission of Shell International Trading and Shipping.)

Tank Cleaning Procedures

*8.22. The Inert Gas Scrubber Tower draws up the gas from the Bottom Through a Water Seal.
(By kind permission of FGI Systems Ltd & John Mills (Photographer) Ltd.)*

*8.23 Internal View of the Scrubber Tower Looking Upwards through the Packing of Baffle Plates.
(By kind permission of FGI Sytems Ltd & John Mills (Photographer) Ltd.)*

Supertankers Anatomy and Operation

A number of different designs of inert gas systems are available, as well as a variety of methods of pumping gas into the tanks. FGI Systems, of Leatherhead in Surrey, specialise in inert gas system design (which they have acquired from Howden's). They also manufacture and service the VLCC and wider tanker trades. Their equipment introduces boiler exhaust flue gas, through deck lines over the cargo as it is discharged or after loading. It operates via a closed system so that all oxygen is totally excluded. To do this, the IGS consists of two main components. There is the plant contained in the engine room whose function is to draw, cool and purify the gas in non-hazardous conditions. (Diag. 8.9). The second component is concerned with distributing the inert gas to the tanks, an operation which is potentially more dangerous. This latter function is examined later. Amongst other important features, the system must be capable of regulating the gas, so that it is delivered in the quantities required as it is needed, and to prevent any oil gases existent within the lines or tanks back flowing to enter into the system itself. The final requirement demands that, on occasions when the system is used for tank venting, gas blowers must be used to introduce air into the pipes and tanks. It has been shown, in Chapter 5, that close control and monitoring of the IG system is regulated from the Cargo Control Room (CCR), with an additional monitor fitted in the wheelhouse of all VLCC's.

Using flue gas implies that the gases will be hot, wet, and dirty with soot and other particles. The first requirement, therefore, is to prepare the gas so that, when it is introduced into the tanks, it will be in a more refined condition and, vitally important, that it will be of a known quality that is sufficient to reduce oxygen to the point where combustion is impossible, regardless of the hydrocarbon concentration within the tank. The first stage in this process requires a remotely operated butterfly uptake valve to be fitted into the flue pipe in the base of the funnel. (Diag. 8.10). This provides the tapping-in point necessary to draw-off the hot gas and also isolates the system from the boiler when it is not in use. The valve has to be extremely robust to prevent any possibility of gas entering the scrubber and causing corrosion and it is invariably constructed with a hermetically sealed double disc. The gas is then piped into an adjacent Scrubber-Tower (Photo. 8.22), which is a rectangular tower mounted on a broad base tank, where it undergoes a series of three refining processes. The gas enters at the bottom of the tower through a submerged inlet water seal and is drawn upwards through a packing of rhomboid-shaped water soaked baffle plates. (Photo 8.23). The water is forced, by a powerful 290 tonnes per hour pump, from the sea-water intake, although the quantity of water will vary with the system used, into the top of the tower and is distributed evenly throughout the packing by a series of very high capacity course water spray nozzles. The water cascades downwards through the tower into effluent trays in the base tank which drain overboard via a discharge valve taking with it virtually all solid particles, such as soot and any alien pieces. Up to 99% of the sulphur dioxide is also removed in this process. The entire interior of the scrubbing tower, including all packing, is protected against the effects of corrosion and contamination as, although

Diag. 8.10 The Flue Gas Valve Provides the "Tapping-In" Point and Isolates the IG System from the Boiler on occasions when the system is not being Used. (By kind permission of Fredrikstad IGS.)

8.24. A Scrubbing Tower and De-Mister Unit seen from Different Angles …
(By kind permission of FGI Systems Ltd & John Mills (Photographer) Ltd.)

8.25. … (2).
(By kind permission of FGI Sytems Ltd & John Mills (Photographer) Ltd.)

Supertankers Anatomy and Operation

8.26. The Main IG Fan and ...
(By kind permission of British Petroleum Shipping.)

8.27. ... Motor Assembly Unit.
(By kind permission of FGI Sytems Ltd & John Mills (Photographer) Ltd.)

the quantities of sulphur dioxide are small, the gas is transformed in the water thus releasing sulphuric acid. The inside of the tank is treated, generally, with a fibre-glass reinforced epoxy resin coating. Sometimes rubber is used as a protecting layer. The sea water pipes, nozzles and valves are often made of Incoloy. It is interesting to note that the over board discharge pipe can be made of stainless steel and, in these cases, is one of the few non domestic items of this material to be found on board ships. There are two alarm systems fitted to most scrubbing towers. One, on the sea-water line which becomes activated if, for example, the water flow drops below an accepted limit or, if any temperatures rise above the norm. In these instances, the water supply is automatically discontinued and the entire system shuts down. The sea-water line is fitted also with a pressure gauge and a flow indicator. Another alarm is fitted at the base of the tower which alerts engine-room personnel in an emergency. The cooler, cleaner, saturated flue gas then passes into a demister unit, at the top of the scrubber, which subjects it to a swirling motion so that any water residue is thrown clear of the gas, along with any occasional particles that might not have been removed as the gas passed through the scrubber. Only the Howden IG System, in fact, incorporates this type of demister as other systems use mesh demister pads. Two different views are shown of the scrubber and demister (Photo. 8.24 and 8.25) as these particular items of equipment were assembled at FGI's factory before despatch to a ship. The pipework and valves can be seen clearly.

The next stage is for the now refined gas to be piped through inlet valves from the top of the scrubber into fans or blowers before entering the pipe lines on the main deck. (Photo. 8.26 and 8.27). Normally, the most common arrangement is for two blowers each of full capacity to be used. One can act as a standby to the other but, occasionally as an alternative, two blowers of 50% capacity operate in parallel. If one fan is out of service, however, this will have the effect of reducing the rate of discharge of the cargo. Fans aboard turbine driven VLCC's were steam driven, but nowadays, they are usually both motor driven around 115 hp — and, at 12,500 M³/hr. Sometimes one is used whilst the ship is at sea to meet the requirements of slower 'topping-up' purposes. The blower has gas valves so that all operations, using this motor only, can be used remotely from the IG equipment control room. The fan serves a vitally important function because it ensures that the gas remains constant and stable throughout its entire flow. The gas is now ready to be pumped into the tanks but, before this occurs, it has to pass through the gas regulating, or pressure control, butterfly valve which, to help maintain a constant pressure to the tanks, may be operated either manually or remotely. Often, an additional valve is fitted for use when the regulating valve is closed. This is a recirculating valve which discharges the inert gas into the atmosphere via a pipe inside the funnel casing. This valve operates in conjunction with the main regulating control valve. It is used also when starting up the IG system in order to stabilise and check the oxygen content of the gas before passing the gas to the tanks. On older systems, the valve was fitted just before the main-deck tanks started, and the 'hazardous' area of inerting commenced, and it led the gas through a submerged inlet into the deck water seal which was situated on the main deck immediately forward of the accommodation block. (Photo 8.28).

The second part of the IG System is the hazardous area that is concerned with distributing the inert gas to the tanks. The important function of the deck seal is to prevent any reverse flow of hydrocarbon gas from the cargo tanks to the boilers and engine-room machinery spaces so that it can never enter the non-hazardous area. It is useful also in the event of fan failure or, if plant shut-down happens for any reason, then the gas inlet pipe into the seal is immersed in water to a depth that it greater than the pressure in the top half. There are a number of different kinds of seal in use. In the older type, the top part consisted of a drop tank through which the water was piped, via a number of hydraulically operated valves, into the larger lower seal tank. The following photographs show two different views of a typical deck seal (Photos. 8.29 and 8.30) from which the inlet and outlet pipes can be clearly seen. Although some operate with dry seals, the majority rely upon the provision of water, taken from the

Supertankers Anatomy and Operation

8.28. The Deck Water Seal forward of the Accommodation Unit on a VLCC.
(By kind permission of Mobil Ship Management.)

8.29. Two Views of a Deck Seal in the
Manufacturer's Shop ...
(By kind permission of FGI Sytems Ltd & John Mills
(Photographer) Ltd.)

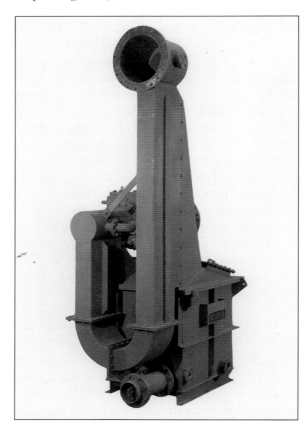

8.30. ... Awaiting despatch to a Tanker.
(By kind permission of FGI Sytems Ltd & John Mills
(Photographer) Ltd.)

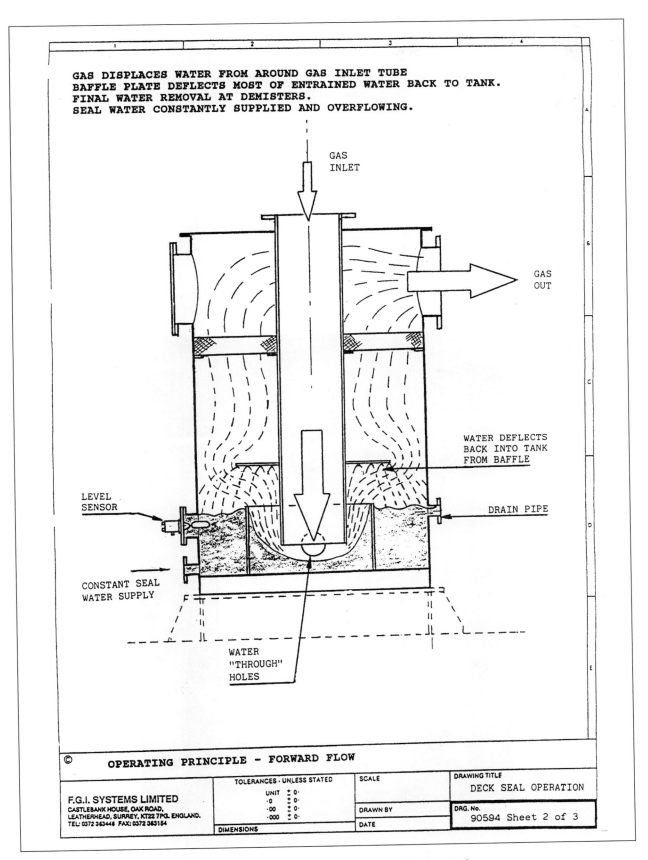

Diag. 8.11 The Operating Principle of a Deck Water Seal.
(By kind permission of FGI Sytems Ltd.)

sea-water main uptake in the engine room, in order to be effective. The operation of the water through the seal can be seen clearly in the diagram above. (Diag. 8.11). The seal often consists of a mild steel rubber-lined gas duct that is connected to an external reservoir. Again, as with the scrubber tower, the internal surfaces of both chamber and tank are coated with bituminous epoxy, of about 300 microns, for protection against corrosion. A steam heating coil is incorporated to prevent freezing on those rare occasions when the IG system is used in extreme cold weather conditions. A low level float switch is fitted so that when the water drops below a certain depth in the reservoir an alarm is raised. The latest model in wet type deck seals (Photo. 8.31) is shown undergoing tests before despatch.

The inert gas is forced through the water from the deck seal into the deck distribution system. The following plan (Diag. 8.12) indicates the pipe lines and various components which take the gas from the deck seal along the main deck. Considerable safeguards are built into the system in order to prevent accidents. The first of these is a main non return isolating valve whose main function is to vent any gases that may remain in the pipework between the deck seal and the main regulating valve. It prevents also any hydrocarbon gases that may have got through the deck water seal from seeping into the system. A flame screen is also incorporated that is always open when the IG system is not in use. (Photo. 8.32). The valve performs the

8.31. The Latest design of a wet Type Deck Seal Undergoing Pre-Despatch Tests.
(By kind permission of FGI System and John Mills (Photographer) Ltd.)

Tank Cleaning Procedures

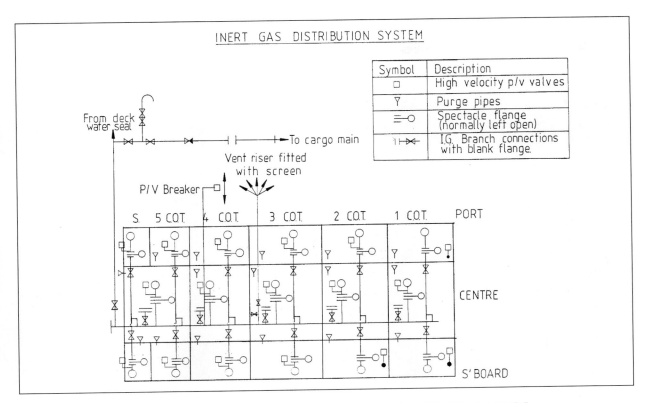

*Diag. 8.12 The IG Deck Distribution System on a Modern 290,000 sdwt VLCC.
(By kind permission of Shell International Trading and Shipping and Paul Taylor.)*

*8.32. The Main IG Non-Return Vent Isolating Valve aboard a 215,000 sdwt Tanker.
(By kind permission of British Petroleum Shipping.)*

Supertankers Anatomy and Operation

8.33. IG Main Piping System along the Main Deck of a VLCC ...
(By kind permission of British Petroleum Shipping.)

8.34. ... with Branches Serving the Various Cargo Tanks. (By kind permission of British Petroleum Shipping.)

8.35. High Velocity "Bullets" or Pressure Vacuum Valves Serving Each Cargo Tank.
(The Author)

8.36. IG Purge Pipe on a Very Large Tanker.
(By kind permission of British Petroleum Shipping.)

258

8.37. High Pressure Vacuum Valve on the starboard Side Wing Tank.
(The Author)

8.38. Liquid Filled Pressure Vacuum Breaker, Tested
and Awaiting Despatch.
(By kind permission of FGI Sytems Ltd & John Mills
(Photographer) Ltd.)

8.39. Oil-Filled Pressure Vacuum Breaker
fitted into the IG System on a VLCC.
(By kind permission of British Petroleum Shipping.)

Supertankers Anatomy and Operation

8.40. The Chief Officer shows a Deck Cadet how to Analyse the CO2 Content of IG with a FYRITE Indicating Measure.
(By kind permission of British Petroleum Shipping.)

8.41. A Vent Riser Fitted to the Midships Port Derrick Mast Carries the IG out of the System.
(By kind permission of Mobil Ship Management.)

vital function of preventing hydrocarbon gas from seeping back into the scrubber and possibly the blower fans. The inert gas is then blown into a 700 mm gas main pipe, the size of which may vary, depending on the size of the system, which passes along the maindeck to lead-off into each of the wing and centre cargo tanks. (Photos. 8.33 and 8.34). Obviously, there is no need for the permanent ballast tanks to be connected to the system for inerting so their piping system remains isolated.

The inert gas is piped into tanks that are full of air which, of course, contains oxygen. On most Supertankers, the gas enters under sufficient pressure to cause a turbulence that forces it to the bottom of the tank from whence it mixes with the air ultimately to dilute it completely. The expelled air can exit from the tank via a purge pipe or, more usually, a high velocity pressure vacuum valve, or 'bullet', as they are known for obvious reasons, at the top of the tank. (Photos. 8.35, 8.36 and 8.37). An alternative method of inerting, used on some tankers, is to inject the gas into the tank by less forceful means which displaces the air to the bottom of the tank from whence it exists, again via a purge pipe or PV valve, but one which extends from near the bottom of the tank. The risk of over-pressurisation in tanks and system lines is an ever present danger and, in order to eliminate this, pressure vacuum valves and a gas PV breaker is introduced, which may be liquid filled if the size of the PV valves permit this. (Photos. 8.38 and 8.39). The first of the photographs shows the PV in the manufacturer's yard whilst the second indicates how it is fitted into the system aboard a VLCC.

It is a common practice for the PV breaker to consist of a glycol/water mixture to the order of 25%/75% respectively. The glycol is incorporated in order to prevent freezing and often a minute amount of lubricating oil can be added which forms a thin film on the top of the mixture and helps prevent evaporation. The breaker acts also as a pollution preventer in the event of the breaker blowing thus causing cargo to flood over the deck. An oxygen analyser is invariably added to the system, the function of which is to monitor that the correct quality of inert gas is being introduced into the system. Should the oxygen concentration deviate from a certain predetermined level then an alarm is activated and the system is automatically tripped. Portable monitors are also available for use in the cargo tanks and, in the following photograph, a chief officer aboard a 215,000 sdwt VLCC is shown instructing a deck cadet in the procedure for checking the oxygen levels of inert gas with a portable FYRITE indicator. (Photo. 8.40). It is interesting to note that a leading tanker operator wisely instructs his personnel to beware of the danger of using the correct glycol additive warning against using in error the 'other glycol', a form of paint stripper, so that this is not introduced instead.

The inert gas is carried out of the system lines by a 'vent riser' that is fitted with a spark screen. It is invariably a few metres in height and is mounted on to the mast tops which support the derricks and carries it well clear of the main deck. (Photo. 8.41). The officer servicing the riser gives an indication of the dimensions involved.

8.13. PORTABLE FANS

The cargo tanks are normally kept in an inerted condition permanently, but occasionally it is necessary to enter some of the tanks either for inspection purposes or to carry out repairs. It is also essential for the VLCC to enter dry dock every thirty months for a survey. Any of these occasions indicates that either all, or a set, of the tanks have to be gas freed. There are two main methods of doing this: either by using portable fans or the ship's inert gas system itself. There are a number of powerful portable models on the market which can be used to create a flow of fresh air throughout the tank. They have a mutual facility of either extracting or infusing air, thus dispelling any possible residual vapour pockets and eliminating the mustiness in the tank caused by the inerting process. The portable models are driven either by air, hydraulics or water from the ship's deck lines. They are surprisingly lightweight, to the order of 40 or so kilograms, and measure around 700 x 600 mm, which makes one man operation quite feasible. They are usually employed when only a tank or set of tanks need to be gas freed. The method of isolating the tanks is to close the spectacle flanges, (See Diag. 8.12),

8.42. A "JETSAN 125" Portable Gas Freeing Fan.
(By kind permission of Dasic Marine.)

8.43. Portable Gas Freeing Fans.
(By kind permission of Dasic Marine.)

in the IG distribution system covering the tanks to be entered. The above photographs offer an idea of the construction of gas freeing fans. (Photos. 8.42 and 8.43). Nowadays, it is possible to close the boiler uptake valves and use the ship's own IG System to gas free the tanks with air. Prior to starting the system, the blank flange is removed from the gas outlet pipe from the scrubber and the fresh air inlet valve opened.

8.14. EFFECTIVENESS OF C.O.W. AND I.G.S.

The effectiveness of crude oil washing can be seen in the following range of photographs which make interesting comparisons with those used to commence this chapter which showed the range and extent of the sludge. (Photos. 8.44, 8.45, 8.46 and 8.47). The areas which were covered can be seen to be completely clean and, one feels, would quite satisfy the inspectors as they examine the bottom of the tank and the stringers. (Photos. 8.48 and 8.49).

The use of COW and IGS have undoubtedly revolutionised the safety and efficiency of tank cleaning operations, with a corresponding benefit regarding the saving of time and manpower. In keeping with virtually all tasks that are performed at sea, however, considerable caution and attention must be paid to the regulations and operating procedures that have been tested, proved and laid down as safe practices. There are a number of recorded incidences where momentary lapses, by totally responsible officers during the ordinary course of duties whilst loading or discharging, have led to potentially dangerous situations. Whilst taking on cargo at an Iranian Gulf port on one occasion, for instance, a valve was not closed completely with the result that about ten tonnes of crude oil was allowed to escape over the main deck before the incident was noticed. The flush main deck on all VLCC's all too readily permits such accidental discharges

Tank Cleaning Procedures

8.44. *Series of Photographs showing Crude Oil Cargo Tanks aboard a VLCC following a Successful Tank Cleaning Operation.*
(By kind permission of Victor Pyrate.)

8.45. ... 2 ...
(By kind permission of Victor Pyrate.)

8.46. ... 3 ...
(By kind permission of Victor Pyrate.)

8.47. ... 4 ...
(By kind permission of Victor Pyrate.)

8.48. Cargo Tank Inspections indicate the Efficiency of Tank Cleaning Operations ...
(By kind permission of Victor Pyrate.)

8.49. ... and Prepare for Inspections in Dry Dock Prior to commencement of a New Voyage.
(By kind permission of Victor Pyrate.)

to flow over the ship's side back into the Gulf from whence they came. The resulting heavy fine imposed on the master was not greeted by him with particular enthusiasm, nor by his insurance company.

Similarly, when using inert gas, there can be no room for 'short cuts' because, if it is not used correctly, the system can prove disastrous. There have been a number of incidents over the years that have been examined by the Marine Accident Investigation Branch of the Department of Transport, which includes similar accidents that occurred aboard two Supertankers using IGS whilst discharging crude oil.

Chapter Nine

DECK DUTIES

9.1.   INTRODUCTION

9.2.   PAINTING

9.3.   ALLOCATION OF DUTIES

9.4.   LIFEBOAT DRILLS

9.5.   FIRE-FIGHTING EQUIPMENT

9.6.   STRETCHER PARTY EXERCISE

9.7.   SOME GENERAL OBSERVATIONS

9.8.   HELICOPTER OPERATION

9.9.   PUMP-ROOM BLOW OUT

9.10. SOME UNUSUAL DUTIES FOR OFFICERS, INSPECTIONS

9.11. SEA-TRIALS AND THE "HANDING-OVER" CEREMONY

Chapter Nine

DECK DUTIES

(Introductory Remarks – Routine Maintenance Painting – Weekly Meetings – Lifeboat Drills – Fire-Fighting Exercises: Foam Blanketing – Use of Sea Water – Cargo Tank Stretcher and Rescue Exercise: Incidents of Cargo Tank Rescues: Successful and Tragic – Deck Work on Passage – Helicopter Operations and Practice – Officer of the Day – Handing Over Inspections – After Peak Inspection – Paint Film Testing – Handing Over Ceremony and Celebrations.)

9.1. INTRODUCTION

The range of deck work performed by officers is varied and includes supervision of many duties carried out by the crew which should have been undertaken by the officer whilst he or she was a cadet. Such work can range from rigging of gangways, inspection and maintenance of deck gear and equipment, to taking charge of helicopter or docking operations. Where appropriate, many of these duties are described in the chapter most appropriate and relevant to the task, leaving only those which do not fit readily elsewhere to be considered here.

9.2. PAINTING

Much of the chief officer's time, when one of the 'early' generation of VLCC's was on passage, would have been taken up with organising the crew-members in the fight against weather and sea/air effects. It remains an essential, full-time and never-ending job and, because of the vastness of the surface areas involved, it is rather akin to painting a moving apocryphal Forth Bridge. There is not only the anti-element to consider even though, as discussed in Chapter 3, a great deal of this has to be outside the control of the ship's officers. The functional Supertanker is invariably one that looks clean and smart, particularly in major areas of concern such as main deck, superstructure and funnel. These require various levels of maintenance virtually before the maiden voyage has been completed up until the time when the VLCC is finally scrapped. The ship, after all, is the home of the officers and crew for the duration of their time on board whilst, for the owner, a pleasing appearance is not only good publicity, but helps reduce repair costs.

As it was explained briefly, in Chapter 2, ships are made from various types of steel which means that deck scaling, paint chipping and scraping and then the interminable undercoating of primer followed by painting, become the norm. The main deck areas are those upon which a ready eye has to be kept because this is the main working space on the ship. Virtually everything and everyone come into contact with the main deck; cargo is worked on it, berthing and mooring duties performed there as well as countless routines undertaken, so it is the most used and abused part of the ship. It requires therefore the maximum maintenance and, in the days when a larger number of deck or general purpose crew were available than is often the case

Supertankers Anatomy and Operation

*9.1. Crew Members Scraping and De-rusting Part of the Main Deck.
(By kind permission of British Petroleum Shipping.)*

today, this was a major deck officer's task. Current practices are often to put aboard a shore crew, who may well be ex-mariners, who live on the ship for periods of up to six months working specifically on the routine cleaning and painting necessary. Rust is the main enemy for the seafarer and when it is noticed in the 'pimply' stage, then little more is necessary than a light wire-brushing and a protective layer of paint. In the picture of crew working at these tasks, (Photo. 9.1), the condition has deteriorated to the extent that an hydraulically operated scaling machine has to be used. A member of the crew operates the scaler whilst the deck scraper and chipping hammer are used by another deck hand, with a third on the inevitable broom and shovel, tidying the deck. Due regard needs to be taken of the air temperature, in order to avoid possible condensation of the places to be treated, and the usual preparations made to ensure that adequate cleaning takes place before the painting is commenced. The specific area of the ship to be painted, and the type and quality of the paint used, will determine the thickness to be applied, but generally weatherdecks receive a 200 micron epoxy-based coat on top of a 50 micron primer. Little more than a "holding operation" can be practised normally at sea because of the extensive areas requiring attention. All that the Mate can do practically is to be on guard for first signs of deterioration and correct to the best of his ability and manpower. Other sections on deck exposed to the sea-air which need attention are the railings, pipes, foam monitors and stands as well, of course, as the external coverings of deck machinery.

9.3. ALLOCATION OF DUTIES

The Captain's weekly meeting, between the four senior officers; Master, chief engineer, Mate and second engineer determine interpretation of shipboard policy, based on company's regulations, where decisions are made concerning the efficient and smooth-running of all areas

Deck Duties

*9.2. Some of the Ship's Officers in Discussion at the Weekly Planning Meeting.
(By kind permission of British Petroleum Shipping.)*

in the ship. Other officers may well be co-opted as the occasion demands, (Photo. 9.2), especially when decisions have to be made that affect directly mid-management officers, or even involving all officers, such as planning, loading or discharging of cargo. Another example of the kind of decision made concerns distribution and delegation of duties regarding exercising of emergency drills. These consist of possible variations in a number of duties that are practised weekly on a Sunday whilst the ship is at sea. Generally, there is far too much going on whilst in port or alongside although, to conform with SOLAS Regulations, the lifeboats have to be lowered into the water on occasions, to ascertain that the lowering gear is working efficiently. The more modern 'free fall' lifeboats, frequently found today on VLCC's, have also to be launched at least twice a year in order to conform to SOLAS. From a safety aspect, these exercises are best undertaken whilst the ship is in port. Additionally, at the meeting, an emergency situation is devised and a decision made regarding that part of the ship to be involved, and which officers should be delegated to take charge in one of a number of simulated activities. This is an ideal chance for junior officers and cadets to be given responsibility showing their ability in leading the crew to participate effectively as a team.

9.4. LIFEBOAT DRILLS

During the weekly lifeboat drill, lifejackets are worn by everyone involved and the opportunity taken, (Photo. 9.3), for the buoyancy aids forward to be given an airing and exchanged for others in the ship. A number are always kept in the Bosun's stores forward as an emergency measure which, if not exchanged occasionally, would become musty and perhaps have their effectiveness impaired. At the beginning of the voyage one of the first duties undertaken by either the Mate or the third officer is to draw up a lifeboat drill delegating all personnel on board to one of the two boats and to assign duties as relevant. 'Boat Drill', or 'DTp Sports'

Supertankers Anatomy and Operation

9.3. Deck Cadet Carrying Lifejackets from the Fore-Peak Store to the Boat Deck for Lifeboat Drill.
(By kind permission of British Petroleum Shipping.)

9.4. Lifeboat Drill is a Serious Occasion on all Ships.
(By kind permission of British Petroleum Shipping.)

9.5. Lifting Gear for the Lifeboats being tested by Lowering to Embarkation Deck Level.
(By kind permission of British Petroleum Shipping.)

as it may be more flippantly known, in an effort to make light of the concentrated effort involved, is *always* treated as a very serious occasion. In the tropics, the shorts and casual deck shoes customarily worn for watches and relaxation, are exchanged for working clothes and hard hats, with sensible, hard wearing and reinforced shoes. (Photo. 9.4). The boats are invariably lowered to embarkation deck level. (Photo 9.5). Stores and all gear are checked, and the engines tested by running the propeller for a brief while. This exercise is, of course, an excellent opportunity for the crew to practice drills in a leisurely fashion which prepares officers and crew, by familiarity, for the occasion when the boats may have to be lowered in a genuine emergency. It is at this time also that the efficiency of the officers assigned to looking after the boats during the voyage comes under close scrutiny from the chief officer and, perhaps, even more so by the Captain.

9.5. FIRE-FIGHTING EQUIPMENT

Although VLCC officers do not go about living nightmares regarding the possible hazards of a fire on board, clearly all officers and crew have to be mindful of this very real danger. This is particularly true on board an unloaded VLCC where, it has been seen, there may well exist a very real potential for explosion. Personnel have to be on guard against causing sparks, and mindful also of the places where they choose to smoke, until such precautions become very much second nature. It is the security that arises by conforming to these SOLAS precautions which gives rise to a degree of equanimity over the possibility of a fire occurring and helps all to live with the threat. Immediately following boat-drill, therefore, an exercise is always practised on VLCC's which takes the form of a fictitious fire in any part of the accommodation, engine-room or deck housings. The policy on many Supertankers is to use the ship's safety plan in order to work selectively through the various appliances recorded so that eventually all equipment receives a regular testing, other than those alarm systems which are examined on a daily or weekly basis during the normal course of routine inspections. The plan gives a highly detailed and comprehensive indication of the number and accessibility of all fire-fighting equipment, including the positions of hose points, (Photo. 9.6), indicating the type of equipment likely to be found at any particular place. All fire points, foam smothering lines and fire extinguishers are shown, together with information regarding the numbers and types of the latter. There are also breathing apparatus sets in the ship, (Photos. 9.7 and 9.8), consisting of compressed air tanks, breathing masks, coils of rope, the inevitable but extremely useful hard

9.6. Fire Hoses are Placed at Strategic Points on Board.
(By kind permission of Shell International Trading and Shipping.)

Supertankers Anatomy and Operation

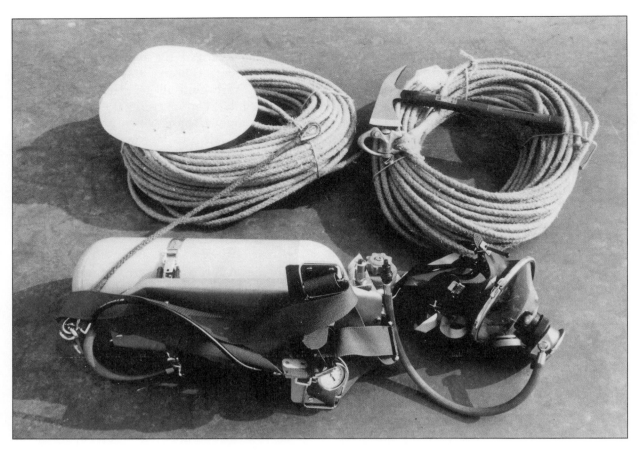

9.7. *Fireman's Breathing Apparatus Aboard a First Generation VLCC ...*
(By kind permission of Shell International Trading and Shipping.)

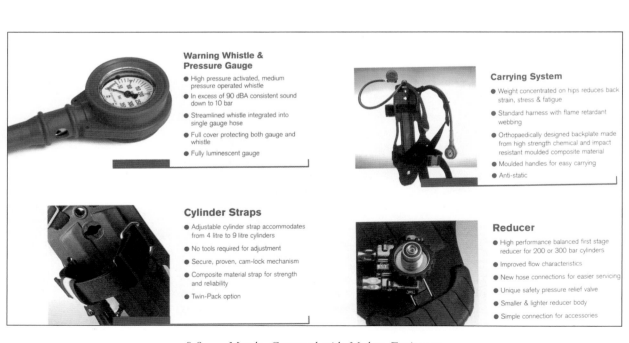

9.8. *... May be Compared with Modern Equipment.*
(By kind permission of Draeger Ltd.)

9.9. Instruction in the Correct wearing of Compressed Air Breathing Apparatus for Entering Smoke or Gas Filled Compartments.
(By kind permission of British Petroleum Shipping.)

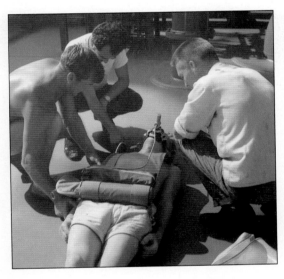

9.10. Exercise in the use of 'Rescuepak' Oxygen Resuscitating Equipment.
(By kind permission of British Petroleum Shipping.)

hat, and an axe. This equipment is also worn regularly, (Photo. 9.9), during exercises so that all personnel soon learn how to do this as a matter of course. It is not always the case that the poor long-suffering deck/catering boy or cadet, where these are still carried, who is chosen to demonstrate the gear as, in an emergency, it is very often (and rightfully so), the officer who will find himself going into the danger area. Oxygen can, of course, be administered independently of the breathing apparatus in appropriate conditions. (Photo. 9.10).

Simultaneously with the fire drill, two other drills are normally being carried out, with the officers taking it in turns on each occasion to rotate with their permanently assigned parties of PO's and ratings. The first of these are fire monitor practices. Run from a foam tank, situated in the fore-part of the main deck accommodation, (Photo. 9.11), the lines serving the monitors are supplied from one of the service pipes running from the engine room and emerging forward of the accommodation. The line carries to the monitors a thick jet of foam that can

Supertankers Anatomy and Operation

9.11. Foam Storage Tank Aboard a VLCC.
(By kind permission of Shell International Trading and Shipping.)

9.12. Testing the Foam Jet from a Main Deck Monitor.
(By kind permission of Shell International Trading and Shipping.)

be directed to all parts of the main deck. (Photo. 9.12) and whose function is to blanket the fire by smothering the flames. The monitors can be operated also from the water-main and the very powerful jet produced can be used literally to push the fire area over the ship's side out of harm's way. I recall (unwisely as it happened) allowing my exuberant young deck boy to use the water function of the monitor during one of our drills. Initially, he sensibly kept the barrel pointing downwards as instructed but, as the pressure increased, lost control of the monitor and as it swung upwards nearly took away the radio aerials and then, swinging in order to

276

Deck Duties

9.13. A liberal dose of Foam can also be administered by portable hoses to those areas difficult to reach by the Monitors. (By kind permission of British Petroleum Shipping.)

regain control, totally drenched the officer and crew who were operating the next monitor forward. Needless to say, neither he nor I were "flavour of the month" for a while. He was nearly lynched by the ratings, whilst the incident cost me a couple of beers, when next we were in port. It is possible for those areas of the ship which cannot be reached by the monitors to receive a liberal coverage by portable hoses operating of course, again from the water or foam lines. (Photo. 9.13). Two crew members are required to control the powerful jet emitted and they would need to wear appropriate gear for protection and safety purposes. In this shot, the area selected for exercise is that within the vicinity of the manifolds.

9.6. STRETCHER PARTY EXERCISE

The final drill practised weekly is perhaps the most difficult in real life to perform. This is the stretcher party exercise using the Neilson safety stretcher which is designed to lift a casualty from any area within the ship in which he may have become incapacitated. The photographs, (Photos. 9.14-9.18), show a 'casualty' secured inside a stretcher at the bottom of a cargo tank. All concerned in the drill have their breathing apparatus fitted. The arms of the 'casualty' are strapped safely inside, alongside the legs, so that no movement is possible, and the feet are supported by a rope toggle. The following series of shots, taken aboard a Supertanker during her trials, shows the problem that could be experienced, and one that is frequently practised, when the casualty occurs in a part of the ship that is hard to access. It results in a particularly difficult rescue operation. I recall Captain Colin Graham relating to me, as we discussed this Chapter, an incident which occurred when he was a deck cadet serving on board a 60,000 sdwt tanker, which has of course cargo tanks considerably smaller than those on a VLCC. A crew member had sustained an accident in a tank, an examination revealed he had broken his right leg and received injuries to his back and skull. After administering morphine to

Supertankers Anatomy and Operation

9.14. ... 9.15. ... 9.16. The Drill to Rescue a Casualty from a Cargo Tank is both ...
(By kind permission of Shell International Trading and Shipping.)

9.17. ... *dangerous and difficult to perform, requiring patience and ...*
(By kind permission of Shell International Trading and Shipping.)

9.18. ... *skill on the part of all concerned.*
(By kind permission of Shell International Trading and Shipping.)

the unfortunate man, it took over three hours of careful and gentle manoeuvring to lift him onto the main deck. It is good to record that he subsequently received hospital attention and made a complete recovery, but this real life incident emphasises the considerable difficulty in undertaking rescue operations within the confines of the much larger tanks on a VLCC.

In the sequence of photographs, the 'casualty', in this instance a straw-filled dummy, has met with an accident in the lower part of one of the tanks. He first has to be strapped into the stretcher, (see previous photos. 9.14-9.18), a manoeuvre hampered both by the rescuer's breathing apparatus, and restricted and cramped movement within the very close confines of tank obstructions. The stretcher and patient are then hauled up the recess ladder and through manholes to a position whereby the top rope support of the stretcher can be attached to a runner lowered from the main deck tank entrance often operated from one of the portable gantries used normally to lift out sludge from the bottom of a tank. (Photos. 9.15 and 9.16). Both stretcher and patient are then winched through the main tank area (Photo. 9.17) until they can be lifted out onto the main deck and transported to safety. (Photo. 9.18). The sequence requires considerable patience and dexterity with strong nerves necessary on the part of any conscious patient and rescuers. I recall seeing an accident occur on a dry cargo ship in Santiago de Cuba when a docker had an accident on top of cargo near the bottom of a hold. The stretcher rope parted (someone's knots were faulty) as he neared the trunking of the hatch and he fell to his death. The Cuban authorities left him on the stretcher, covered by a blanket and seeping blood into the scuppers for over five hours, before they sent a van to take him to the mortuary. The final stage of many rescue exercises is a renewing of oxygen from the resuscitation equipment in which the casualty having experienced his manhandling through manholes and up ladders, followed by his lift through as much as 25 metres of tank, is given a well earned shot of oxygen. Although, today, many ships are fitted with lightweight aluminium harness type stretchers, instead of the Neilson, the inflexibility of these does not in any way make the rescue operation easier.

9.7. SOME GENERAL OBSERVATIONS

Deck officers usually work on deck for a few hours when they are not on the bridge. The range covered varies considerably dependant upon the needs of the ship and can include working directly with the crew where duties are involved that are out of the ordinary run of deckwork normally supervised by the bosun as senior G.P. rating. This is particularly the case

9.19. Officer's Deck Work can be spent looking for Hair Line Cracks in any pieces of equipment, such as the Anchor Shackle. (By kind permission of Ewan Muldane.)

when no deck cadets are carried. On an 'early generation' VLCC the duty mate concerned might be given a hand to assist with servicing of the Whessoe gauges for instance, entailing opening the front and lightly oiling the runners. There may well be adjustments to the deck gear, which are outside the scope of the able seaman assigned to this duty, or which the chief officer decides would be better served by mates than crew. The taking of ullages and temperatures, and the soundings of empty tanks cofferdams and pump room, with the findings recorded on a board situated in the wheelhouse, may well be taken by a mate, especially if the pumpman or crew are engaged in the tanks or elsewhere on other momentarily more immediate duties. If a shifting of some ballast is necessary, whilst the ship is under weigh, the officer may well be called to officiate in the cargo control room or pump room itself. (Refer to Chapter 6). Work might be undertaken in the lifeboats or drying hoses following a fire drill, to name just a few examples which can cause the mates to 'turn-to' for a couple of hours when on watch below. A specific duty is that of 'keeping eyes open' on all occasions and at all times looking for 'anything unusual' that appears to be out of the ordinary. These, like the pre-handing over inspections, that will be examined later, cover an extremely wide brief. They can include, but not exhaust, checking pipes, connections, runners to the derricks or cranes aft and midships, looking in these instances for loose connections at blocks that might be due for a routine oiling. Traces of the ubiquitous rust, virtually anywhere, hairline cracks appearing in any aspect of equipment, (Photo. 9.19), any rope lines becoming unravelled and in need of a splice such as the anchor shape where this is of the basket type, or flag halyards, for example. The list is endless and ingenious, but still needs to be done. I used to carry a rough book around the deck and record any such defects or deficiencies for later attention.

9.8. HELICOPTER OPERATION

A routine deck duty nowadays is for the chief officer to cover helicopter operations. As in all drills, safety factors involve strict adherence to IMO regulations. The Master, of course, remains in total charge during the approach, landing or hovering, and departure of the helicopter and usually assumes his command from the wheelhouse from whence he has an overall view of the events taking place. He is in direct VHF communication with the chief officer on deck, not only to be informed of the progress of events, but because there may exist also a blind sector between the crew working on deck and the wheelhouse view. The International Chamber of Shipping, (ICS), see Appendix, has published guidance to masters engaged on helicopter operations and, on page twenty-three of their manual, "Guide to Helicopter/Ship Operations", made the following comment when considering different types of ships likely to be involved in this work:

*"Large tankers are among the few commercial ships able to provide a landing area for the largest helicopters in normal marine service and they will often be able to provide a winching area on the opposite side of the ship in addition to a landing area. When landing areas are provided the preferred mode of operation is **always** to land the helicopter. This is quicker and much less hazardous operation than winching. It therefore puts both ship and helicopter at risk for the shortest time."*

Undoubtedly, landing the helicopter would be the rule followed in even near adverse weather conditions, and on nearly all occasions when personnel, such as relieving officers/crew or pilots are to be transferred. When both the sea is flat calm, and the wind is around force one or two, use of the winching area is often a preferred and safe method of transferring stores and, in these conditions, it would be rare for the helicopter to land on the deck. Indeed, the manual recognises the legitimacy of winching operations:

"With the increasing use of helicopters for routine operations with ships, it is strongly recommended that where it is impracticable to provide a prescribed landing area a clear winching area is provided over which a helicopter can safely hover while winching people or stores to or from the ship."

Supertankers Anatomy and Operation

9.20. Helicopter off Capetown Lowers Supplies over the Winch Area.
(By kind permission of British Petroleum Shipping.)

Whichever way is considered the most expedient for the circumstances and conditions, strict criteria apply to the designation of the 'landing pad', with its familiar and distinctive "H" on the main-deck forward of the manifolds, or the less familiar 'Winch Only' instruction covering the winching area. Obviously, both must be clear of all radio aerials and other shipboard obstructions. The wind direction is of considerable significance to both the pilot of the helicopter and the deck side of the operation. This is because, following the guidelines for helicopter operations, the ship has to be positioned with the relative wind approximately thirty degrees on either side of the bow. The ship normally flies a wind sock from the foremast in order to assist the Master to manoeuvre into the correct position to assist the helicopter. When winching, once the helicopter is overhead, a 'static line' is lowered, and permitted to touch the main deck, in order to permit discharge of any electrical build-up. This is perfectly safe owing to the totally enclosed nature of the tanks and there have been very few recorded incidents affecting any aspects of working with helicopters. The taking of ship's stores off Capetown, for instance, is particularly common-place and, indeed, has been for the last thirty years or more. The shot of stores being taken on board, (Photo. 9.20), by the crew shows those already unloaded being carted away and the net containing the next drop being lowered to the deck. The officer is shown in the white topped uniform cap supervising the proceedings. He not only informs the bridge of the progress of events, but acts also as a relay, through the Captain to the pilot of the helicopter, as usually that which is happening on deck is outside the pilot's line of vision.

9.9. PUMP-ROOM BLOW-OUT

Other duties occur as circumstances dictate, and these can be many and varied. I recall an occasion when one of the cargo pumps blew-up. We were homeward bound from the Arabian Gulf towards Europe, about 1900 hours, when the ship was shaken lightly by a noticeable explosion. Our immediate thoughts were of an incident in the engine-room, but a quick 'phone call eliminated that source and only added to the confusion as they asked us what had happened. We thought then of a possible blow-out from one of the deck pipes, but careful inspection showed nothing amiss. The only place left to examine was the pump room and a glance in there revealed that the ladder and trunking leading to the bilge itself was full of water, to a level, of course, with the ship's draught.

After the engineers had pumped out the area, the 'deck duty' of the officers for the next few days was to organise some of the crew in cleaning, with permoglaze fluid and rags, as well as patience and elbow-grease, the entire area, although we co-opted the two deck boys to lift the residue from the pumproom by use of the hydraulic winches mentioned previously. We collected the pieces of the pump and loaded these into crates that were eventually off-loaded in Coryton into a motor barge, the "LADY KITTY", for onward transmission to Heathrow and hence back to the manufacturers in Sweden. It is interesting to relate that, just a brief while ago, I saw the barge discharging machine parts to a VLCC at the same Thameside berth. It was vaguely comforting to know that this sterling work horse had not been 'put out to pasture', but was still functioning.

9.10. SOME UNUSUAL DUTIES FOR OFFICERS, INSPECTIONS

Deck officers are occasionally called on board VLCC's to act as officer of the day (OOD). This is an unusual duty and arises from those rare occasions when the Supertanker is alongside and no cargo is being worked. This happened once only with me and that was in

9.21. Welding defects would have to be Corrected ...
(By kind permission of Jotun Marine Coatings.)

9.22. ... Before the Handling Over Ceremony.
(By kind permission of Jotun Marine Coatings.)

Milford Haven when the ship was taken alongside the berth, but the tank capacity ashore was insufficient to take our entire cargo in one drop which was, on that occasion, the preferred method. We were kept alongside for twenty-four hours which the two most senior mates split between us, each doing a twelve-hour stint. The duty consisted of keeping an eye on the moorings and the deck man, and generally being on call for any emergencies that might arise, with particular emphasis on crew safety going ashore, and returning, and fire prevention.

Any officer who spends a period standing-by a VLCC during (to him) that 'grey area' when the ship is between the final stages of fitting-out, and the handing-over ceremony between the builders and the owners, his employers, is in a very privileged position. This is because he has the opportunity during that period of working the vessel to seek out in virtually undisturbed leisure every 'nook and cranny', if these can exist aboard ships, without the ultimate responsibility for the labour performed or the pressures that will later colour his every duty action. He has, nevertheless, to perform the tasks given responsibly because, notwithstanding the lack of pressure, that which he has to do is important. One of his roles is to go around the areas of the ship delegated to him by the chief officer with a large piece of chalk and his deck note-book looking for any defects that might need correction before the ship can function effectively. The chief officer selects also his own sections and the aim is to cover as much as possible of the deck and tank areas of the entire ship. The world is made up of imperfections and Supertankers are no exceptions either in ship or associated personnel, be they shoreside workers or seagoing officers and crew. Mistakes can be made, but damage inflicted is very often outside the direct scope of human error. The officers' tasks are to note any combination of these defects so that they can be rectified. The kind of malfunction to be looked for covers a vast selection, ranging from spindles on deck which lack wheels, to extensions of metal frames that obstruct entrance hatches thus preventing access. The defects are not always obvious. They do not attract attention to themselves and do actually need to be looked for. In the bottom of a tank it is extremely easy to miss weld defects, particularly in conditions of poor lighting. (Photos. 9.21 and 9.22). Sharp edges that have been overlooked during building, for instance, and where the toe of a weld may have been missed. Other defects for which the inspecting officer should be on guard include pipes that have been fitted too close to the ship's sides or frames, incomplete transversals, webs or stringers, faulty tank top openings or water-tight doors that malfunction, as well perhaps as vertical inspection manholes which have not been cut into their frames. These are examples of defects experienced on ships over the years. Sometimes, plates become buckled owing to uneven or unexpected stress areas placed upon the hull during launching. The following photographs of a large VLCC, for instance, (Photos. 9.23 and 9.24), indicate other defects likely to be encountered during inspections. Here, minor damage had been sustained in two different areas of number four centre tank with a corresponding partial collapse and distortion of plating and buckling of verticals. Any welding defects should also be noticed and marked. It is really a case of *"looking to see without really knowing what you are likely to find until you see it"* — if I might put it so!

During the final fitting-out stage inspections undertaken, for example, as a very young and new deck officer to VLCC's, I visited the fore-peak tank and went up into the bulbous bow. The area, as I recall, was very moist and warm. It was totally dark and the rails, ladders and deck were rather slippery. This latter hazard was due no doubt to some condensation in the tank. I had to be extremely careful wandering around the entire area and relied totally upon a specially powerful non-sparking safety torch, supplied by the Lindo Yard, together with a set of spare bulbs and batteries.

I remember inspecting other areas and continue to recall, with some misgiving, vaguely disturbing experiences. I had to visit the port-side section of the after-peak tank running down the shell plating at the ship's stern. In comparison, the enclosed cellular section was not too dissimilar to the inside portion of an old-type squash box with its honeycombed lattice-work

9.23. *Internal Cargo Tank Damage Caused by the VLCC's Launch ...*
(By kind permission of Shell International Trading and Shipping.)

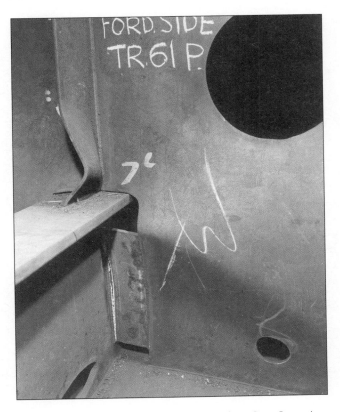

9.24. *... and Discovered During Pre-Handing Over Inspection.*
(By kind permission of Shell International Trading and Shipping.)

Supertankers Anatomy and Operation

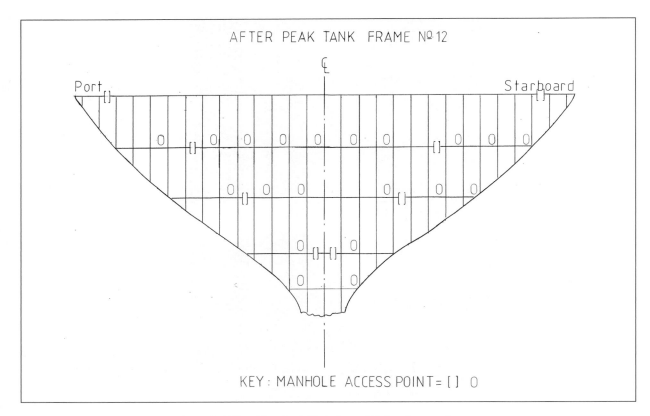

Diag. 9.1 The After Peak Tank is a Cellular Construction ...
(By kind permission of Mobil Ship Management and Paul Taylor.)

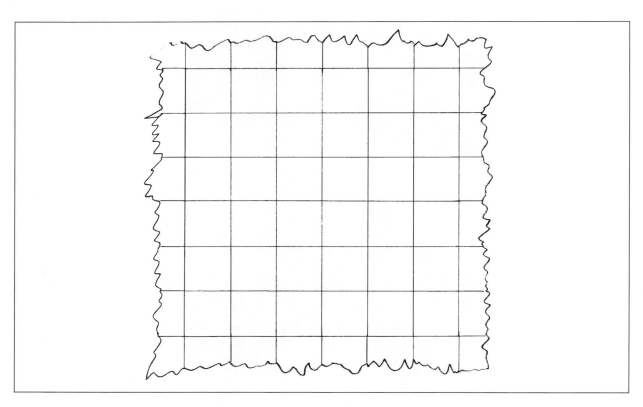

Diag. 9.2 ... as seen in the Plan View ...
(The Author and Paul Taylor.)

Deck Duties

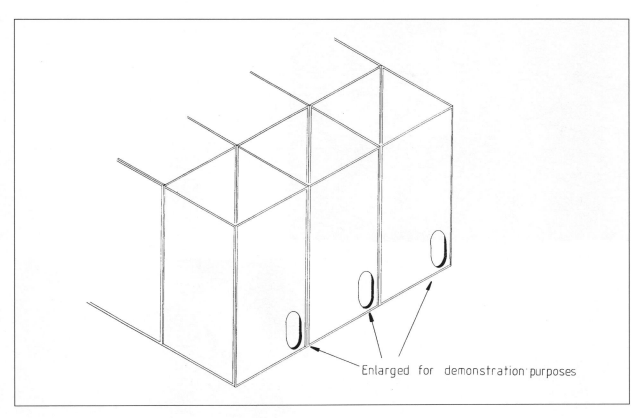

Enlarged for demonstration purposes

Diag. 9.3 ... and Isometric View with ...
(The Author and Paul Taylor.)

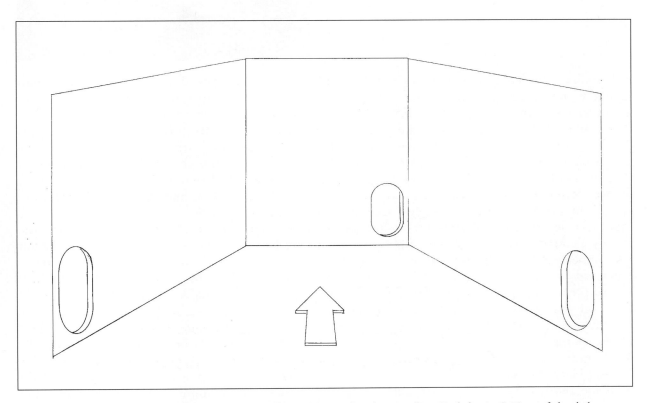

Diag. 9.4 ... Manholes situated in all traversals and longitudinals, as well as Ratholes in Sections of the deck.
(The Author and Paul Taylor.)

9.25. All Cellular Tank Spaces are Cramped ...
(By kind permission of Jotun Marine Coatings.)

9.26. ... and Difficult to Access.
(By kind permission of Jotun Marine Coatings.)

partitioning that acted as separators to twelve bottles, if this can reasonably be imagined. Inspection manholes had been cut into the verticals of each transversal and longitudinal section as well as appropriately named 'ratholes' at intervals into the deck plating on each of the different levels descending to the bottom of the tank. (Diags. 9.1, 9.2, 9.3 and 9.4). Spaces within the confines of all cellular constructed tanks which have to be inspected are extremely cramped with the ratholes and manholes awkward to reach and difficult to access. (Photos. 9.25 and 9.26). I recall that the confines inside the tank were completely dark (even blacker than the forepeak) so that, on wandering around inspecting, I soon felt a sense of disorientation and total loss of direction. As I bent my body into the angles necessary to descend each level and went deeper into the tank recesses so the atmosphere became increasingly claustrophobic. The musty tang of the freshly painted surfaces added a slightly nauseous ingredient to the eeriness of the task. It felt also unbelievably lonely and I went with extra care in case of accident. The idea of breaking a limb and having to be extracted from the bottom layers of the tank was not a completely re-assuring contemplation for the imagination and thoughts of a possible difficult tank rescue, of the kind that was described earlier, were at the back of my mind causing me to proceed very carefully and slowly. I found a new use for the chalk soon after entering the top of the tank. Realising the job with which I was to be confronted, I drew arrows on the bulkheads with the heads pointing towards the entrance and, to put my mind at ease, occasionally repeated the adage "Head for Safety". This simple action also helped me to retain my sanity. I had brought with me two torches on this particular inspection along with a number of spare batteries and bulbs all of which contributed to my sense of well-being and gave me the impetus to make a more thorough inspection than otherwise I might have been tempted to perform. Another task

9.27. Modern Digital Paint Film Thickness Gauge. (By kind permission of Sheen Instruments.)

given to me was to test the paintwork throughout the tank and everywhere else which was inspected on the Supertanker for that matter, endeavouring to detect inadequate levels of paint layers, and any uneven applications of thickness in the paint coats. To do this, I was armed with a metal paint gauge made by Sheen Instruments – a very effective device that gave an accurate indication of the thickness in microns of the paint against a number of different surfaces. A modern digital model is shown in the following photograph. (Photo. 9.27). In the Mate's office resided a book of specifications provided by the paint manufacturers' which indicated the requisite thickness that could be expected in any part of the ship. It is worth recording that, over the maiden voyage, we deck officers tested between us not only the entire main-deck area, but every other part of this massive ship, and found no discrepancies. The inspection of this tank seemed to take a lifetime and I felt decidedly ill when I almost blindly followed my arrow-heads to emerge safely and with some considerable relief, into the fresh air. The Captain's laconic statement to the effect that he and the chief officer had wondered where I had got to, and that they were seriously contemplated initiating a search party, did not encourage me to disclose much of my experiences. Today, of course, with the more powerful VHF personal radio sets and enhanced IMO/SOLAS regulations, which govern more rigidly the entry by officers and crew into enclosed tank and other spaces, my experience would itself not have been permitted. An officer would now be stationed at the entrance to the tank who would be in constant radio contact every ten minutes or so confirming progress and monitoring safety. But, that was how things took place aboard many of the pre-MARPOL/ISGOTT VLCC's.

9.11. SEA-TRIALS AND THE "HANDING-OVER" CEREMONY

More pleasant pre-trial inspections were to go inside the trunking leading to the rudder pintle and top in order to examine the fragile-looking lubricating oil pipes, testing the paint on the way of course, and going up into the funnel and emerging onto the veranda at the top.

Our duties during sea-trials were to understudy the Danish officers and become familiar with what was going to be 'our ship', in my case for the maiden, second and a number of subsequent voyages. It was a sound experience because the Supertanker was really put through her paces – all cargo and ballast tanks were tested with sea-water, every piece of equipment was exercised and the times taken for the ship to turn under full helm, and perform all engine revolutions, noted for future reference. The handing-over ceremony, when a ship is accepted into the owner's fleet, is always a momentous occasion and a cause for great celebration. The following photographs indicate the significance of this event. There is always a handing over party, hosted by the builders, to which the Captain and all officers are invited. Special catering arrangements, invariably of an excellent standard, are provided, (Photo. 9.28), and, certainly the party held for the Supertanker on which I served initially was lengthy and thoroughly enjoyable. Owing to a mistake by

9.28. The Handing Over Ceremony is Always a Significant Occasion.
(By kind permission of Shell International Trading and Shipping.)

the photographers, my photo at the party was mislaid and, for some reason, I received a shot of our Master, Captain "Tommy" Agnew instead. (Photo. 9.29). Captain Agnew was an ex-commodore from Esso tankers and had served in tankers throughout his entire professional maritime career advancing, along with the tonnage, also to reach his ultimate achievement. He was an excellent captain for whom I had the highest respect and, indeed affection, enhanced as I came to know him better during the long sea-watches when he invariably popped-up on the bridge for a cup of cocoa and a chat. He was by far the best Master under whom I served – an accolade that puts him in extremely high esteem – and it is a privilege to dedicate this book, on the class of ships which he loved most, to his memory. Quite astonishingly for a seaman, he never used any bad language, but you could always tell when his "feathers were ruffled" for he would say *"Oh, my goodness, Mr Solly — this ... x ... is (whatever was happening) ... "*: a case then of standing-by and, if an individual was involved, waiting for the unfortunate to receive a quiet blistering that was all the more effective for its mild even – toned delivery. He was a magnificent seaman and a truly good man. I was perhaps not so sorry over the mistake that left me with a copy of his photograph instead of my own.

9.29. Captain 'Tommy' Agnew at Handing Over Ceremony Party. (By kind permission of J. Schou, Odense.)

Chapter Ten

BRIDGE WATCHES AND SHIP HANDLING

- 10.1. MASTERS AND DECK OFFICERS DUTIES
- 10.2. PASSAGE PLANNING
- 10.3. PREPARING THE SHIP, DUTY PERIODS
- 10.4. BRIDGE DUTIES, VIEW FROM THE BRIDGE
- 10.5. USE OF RADAR
- 10.6. COLLISION AVOIDANCE
- 10.7. PLOTTING TIME
- 10.8. VISUAL LOOK-OUT
- 10.9. AUTOMATIC RADAR PLOTTING AID – ARPA
    - 10.9.1. Parallel Indexing
- 10.10. CELESTIAL OBSERVATION
- 10.11. SIGHT BOOK
- 10.12. NOTICES TO MARINERS, BUOY MOVEMENTS
- 10.13. COASTAL WATCH NAVIGATION
    - 10.13.1. Unusual Current Surge
- 10.14. HANDING OVER THE WATCH, THE LOG-BOOK
- 10.15. ENTERING HARBOUR, THE PILOT
- 10.16. MANOEUVRABILITY, TURNING, STOPPING
- 10.17. FOG
- 10.18. SQUAT, INTERACTION

Chapter Ten

BRIDGE WATCHES AND SHIP HANDLING

(Master and Deck Officers — Duties: Leaving Port — Lookout — Arcs of Visibility — Binoculars — Radar: Elementary Plotting Theory — Use for Collision Avoidance — Limitations of the Set — Automatic Radar Plotting Aid ARPA — Deep Sea Navigation: Celestial Sight and Amplitude Workings — Coastal Navigation: Chart and Publication Corrections — Position Fixing — Passage Extracts — Pilot Comment — Tidal Rip off Ruytingen Shoal. Relieving the Watch — Harbour Duties: Taking and Working with the Pilot on Board — Movement Book — Blind Sectors — Angle of Depression. Elementary VLCC Handling Practices: Deep Sea — Emergency Turns — Methods of Stopping Practices in Conditions of Restricted Visibility — Potential Difficulties Due to Squat and Interaction.)

10.1. MASTERS AND DECK OFFICERS DUTIES

By convention, but more importantly by law, the Master is in command of his ship at all times and has been held legally liable even when taking obviously essential periods of rest and sleep. He has, therefore, to rely on responsible and professional deck officers to support him, and to reserve the right — for his own peace of mind and protection — to recommend for dismissal those of his subordinates who are found wanting for various reasons of totally unsuitable temperament or incompetence. The picture is generally not too gloomy. Most deck officers *do* fit into the positive category and, because many aspire to command, they possess interest in their work and take a responsible view of their role in the hierarchy. Perhaps the over-riding lesson any Mate can learn, invariably as a deck cadet, is that of knowing **when** to call the captain — and to do so in sufficient time enabling him to come to the bridge, possibly from a deep sleep, assess the situation yet still have a chance either to offer appropriate advice or take the necessary action. A "rule of thumb" application might generally determine that, if the navigating officer even thinks about calling the "Old Man", then that is the time to do so!

The duties of a deck officer in keeping a sea watch aboard a VLCC, with the essential shiphandling that this involves, are very similar to those of a deck officer navigating any merchant vessel and cover a number of different areas. Clearly, navigation and collision avoidance, in both Deep-Sea and coastal situations, are paramount along with duties performed before and during the operations of coming alongside and leaving port, taking and/or dropping a pilot, and during adverse weather conditions. In each of these circumstances, the officer-of-the-watch has specific tasks to perform. Only a brief outline may be offered, however, because each has to be approached in the light of a number of different conditions and circumstances, and it is impossible to offer comprehensive examination.

Supertankers Anatomy and Operation

10.2. PASSAGE PLANNING

The second mate is in charge of the navigation of the ship and would normally discuss with the Master the passage plan to the destination port and learn his requirements. Usually these are pretty basic: to ascertain that the VLCC is kept a suitable distance off the coast with due regard to safe navigation and national and international regulations. Twenty-five miles is now the distance off salient points whilst rounding the South African coast for example but, as Captain Colin Graham explained to me, when making a passage off Cape Agulhas, which is only twenty or so miles from the coast, VLCC's are allowed briefly to modify the legal requirement by steaming North of the Alphard Bank, for two hours and about five miles distance, in order to keep within the Summer Load Line Zone. The stipulated 25 miles distance not only complies with the South African Government's Movement of Supertankers' requirements, but gives also the advantage of the Aghulas current which sweeps up the East coast. Other than that, he would have a free hand *within the IMO Regulations*, of course, to formulate a detailed passage plan, and then to act reasonably independently in setting his courses navigation-wise, prior to talking these through with the Captain. It is always interesting to prepare the charts overall and then reduce these to the various large-scale ones in the folio's consulting the various Admiralty Pilot publications and tidal stream atlases. It is quite a time-consuming job and one which has to be fitted-in with any cargo deck watches although, usually the previous deep-sea crew are responsible for discharging 'their' cargo before handing over to the joining crew who will ballast the ship before her departure for the oil-fields and a new cargo. Obviously, the immediate passage charts are priorities, ascertaining that very recent corrections from Admiralty Notices to Mariner's are taken into consideration. Practices vary enormously, but this task was aided enormously by one company with whom I sailed because the superintendents made the corrections available at the hotel where we all met immediately prior to joining the VLCC. This enabled a preliminary copy of the directly relevant amendments to be written into a bridge work book even before joining the ship.

10.3. PREPARING THE SHIP, DUTY PERIODS

The deck officer, in preparing the bridge for departure once the pilot has reported on board, orders water, air and steam on deck, takes weather forecasts and the ship's draught, switches on and checks the echo sounder, tests the steering gear and puts the engine-room telegraph on 'Stand-by'. He would also hoist the appropriate flags, synchronise clocks with the engine-room, thus ensuring that a common accurate time is being kept for records, check the gyro compass and repeaters and generally assist to prepare the navigational side of the ship for sea. The chief officer, in consultation with the Captain, checks the state of the Inert Gas System, puts the crew on stand-by and then goes to his station forward. Meanwhile, the second mate would go aft, so that each officer prepares his sections by 'singling-up' those mooring wires not essential for keeping the ship hove taut alongside the jetty. It is customary, following consultation with the pilot before going to stand-by on deck, to confirm his wishes for casting-off the moorings. Depending upon the state of the wind and tide, most ships keep out a back-spring and sometimes a breast wire at each of the fore and aft ends along with head and stern wires.

On leaving port, with the Supertanker bound for her destination, a steady routine is soon established ruined, those 'sea routine' fanatics would say, only by *having* to call in at the next port. The deck officer's duties on the bridge continue to be divided into periods of four hours between three officers over the day such that each man spends a total of eight hours, in two four-hour periods, on watch with, for the keen officer who wishes to learn as much as he can about his profession, a couple of hours working on deck. Some companies prefer to vary this arrangement — one leading VLCC operator, for instance, allows their masters to organise the deck officers' day into periods of six hours for bridge work, two or three compulsory hours on deck duties, with the remainder of the time for rest and recreation. We tried this as an experiment, for one leg of a voyage to see how it would work for us, but found that the amount of spare time was excessive and, by mutual agreement, very soon reverted to the traditional watchkeeping pattern.

10.4. BRIDGE DUTIES, VIEW FROM THE BRIDGE

The fundamental over-riding Bridge duty should always be that of look-out — both for shipping and any other potential threats to navigation. Whilst most container ships and bulk carriers offer the officer of the watch a clear line of view, a considerable number of the older type of conventional cargo-ships, together with other traditional dry-cargo carriers that have along the foredeck a number of cranes as well as fore and main masts, continue regular world-wide trading. Even a short while ago, in the summer of 2000, one of these ships was

10.1. Differences in the views from the Bridge aboard a dry cargo ship ...
(The Author)

10.2. ... and a VLCC.
(The Author)

seen outward bound in the Dover Strait with a forest exceeding twenty "sticks", or masts and derricks. As many of these extend above the height of eye from the wheelhouse, it means that the view from the bridge is comprised of a number of 'blind sectors'. The Supertanker, even when fully-laden, gives a height from the bridge, above the water-line, to the order of 28 metres which, when the ship is in ballast, can increase to around 40 metres. Thus, the view from the wheelhouse is virtually unhindered. (Photos. 10.1 and 10.2). Other shipping and dangers can readily be observed, in good visibility, although this would never negate the need for constant movement around and between the wheelhouse and bridge wings. The conventional image and look-out role of the duty mate is epitomised in the following photograph of an officer-of-the-watch in the wheelhouse of a VLCC (Photo. 10.3). It represents the kind of scene at which any navigator could look and empathise. To pick up binoculars for a closer sighting is definitely a reflex action on the bridge in order to obtain a concentrated look at whatever has attracted attention in order to make a value-judgement preliminary to action. Generally, the 7 x 50 glass is the best for a clear-angle view and has been found to possess sufficient strength for efficiency and reducing potential eye-strain when (as often happens) the "bino's" are used over a long period. There is a compactness about the layout of this wheelhouse (compared with others examined in Chapter Five), considering the 270,000 sdwt size of this ship. This enables essential controls, necessary for identification and subsequent manoeuvrings in collision avoidance situations, to be readily available for convenient one-man operation without too much wheelhouse movement between them, which is often the case. From the manner in which the officer is standing near the radar, it is indicated, perhaps, that a probable target on the PPI (Plan Position Indicator or picture on the screen) has invited the closer visual look.

10.3. The Conventional image of the Officer of the Watch relying on the all important visual look out as well as the aid to collision avoidance of radar.
(By kind permission of British Petroleum Shipping.)

Bridge Watches and Ship Handling

10.5. USE OF RADAR

Radar was always intended as an *aid* to navigation in the same way as any other electronic devices placed on the bridge. It was never intended as a replacement for the efficient 'mark one eyeball' of visual observation but, correctly used, it has always been an excellent means of re-assurance and information with both navigation and collision avoidance. On a Supertanker in ballast, the scanner is over fifty metres above sea-level which means that coast-lines start to appear at the maximum forty-eight mile range. They would be too inaccurate for position fixing at that distance, but they certainly helped identify landfalls, following an ocean passage, by making approximate chart comparisons. On the five/six day run, for example, from off Monrovia to south-west Africa, there is a quiet feeling of deep satisfaction in being able to telephone the master, informing him that Cape Columbine light — the landfall light some fifty miles north of Capetown — has been picked up on the radar, fine on the port bow, more-or-less where and when expected. In the passage up the African coast, from East London to Durban, radar gives considerable re-assurance regarding position-fixing in an area where the coastline looks very much the same for a distance in excess of one hundred miles.

10.6. COLLISION AVOIDANCE

On collision avoidance, the height of the scanner enables the echo-return of most ships to be traced at around seventeen miles. This gives more than adequate time for collision risk to be assessed and, if determined to exist, for an alteration of course to be made so imperceptibly that virtually nobody else on the ship (except perhaps the Master) is aware that this has happened. It assists the detection of suspected close quarter ships — which can be defined as any vessel that is likely to close own vessel's course-line ahead at a distance of two to three miles or less, (depending upon circumstances) so causing a radar plot to be commenced. The information

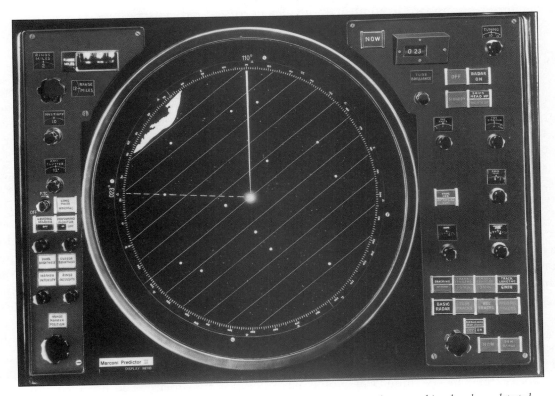

10.4. The Basic Radar PPI gives no indication regarding targets except that something has been detected. (By kind permission of Marconi Marine.)

Supertankers Anatomy and Operation

essential in determining risk of collision, before a professional judgement may be made as a prelude to deciding to stand-on or to make an alteration of course, is the course and speed of the other 'target' ship: the nearest that she will come to our course-line and the time that this will occur. This is the immediate information that would normally be requested by and given to the Master if he is on the bridge and has taken over the watch.

On glancing at the PPI, it is not possible to make any kind of assessment regarding the target which is seen. Indeed, one of the greatest dangers of radar interpretation remains that of acting on assumptions largely because these can always be determined false. In 'olden days', the maritime courts determined that a number of collisions were actually caused by both ships having radar, and what was deemed paramount to the making of little more than a number of 'inspired guesses'. These 'radar assisted collisions', as they became known, would probably not otherwise have happened if the ships had not been so fitted, *or had been using the set correctly*. The accompanying photograph gives an indication of the 'raw' radar, as for convenience it may be termed, representing the PPI that would be seen on first switching-on the set. (Photo. 10.4). The display indicated is called 'Relative Motion' which means that all targets observed are relative to the course and speed of the observer's ship. This is a popular alternative to the 'True Motion' display (mentioned later) because it is the mode which helps ascertain the course and speed of the other ship. It helps also to actually determine if the target is stationary or running at reduced speed, as well as giving the distance and time of her closest point of approach (CPA and TCPA), the key information essential for the planning of collision avoidance action. The range on this set is twelve-miles so that the distance is two miles between the parallel lines of the mechanical cursor. The ship's course is 110° (T), indicated by the heading marker, and the land returning the echo, around a mean at 070° (T) is Beachy Head on the Sussex coast. There are twelve targets indicated by light returns, or echoes, on the PPI. The set is an anti-collision course PREDICTOR devised by Marconi and is very much the type of set found at sea on VLCC's (amongst other ship classes) in the 1980's. They proved robust, by being able to put up with pounding of the ship in heavy seas, and totally reliable and, indeed, many remain in use aboard a number of smaller deep sea and coastal ships.

10.7. PLOTTING TIME

It is impossible to determine in any way shape or form what each echo could represent. They could be navigational buoys with radar reflectors, as each target is very strongly 'painted' up to, remember, a distance representing twelve miles. They could be small steel fishing vessels beam-to, hence returning the maximum energy: They could be large container ships — or even (possibly, at extreme distance), Supertankers seen end-on, thus presenting the minimum aspect. They could be coasters or ferries or naval vessels. Additionally, they could be stationary or moving at anything between five knots for a cabin cruiser or thirty for a large container ship. Additionally they could be heading in the same direction, the opposite direction from our own vessel, or, if engaged on fishing and following a shoal, on any course imaginable. On initial glance it is impossible to tell. In very restricted visibility, it may well prove impossible for all of them ever actually to be identified visually. Certainly, the only way to determine potential movement with accuracy is to plot — or, in the case of this radar, to use the controls on the bottom right-hand side, to predict possible movements. The next photograph in this sequence of three (Photo. 10.5) indicates the *predicted* positions in nine minutes with the distance between the dots on each target representing an interval of three minutes. Six minutes is the "plotting time" usually selected because it represents one tenth of an hour — such that, if the distances covered in six minutes are multiplied by ten, so the speed of the ships involved can be found. Thus 1.6 miles distance equals a speed of sixteen knots and, similarly, if a ship's speed is twenty knots, then in six minutes it would have covered two miles. The use of a nine minute predicting vector allows a fifty per cent time margin or 'assessment time'. As our ship moves ahead on 110° (T), it 'drags' every echo in the same direction. It would be unwise to make further

Bridge Watches and Ship Handling

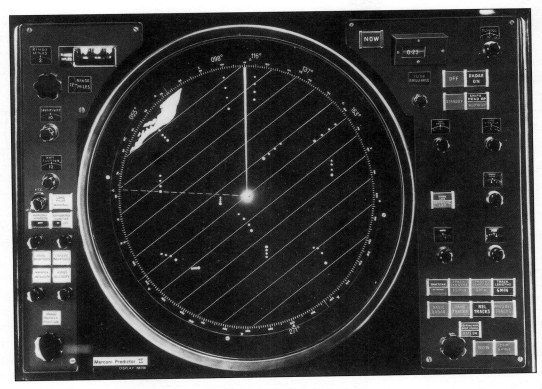

10.5. *The Predicted Relative Motion of targets over a period of nine minutes. Each "Dot" represents a three minute interval.*
(By kind permission of Marconi Marine.)

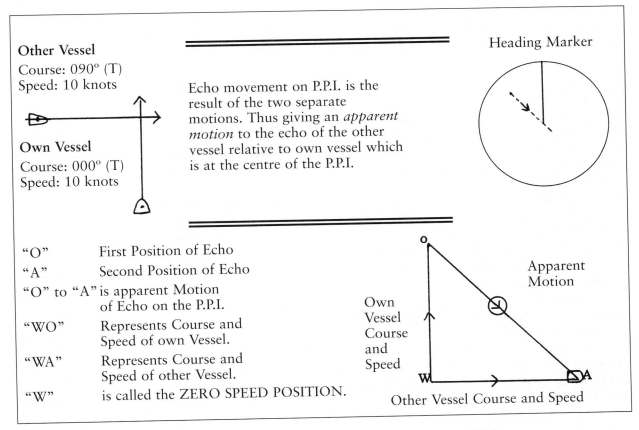

Diag. 10.1 *The Formation of the Relative Triangle to give Apparent Motion.*
(By kind permission of School of Navigation, London.)

Supertankers Anatomy and Operation

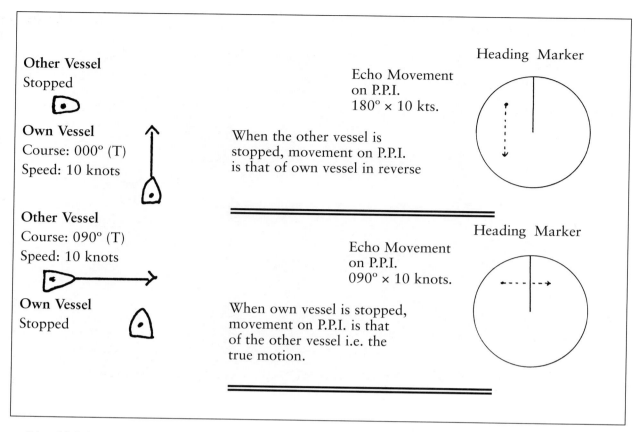

Diag. 10.2 Stationary Targets "appear" to be dragged across the PPI in a direction equal and opposite to the ship's course and speed but, when Own Ship is Stationary, a True Motion PPI is obtained.
(By kind permission of School of Navigation, London.)

deductions, upon which collision avoidance action might be taken, without plotting the targets on the PPI. This can be done in one of three ways: by chinagraph pencil on the screen, by a special mechanical plotting device such as the 'RAS Plotter', or by manual plotting using a plotting sheet, parallel rules and, in order to attain greater accuracy, a very sharp pencil. The combined courses and speeds of both ships under weigh resolve themselves into vector quantities where the resultant of the relative velocity triangle so formed represents the true course and speed of the target ship. (Diags. 10.1 and 10.2). The theory and practice of radar plotting remains today as an examination requirement for navigating officers. Generally, with manual plotting, after the initial range and bearing taking, a 'check' bearing and range would be noted after three minutes so that an indication might be made as to possible movements. The final range and bearing would be taken three minutes after the second, thus giving the third dimension of the triangle in a six-minute interval. By extension, an indication of the target's course/speed and the distance/time of her nearest approach can be assessed. (Diag. 10.3). The entire plot is dependant upon both the target ship and our own vessel maintaining their relevant courses and speeds. Later plotting stages can determine if this has in fact happened and, if necessary, a new plot can be commenced. In a similar way, future movements of our own ship can be assessed before the action is taken. These diagrams were taken from a radar plotting manual originally offered as supporting material for a radar course undertaken at the City of London Polytechnic's (now Guildhall University) School of Navigation – now, alas, no more.

The final photograph (Photo. 10.6) is in the 'true mode'. This is the actual situation that is taking place and represents a disposition of the type that could, perhaps, be seen from an aircraft. The positions of the targets are predicted again for nine minutes and a clearer

THE RELATIVE PLOT

The plot represents both the PPI and the situation outside.
Always plot True Bearings, even with a Head Up Display.

Own Vessel's Head
Course 320°
Speed 15 kts.

Six Minute
Plotting
Interval

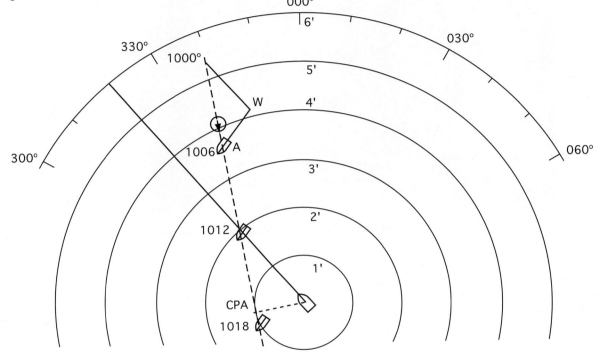

"CPA" Nearest Approach of 1 mile is about 1017.

"WA"   Other Vessel's Course and Speed is 221 (T) at 10 knots.
       Own Vessel Remains at Centre of PPI and Plot.

"O"   The position of the echo on the PPI at the time of the first observation. If target has no speed of its own, our own vessel's motion will draw it across the screen in the reverse direction of our course and at our speed to position W.

"W"   The zero speed position in the plotting interval.

"A"   The position of the echo on the PPI at the time of the second observation. If "A" has not fallen on "A" the other vessel must have course and speed. This is found by joining W to A.

"WA"   That represents the other vessel '8 true course and speed.

"OA"   The apparent motion of the echo seen on the PPI and if neither vessel alters course or speed the echo will continue this movement across the screen. The other vessel may be visualised as tracking crabwise down this line, 1012 and 1018 are forecasts of the position of the other vessel.

Diag. 10.3 A Completed Relative Motion Radar Plot showing CPA and TCPA.
(By kind permission of School of Navigation, London.)

Supertankers Anatomy and Operation

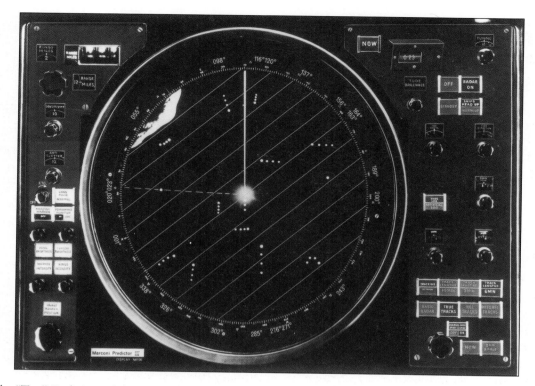

10.6. The "True" Predictions of the Targets akin to the view likely to be obtained from an aircraft looking down onto the Scene. (By kind permission of Marconi Marine.)

picture is obtained, at a glance, of the intentions of the targets. The CPA and TCPA, however, are not indicated and would again have to be found from plotting and it is perhaps, to some navigators, the reason why the Relative Motion plot would be preferable.

10.8. VISUAL LOOK-OUT

It is again worth stressing that the availability of radar in no way detracts from the essential importance of keeping an effective visual look-out — both from the officer-of-the-watch and, when available, the duty able-seaman. I have heard it expressed, in tones of extreme concern from a number of serving VLCC masters, that many young navigating officers today spend far too much time over the radar, sometimes at the expense of a more frequent visual collision avoidance look-out. Certainly, as this appears to be the case in fact, it denotes a very worrying trend because not all ships and other objects sighted are necessarily detected by radar and hence appear on the PPI. Sometimes if they are constructed largely of wood they return a weak echo or even no echo; at other times they are caught up in sea-clutter, which is the return of wave-crests — usually in the most dangerous sector, namely immediately dead-ahead of the course-line. This is especially likely to occur during wind conditions in excess of force four, or about 16 knots on the Beaufort Scale. There remains no substitute for the 'Mark One Eyeball' and the 'extra computerised brain' to which it is attached.

10.9. AUTOMATIC RADAR PLOTTING AID-ARPA

Developments in the 1980's led to the application of computer software to radars which solved the WOA triangle, and its forecast predictions concerning heading alterations and speed adjustments of own and target vessels. Automatic Radar Plotting Aid — or ARPA, as it is

Bridge Watches and Ship Handling

known — became a mandatory fitting for all Supertankers of whatever age, in September 1986, whilst attendance at an ARPA course became integrated into the professional certificates of competency of Masters and Mates at all grades. Certainly, by use of ARPA, considerably more targets could be plotted than would be possible by even the most proficient and competent manual plotter with up to twenty tracks catered for between distances of one quarter and up to twenty miles. Numerous radar set models, each with its own strengths and comparative weaknesses, have been produced, offering a variety of display modes — each, however, conforming to IMO regulations. By use of a golfball, or mouse, targets can be highlighted and immediate collision avoidance information obtained of the selected ship, including forecast movements and own vessel trial manoeuvrings. The system proved a breakthrough in radar technology, the further developments of which will be examined in greater detail (Chapter 16), when discussing briefly computerised techniques applied to navigation.

10.9.1. Parallel Indexing

The late 1970's saw also an additional use to which radar could be put as an aid in the field of safe navigation. This was a device called "parallel indexing" which enabled the officer to construct lines onto the screen with chinagraph pencil, using the mechanical reflection plotter provided on nearly all of the older sets, so that an entire passage could be built up in a series of stages. Indexing used the basic plotting theory that all stationary targets on a Relative display are 'dragged' across the PPI in a direction equal and opposite to own ship's course and speed, so that when the lines are drawn of stationary objects, parallel to the ship's heading marker, so the progress of the ship relevant to that object may be continuously monitored. The object should take a reciprocal course along the marked line: clearly any minor deviation from the track would be as a result of the set and drift of the current, enabling this vector also to be determined and checked against the chart's tidal data. The following diagrams outline some theory of the practice. (Diag. 10.4). A Course of 160° (T) is set, for example, to pass 3.9 miles off North Foreland, the extreme northerly tip of the Kent coast, keeping to seaward of the buoyed channel. A parallel line is drawn on the chart, after the passage area has been scrutinised carefully for safety. Once this has been satisfactorily established, it is possible to transfer the course line to the radar PPI, either by using chinagraph pencil plotting or by using the mechanical bearing cursor on the set or electronic variable range marker — or a combination of both. The next drawing indicates how this should look. (Diag. 10.5). The variable range marker has been set to 3.9 miles and the mechanical cursor to 160° (T), the ship's course. The expected apparent motion can be seen in the northerly 'ghost' drawings indicating the position as the coastline is dragged in a direction that is reversed to the ship's course. Any deviation from the course, due to set and drift of the current, as the most likely instance, would be immediately detected as the coastline drifts away from the variable range marker's safety margin line.

The device enabled safer passages to be planned by monitoring ship handling and pilotage in restricted waters such as narrow channels, traffic separation zones, port entrances and areas of rock outcrops or wrecks and shoals. Margins of safety in all situations could be designed, as well as approaches to anchorage's and search and rescue 'box' searches integrated.

10.10. CELESTIAL OBSERVATION

Out of sight of land, which on deep-sea trading VLCC's is the norm, the officers' duties are concerned with fixing the ship's position by electronic, or celestial observations (or a combination of both) of, in more-or-less order of selection, the sun, the five navigational planets (Venus — Jupiter — Mars — Saturn and Mercury), the 52 selected navigational stars

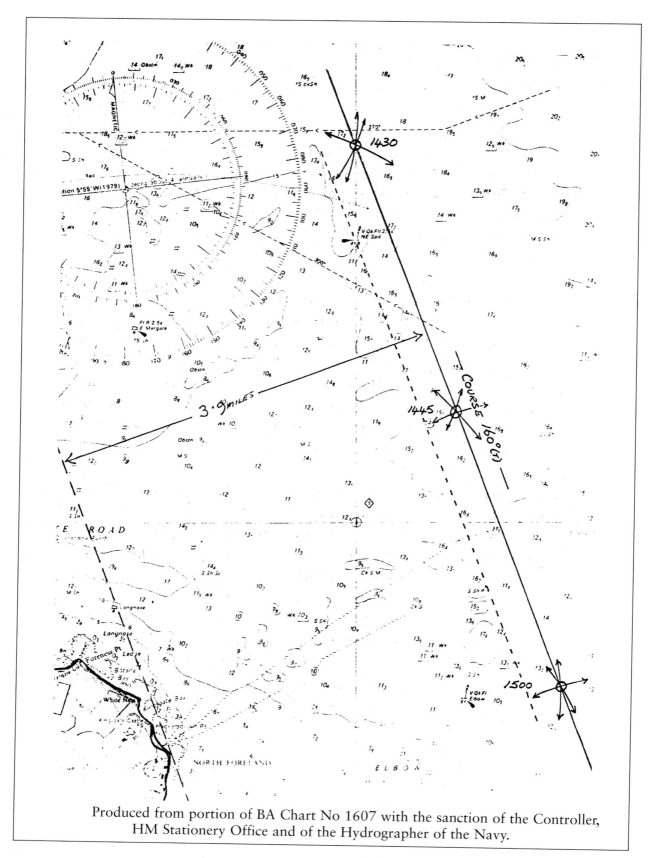

Diag. 10.4 The Theory of Parallel Indexing as an aid to Navigation …
(By kind permission of School of Navigation, London.)

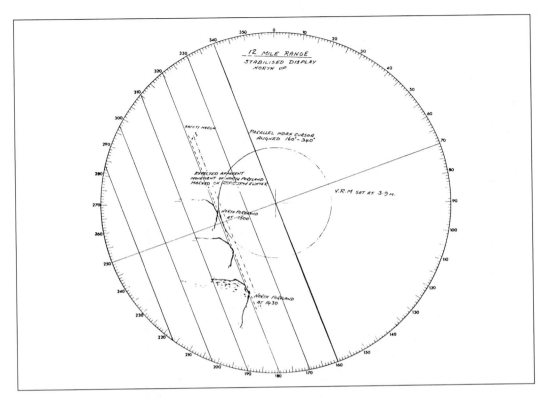

Diag. 10.5 ... and the Practical Application as it would appear on the PPI.
(By kind permission of School of Navigation, London.)

and the moon. Although the latter is certainly the closest as a satellite of the earth, the moon is not always the navigator's first choice for his Sight. This is because the various corrections for parallax tend to make it a mathematically 'untidy' selection (and hence the more prone to error), especially when, as often happens, more convenient choices are readily available. Some of the stars have magnificent names — Altair, Procyon, Rigel Kentaurus, Miaplacidus, Kochab, Eltanin, Adhara, Spica, Formalhaut ... to name just a few whilst, if boredom led to a 'winding-up' of the Master, the devilish-minded mate could always use an unabbreviated Zubenelgenubi to cut-across two lines of the Azimuth Book.

10.11. SIGHT BOOK

The following extract from a bridge note-book which I kept at sea indicates the progress of four bridge Watches over a period of two days. (Extract 10.1). The Supertanker was en passage between Capetown and Sete (in the Gulf of Lyons near Marseilles in the Mediterranean) and, over the period, was between a position about 800 miles off Lobito to about 1000 miles off the Mouth of the River Congo. The routine can be clearly observed: the steering motors were changed about the same time every morning, the current practice on that ship, followed by the general pattern of sight taking. Whoever was on watch would shoot the sun 'around 0900 hours, dependant upon the cloud thickness. On each of the days considered, the time was nearer 1030 hours in order that the sun would be sufficiently high to provide a position line by one of the conventional methods. The Marcq St. Hilaire method was the one I favoured. Even though I prepared the research, the records show that it proved too overcast, on this occasion, to obtain a latitude by meridian altitude of Venus (as a morning planet) in order to obtain a position fix instantaneously. This was not really so important, but served as a bonus if it could be taken because the sun might well be obscured by cloud when the time came later for taking a meridian altitude of the sun. This was usually when the sun was directly overhead, around 1200 hours, or whatever the local time happened to be as determined by the longitude of the

Extract 10.1 Typical Entries in a Bridge Book recording four Watches over two days.
(The Author)

Bridge Watches and Ship Handling

10.7. The 2nd and 3rd Officers engaged on 'Shooting the Sun' in order to obtain the ship's Noon Position. (By kind permission of British Petroleum Shipping.)

10.8. Young Deck cadet engaged on Sight Reduction work in order to obtain a Celestial Position Fix. (By kind permission of British Petroleum Shipping.)

ship. On this occasion in June, at an approximate Longitude 0° 22'.0 West for instance, it was 1300. The object of the time interval between 0900 and mid-day sights was to allow a good run of the ship before obtaining the second position line – hence, a fix of the ship's position at noon, for the day's run. The photograph of the second and third officers engaged on the traditional 'Shooting the Sun', (Photo. 10.7), illustrates the practice of a traditional art and science that was quick, less than thirty minutes from sight to a fix, sufficiently accurate for practical purposes and a reliable method of position fixing. This was useful particularly on those occasions when a total electrical failure effectively destroyed the satellite and other long-range methods of position fixing. It is encouraging to note that the Marine Safety Agency of the Department of Transport continue to include the process of sextant observations, and their subsequent calculations and plotting, within the deck officers' examinations.

A young deck cadet (Photo. 10.8) can be seen going through the systematic process of turning the corrected sextant altitude of the sun into calculations which will lead to an ultimate position line. The book by the boy's right arm is Norie's Nautical Tables of calculations.

2ND MSH ☉	3RD MSH ☉	VENUS ON MERID. PASS
Run @ 15 hm on 0:20⁴ from 28'⁷ ITP	from 08'² ITP	MP 0952.3
DR Pos: 55 = 13/84, 18°32'.75 18°31'.5, 39°10'.E	Long 0236 (39°E)
LMT 09¹⁵ 139°03'.8 E 09³⁸	GMT 06⁴⁷
Chron: 5-07-15-16 s 07-38-15	Long 0200
Ob alt ☉ 38°25'.6 41°41'.4	LMT 0847
N. 22° 30'.8	N 22° 30'.9	
285° 25'.0	285° 25'.0	ob. alt
3° 49'.0	9° 33'.8	Dip
289° 14'.0	294° 58'.8	apt alt Too
39° 03'.8	39° 10'.0	T. Corr.
328° 17'.8	334° 08'.8	apr. T. alt O'CAST
31° 42'.2 E	25° 51'.2 E	alt. Corr
0.54 N	0.69 N	Tr. alt
0.81 N	0.94 N	T Z D
1.35 N	1.63 N	Decl
N 38°.1 E	N 32°.6 E	LAT.
038°½(T)	032½°(T)	
2.87280	2.69935	
T.97684 18°32'.7 S	T.97691 18°31'.0	
T.96557 22°35'.8 N	T.96556 22°30'.9	
2.81521	2.64182	
0.06535	0.04384	
0.12298 41°03'.5	0.12283 41°01'.9	
0.18843	0.16667	
38°25'.6	41°41'.4	
5'.7	5'.8	
38°31'.3	41°47'.2	
51°28'.7	48°12'.8	
51°27'.2	48°11'.4	
INT: 1'.5 (AWAY)	1'.4 (AWAY)	

Extract 10.2
(a) An insight into the calculations ...
(The Author)

Extract 10.2 (b) … necessary to obtain the Noon Position.
(The Author)

3rd June	2nd June	1st June	31st May	DA
* KAUS AUSTRALIS	* ALIOTH	* ALPHECCA	* HADAR	BODY
RT→CT	RT→CT	RT→CT	RT→CT	RUN
10°05.5	04°38.5	01°10'N	06°08'N	DR. LAT
43°40'E	46°08'E	49°35'E	52°11'E	LONG
3-17-44-21	2-17-27-45	1-17-38-15	31-17-32-35	Chronom
206°	205°	208°	210°	Should. Gyr
232°	221°	221°	220°	Mag
5½°W	3°W	2°W	1°W	Vartn
123°	356½°	044½°	176°	Gyro Body Bearing
* KAUS AUSTRALIS S.34°27'.8	* ALIOTH N 36°06'.0	* ALPHECCA N 26°48'.0	* HADAR S.60°15.3	Body = Decl
146°41'.0	145°46'.9	144°47'.7	143°48'.6	GHA T
11°07'.1	6°57'.4	9°35'.3	8°10'.1	Incr
157°53'.1	152°44.3	154°23'.0	151°58'.7	GHA T
84°22'.2	166°45'.9	126°35'.4	149°29'.0	SHA *
242°15'.3	319°30'.2	280°58'.4	301°27'.7	GHA *
43°40'.0	46°08'.0	49°35'.0	52°11'.0	Long
285°55'.3	365°38'.2	330°33'.4	353°38'.8	−360°
74°04'.7 E	360°	29°26'.6 E	6°21'.2 E	LHA *
—	5°38'.2 W	—	—	=
0.05 N	0.83 N	0.03 S	0.94 S	A
0.71 S	15.20 N	1.03 N	15.60 S	B
0.66 S	16.03 N	1.00 N	15.54 S	C
S.57°.0 E	N.03°.6 W	N.45°.0 E	S.03°.8 E	a_s
123°	356½°	045½°	176½°	=
123°	356½°	044½°	176°	Gyro Bearing
NIL	NIL	½° LOW	½° LOW	GYRO ERROR
206°	205°	208½°	210½°	Corrected Gyr C
232°	221°	221°	220°	Mag. Co
26°W	16°W	12½°W	9½°W	Comp. Error
5½°W	3°W	2°W	1°W	Vartn
20½°W	13°W	10½°W	10½°W	DEVTN
123°(g)	356½°(g)	044½°(g)	176°(g)	Comp. Brg
26°W	16°W	12½°W	9½°W	Comp. Error
149°(m)	012½°(m)	057°(m)	185°(m)	Stard. Brg
NIL (p)	NIL (s)	½° LOW (p)	½° LOW (p)	
20½°W	13°W	10½°W	10½°W	GYRO ERROR

Extract 10.3 Calculations necessary in order to determine a Compass Error from Star Sights. (The Author)

Bridge Watches and Ship Handling

10.9. *Navigating Officer preparing to take an Amplitude of the Setting Sun in order to determine any Error of the Compass. (By kind permission of British Petroleum Shipping.)*

He is writing into his sight record book, whilst the book partially hidden by the photograph frame, is the Nautical Almanac. The blue-covered book alongside is the master's Night Order Book into which are written the standing orders governing the navigation of the ship during darkness when the master would generally be below. It is customary for each officer to read and sign the relevant page indicating his awareness of the Captain's requirements. In the extract from my bridge book, the run for the day was 360 miles, which is about the norm for VLCC's, and the comments indicate a quiet watch was spent with few other ships about and a fine morning. The chief officer and second mate would usually take stars during their respective watches so that a number of fixes of the ship's position would be obtained over the day. In order to check the reliability of the gyro compass, an amplitude was generally taken at the moment of sunrise/sunset. A navigating officer, (Photo. 10.9), prepares to take his bearing as the sun closes the horizon one evening during his watch. It is normally the practice to take a bearing of the heavenly body simultaneously whilst taking the sight in order to provide a check on the gyro compass during each watch. The extract from my sight book (Extracts 10.2(a), 10.2(b) and 10.3) indicates something of a "taster" concerning the calculations involved both in working a sight and azimuths/amplitudes of the sun/stars.

10.10. *Chart Correcting has always been a concentrated and time consuming job to the Second Officer. (By kind permission of British Petroleum Shipping.)*

10.12. NOTICES TO MARINERS, BUOY MOVEMENTS

One of the duties which befalls the lot of the navigating officer is to come up into the wheelhouse, during his off-watch time, and to work in the navigating side of the bridge engaged on chart and Admiralty sailing direction corrections from the weekly published Notices to Mariners. These weighty booklets notify permanent and temporary amendments to the various folio's of charts carried aboard all ships, and each one covers all navigable waters world-wide. An example of this very painstaking and time-consuming job is shown in the accompanying photograph (Photo. 10.10). The officer is preparing to place a trace showing a considerable movement of buoyage at the northern tail end of the bank marked by the Varne light-vessel, on Chart BA 1892, which is situated in the Dover Strait about six miles off Folkestone. The number of green buoys in this area indicates, quite possibly, a rather notorious collection of wrecks that occurred, some years ago now, when one ship went aground on the sands and two others collided with her. The green buoys would, before the IALA changes in 1980, have indicated wreck buoys. Other Notices may well cover changes in light characteristics to landmarks or buoyage, changes in fog signals or shoal areas — new constructions to harbours, river jetties or docks — the list is endless. As a rule only the areas in which the ship is immediately trading are amended initially then, as time permits, other corrections applied.

10.13. COASTAL WATCH NAVIGATION

Whilst the ship is in coastal waters, the responsibility for keeping a close look-out remains supreme especially as the inevitable build-up of shipping occurs. Day and night, whilst under way, the navigating officer fixes the ship's position, between every ten and thirty minutes, depending upon the proximity of the coast. An established pattern, although various shipboard practices vary, would be for a visual fix to be taken of three prominent and reasonably spaced points of land, ideally not less than 30° apart, the distance of each checked by radar, and then the position compounded by a Decca, or other electronic navaid fix. Thus, the relation of the Supertanker to her course-line would be monitored by three different methods each serving as a check to the other. This would lead to a pattern of two sets of fixes against each time placed on the chart, as invariably the electronic fix was always plotted last and would, as a consequence of the movement of the ship, appear at a constant distance slightly astern of the visual/radar fixes.

Sperry Marine Systems, in the American magazine, "Sea Technology", in January 1995, made an extremely interesting survey which preceded the introduction of their "VISION 2100" Integrated Bridge Navigation and Safety Control System, that is examined briefly in Chapter Sixteen. They determined the extent to which: *"Casualty data ... (had) ... clearly pointed to human error as the main causative factor of marine casualties."* In an accompanying table, (Extract 10.4), reproduced with permission, Sperry's researchers considered specific reasons why grounding on ships occurred, and considered also some of the factors that had led to incidences of ramming between ships. The Table speaks for itself and reinforces the need for concentrated vigilance so far as the Bridge functions of navigation and collision avoidance are concerned.

The chief officer in the accompanying photograph (Photo. 10.11) is seen measuring against parallel rules the distance of a coastal point. The main radar, in the extreme foreground, is switched-on, as the slight trace of the heading marker indicates — at around 340° (T). The round glass-covered box, on the left of the radar, encloses the ship's chronometer. Very close waters, such as the Cap Gris Nez part of the Dover Strait, would be an example of when positions are recorded onto the chart every fifteen minutes. As a rule, I would notify the Master once we had Boulogne abeam that we were closing Gris Nez and inform him of the density of the traffic. We would always be coming in partially laden in the shipping Traffic

Sperry Marine Inc.

Causes of Groundings Contributable to Navigation Error

Primary Cause(s) of Groundings	No. of Groundings	Primary Cause(s) of Groundings	No. of Groundings
Didn't keep informed of position although navigation aids were available.	8	Informed incorrectly by pilot that buoy was off-station. Used it to navigate.	1
Determined erroneous position/course although navigation aids were available.	1	Didn't wait for pilot in safe area. Navigation aids were available.	4
Erroneous position - Conning Officer not fully licensed. Didn't use available navigation aids.	1	Didn't wait for pilot in safe area. Misjudged set. Navigation aids were available.	1
Didn't keep informed of position, then turned on wrong buoy. Navigation aids were available.	2	Misjudged set or drift in a maneuver.	6
Misjudged set, thus didn't know position. On watch over eight hours. Navigation aids available.	2	Maneuver too close to edge of wide passage. Navigation aids were available.	1
Inaccurate position in poor visibility.	2	Made turn too close to edge of wide passage and barge sheered. Navigation aids were available.	1
Couldn't determine position due to aids to navigation failure. Didn't wait for pilot.	2	Inaccurate position in aiding vessel.	1
Didn't keep informed of position. Gyro failed.	2	Used buoys to navigate. Failed to enter buoy changes on charts.	2
Inaccurate position in poor visibility. Radar failed/unreliable.	2	Radar unreliable due to weather conditions.	1
Radar failure.	3		

Causative Factors for Rammings

Causative Factor Involved	No. of Rammings	Percent of Total
Failure to maintain proper lookout.	3	50
Conning Error - poor maneuvering.	2	33
Navigation Error - poor navigation practice.	1	17

Selected Causative Factors for Groundings

Causative Factor	No. of Groundings of which Factor is Involved	Percent of Total Groundings in Which Factor is Involved
Navigation Error (e.g., erroneous position) - all causes.	40	72
Navigation Error - poor navigation practice.	21	38
Navigation Error - inoperable or malfunctioning equipment.	9	16
Navigation Error - lack of charts.	5	9
Conning Error (i.e., poor maneuvering) - all causes.	10	18
Conning Error - misjudged set.	6	11
Didn't wait for pilot or didn't wait in safe area.	7	13

NOTE: Some cases involve more than one of these factors, and some cases involve unique factors not listed above.

Causes of Ramming

Primary Cause(s) of Rammings	No. of Rammings
Didn't keep informed of position although navigation aids were available.	1
Misjudged set or drift in a maneuver.	2
Failed to maintain proper lookout.	2
Failed to maintain proper lookout when radar was not usable due to weather.	1

Extract 10.4 Factors researched as Causes of Grounding and Ramming between Ships. (By kind permission of Sperry Marine.)

Extract 10.5 Bridge Book Entries show two busy Watches in the English Channel and Dover Strait. (The Author).

10.11. The Chief Officer engaged on Chartwork during his 4-8 Watch on the Bridge. (By kind permission of British Petroleum Shipping.)

Separation Zones near to the French coast and he, as a prudent and responsible mariner, would invariably come to the bridge. We would work the familiar routine that, even when I became a senior officer and found myself on watch in this vicinity, I would take the main radar on the twelve-mile range and feed to the Captain information concerning the targets bearing and intentions as these had been plotted to date. He would pick these up on the Secondary radar at six miles and give me any alterations of course that might be proved necessary for collision avoidance purposes. I would also keep track of the ship and plot our position.

An extract from one of my deep-sea bridge-books gives an indication of the work that can be involved during the coastal passage of a VLCC. (Extract 10.5). The passage covered by two watches over one day, 4th July, was from rounding the Casquets Light in the English Channel until I next came on watch when we were in the Dover Strait. The extract gives a clear record of fixes from visual, radar and Decca plots, plus an indication of shipping density. It shows an alteration of course at 11.24 hours, slightly to port, so opening the passing distance with a Russian ship, the m.v. "AKADEMIC KRYLOV", that was being overtaken on our starboard side. As she veered more to starboard of her own course-line, we were able to revert to our original course eleven minutes later thus passing in perfect safety. Both watches proved extremely busy, with the large, but not unusual, figure of over seventy targets at one time on the 12 mile radar range. To me, the Dover Strait was always an interesting passage to navigate and was one in which I had frequent experience in power-driven pleasure craft as well as coasters. To do so aboard a Supertanker was non-threatening, but challenging. I am reminded of the experiences of a Captain A. Knowles — a professional pilot specialising in handling VLCC's in these waters. He wrote in an article entitled: 'Practical Navigation in the Dover Strait', published by the Nautical Institute's Dover Branch, on page 25, about some of the hazards associated with Supertanker handling in these very restricted and busy stretches of water:

"From the Sandettie Light Vessel, through the deep-water channel to the North Hinder is considered the most hazardous, the vessel being in close proximity to the banks, shoal areas and the South-West lane. It is also the area where sandwaves are predominant and tides erratic. Room for manoeuvre is severely restricted and no large deviations from the track can be considered until passing north of the Sandettie".

10.13.1. Unusual Current Surge

I recall navigating a dry-cargo ship in these waters and, very unexpectedly, meeting a tidal rip that came across the tip of the Ruytingen Shoal and set the ship towards the Sandettie Bank, shortly after we had altered course from 043° (True and Gyro) to 062° (T). (Diag. 10.6).

Supertankers Anatomy and Operation

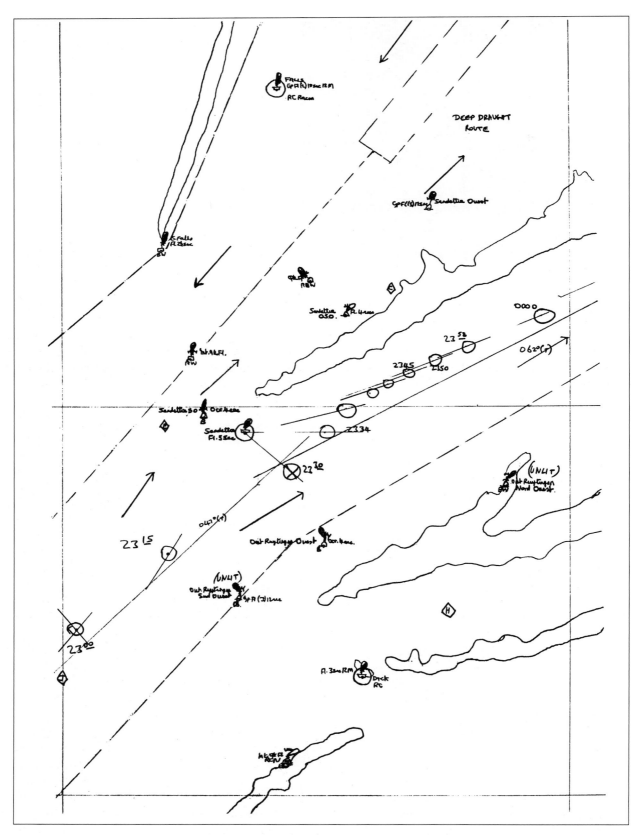

Diag. 10.6 An Example of a Tide Rip encountered by the Author whilst Navigating the mv. "GOLD VARDA" off the Sandettie Light Vessel.
(By kind permission of HMSO and the Author)

The ship adopted a distinctive 'crabbing' motion under the influence of this very powerful surge in current. We had to react extremely quickly in order to recover from the initial effects of shock and realisation of what was happening to the ship. This is the kind of situation where the benefits of professional Simulator training more than repay the financial investment involved. We found that our minds went 'into over drive' and led to a quickly and accurately reasoned bold alteration of course to starboard in order to counteract the immediate conditions. We then had to carry about five degrees of starboard helm in order to make good the True Course. I noticed that the effects of this, from the time of detection until we could make our charted Course of 062° (T&G), was not far short of twenty minutes: a lapse which proved a quite nerve-wracking experience.

10.14. HANDING OVER THE WATCH, THE LOG-BOOK

Handing over the watch at the end of four hours' continuous, and often extremely busy duty is a welcome time, and generally, it is not until the duty man had been sent below to give the relieving officer a shake, or a telephone call in preference, that the mate realises just how tired he has become. It is invariably a responsible, but relaxed occasion, particularly if the relief is on time. The change-over is made smoothly, and the process of "coming-too" assisted considerably, if the duty officer has the kettle boiling and coffee made, whilst his colleague adjusts his night-vision, and invariably also his liver, often disturbed and irritated by the move from sleep into attentiveness at four o'clock in the morning. Invariably, because considerations regarding the possible risk of collision are paramount, the first question **always** concerns traffic, how many ships, how they are heading, what forecasts have been made regarding their movements? Questions relevant for those ships observed both visually as well as detected by radar. The navigation, particularly the position of the VLCC, comes virtually simultaneously along with any future passage/course alterations likely to be encountered in the next four hours. I have never forgotten, when I was a second mate, the peevish chief officer who relieved me, and his feelings, by voicing the rhetorical question which asked why I always planned that his watch should continually have the major alterations of course? As I was leaving the wheelhouse anyway I was able to accept his mood with scarcely concealed humour. Other areas covered include referral to the master's Night Order Book and statements of a general nature regarding weather and visibility, functioning of equipment, availability of quartermasters and look-outs, etc., unless circumstances necessitate more detailed explanations. Having made certain that the relieving officer is fully *au fait* with the situation, and ready to assume the watch, the Logbook would be entered-up covering the previous four hours and then a quick retirement made for a well-earned rest before a prolonged conversation could develop, one that would enliven *his* watch, whilst keeping me from my bunk! Most mates aboard Supertankers use one of the conventional series of log-books to record the events that have transpired during the course of their period of duty. The 'South Tower' Deck Log, published by Ord Printing Company of Hartlepool — and produced with their kind permission, indicates a representative page of a typical Chief Officer's Log Book. (Extract 10.6). The ship's Master and chief officer would sign the written-up entries usually on a twice or thrice weekly basis for, after all, both are in attendance on the bridge every day, with the chief officer usually keeping a navigating watch, enabling all entries to be monitored regularly. Under SOLAS regulations, the Log-Book is a legal document and would be asked for in the event of an incident. In such cases, each entry over the entire voyage, including all days prior to the event, would be studied carefully to gain an impression of the way in which duties had been conducted. The recording of the incident itself would, of course, also be given very close scrutiny. On occasions, extracts from the Log may be copied and transmitted to interested parties. When navigating in restricted visibility, an entry is generally made to the effect that Department of Transport regulations were complied with over the duration of the time spent in fog, thick mist or heavy squalls and other conditions similarly affecting the ability to keep an effective lookout. The all-important Course Recorder sheet and the Compass Error Book would also be signed at the end of each watch.

Extract 10.6 Page from the "CHIEF OFFICER'S LOG BOOK" widely used at sea as a record of Bridge Watchkeeping which becomes a legal document.
(By kind permission of Ord Printing.)

Bridge Watches and Ship Handling

10.15. ENTERING HARBOUR, THE PILOT

Bridge duties whilst entering harbour, be that in a port or, more frequently with VLCC's, at a jetty or oil island, are fundamentally the same and the only variations, from the ones described earlier in this chapter on leaving, will be cited. For instance, water, steam and air are still ordered on deck and navigation equipment not already in use would be made ready. The Captain of course is present with the pilot, but there is invariably a duty officer in attendance assisted by a deck cadet, if one happens to be available and is not placed, for steering experience, at the wheel. The mate's function is to maintain navigational position fixing, advising the Master and/or pilot if he notices anything untoward. Generally, this is unlikely in European waters, where the standard of pilotage is excellent, but the situation might call for closer attention to detail whilst working some ports in the world. As the ship closes the pilot station, the mate on watch would notify the engine-room of the situation. He would also hoist the appropriate International Code flags, the letter G indicating the requirement for a pilot, which he would change to H, 'I have a pilot on board' once this has occurred. Entering the pilotage zone itself, to drop or take on a pilot, is a time to be on maximum alertness because other large ships may well be in the area doing precisely the same thing and also because at slow speeds VLCC's are not always consistently quick in answering the helm. Invariably, the zone is in coastal waters with a high density of shipping that can, and often does, include warships, fishing vessels, coasters, ferries, tugs with tows, single tugs, yachts and pleasure craft. Notwithstanding its size, the supertanker has to conform to the Collision Regulations in the same way as every other ship. The photograph that follows shows a radar on board a ship, in the centre of the screen, that is heading in the dredged channel out of Liverpool. (Photo. 10.12). The set is on the 24 mile range, with the rings each four miles apart and the ship is probably heading for the pilot vessel to drop her river pilot. It could be questionable that the twelve mile range might have been more appropriate for the conditions and circumstances, but possibly

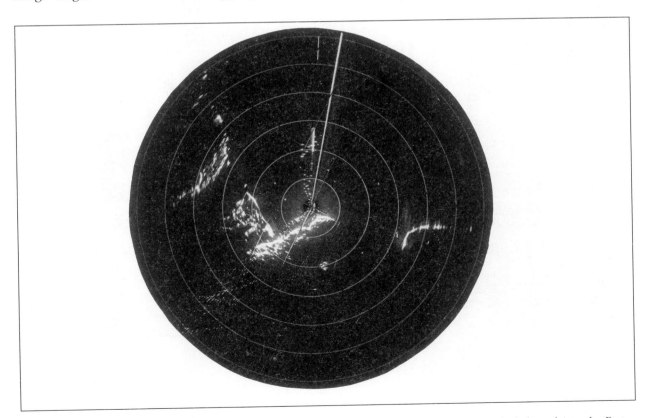

10.12. The Port of Liverpool showing the RACON indicating the Bar Light Vessel and the dredged channel into the Port. (By kind permission of Kelvin-Hughes.)

Supertankers Anatomy and Operation

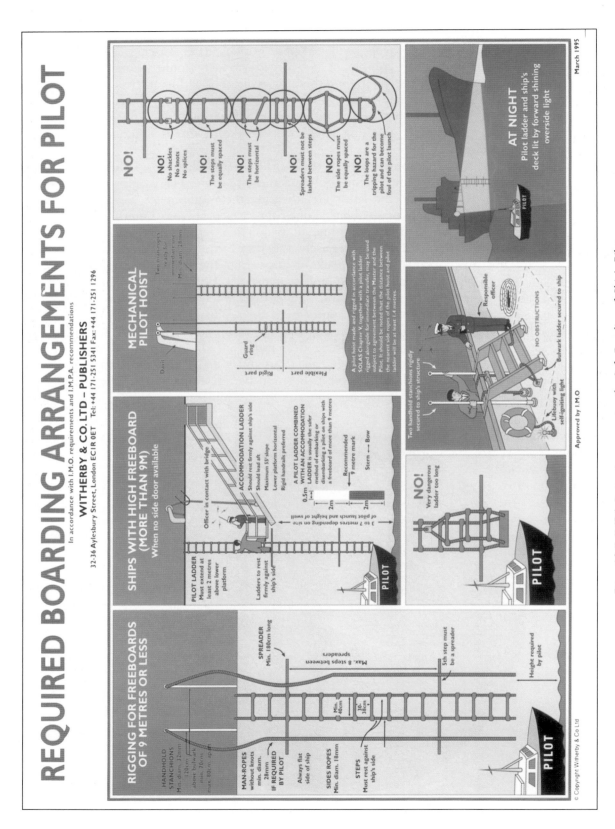

Diag. 10.7 IMPA Instructions concerning the Safe Boarding of Ship's Pilots.
(By kind permission of International Maritime Pilots' Association.)

it has been on this range for 'demonstration' purposes. There is a preponderance of shipping, around the eight mile range which denotes the ending of the channel (at about five miles), within the area of the Bar Light Vessel. The latter is indicated by the oblong blaze, called a RACON, and it is used to show the precise position of the light ship, an essential prerequisite in identifying this important navigational aid easily and quickly from amongst the surrounding shipping. The channel leading into Liverpool is clearly shown, along with the outline of the bay, and the dark areas indicate the blind sectors that have been caused by predominant land shadowing the other side of which is consequently out of the line of sight of the scanner.

The duty mate will occasionally con the ship, under the supervision of or, more testing, under the observation of the Captain until the pilot launch is closing. He will go to the gangway to see the pilot on board safely, taking his duty AB with him to look after the pilot's gear. The carriage of a pilot is compulsory for all Supertankers, but his presence on board does not relieve the Master or his officers from their responsibilities. He acts in an advisory capacity, excepting only in the Panama Canal (not that any VLCC would ever be able to use the system of locks in that waterway) and in all dry-dock sills. Taking the pilot onboard is a potentially dangerous operation, especially if a heavy sea or swell is running, and the International Maritime Pilots' Association, in conjunction with IMO, has laid down very specific rulings designed to offer him maximum protection. (See Diagram 10.7). After escorting the pilot to the bridge, the Captain would notify details of the ship's characteristics thus assisting the pilot, and discuss the passage plan to the berth. The duty mate may request the steward for a tea/coffee tray with sandwiches for the pilot — common courtesies that, in many companies, remain part of the way of life, and are invariably reciprocated by the pilot bringing with him copies of some of the daily newspapers. The deck officer's main function from thence would be to prepare and keep the Movement Book. This is a legal record of all engine-room telegraph and ship movements undertaken up to when the ship is in position and tied-up alongside etc. Times and events are recorded, including the pilot's name and details of any tugs taken, including their name and whether or not they were merely in attendance or were made fast, again, with times and positions relative to the ship's length. Included would be times of the first line ashore, which indicates when the ship legally becomes alongside, even if not yet fully secured, and a record of all telegraph movements. The mate would relay any telephone messages and possibly VHF commands to the other duty mates forward and aft. The pilot would communicate directly with his tugs telling them the positions to adopt and giving engine revolutions for them to follow. Usually, the third officer would be on the bridge, but most Captains and chief officers, with an experienced third, generally allow him to widen his knowledge by going aft or on the fo'c'sle.

Although not such a serious problem deep-sea, because of the wider arc of visibility and the general absence of close-quarter situations, a considerable blind sector exists forward of

Draft (Metres)	TRIM BY STERN (METRES)						
	0.00	2.00	4.00	6.00	8.00	10.00	12.00
4.00	579	684	825	1020	1312	1793	2738
6.00	537	637	769	954	1230	1684	2576
8.00	495	589	714	888	1147	1575	2414
10.00	454	542	659	822	1065	1466	2253
12.00	412	494	603	756	982	1357	2091
14.00	371	447	548	689	900	1247	1929
16.00	329	399	493	623	818	1138	1767
18.00	287	352	438	557	735	1029	1606
20.00	246	304	382	491	653	920	1444
22.37	197	248	317	413	555	790	1252

Table 10.1 Table of Watchkeeping Officer's Obscured vision according to the various stages of draft and trim of the VLCC. (By kind permission of Shell International Trading and Shipping.)

Supertankers Anatomy and Operation

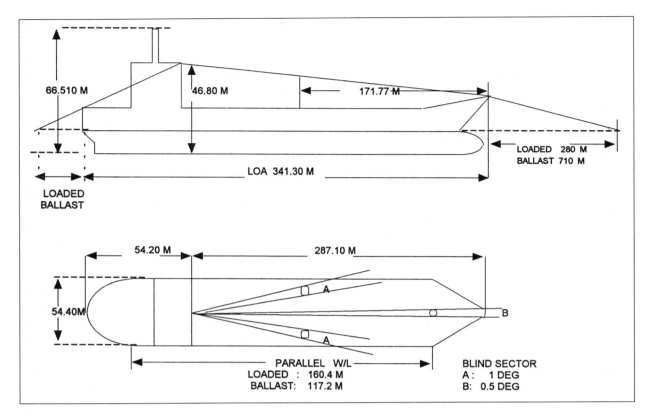

Diag. 10.8 Diagram showing the computation of the Blind Sector ...
(By kind permission of Mobil Ship Management.)

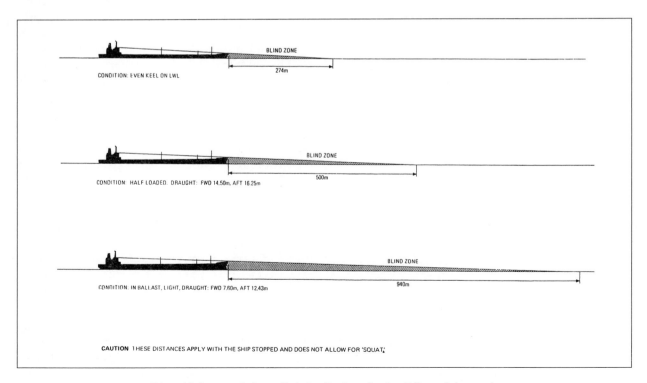

Diag. 10.9 ... and the realistic implications for the Officer of the watch.
(By kind permission of British Petroleum Shipping.)

a Supertanker due to the angle of depression from the officer on the bridge, extending over the fo'c'sle head, and beyond the bows, which could have serious consequences in restricted waters. The extent of this region of obscured vision may best be appreciated by reference to tables (Table 10.1). If the ship was trimmed by the stern at 6.0 metres, for instance, at a draught of 12 metres, then there would be a distance in which the officer could not see any movements of smaller ships or boats, or even buoys, for the quite considerable distance of 822 metres. The accompanying drawings (Diags. 10.8 and 10.9) indicate a 270,000 sdwt VLCC in various stages of her loading and illustrate quite dramatically this potentially dangerous situation. Duty officers and visiting pilots would need to be made aware of such limitations so that they could take them into consideration when ship-handling, particularly in highly congested waters, mooring operations and conditions of restricted visibility. Sometimes the answer resides in stationing an officer or a cadet on the fo'c'sle head in VHF communication with the bridge.

10.16. MANOEUVRABILITY, TURNING, STOPPING

Undoubtedly impressive when seen from the shore and, indeed, when looking out of the wheelhouse windows, the sea is a great leveller and it is soon found that Supertankers handle remarkably like most other ships. The thing which has constantly to be borne in mind, until it becomes second nature, is that the pace of all movements is considerably slower and, therefore, much forward thinking needs to be done when considering any helm orders necessary to shift the VLCC from her course for whatever reason. In my experience, many of the horror stories which abound concerning manoeuvrability, within marine circles as well as without, are simply not true and it most certainly does not take distances of twenty miles to stop the ship, extending over periods in excess of one hour. Fully laden, and steaming at the VLCC optimum speed of around 15–16 knots, and, very importantly, depending upon the loading condition, weather and tides, amongst a number of other minor but potentially cumulative variables, the ship could be stopped within about ten ship lengths, or 2.0 miles, at a time of around 15 minutes. During this period she could make a lateral displacement from her course of about three quarters of a mile. Like all steering systems, those on VLCC's have a maximum hard over helm limit of about 35 degrees, to each of port and starboard, and certainly the helm would be used effectively, in conjunction with stopping the engines, if such an emergency situation arose. It is worthwhile remembering that, with steam-turbine engines, it is not possible to ring on full-astern from full-ahead, through stop, with effect, until the propeller has completely ceased turning. The manoeuvring data of an average VLCC, (Diag. 10.10), based on her fully-laden sea-trials, shows the factors involved in working the telegraph from full ahead to full astern on that steam-turbine VLCC, and indicate that the astern revolutions can be started in less than one minute from stopping the engines. The subsequent halting of the vessel occurs just over twenty minutes later. Generally, in the ordinary course of reducing speed on a VLCC, the times to stop her in various conditions of loading would, as discussed, be determined by a number of factors. A "typical" manoeuvring time and distance travelled for a VLCC may be assessed from the following table. (Diag. 10.11). As mentioned in Chapter 5, when a sectional analysis of a Supertanker was made, the propeller and rudder indicators on the bridge and each bridge-wing (Photo's. 10.13 and 10.14) are invaluable aids when handling the ship, enabling the situations of helm position and engine revolutions/speed of propeller turn, to be seen at a glance.

If, because of an emergency, it becomes necessary to make a complete round-turn by keeping the helm hard over, she could (subject to the proviso's already mentioned) make the turn in between five to seven ship's lengths, or about one nautical mile or so. The following drawings indicate the turning circles of a large Supertanker, (Diag. 10.12), from which it can be seen that the diameter of the turning circle to port is tighter than that to starboard. This is because nearly all ships of this class have a 'clock-wise' right-turning propeller which means that there is a natural tendency for the ship's head to swing to port, thus the "propeller thrust" can be used to advantage to make a more effective turn. It is for this reason that the preferred

Supertankers Anatomy and Operation

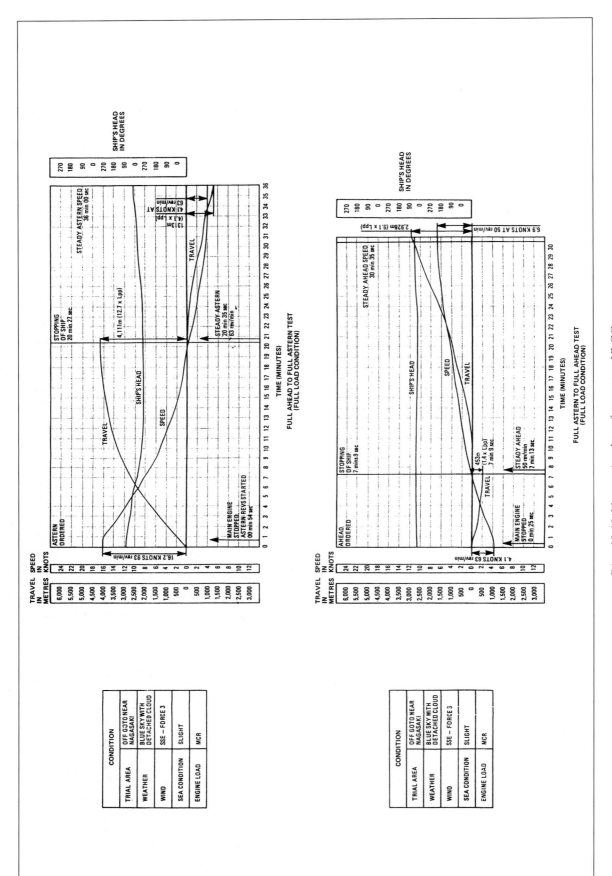

Diag. 10.10 Manoeuvring data of an average VLCC.
(By kind permission of British Petroleum Shipping.)

TIME AND DISTANCE TO STOP

	LOADED CONDITION		BALLAST CONDITION	
	TIME	DISTANCE	TIME	DISTANCE
FULL SEA SPEED	19.3 MINS	2.45 MILES	9.6 MINS	1.35 MILES
FULL SPEED	22.5 MINS	2.02 MILES	11.3 MINS	1.13 MILES
HALF SPEED	28.1 MINS	2.01 MILES	14.1 MINS	1.13 MILES
SLOW SPEED	33.6 MINS	2.00 MILES	16.9 MINS	1.12 MILES

MINIMUM STEERING SPEED
NORMAL LOADED CONDITION   2.2 KNOTS
NORMAL BALLAST CONDITION   1.7 KNOTS

*Diag. 10.11 Time and Distance taken to stop an average sized VLCC.
(By kind permission of Mobil Ship Management.)*

*10.13. Speed Indicator mounted outside the Wheelhouse on a VLCC.
(The Author)*

*10.14. View from the Bridge Wings showing the close manoeuvring controls above the Wheelhouse Entrance.
(The Author)*

Supertankers Anatomy and Operation

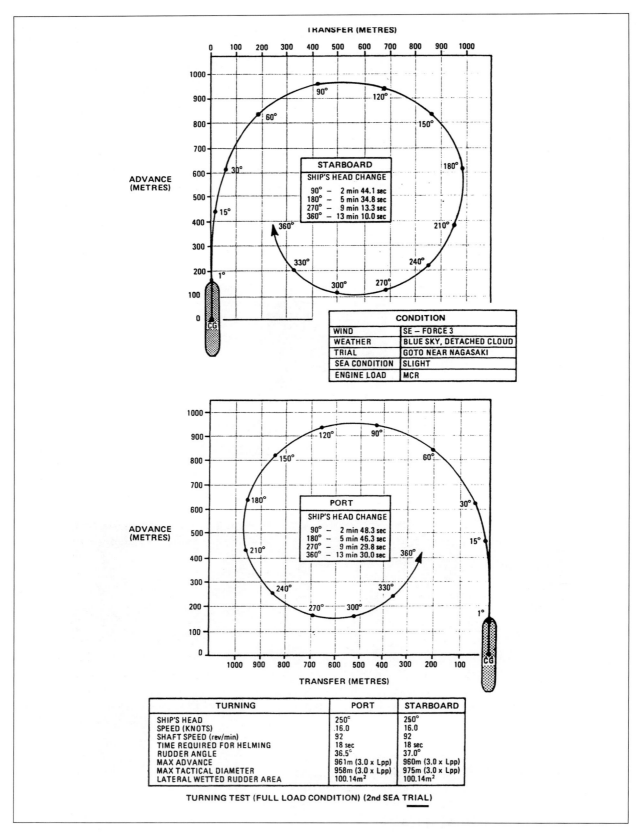

Diag. 10.12 The "Turning Circle" of a VLCC invariably tighter to Port than to Starboard owing to the assistance given the turn by propeller thrust.
(By kind permission of British Petroleum Shipping.)

side to use in jetties or berths is the port side. It follows that an appropriate kick to starboard, at the requisite moment when going astern, helps fit the ship snugly alongside the wharf. There is another method used to stop the Supertanker in an emergency which is called 'rudder cycling'. If the VLCC is steaming at normal revs — around 90 rpm to give a speed of 16 knots — then by deviating the ship's head from her course-line a reduction in speed would follow. Certainly, we experienced this phenomena every time we altered course on the VLCC's on which I served. It follows therefore that by alternatively swinging the rudder from port to starboard, and simultaneously gradually reducing the engine revolutions, a slowing of speed and eventual stopping of the ship can be effected. The two methods of manoeuvring are generally adopted both deep-sea and in restricted waters, along with important qualifications, and subject to due allowances that would have to be made to suit modifications necessary for any specific VLCC. Given appropriate regard to existing circumstances and conditions regarding other shipping and their actions, such emergency manoeuvres, however, should prove rarely necessary. Being fortunate, I suppose, in never having encountered on VLCC's the notorious 'Levantine left-hander', that has had its officers and crew shaken out of the trees from some obscure jungle, I cannot recall having to take any such dire action during the whole of my time aboard ships of this class. Dry cargo ships, for some reason were a different matter, admittedly on rare occasions. In coastal waters, a Supertanker will react more slowly indeed, sluggishly even, and this element needs also to be taken into account when using the helm.

10.17. FOG

Notwithstanding the modern electronic aids provided on Supertankers and, normally, the responsible approaches adopted by their officers, fog at sea is an occasion which puts masters and mates on their mettle. There is rarely need for panic and over-stress, but certainly those on watch need to be on the alert more than usual. Again, horror stories abound of ships suddenly jumping at the VLCC from out of a dense bank giving all and sundry heart-attacks by the score but, again, reality is different. First of all the Officer of the Watch has to ensure the safety of the ship by notifying the engine-room, putting the telegraph on "Stand-by" and reducing to a safe speed for, as the Collision Regulations state: *"the prevailing circumstances and conditions"*. He would then commence sounding on the ship's sirens the regulation fog-signals of one prolonged blast at intervals of not more than two minutes indicating a power-driven vessel under way. If they were not already operational, or on stand by, then the radars would be activated. Traffic conditions and proximity would determine whether or not both sirens or only just one would be used. The foghorns are very powerful, can be heard for considerable distances, and would virtually deafen any hand posted within their vicinity. The navigation lights would have been switched on and extra look-outs posted as necessary and, if a hand were to be placed on the fo'c's'le, he would have been provided with a bridge-linked VHF set, and use made only of the after foghorn. The officer would, by then, have notified the Master that his ship is about to enter a fog-bank. Invariably, if Deep-Sea, the state of the shipping detected visually beforehand and checked on the radars would also be reported, assisting the captain to decide if it were necessary for him to appear on the bridge immediately or if it would be more appropriate for the duty mate to keep him informed of any developing situation. In coastal waters, the speed of the ship would *always* be reduced and the Master would generally be attending in the wheelhouse. Regular radar plotting would have been used on the 'early generation' of VLCC's and, in very extreme conditions, the anchors cleared for possible use.

10.18 SQUAT, INTERACTION

Particular instances of shiphandling of which the VLCC mate needs to be aware are those of squat and interaction. The former condition occurs when the Supertanker is under way in shallow, and perhaps more confined, waters than the open sea. She has a tendency to sit lower in the water and often requires more helm before she commences to alter course. Very often one of the initial signs is a fairly pronounced increase in vibration as the ship goes ahead. The answer invariably is to reduce the engine revolutions slowing the speed of the ship. Certainly, I can recall only one noticeable occasion experienced when encountering squat, largely because we regulated speed automatically in the knowledge that it could occur, reduced speed accordingly and notified the Master of the action taken.

Interaction takes place when the ship is manoeuvring in close proximity with other vessels in rivers or dredged channel is particularly when passing them or overtaking. If squat is caused, basically, by forces tending to pull the hull towards the river bottom, so interaction occurs as the effect of similar forces which pull the VLCC towards and/or away from other objects to which she passes too closely. Often excessive speed is a vital factor causing interaction. When overtaking another ship, for instance, the (often) larger VLCC as she approaches will force the stern of the other vessel away from her own head, so that the other's bows turn across the bows of the Supertanker, then, as the bows of each close, so the bulk of the other is pushed away. As the VLCC continues to pass, there would be a tendency for the sterns of each to be attracted, so that the stern of the smaller would be sucked into the propeller area of the larger ship. In either instance, the danger to the other vessel is potentially quite serious. In practice when there is a likelihood of this situation arising, especially when the VLCC is in ballast, the pilot and master would have to take considerable way off their ship and allow the other to draw ahead. In an instance when I observed such a situation building-up, the other ship on seeing us bearing down upon her shot off well to the starboard side of the channel, which she was able to do easily within her draught and the state of the tide, where some would argue she should have been anyway, thus leaving us open to pass without further reduction of speed or alteration of course. This action by the other vessel showed prudence if not common sense also, within the ordinary practice of seamen, but did not relieve us of our obligation in law as an overtaking ship. The alternative situation in which interaction between ships is likely to be encountered is a meeting between vessels in a narrow waterway head-on or nearly so. By the Collision Regulations, each ship has to keep to her own starboard side of the channel as closely as possible, but again in practice, the Supertanker would be proceeding at a safe speed, determined by the pilot and master and often specified by harbour by-laws. She would also be working within the appropriate regulations of the Rule of the Road and, under such circumstances, be *"exhibiting a cylinder or the three red lights in a vertical line at night-time"* indicating her deep draft and notifying other vessels of her inability to deviate from a dredged channel or the centre of a river. Generally, the pilot would notice other ships that were coming perhaps too close and be open to taking avoiding or warning action, even to the extent of using the normally compulsory attending tug. For many years now, certainly since the early 1950's, some of the more enlightened and responsible port vessel tracking authorities use a harbour service launch to precede the path of the Supertanker, particularly when she is entering the port fully or partially laden, keeping other craft out of the way. This trend, however, is not reflected by the authorities in every port which handles VLCC's. Regarding other ships, very close attention to a common sense interpretation of Regulations has to be paid, for, so long as each ship does what is legally required of them, then very few problems are usually presented. The other condition when interaction is likely to be encountered is when the Supertanker is closing river banks or entrances to harbours. Again, by virtue of their size, it is unlikely — but not impossible — that they would close banks sufficiently for this to happen. If it did, then invariably, the answer would be for a reduction in speed and possibly to use the services of tugs.

Undoubtedly, many ideas involved in handling Supertankers by the deck officers are shared, but these can be succinctly summed-up in the rules directing the need for a responsible and professional attitude on the part of those in control, and intelligent use made of speed when appropriate for the existing circumstances and conditions in which the VLCC is steaming.

Chapter Eleven

ANCHORING/BERTHING/LIGHTENING

11.1. INTRODUCTION

11.2. ANCHORING

11.3. LYING AT ANCHOR, THE ANCHORAGE

11.4. BERTHING, THE USE OF TUGS

11.5. MOORING AT JETTIES

11.6. THE USE OF DOPPLERS: SONAR AND RADAR

11.7. MOORING ROPES AND WIRES

11.8. ADJUSTING POSITION TO THE CHIKSANS

11.9. MOORING EQUIPMENT

11.10. DEPARTURE AND DROPPING THE PILOT

11.11. SBM'S

11.12. LIGHTENING AND SHIP-TO-SHIP TRANSFER

Chapter Eleven

ANCHORING/BERTHING/LIGHTENING

(General Introductory Comments — ANCHORING: Letting go — Amount of Cable Used — Bringing up to Anchor — Anchor Bearings and Watch — Weighing. MOORING: Preparations from Deep Sea — Time Factors — Tug Work — Sonor Doppler-Radar Doppler — Deck Preparations — Mooring Ropes and Wires — Departing Berth. OFF-LOADING OPERATIONS: SINGLE BUOY MOORING: Buoy Descriptions- — Methods — Single Point /Pipe Mooring. LIGHTENING: Operation Details — Photographic Talk Through).

11.1. INTRODUCTION

In the words of Rule 3, 'General Definitions', Section (f), of the 1972 IMO-based Collision Regulations: *The word "underway" means that a vessel is not at anchor, or made fast to the shore, or aground.* The Rule implies, therefore, that the range of manoeuvres required to bring a VLCC to rest when she is in ballast, or in a fully/partially loaded condition, are really quite limited. The ship is either in an anchorage, berthed to a jetty, alongside at an oil terminal, or she is brought up to a mooring buoy. Clearly, the former conditions constitute the Supertanker being "made fast to the shore", whilst being "aground" means that she is 'on the putty', as the expression goes, with its implications of ill-intent by Act of God or, far more likely as the Courts testify, to mis-act of man. The latter, for the purposes of this chapter can be ignored, but each of the other manoeuvres is briefly examined. Throughout all operations, strict attention is always paid to SOLAS requirements concerning environmental protection.

11.2. ANCHORING

The process of anchoring a VLCC is generally quite straightforward. This is because very often Supertankers arrive at a specially designated deep-water anchorage. The main advantage of 'specially reserved' areas means that, in the run up to anchoring, the Master can generally make an unhurried and unhindered approach. This of course does not relieve him of his obligations to Rule 5 of the Collision Regulations, which emphasises the importance of maintaining *"a proper lookout by sight and hearing as well as by all available means in the prevailing circumstances and conditions"* and thus being constantly mindful of the existence of other shipping and their potential movements. This is particularly important when these very large ships approach a non-designated anchorage. Other precautions usual for anchoring any vessel would be taken into consideration, such as the nature of the seabed and its potential holding properties, depth of water, weather and tidal stream conditions, etc.

Usually, the port anchor would be chosen from which to swing and, as a rule, this would be 'walked out', that is allowed to run from the hawse pipe a little way in order to

Supertankers Anatomy and Operation

11.1. The Chief Officer on Fo'c's'le Duty is in constant Communication with the Pilot and Master by VHF Radio. (By kind permission of British Petroleum Shipping.)

ascertain that the cable had not become jammed in either the spurling or hawse pipes, an uncommon situation on Supertankers, but always a possibility. Once the ship is in the position chosen by the Master the order to let go would be given over the walkie-talkie (Photo. 11.1) and the officer in charge on the fo'c'sle would pass-on the instruction to the ship's carpenter (or 'Chippie') on the windlass. Invariably, the carpenter and deck officer would have discussed their plan and orders beforehand so that there would be no ambiguity in the event of any kind of emergency. It is for this reason that on most ships, the words 'let go' are generally restricted purely to anchor work. The amount of cable paid out would vary according to circumstances. I note from voyage records that, on one anchoring off Europoort we brought-up the ship in six shackles, or, 165 metres of cable in 31 metres of water. Thus, a ratio of between 152 and 183 metres of cable to each 30 metres of depth, depending upon the weather conditions. The shackles are marked with white paint on their actual position both on the cable and, for easier identification, the adjacent links on each side. The shackle is logged as being either 'in the water' or 'on deck', depending on the position of the cable mark. In this instance, the sixth shackle was on deck which, in practical terms, meant that the mark was slightly below the hawse pipe, thus we had 12.2 metres or so of cable out of the water. It would often take the best part of a shackle, (27.5m), before the anchor touched bottom.

Just prior to letting-go, the ship's engines should be put astern so that the cable is led out along the sea-bed and will not land in a heap on top of the anchor, with the possibility of fouling. As the cable slows its descent, following the initial momentum, the engines should be stopped and the ship allowed to take up a natural position determined by her draft and the strength of tide and wind. Invariably, she lies in the direction of the strongest of these elements but often, when in ballast, there will be a tendency for this to be in the direction of the wind due to the massive expanse of hull, superstructure and funnel. The remainder of the cable will then be allowed to run until a steady strain is noticed, the brake is then applied until the strain comes off the cable. A few minutes after the weight is felt to come off, and an even strain on the cable is experienced, indicates that the anchor has 'bitten' into the sea bed. Then the ship is said to be 'brought-up' to her anchor. The time for this to happen varies, but can be anything between eleven and thirty minutes depending on how long it takes the VLCC to settle. This period is generally longer when the ship is fully laden. To offer comparative times, for instance, I notice from records that, during other anchoring operations: in Ras Tannurah the port anchor was used, with three shackles on deck that took ten minutes to bring-up: in the Southend and Warp deep-water anchorage, we used the port anchor with five shackles on deck and this took twelve minutes. In Sete, the starboard anchor was used with six in the water, having taken 25 minutes to bring-up, this was due to strong winds and quite heavy seas.

11.3. LYING AT ANCHOR, THE ANCHORAGE

Once the ship is reported as brought-up, the duty mate on the bridge would take visual anchor bearings and, almost invariably, fix the ship's position additionally by Radar, Decca Navigator, or a similar electronic device. The officers then stand-down apart from the officer-of-the-watch who undertakes anchor watch duties. This means that he has to remain on the Bridge for the normal four-hour period of duty as the ships' officers usually keep sea-watches during these periods. The prime duty is to check that the ship is maintaining her anchored position, particularly during the initial stages when the possibility of dragging the anchor could most reasonably occur. A change in weather conditions may also lead this to happen. The second, equally important, role is to keep a look-out for approaching shipping. Whilst avoiding action is not easily undertaken, certainly a 'waking-up' signal on the ship's siren could be made in the event of any other vessels approaching at a rate and distance deemed potentially dangerous. An incident, just a few years ago, occurred when a coaster ran into the side of an anchored VLCC: the mate-on-watch on this dry cargo vessel had fallen asleep in the wheelhouse through sheer exhaustion, apparently having been on continuous passage watch for eight hours. The first he knew that the accident had happened was when he was thrown across the wheelhouse following the impact of the collision, something of a rude awakening. Luckily, the coaster hit one of the frames of the VLCC, otherwise the repercussions could have been very serious. An effective VHF listening watch, for any movement orders for the vessel from the pilot station, needs to be undertaken. An eye is usually kept on any men working on deck, especially those who might be painting in the vicinity of the ship's rails, or rigging the gangway as a preliminary to going alongside, or preparing this in conjunction with the pilot ladder. Any over-side discharges should also be noted, looking particularly for contaminants.

The anchorage off Rotterdam, a leading European oil-port, is about forty miles from the berth in Europoort and, at this kind of distance, the pilot would come to the ship by helicopter. (Photo. 11.2). Very often there are two pilots, a senior captain, who performs the operation, accompanied by a junior pilot who might be training for the senior job. The latter invariably covers the steering, either by doing so himself or supervising the deck cadet or quartermaster. This is especially so in the later stages as the ship approaches the berth. He would also do any administration, and sometimes the VHF communication with the tugs. The ship's officers remained in charge of the navigation of the ship with the entry in the log-book "VTMOPA" or, 'Vessel to Master's Orders and Pilot's Advice'.

Other than in an emergency, Supertankers invariably anchor because they are awaiting a tide or for a berth. In the case of the former, the delay is generally a matter of hours but, in the latter, it might well be a few days. When orders to go alongside are received, the duty mate

11.2. When VLCC's are anchored far from land, the Pilot usually comes to ship by Helicopter.
(By kind permission of British Petroleum Shipping.)

goes on to the fo'c'sle to weigh anchor. The depth generally prevents the anchor from having attached to it a marker buoy, which is the practice frequently adopted in smaller ships so that the actual position on the sea-bed can be seen more readily. This is a useful device because, on manoeuvring away from anchorage, it is important to see the direction in which the cable lays in relation to the anchor and the lie of the ship. This direction is notified to the bridge and very often the Master brings up the ship to dead slow ahead in order to facilitate weighing the anchor. It is not unusual for the cable to run under the ship, apparently in the most unlikely directions, as the VLCC moves to strong tidal ebbs and floods, and the varying wind strengths. One of these forces usually tends to dominate at any particular time and, when the ship has been anchored for a few days, the freely-suspended swing around her cable can be equally as haphazard. The swinging circle of a supertanker around her anchor can be quite considerable and often as much as half a mile bearing in mind the length of the ship itself and the cable to which she is lying. Invariably, dependant upon the amount of cable paid-out, it takes about twenty-five minutes to bring the cable 'up-and-down' in a vertical line, and to then house the anchor. It is always washed as it comes through the hawse pipe.

11.4. BERTHING, THE USE OF TUGS

Mooring a Supertanker alongside remains a skilled and time-consuming job. A notification of the ship's intentions is always passed to the engine-room so keeping them in the picture as events are planned, developed and, amended as necessary. If arriving at a port, the practice normally is to radio in at least twenty-four hours beforehand and keep the berth informed of progress as the ship continues on passage. The radio message indicates also details of the cargo and a suggestion regarding the ordering of tugs. The first action, once the pilot has boarded, is to discuss with the Captain the berthing plan. Whilst the Master often knows the port conditions due to previous visits, the pilot's up to date and often wider knowledge of local variations is always taken into account. In the light of the ship's draught, and the power ahead and astern of available tugs, a decision is made concerning the number which will be used. The longest period I can recall spending on berthing, from the initial time approaching the jetty until we were actually in position alongside and finally tied-up, was seven hours. We commenced, in Kharg Island, at 0700 and finally stood down at 1400 hours. This was an exception and was caused by a combination of tug shortages, thus necessitating use of a couple of tugs that possessed a reduced bollard pull, and thus were not so powerful as the harbour ones normally available. On this particular mooring we had also to cope with adverse weather conditions, in the form of strong off jetty winds, and a 'difficult' mooring party on the jetty and dolphins. The berthing crew were in the middle of a dispute concerning overtime payments and were thus 'going slow'. The circumstances proved a rather tiresome combination of events.

When about two or three miles off the berth, a single tug comes out to the VLCC and makes a line fast for'd thus acting as the attendant tug. It is generally the practice to use one of the tugs wires to make it fast and not the ship's lines. The shot of a VLCC owned by a major shipping company (Photo. 11.3) shows the attendant tug coming out to meet her charge. She remains 'in attendance' with the ship and can be joined by up to five others. Invariably, Supertankers are turned in the river before coming alongside so leaving the ship's head facing the sea in order to facilitate eventual departure. They berth just before the ebb tide commences, to help keep a tighter control of movement, often with just a few metres underkeel clearance between the bottom of the hull and the seabed.

On approaching the berth, the tugs are manoeuvred by the pilot to the best positions, again in the light of tide and weather, (Photo. 11.4) and the final run-in to the jetty is made. The disposition of the tugs here is quite significant for each one has an important part to play in drawing a different Supertanker (owned by the same Company) to her berth. The power

*11.3. An Attendance Tug approaches a VLCC as the Ship nears her Berth.
(By kind permission of British Petroleum Shipping.)*

*11.4. Tugs on the Port Quarter of a VLCC take the strain to lead the Ship into her Berth.
(By kind permission of British Petroleum Shipping.)*

exerted by the tugs is indicated by the wash from their propellers, their stems are really down into the water as they exert maximum pull on the large fully-laden ship. The tug on the right-hand side of the three is currently dominant because it can exert the maximum leverage forward of the ship to pull in the stern to the jetty. The centre tug is more restricted in its ability to manoeuvre, but will have maximum leverage when the ship needs pulling astern, whilst the third, left-hand side tug, although performing a minor role in the present situation, will also be highly effective later in drawing in the stern of the ship. Today, tugs possess such enhanced bollard pull that a 300,000 sdwt ULCC may be berthed quite easily using only two or three tugs, either made fast or in attendance.

11.5. MOORING AT JETTIES

The angle of approach can be up to thirty degrees, but sometimes the VLCC is brought alongside parallel to the jetty and then pushed into position by the tugs. The approach speed is crucial. If it is too slow then an unnecessarily prolonged operation could ensue, too fast and there could be a danger of demolishing the jetty. Many jetties continue to be made from greenheart timber, which is very strong and durable and has the resilience to withstand a fair amount of kinetic energy, but the impact of being hit by the momentum of a 300,000 dwt Supertanker, if she is moving much above the critical approach speed, might well prove too much. We hit a jetty on one occasion coming alongside, with about one third full load on board, and I saw distinctly, from my position aft, a chunk of concrete about three square metres fall into the water from underneath the jetty. When the Master and I inspected the jetty, once the ship was in position alongside, this showed clear evidence that the piping had been moved slightly out of alignment. It was a few centimetres out of true. Generally, the approach speed requires a degree of accuracy far too sensitive for the human senses to assess. Technological methods have been devised to help. There are two electronic aids open to the seafarer, both of which are activated for the final quarter mile approach when the speed should be about one quarter of a knot, or between approx 5 to 7.5 metres per minute, which reduces to about one metre per minute for the ultimate stage. Both methods work on the Doppler shift in frequency principle, proportional to the relative motion of the ship.

11.6. THE USE OF DOPPLERS: SONAR AND RADAR

Sonar Doppler is a robust high frequency device, operating around 300 kHz, which has the advantage of being a self-contained unit devoid of any shore contacts. It does not suffer problems of vocal ship's radio/VHF interference and the errors of accuracy in Doppler shift are perfectly acceptable. Some users of the early type models found problems with random noise effects on accuracy of the readings caused, they believed, by the ship's engines turning-over albeit gradually. Perhaps more interference, however, is caused in this respect by the noise of the engines of the powerful tugs ranged along the length of the ship. The wave is often subject to absorption and refraction energy loss, and dirt on the transducers, marine growths and air bubbles also play a detrimental part.

Doppler Radar, on the other hand, is sometimes regarded as a more favourable alternative. This devise operates on the ultra-high frequency bands at about 14 Ghz, but is not a self-contained system. It relies on a transmitter on the jetty, and a receiver on the bridge usually brought on board by the docking pilot. It remains essentially a line-of-sight system, very often owned by the jetty or oil company/owner. The fact that it is portable and transferable is useful as also it gives direct, see-at-a-glance readings.

11.7. MOORING ROPES AND WIRES

The crew meanwhile working on the main-deck (Photo. 11.5) are busy under the supervision of their officers preparing the moorings for winches in readiness to be sent to the shore by

11.5. *Moorings being prepared on the Fo'c'sle of a VLCC.*
(By kind permission of British Petroleum Shipping.)

Anchoring/Berthing/Lightening

*11.6. Nylon Rope on its Hawser Reel in the Bosun's Store For'd.
(By kind permission of Shell International Trading and Shipping.)*

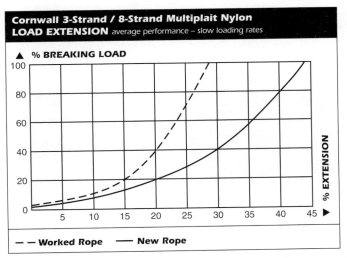

*Graph 11.1 Worked Nylon Ropes, when properly cared for, continue to offer excellent service aboard very large tankers.
(By kind permission of Marlowe Ropes.)*

the mooring boat. It has been shown already, in Chapter 4, that the mooring wires aboard VLCC's are kept permanently housed, originally on split drum self-tensioning winches, that are fitted with drum ends enabling synthetic mooring ropes to be handled during docking operations. Chapter 5 indicated that the nylon rope hawser reels (Photo. 11.6) are kept in the Bosun's store forward and the Steering Compartment Aft. These reels are self-winding and those used aboard Supertankers usually have a length of 420 metres with a Size 10, or 80 mm, 8-strand multiplait nylon rope, although double braid ropes are often used for additional strength. The drawing, reproduced from Page 70 of the OCIMF Mooring Equipment Guidelines, indicates the structure and differences between the various strands of rope. (Diag. 11.1). Synthetic ropes have considerable advantages for use on board very large ships over natural fibres. They are very strong, possessing twice the strength of sisal, and have a greater degree of elasticity that enables a considerable amount of shock to be absorbed during docking. The following table and graph give an indication of the breaking strains in nylon. (Table 11.1). Certainly, the graphical representation, (Graph 11.1), illustrates the load extension permitted by new ropes and shows that, when worked and sensibly cared for, the percentile breaking load remains within acceptable limits aboard very large tankers.

Supertankers Anatomy and Operation

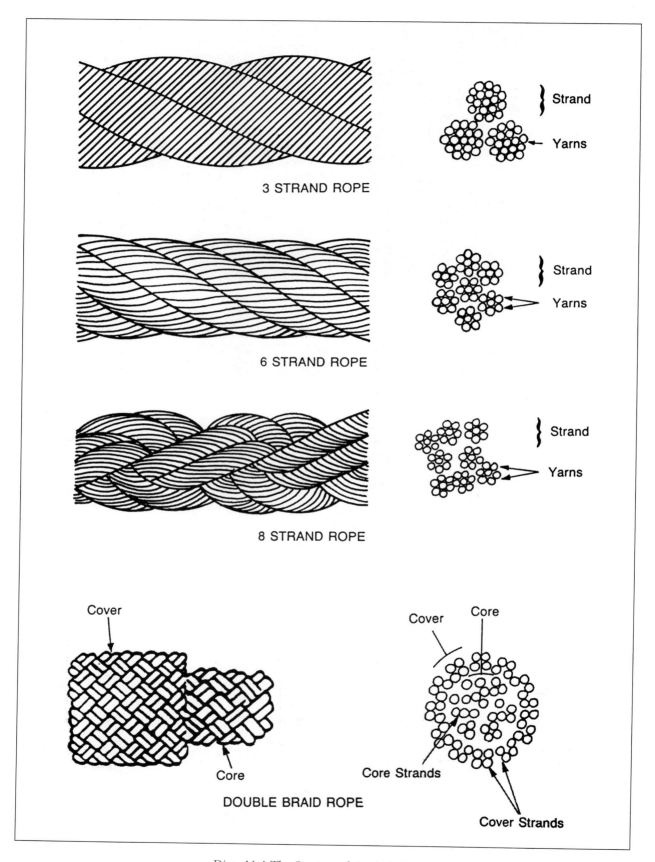

Diag. 11.1 The Structure of Synthetic Ropes.
(By kind permission of OCIMF).

Anchoring/Berthing/Lightening

		Cornwall 3-Strand and 8-Strand Multiplait Nylon			Superline Circular Braided Nylon		
Dia. mm	Size No.	Nominal Mass kg/100m	Minimum Strength tonnes	Average Strength tonnes	Nominal Mass kg/100m	Minimum Strength tonnes	Average Strength tonnes
6		2.37	0.75	0.83			
8		4.20	1.35	1.49			
10		6.50	2.08	2.29			
12		9.40	3.00	3.30			
14		12.8	4.10	4.51			
16	2	16.6	5.30	5.83	15.2	7.35	8.09
18	2¼	21.0	6.70	7.37	19.2	19.14	10.1
20	2½	26.0	8.30	9.13	23.2	10.7	11.8
22	2¾	31.5	10.0	11.0	28.1	12.9	14.2
24	3	37.5	12.0	13.2	33.6	15.7	17.3
28	3½	51.0	15.8	17.4	45.7	19.9	21.9
32	4	66.5	20.0	22.0	60.1	25.2	27.7
36	4½	84.0	24.8	27.3	76.2	35.7	39.3
40	5	104	30.0	33.0	90.9	39.9	43.9
44	5½	126	35.8	39.4	110	48.3	53.1
48	6	150	42.0	46.2	129	58.8	64.7
52	6½	175	48.8	53.7	152	68.2	75.0
56	7	203	56.0	61.6	180	78.7	86.6
60	7½	233	63.9	70.3	207	90.3	99.3
64	8	265	72.0	79.2	229	103	113
72	9	336	90.0	99.0	288	129	142
80	10	415	110	121	361	161	177
88	11	502	131	144	437	195	215
96	12	598	154	169	520	232	255
104	13	700	180	198	610	271	298
112	14	810	210	231	705	308	339
120	15	930	240	264	810	353	388
128	16	1060	270	297	925	394	433
136	17	1197	305	336	1045	444	488
144	18	1340	340	374	1170	488	537
152	19	1493	379	417	1306	544	598
160	20	1660	420	462	1450	590	649
168	21	1830	460	506	1590	650	715
176	22	2010	510	561	1750	697	767
184	23	2190	560	616	1915	762	838
192	24	2390	620	682	2100	840	924

Minimum strengths illustrated for Cornwall are in accordance with BS4928/85. Minimum strengths based on an approximate 98% confidence factor. Please contact us for in-depth technical details.

Table 11.1 Table Indicating the Breaking Strain of Nylon Ropes. (By kind permission of Marlowe Ropes.)

Nylon ropes are, to all intents and purposes, practically waterproof and appear to be little affected by the heat. The advantages far outweigh practical problems encountered by the deckhands who have to work with them, for they tend to be very heavy when wet and rather difficult to manipulate. From the owner's consideration, they are expensive to buy and replace, even though they have a life span which makes them economically viable. During my time, certainly, I never recall any ropes parting, as I have experienced on dry cargo ships, and they never required splicing or replacing. This was probably just as well as the job of splicing on board warps of these dimensions would be virtually impossible for the resources of most VLCC's.

The wires used in mooring are often of 137mm circumference and 280 metres in length. They are made from galvanised round strand wire, which has six strands in equal lay around a 41 mm steel core. The steel core is often used in preference to a fibre core because the heat generated in the storage on the drum, especially when the ship is in tropical waters, could lead more readily to the rotting of a rope core. Galvanised wire is used owing to its greater resistance to corrosion. So long as care is exercised when winding in the wire, to make certain that it does not become too unevenly coiled on the drum, maximum life can be obtained from the wire. The idea of mixing ropes and wires is a matter of concern to some serving captains. Inevitably, it is totally inconceivable that ropes only are used as hawsers for mooring purposes, but many Masters dislike the idea of a mixture of ropes and wires and would prefer their VLCC's to be fitted only with wires. There are arguments on both sides. It is one of the 'personal preferences' referred to in the Preface. Manually operated winches are now in use on all VLCC's. Self-tensioning winches, when they first appeared on VLCC's, were supposed to adjust themselves automatically against the rise and fall of the tide, but they proved so lethal on VLCC's, especially in very strong tidal conditions, that they have now been banned completely. They often ran out the wire when the current was strong, but were not always effective when the current slackened. This had a tendency of allowing the ship to move along her moorings placing undue strain on the chiksans. Captain Graham, in fact discussing this issue, recalled an incident, on a smaller tanker, when one of the self-tensioning winches let out the mooring to an extent that a chiksan was pulled over. A chain reaction was then created which led each of the VLCC's mooring lines successively to part.

11.8. ADJUSTING POSITION TO THE CHIKSANS

There is very little that can be done to prepare the ship for cargo discharge until she is alongside as the ship must be kept in a safe condition whilst the tugs are in attendance. All tanks have to be kept securely fastened as occasionally a tug might emit a spark from its funnel or exhaust. The photograph of a VLCC approaching parallel to her berth shows clearly

11.7. A VLCC in the final stages of approaching her berth at Lavera, France.
(By kind permission of British Petroleum Shipping.)

Anchoring/Berthing/Lightening

11.8. *Tugs keep a VLCC in position alongside the berth until all moorings are in place.*
(By kind permission of British Petroleum Shipping.)

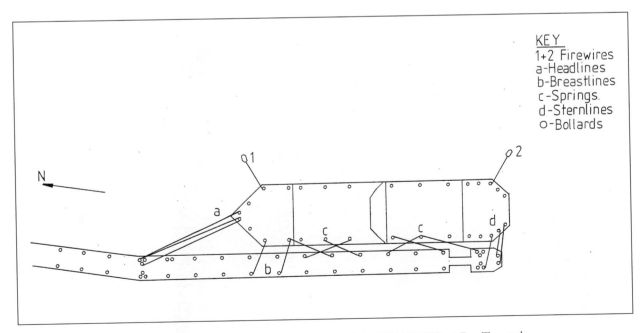

Diag. 11.2 *Mooring Diagram of VLCC RANIA CHANDRIS at Ras Tannurah.*
(The Author and Paul Taylor.).

Supertankers Anatomy and Operation

a number of the stages involved. (Photo. 11.7). Up forward, by the end of the jetty, a tug takes the strain on a hawser, keeping the Supertanker's head in the position required by the pilot and Master. A mooring boat has taken a breast rope just aft of midships to the bollard on the jetty that would probably act ultimately as the for'd back spring, warping-in the ship. When she is finally brought alongside there may well be a period of adjusting along the berth in order to line up directly the manifolds on the ship with the chiksans, or loading arms, ashore. Once this alignment has been effected, a note is made in the Movements Book that the ship is 'in position alongside', and the time noted. A number of tugs will then keep her into the berth by pushing along the length of the hull against the ship's frames, in positions indicated by narrow white vertical lines, thus avoiding the danger of puncturing any tanks. The tugs will remain on duty until sufficient mooring lines have been put ashore for the ship to remain in position unaided, as indicated in this photograph of the same VLCC awaiting additional wires. (Photo. 11.8). Moorings are laid in accordance with the requirements stipulated by IMO, but could very often be of the following pattern, (Diag. 11.2), which was a sequence that we used during a berthing at Ras Tanura. The springs are normally put ashore first to allow the ship to surge alongside. These are followed by the first of the head and stern lines, then the breast ropes, necessary to keep the VLCC well into the jetty, and then the other ropes and wires added until sufficient are put out, along the entire length of the ship. The number being determined by the weather and sea conditions, particularly the direction and strength of wind and current. The fire wire is lowered off the off jetty side bow and quarter and adjusted to keep it within a couple of metres of the water line so that, in an emergency, a tug can quickly take the ship off the jetty.

11.9. MOORING EQUIPMENT

There are usually seven lines forward, and the same number aft. These might well

11.9. A Mooring Dolphin at Number Four Jetty, Coryton, River Thames.
(The Author)

consist of three nylon/polypropylenes which, because of their elasticity, act as back-springs and breast ropes, with two wires to act as stern moorings, one of which shall be paid out over the opposite quarter to hold the stern more effectively. Another set, consisting of a wire and a rope act as additional breast ropes to keep the vessel from surging too much on the jetty. The back springs, for'd and aft, are always the first ropes ashore and are used to ease the ship alongside. The breast rope/wires and head/stern wires are then put out. Obviously, the positioning of moorings varies according to the facilities provided by the jetty. In the River Thames at Coryton, for instance, a number of the fore and aft wires would be attached to dolphins. These were isolated mooring pontoons extending beyond the jetty. (Photo. 11.9). I found, during berthing operations, that the ship was really too lengthy for the facilities available because the mooring wires, especially those extending as head lines forward of the vessel, were stretched almost to the limit of their capacity leaving very little leeway if it had been required. The tides in the Thames run at quite an appreciable rate and it is necessary to check moorings continuously during the full ebb and flood, making certain that the ship is kept well in to the berth.

11.10. DEPARTURE AND DROPPING THE PILOT

The procedure for leaving a jetty is rarely a problem. A tug is taken aft and occasionally another forward to assist, both with steering, and to help the ship's engines pull the VLCC off the berth. This/these are then cast-off and the ship proceeds on her way. From records, on leaving Las Palmas on one occasion, I noticed that checking the crew were on board and then turning them to for stand-by, was the first duty. The pilot boarded at 0154, after which the gangway and pilot ladder were lowered in preparation for his eventual departure. We rung stand-by engines at 0248, took the tugs fore and aft at 0250, and gently manoeuvred off the berth and out of the harbour entrance. We dropped the pilot at 0400 from whence I proceeded on watch until 0800. Most departures took approximately one hour.

Hydraulic pilot hoists were features aboard the early generation of supertankers due to the fifteen metres or more distance from the water line to the main deck when a VLCC was in ballast. These however proved too dangerous and, after the sad death of a pilot whilst using a hoist, they were immediately discontinued. The current practice allows the gangway to be lowered as conveniently as possible and then a pilot ladder continues the distance to the water line. When the ship is loaded, of course, the freeboard is only a matter of a few metres and so the pilot ladder only is used, in the manner conventional to all ships that are not fitted with a lower gun port door in the ship's side.

11.11. SBM'S

It is possible for fully laden Supertankers to proceed up the English Channel on the French side, thus keeping to the Traffic Separation Zones, to go into Europoort, leaving the lanes on the English side for outward bound ships. There is, however, even in the deeper waters on the French side of the Dover Strait, not a great deal of under keel clearance and the majority of masters would prefer not to bring fully laden, 98% capacity VLCC's into these waters. It is for safety and environmental reasons, as much as preference, that many Supertankers call-in at another port for partial discharge, or lighten ship by discharging partially at a single-buoy cargo mooring, or off-loading into a smaller tanker.

Supertankers of a fleet with whom I served would invariably call in at Le Havre or Antifer for lightening although, on occasions, we would go to various Mediterranean ports including Sete, a French oil terminal opposite Marseilles, in the Gulf of Lyons. There we would discharge at the Frontignan Single Buoy Mooring (SBM) which was situated some 3.6 miles from the harbour, in about thirteen metres. On one voyage, we picked up the Sete Fairway Buoy at

0800, at a distance of 5.2 miles and, twenty-five minutes later, swung the ship to enter the Fairway proper, so commencing the twenty-minute approach along the dredged channel to take-on the pilot at 0900. It was too rough for us to pick up lines from the SBM and thus we had to anchor until the fresh ESE gale abated from its Force 8 winds. This was not until 0500 the next day. We eventually took on the pilot and turned very slowly into the now Force 2/3 wind to pick up the floating mooring lines that are provided at the buoy.

The SBM's, although of various sizes, are usually substantially constructed and are fixed firmly to the seabed with a combination of sinker and grapnel type anchors, similar to those used for mooring light vessels, whose mushroom shape digs firmly into the mud and are capable of withstanding considerable strain. The moorings are so constructed that they enable the buoy to swivel to the changes of tide so that the VLCC made fast will take up the pressure on the mooring ropes and swing with the appropriate condition of tide. The photograph, (Photo. 11.10), shows a VLCC made fast to a SBM and gives a good indication of how the ship and buoy are connected. The OCIMF book on Single Point Moorings offers universally accepted recommendations governing the fittings and equipment of the buoys. A service launch and berthing master would normally be in attendance to assist the VLCC to pick up the moorings. On board some ships, the two bow chains are made fast by means of a Carpenter Stopper and led by a docking winch through to the windlass thus ensuring a sufficiently powerful attachment that will withstand gale force winds. In more common use, as it was shown in Chapter Four, is the AKD Stopper which performs the same function of securing firmly the mooring wires, but is less cumbersome to use, and more reliable. The oil-pipelines are flexible and float on top of the sea. They are picked up by the launch and connected to the cargo manifolds once, of course, the ship has been moored. A submerged cargo pipeline runs through the centre of the SBM and leads to the refinery, which can be a considerable distance inland, and through which the oil can be

11.10. VLCC at a Single Buoy Mooring (SBM).
(By kind permission of British Petroleum Shipping.)

discharged. The SBM is not to be confused with the Single Point or Pipe Mooring (SPM). This type is far more robust and substantially constructed and is fitted with a helicopter pad. It can be in excess of 70 metres in height, from pad top to the bottom of a base that is constructed upon piles fixed onto the sea bed. The extremely strong construction makes this type of mooring ideal for work in the North Sea, for example.

Whilst working cargo at a SBM, the ship's engines are always kept in readiness and it is a requirement that a responsible crew-member is in attendance at all times. The fo'c'sle head must be equipped with a VHF set linked directly to the duty mate so that immediate action may be taken if there is any indication that the ship might override the buoy. Ship's pumps are used for discharge, starting gradually, and building up to about 10,000 tonnes per hour, although this rate is variable depending on the number and duration of the stoppages which might occur for various reasons throughout the discharge. On the voyage in question, we discharged about 50,000 heavy and 100,000 tonnes of light crude over a period of something like sixteen hours.

11.12. LIGHTENING AND SHIP-TO-SHIP TRANSFER

Lightening into a smaller tanker is very much a viable alternative, so far as the deck officer is concerned, if not always to the owner. This operation always takes place with both ships initially under way and, when made fast alongside, with the larger ship only at anchor. Both ships have to be under way initially, because if one ship was at anchor, with the other approaching alongside her, it would prove extremely difficult for them to make fast. The momentum gives a tendency to swing around the anchor cable of the vessel being brought alongside. It is far easier for both ships to be moving ahead very slowly. Coming into the United Kingdom, we used to operate in Lyme Bay, anchoring in a position about 7.5 miles off Berry Head. Again, notification through the ship's owners/agents to the oil company operating the lightening vessel, and later, VHF radio contact with the ship direct, would ensure that both ships met in the agreed place at the agreed time.

From records, I noticed that on one operation, coming in from deep-sea, we rang on 'End of Passage' (EOP) at 0700 hours, 9.4 miles off Start Point. At 0817, we made direct VHF contact with the 50,800 dwt "ESSO CARDIFF" and came-round to an agreed course of 270° (T) and allowed the lightening ship, with her greater manoeuvrability, to draw alongside. The difference in sizes of the two tankers is quite impressive. (Draw. 11.1). By this time, we were both proceeding at a speed of 4 knots. At 0922, we took the first lines from the Esso tanker and by 1020 both ships were in position alongside and made fast. Fifteen minutes later, our larger ship let go the port anchor and by 1050 had brought up the ship to 6 shackles in the water. I notice that the wind was westerly-hence our course to steer directly into this, at a steady Force 6/7, gusting 8. The ship's head anchored at noon was 278° (G). We used the pipeline belonging to the "ESSO CARDIFF", in order to make the connection between the two ships, suspending this by use of the derricks of both tankers. The idea was to keep the pipeline away from all contact with the ships' rails not so much due to a build up of static electricity caused by the speed of the oil discharge, as both ships would be in an inert gas condition, but in order to stop the weight of pipes and cargo from causing damage. In the following drawing, (Draw. 11.2), the pipeline is shown raised above the deck of the smaller tanker and suspended by the married derricks of each. The second diagram, (Draw. 11.3), indicates the completed operation with the pipeline in the lowered position and the manifolds connected ready for off loading to commence.

By 1340, number 3 manifold was connected and five minutes later, we slowly cracked open the cargo valve and commenced a gradual discharge, building up to a rate of 5,000 tonnes per hour. Cargo pumps numbers 2 and 3 were used to discharge 42,000 tonnes total from numbers 1, 3 and 5 wing tanks. We had a problem with number one master valve on number one cargo line which led to a delay in commencing discharge, and ultimately we used number three

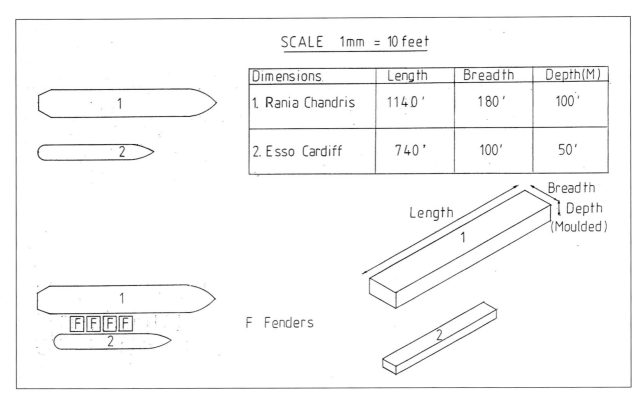

Draw. 11.1 The Relative Sizes of a VLCC and a Lightening Tanker.
(The Author and Paul Taylor.)

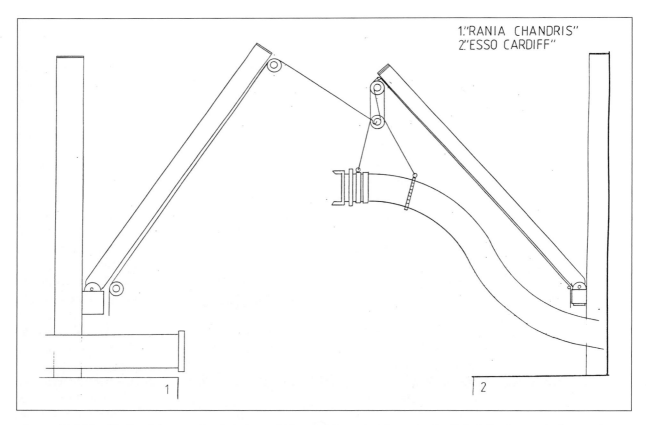

Draw. 11.2 The Pipeline being transferred during a Lightening Operation shown in the Raised Position ready for transfer ...
(The Author and Paul Taylor.)

Anchoring/Berthing/Lightening

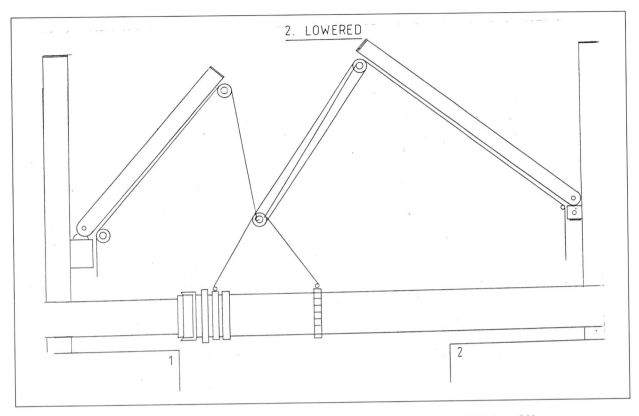

Draw. 11.3 ... and finally in the Lowered Position made fast to the VLCC's Manifold.
(The Author and Paul Taylor.)

11.11. A Swettenham Fender cradled aboard a Lightening Tanker.
(By kind permission of British Petroleum Shipping.)

Supertankers Anatomy and Operation

11.12. The smaller off-lightening tanker approaches the VLCC with fenders lowered in readiness to come alongside. (By kind permission of British Petroleum Shipping.)

11.13. Both tankers proceed slowly under way as the lightening tanker closes the VLCC's Starboard Quarter. (By kind permission of British Petroleum Shipping.)

11.14. With both Tankers in position, the final moorings are made fast. (By kind permission of British Petroleum Shipping.)

starboard manifold. We concluded lightening at 2132 on the same day and, thirteen minutes later, at 2145 hours, cast-off, with the smaller ship leaving us stern first at a very gradual speed and initially small angle. Our VLCC then weighed anchor, shipped the starboard derrick, and proceeded on passage for this particular voyage to Le Havre.

So far as the actual transfer was concerned, we complied strictly with the recommendations of OCIMF. The "ESSO CARDIFF" had been specially adapted for ship-to-ship lightening and was permanently engaged on this trade although, of course, she could soon be adapted to revert to normal trading. As a lightening ship, she was equipped with four special fenders, equally spaced along the length of her hull, called Swettenham Fenders. (Photo. 11.11). They were often foam-filled, measuring approximately 4.5 metres in length and 3.3m in diameter, and were suspended on specially constructed davits thus facilitating quick and efficient lowering and raising.

A series of photographs shows a 215,000 sdwt VLCC, belonging to a leading shipowner, being brought into position for lightening into one of the Company's smaller, 53,000 range sdwt tankers. The sequence offers an ideal practical illustration of the steps involved in this kind of operation. In the first shot (Photo. 11.12) the smaller tanker sets out to approach the larger VLCC in the distance. Once she closes, on a slightly converging course, she makes ready to come alongside, port side to on the starboard side of the larger tanker. (Photo. 11.13). The fenders are already lowered into position from their special derricks and at the appropriate level in the water, whilst the gantry crane is topped and prepared for handling the cargo pipe. The engines of the Supertanker are going dead slow ahead, as can be judged by the ship's wake, and her starboard derrick is topped also in preparation for the receipt of the pipe. This ship is 'well down to her marks' with not far short of twenty-five metres of ship below the surface. A calm sea prevails with just a moderate breeze, ideal weather for such a transfer although, as it has been shown, it would be in only very severe weather, around strong gale force 8 or above, that the operation would probably be aborted. Shot number three, (Photo. 11.14), taken about fifteen/twenty minutes later, indicates how the smaller tanker is virtually positioned alongside as the final checks are being made. Already breast and stern lines have been passed aft from the larger ship, as no doubt have those forward, but all ropes will remain in slack readiness until the manifold connections are correctly aligned between the two ships. By referring to the accommodation structures, an interesting comparison may be made of the beam distances between the two ships for, although the smaller is much lighter in draught,

11.15. Using two cranes in this operation, the pipelines are lowered into position. (By kind permission of British Petroleum Shipping.)

Supertankers Anatomy and Operation

11.16. *With the pipelines still suspended, the final manoeuvring into position adjacent to the Manifolds takes place. (By kind permission of British Petroleum Shipping.)*

11.17. *The Manifolds and Pipelines are Connected. (By kind permission of British Petroleum Shipping.)*

the wing tip of the VLCC towers above that of the other vessel. The funnel and after-mast sizes are also very pronounced. It is noticeable that the gangway of the Supertanker is housed emphasising that, for convenience, as much as possible of the gear belonging to the smaller tanker is used. The next shot, ((Photo. 11.15), shows the two pipelines, tied together to facilitate transfer, being passed between the ships and positioned for lowering to the manifolds. (Photo. 11.16) looks forward, and shows the ships in position, the gangway passed, and one head wire, a breast wire and three back springs forward made fast. Two pipelines have been passed to the VLCC and are suspended with their weight taken by the gantry of the crane. The cargo hook belonging to the runner of the lightening tanker is ready to receive the pipes once the crew have prepared the manifold connections. (Photo. 11.17) indicates the midships load line and draught marks of the smaller tanker, clearly visible in the background, whilst the combined crew fit appropriate flanges and reducing pieces at the manifold connections. In the final photo of the series, (Photo. 11.18), the pipes have been connected and are suspended clear of the rails. It is night time and the cargo transfer has been proceeding for some hours: the lightening tanker is now much lower in the water and the deck levels are almost the same height. Another few hours should see the transfer completed with both ships then proceeding on their respective passages.

11.18. The offloading operation nears completion.
(By kind permission of British Petroleum Shipping.)

Chapter Twelve

TRAINING DECK CADETS

12.1. INTRODUCTION

12.2. DECLINE OF THE BRITISH MERCHANT FLEET

12.3. RECRUITMENT, A FUTURE CAREER?

12.4. RECOGNISING AND ADDRESSING MANNING PROBLEMS, TRAINING

12.5. SPONSORSHIP

12.6. CERTIFICATION OF OFFICERS

12.7. CAREER OPPORTUNITIES

12.8. CADET TRAINING: PORTFOLIO
    12.8.1. Deck Duties
    12.8.2. Bridge Duties

12.9. SIGNALLING, COMMUNICATIONS

12.10. CAREER OPPORTUNITIES ASHORE, DIFFICULTIES ENCOUNTERED FILLING POSITIONS

Chapter Twelve

TRAINING DECK CADETS

(General Comments – International Maritime Organization and Standard of Certification and Watchkeeping Regulations – Nautical College Courses – Early Years of Training – Training Portfolio for Deck Cadets – Traditional Deck Skills – Later Years of Training – Bridge Duties – Distance Learning – Experience in Other Departments – Serious problems examined in Associated Shoreside Industries concerning the Employment of ex-Seafarers due to Previous Restrictions on Training – Success of Various Initiatives Considered to Remedy Situation).

12.1. INTRODUCTION

The deck officer's task is often made easier, and indeed more interesting, if the ship is fortunate enough to have on board one or more enthusiastic cadets. Many reputable British shipping companies consider cadet training as an important function and like to employ young men and women as future officers within the fleet. The aim is to equip such youngsters, as adequately as possible in their early years, to undertake or observe duties likely to be encountered as junior and later senior officers. It is, after all, from deck cadets that are ultimately derived ship's masters.

12.2. DECLINE OF THE BRITISH MERCHANT FLEET

The problem of course, perpetually, is and always has been, in attracting the right youngsters initially by offering them a career when so many more attractive alternatives beckon. The problem was confronted effectively – in a manner that was negative, conclusive and extremely short-sighted – in the early 1980's when virtually all deck and engineer cadet training stopped. Arguably, this was due to the emergence of expanding merchant fleets and an influx of maritime labour belonging to 'Third World' countries and foreign flags, that sometimes drastically under-cut the freight rates as well as officer and ratings' wages. A feeling of uncertainty for the future of the British Merchant Navy under the Red Ensign was created which led, almost within a decade, to the disappearance of nearly all the major 'famously named' shipping companies and the amalgamation, virtually beyond identification, of the remainder. This combination of events led to a *considerable* reduction in the numbers of ships, such that entire fleets of major shipowners (i.e.: Clan Line, Royal Mail and British India, to name just three) simply vanished, many to be "swallowed up" by larger Organisations, whilst the fleets of most other leading companies became decimated, with a correspondingly serious loss in jobs. I think particularly of the Ellerman Group, with whom I initially went to sea who, in under twenty years, saw their fleet decrease from around 80 ships to six: the Shell Group from over 250 to about 30 and British Petroleum falling from in excess of 100 to less than 20 now. These companies were all major employers of cadets, often having at least one and sometimes up to three or four on board every one of their ships with, of course, a large number on leave or attending training colleges. Ellerman's alone at one stage in the 1960's, for example, engaged well over two hundred deck, as well as a large number of engineering, cadets. Not surprisingly, therefore, a sense of unease and instability within the industry was soon engendered that was far from conducive to the recruitment of young people. In the eyes of the general public and, more importantly the school-leaving population in particular, the United Kingdom ceased to have a merchant navy and the opinion existed that there were, therefore, no outlets for a sea-going career. There was an additional major recruiting source of deck and engineering cadets which was the Shipping Federation. This organisation had an office in every sea port in the United Kingdom and was an essential agency for the administration of officers, including cadets, and virtually all ratings. They undertook responsibility for interviewing potential applicants, arranging medical examinations and the referral of deck cadets to the Department of Transport for the new entrant eyesight tests. The Federation issued to all seafarers identity cards and the discharge book which covers the career of the seaman as he was assigned to his various ships. For those cadets who applied to the Federation directly, they would issue the aspiring youngster with list of shipping companies. The boy would select, with advice as required, the trade of his choice and apply to the marine department of his chosen company. They would then interview and invariably sign the cadet to his cadetship or, in some instances an indentured apprenticeship, and appoint him to his first ship. Nearly all British shipping companies recruited from the Federation and, even if appointing direct, would still use Federation facilities for the issue of official documents registering the cadet as a British seafarer. The Federation gradually evolved into the Chamber of Shipping which continues to act as a voice on United Kingdom shipping matters but, with the demise of British registered ships and this financial backing, their resources have become extremely limited and they are now no longer able to offer the same scale of cadet recruitment.

Two decades later there has been, therefore, an understandable reluctance on the part of many schools to advertise or, indeed, encourage the merchant navy as a career for their school-leavers. The result has been inevitable. In their comment, from the January 1997 edition of the magazine 'Lloyd's Ship Manager', the editorial pointed out quite bluntly:

> "Owners and managers are finding it increasingly difficult to recruit sufficient experienced and qualified officers and particularly those with the necessary endorsements on specialist ships such as gas tankers. Of course, this can be said to be at least partly their own fault and is the inevitable consequence of years of failure to recruit and train sufficient cadets to maintain the necessary numbers of qualified staff".

Perhaps a counter argument could suggest that the very reversal in fortunes, the decline in numbers and indeed resources, of leading shipping companies just examined, could legitimately have engendered a 'knock on' effect which, by the adverse uncertainty created in the industry and its immediate future, led to equal instability regarding the approach to employing deck and engineering cadets.

12.3. RECRUITMENT, A FUTURE CAREER?

Whatever happens in the future, the nature of the British Merchant Navy, in the early 21st Century, will differ considerably from any time in its previous history but, of course, to those who can be encouraged to enter the profession this will not be of any particular significance. They will accept the circumstances and conditions as they find them as being very much the norm in the same way that cadets, entering the service in the 1960/70's, accepted the situation into which they were recruited as their norm and paid scant attention to what they had been told about the Merchant Navy in the 1950's and before. Apart from a very few Supertankers that sail under the Red Ensign, the majority of VLCC's are registered today with numerous other countries and fly a variety of different flags. In a similar manner, the type of young person who will find satisfaction with the seafaring way of life, and probably remain sufficiently long enough to obtain a qualification, will possess virtually the same kind of qualities which were necessary over thirty years ago. These personal qualities have been summarised adequately in the brochures of those responsible for advising professionally the potential new entrant. The following extract, quoted with their kind permission, from the leaflet outlining the cadet schemes offered by the Maritime Centre in Warsash serve as good as any alternative descriptions of the requirements found:

"A successful ship's officer requires some particular personal qualities in order to cope with the demands of the profession. He or she will spend several months at a time away from home living in close proximity with other crew members. The ability to cope with the stresses of separation whilst exercising tolerance towards others on the ship is therefore essential. Young officers must be able to accept a higher level of responsibility than would be expected at a similar age in most other professions. They must also possess the leadership qualities necessary to direct the work of others, often under difficult circumstances. Self-reliance, self-discipline, initiative and the ability to work as part of a team are also required as are the commercial awareness and management skills demanded by a competitive modern industry. Experience has shown that whilst high academic attainment is not necessary, a cadet needs to be competent in mathematics and a physical science in order to succeed."

This comment makes an interesting comparison with the approach advocated for potential merchant navy officers, that was published in the appropriate volume of an admirable series of booklets to young people, by Her Majesty's Stationery Office, following material prepared by the Ministry of Labour and Central Office of Information of the Central Youth Employment Executive, as long ago as 1952. Their concluding advice to enquiring entrants offered a message which extends *over a gap of nearly half a century* and reads surprising, or perhaps *not* so surprisingly, similar to the message of Warsash:

"You should consider very seriously, therefore, whether you are suited to the life so that you can make a wise decision. A liking for the work the right personality and a willingness to take the rough with the smooth, rather than just the vague desire to go to sea, will probably help you to find satisfaction in the Merchant Navy".

Certainly, it is worth emphasising that there will always remain a niche for the individualist among deck or engineer officers at sea. The efficient function of the team tasks to be fulfilled however will always rely upon good will, requiring the cadet to possess a temperament which allows flexible social interaction. With the reduction aboard most ships, in crew sizes, there remains a tendency for the life at sea to be a rather lonesome existence anyway. It is certainly possible, if the young officer is so inclined, to spend much of his life actually at sea without seeing *any* body other than those met during brief watch exchanges or encountered at the meal table during set times.

The shipping industry, however, has an additional problem upon which it needs to focus and overcome. Those recruiters for the Merchant Navy have to recognise that they are competing with other industries, commerce, the Armed Services and professions for the same kind of intelligent and well qualified youngster. These organisations could well be seen to be offering, in the eyes especially of the 'A' level entrant and their education establishments, far more attractive packages. In a paper, written especially for this chapter, Mr. Jim Gray of Clyde Marine Training in Glasgow, commented:

> *"It is apparent that young people educated to A-level standard will inevitably go to University or other Further Education establishments. Nevertheless some will not see that route as necessarily right for them and may well enter the Merchant Navy. From a recruitment point of view it is therefore vital that we emphasise the best in a Sea Career and certainly, at this time, job prospects have to be an attraction … ."*

There is some confirmation that this view possesses some credence. As a traditional public school, the college where I am employed in East Kent has, without pressure from a biased member of staff(!), found an awareness and interest on the part of the pupils to move away from our long established record of purely Armed Services scholarships for sixth formers. A gradually increasing number of students are now making serious applications to become deck and engineering officer 'Trainees', as they are now called in the Merchant Navy.

12.4. RECOGNISING AND ADDRESSING MANNING PROBLEMS, TRAINING

It takes around three/four years to produce a competent deep-sea watchkeeping officer and ten/twelve years-plus to create a competent Master. As the year 2002 approaches, following this twenty year period of neglect, a condition exists not of a drastic shortage of officers across all ranks, both deck and engineer, but a genuine 'crisis' state. This has been addressed by the industry in the widest interpretation of the word, but with a limited effect. The National Union of Marine and Shipping Transport, or NUMAST — the officer's union has, for many years, recognised both the nature, depth and extent of the problem and, particularly through their magazine 'Telegraph', has offered the shipping industry leadership by making serious efforts to voice genuine concern and thus urge government departments to recognise the vital need which exists for recruitment of both deck and engineer cadets. Additionally, they have presented policy documents to the government urging direct intervention and, in the late 1990's, they organised a joint conference with Warsash Maritime Centre specifically directed at examining issues concerning not only recruiting but, widening their brief, to include the range of national and international vacancies existing for qualified deck and engineer officers. A leading marine recruitment and training agency, in their presentation to the conference, advised 352 vacancies from UK and foreign companies. These included 36 jobs for masters, 64 vacancies for chief officers, 41 and 32 respectively for second and third mates, plus an additional 179 for engineers, along with 48 shoreside posts they had been approached to fill. This included requests for marine superintendents, general and assistant managers and lecturers. This sole agency had, over the previous year, attracted about 150 cadets to the merchant navy. Bearing in mind that other direct recruiting shipping companies and maritime recruiting agencies collectively had attracted about 450 deck and engineering cadets, there was still left a short-fall of something in excess of about 1000 cadets considered necessary to rectify deficiencies in the industry caused by the lamentable lack of foresight of the 1980's. Another perceptive comment on the situation appeared in a 'Lloyd's Shipping Manager' article, also in the January 1997 edition of this magazine, entitled 'Bitten by the Officer Gap':

> *"Even though efforts have been made recently to increase recruitment of cadets in the UK and numbers have increased slightly, they are still falling well short of the figure needed to maintain even the present levels of British officers."*

In an effort to redress something of a balance, there has been available, for some years now, Government Assisted Funds for Training (GAFT) which were intended originally to encourage all shipping companies to become actively involved and offer, if not recruitment, then at least training facilities and so contribute towards producing certificated seafarers. In keeping with many areas of belated government intervention, including the Development of Certificated

Seafarers (DOCS) in 1994, the intention now is fine but, in the absence of youngsters prepared to consider the British Merchant Navy as a career alternative, the practice remains questionable. Ideas are being mooted also which examine the possibility of extensions beyond initial cadetship and it is planned to incorporate continuous training until senior officer status is achieved. A possible criticism of the funding argues against the money being attached to the shipping company operating the scheme, suggesting that it might be more relevant for this to be assigned to the seafarer, on the grounds that taxpayers' money should be assigned to assisting the training only of a *British* Officer. Opposing this view is the equally valid argument that shipping today is far more of an international business than has ever before been the case, as can be seen in the number of British shipping companies who now have their fleets registered under foreign flags.

12.5. SPONSORSHIP

Unlike marine employment in the past when, as we have seen, all shipping companies had their marine and other departments which were responsible for direct recruiting, cadets today are only assigned to cadetships by a limited number of shipowners. Apart from the initiative of these few leading companies, virtually all recruits are now engaged under a sponsorship system, that is administered through a partnership scheme between four or fifty essentially United Kingdom based principals. Each of the involved has a clearly defined role that is measured against strictly interpreted legal criteria. The scheme is co-ordinated by the Chamber of Shipping in London, working in liaison with one of forty managing agents, the cadet, and a sponsoring shipping company. The cadet is put in touch with a managing agent by his/her careers advisors, or makes independent contact. Although practices vary, often considerably, the agent may undertake the important role of initial interviewing, and arranging for the potential cadet to be medically examined and then sponsored by a shipping company operating within his consortium. He/she would then be enrolled on a professional and academic training course with a nautical college, during which a structured training course would be followed operated by the Merchant Navy Training Board.

The sponsoring shipowner agrees to provide the seagoing training essential for a watchkeeping officer's professional qualification, to pay the cadet a monthly sponsorship grant, provide free board and lodging, and allow leave of so many days for each 30-day period served at sea. With many companies 'standard dress' has replaced the more conventional uniform although this latter, with its traditional merchant navy rank braid, continues to be protected by the relevant Merchant Shipping Act. The revised 1978 IMO International Convention of Standards of Training, Certification and Watchkeeping, known as 'STCW95', provides clearly defined rules regulating the duties of all officers on board ship. Chapter Two outlines the conduct of masters and deck officers. Other sections determine in considerable depth the details reflected in the national examination requirements for deck officers. In the UK, the examining authority for mates and engineers is the Marine Safety Agency (MSA) of the Department of Transport (DTp), whose rightly rigorous standards, give British certificates and officers a quality recognition recognised by virtually all countries in the world.

12.6. CERTIFICATION OF OFFICERS

Deck cadet training follows closely, therefore, one of the schemes whose requirements are stipulated by IMO so that youngsters are closely and very precisely taught prior to 'going for their ticket', as the process of sitting their first Certificate of *Competency* continues to be known colloquially. Deep-sea cadets for some years now have taken the qualification of their rank, so that the third officer would, until very recently, hold a class three certificate, after which he would return to sea until he had fulfilled the sea time requirement for a class two. Sometimes a class four certificate, of the lower grade, could equip a young person for a fourth or even a third mate's post aboard a deep-sea Supertanker. Similar to all regulations, however, it is impossible for nautical certificates to tabulate and regulate for every situation likely to be encountered. Whilst the oral element of assessment endeavours to present theoretical situations to candidates by experienced examiners, all of whom are ex-master mariner's who have held command, considerable trust and responsibility has to be placed upon the newly qualified officer in the ordinary practice of his profession. He has to learn by unobtrusive, but continually supervised, experience at sea. This presents an additional problem. Many junior officers are from the 'third world' countries who are certainly competent, but all too often possess little enthusiasm for promotion beyond the rank of second officer. There is a tendency, therefore, in some companies for a potential blockage to be created between the cadet upon completion of his time and senior officer status. Many leading companies see the wisdom of maintaining their investment and, regardless of increased cost in employing a British junior officer, do so nevertheless. There may well be a 'concentration of minds', concerning the employment of nationals within the industry, as revised STCW95 conventions begin to bite which require a more rigid enquiry into the issuing of watchkeeping qualifications emanating from some Third World countries.

Under Scheme A the deck cadet may, after completing a minimum of about fourteen months sea time, apply for his or her first Certificate of Competency provided an 'approved scheme of training' has been followed. Scheme B has a longer requirement for the completion of sea time, around eighteen months but, depending upon the scheme undertaken, the deck officer cadet trainee may commence his programme with a longer initial induction period in a nautical college selected by the shipping company or agent. The following chart (Table 12.1) indicates the progressive nature of the training offered under both schemes which extend over seven phases and take between three and a half and four years to complete, depending upon the initial academic qualifications offered by the cadet. Young people with GCSE's, for example, serve a longer cadetship than those entering the profession with "A" levels in mathematics and physics. The maritime industry has, since 1996, become involved with the Government National Vocational Qualification (NVQ) scheme, with the intention of providing a standard or norm of qualifications available across all British industries thus, it is argued, assisting the UK to become more internationally competitive. The MSA have agreed, and I quote from notes provided by Clyde Marine Training who are a leading agency: *"that a candidate presenting, say, a properly accredited level 3 NVQ would only have to sit the oral exam to get his ticket"*. The actual sea-time required before a candidate may present himself for MSA examination, I believe, compares adversely with the pre-1980 Regulations that stipulated a rigidly enforced four years' actual seatime, with appropriate remission for pre-sea training and/or "A" levels in mathematics and physics etc. I am not alone in knowing personally a number of pre-examination candidates who had to return to a ship in order to obtain often merely a few days' sea-time necessary to total the full four years. Encroachments into sea-service remain very much a cause for concern with all ships' masters with whom research conversations have been held, and is clearly likely to remain an area of concern with the DTp. In fact, the Clyde Marine conference paper continued by pointing out the importance to the MSA of qualifying sea-time, thus rightly preventing an academically able candidate from acquiring his first watchkeeping certificate *"with only very limited experience on board a ship at sea"*. The vital question of adequate sea-time, **actually served on board ship,** has been emphasised additionally in a report summarised from research undertaken by the Tavistock Institute of Human Relations, which was considered of sufficient importance to be presented by the Marine Directorate of the Department of Transport and published by Her Majesty's Stationery Office as early as 1991. I am indebted to each Organisation for permission to quote from page 18 (Section 5/29) of "The Human Element in Shipping Casualties":

> *"In terms of collisions arising from human fallibility, the prospects for prevention by providing more formal training may be limited. Incompetence generally surfaces as a lack of practised skill, which means that seafarers concerned need to spend more time actually at sea, **rather than to spend more time in sea-schools**".* (It is my emphasis that has been placed on the last statement).

TRAINING STRUCTURE-DECK OFFICER

STAGE	PHASE	LOCATION	DURATION* A	DURATION* B	CONTENT OUTLINE
INDUCTION	1	COLLEGE	6 weeks	12 weeks	Basic safety; familiarisation with the marine environment; basic core skills, CPSC, FF(1) etc UPK for selected Level 2 and 3 units.
	2	SEA	12 weeks	26 weeks	Application of UPK. Collection of evidence as per Portfolio requirements.
TRAINING	3	COLLEGE	15 weeks	24 weeks	Review of progress. Assessment MVO Level 2 units as required. Start HND Pt 1/UPK for MVO Level 3. ENS Part 1
	4	SEA	26 weeks	26 weeks	Development of practical skills and application of MVO Level 3 UPK. Collection of evidence as per Portfolio requirements.
	5	COLLEGE	9/5 weeks	12 weeks	Interim review and assessment. Comp HND Pt 1. GMDSS and FF(2)
		COLLEGE	2 weeks	2 weeks	Bridge Watchkeeping Preparatory Course (Optional)
DEVELOPMENT	6	SEA	25 weeks	26 weeks	Completion of Portfolio. Practical experience.
	7	COLLEGE		2 weeks	Assessment MVO Level 3. MSA orals (first certificate).
		COLLEGE	15 weeks	22 weeks	UPK MVO Level 4 Complete HND Part 2**
PROGRESS	8	SEA	12 months	12 months	Completion of MVO Level 4 Portfolio.
	9	COLLEGE	4 weeks	4 weeks	Assessment MVO Level 4. MSA orals (second certificate). HND**
		COLLEGE	6 weeks	6 weeks	UPK MVO Level 5
	10	SEA	24 months	24 months	Completion of MVO Level Portfolio
	11	COLLEGE	4 weeks	4 weeks	Assessment MVO Level 5. MSA Class 1 orals

* Approximate times. Excluding leave.
** Award of HND either at 'first certificate' or 'second certificate' level, depending on achievement.

Table 12.1 Training Schemes for Deck Cadets are progressive and well structured and include elements ashore and at sea.
(By kind permission of Merchant Navy Training Board.)

There exists potentially an additional problem. In the "olden days" of Second Mates and higher certificates, the oral examiner was able to identify and explore any weaknesses which had occurred in the candidate's written papers. This presumably will no longer be the case?

In practical terms, a deck cadet completing successfully the training period would have acquired a range of recognised vocational and academic qualifications: he/she could from 1997, have an MSA first Certificate of Competency, Class 3 or 4, enabling him/her to practice as a third watchkeeping officer at sea, and a Level 3 vocational qualification in merchant vessel operations. The officer would have also underpinning knowledge for a National Vocational Qualification in Marine Vessel Operations, Level 4. He may well possess also a university accredited Higher National Diploma (HND) at either 'first' or 'second' certificate level in nautical science. This qualification, following Grades 2/1 – or, (when ratified), Mate/Master Class 2, and Command Certificate Class 1, with completion of the additional sea-time requirement, attracts the equivalent academic status of a pass degree.

The young officer would also have a choice from a number of alternative immediate employment options and the opportunity of on-going training in order to reach the goal of Master Mariner. For the academic, there would continue to exist the now long standing options of a Bachelor of Science (Nautical Studies) Honours degree, with extension to a Master's degree and/or a doctorate. For continued professional progression at sea, a Class One, Master Mariner's Certificate is often held by chief officers as well, of course, as the ocean-going Captain. The Department of Transport has, for many years, discontinued its Extra Master's Certificate to those candidates who wished to demonstrate exceptional ability. The uptake for this latter, and very valuable, qualification was found limited, due to the considerable financial outlay involved. This attracted very little assistance and had to be borne by the officer concerned. I suppose the death knell for this advanced certificate came with the university degree qualification that carried enhanced recognition by a wider range of shore-side employers. Perhaps this series of parallel academic/nautical qualifications has something to be said in its favour as sometimes a Master Mariner's certificate has little value, in the eyes of many ashore, who simply have no understanding of what the qualification really means.

12.7. CAREER OPPORTUNITIES

Some cadets are lost to the shipping company or agency when, after qualifying, they move to other seagoing employers, often under the 'grass is greener' syndrome, even if this is found to be disappointingly 'marshy' upon arrival. Others leave seafaring entirely once they have qualified and enter numerous professions with their first certificate, but also with a wealth of experience totally unparalleled by their shore-side contemporaries of a similar age. Inevitably, as with all professions, a much smaller number lose interest along the way without even completing their training.

Although there are only a very small number of ratings taken into the industry each year, there exist excellent opportunities which encourage those who aspire to officer status to work towards advancement. These occur mainly by training ashore, very often through nautical college study, as well as 'distance learning' courses at sea following similar MNTB training. The rating who manages to become an officer 'through the hawsepipe', as it is still commonly known, is certainly worthy of respect. To some extent it takes considerably more determination, and sheer hard work, for a rating to become a mate than falls to the easier lot of the deck cadet with his company and officer based support and his carefully planned training programmes. One of the difficulties of aspiring ratings is actually finding opportunities to study at sea and the potential abuse sometimes experienced from other hands for 'considering himself superior to them by wanting to become an officer'; as one rating rather defensively stated his position.

Third officers are often aged twenty-one years or so, second officers from around twenty-two and Mates sometimes in their middle to late twenties. Command, for a particularly outstanding officer, could well be in the late twenties plus age bracket, but generally the company policies of VLCC operators seem to prefer the thirty year plus officers for this important senior post. The responsibilities of a navigating officer, not only today, but also in the past, are really quite considerable as well as being exciting. A twenty-year old in charge of a 300,000-plus sdwt Supertanker for eight hours per day is a challenge and a responsibility. The coastal trades also have their own system of cadetships and promotions. Deep-sea officers have been known to serve periodically aboard coastal tankers, with varying degrees of successful adaptation, but the traffic rarely goes the other way. There are, then, no hard and fast rules regarding promotions in later years. Clearly, the seniority which comes from length of service and its accompanying sea experience are extremely important. But dare I say, 'being in the right place at the right time' is occasionally useful when an officer of higher rank is required urgently, perhaps through an unexpected incident such as sickness, and there happens to be only the one officer available. In some companies, final year cadets are promoted uncertificated fourth officer for their last voyage before their first certificate examinations in order to take direct responsibility for a deck watch and other duties. When this occurs they invariably understudy the chief officer's Bridge watch and some of the duties of the other mates, such as taking charge on the fo'c'sle, or down aft, during a mooring operation, with the duty officer in attendance, ready to offer a word of advice, as necessary. He can be trusted also with greater responsibilities whilst loading or discharging cargoes, attending moorings etc. in port, all of which are immense 'confidence boosters' at just the correct time in the cadet's professional career.

12.8. CADET TRAINING: PORTFOLIO

Cadets are still required, as of old, to follow professional distance learning study courses whilst at sea. Much of the traditional practical 'hands-on' experience common to all deck cadets over the past years, continues to be integrated at all levels with the essential theoretically college-based courses. The Merchant Navy Training Board (MNTB) publishes for each cadet a Training Portfolio (Deck) which has only recently replaced the Deck Cadet's Record Book which had served some two generations of potential navigators. Like its predecessor, the portfolio is regarded as a very important document designed, in the words of the introductory remarks:

> "It constitutes a major part of the cadet/trainee's (or candidate's) planned training programme. Its completion and presentation to the candidate's assessor is therefore an integral and required element of the Training Programme."

It is interesting to note that there is a wide class inclusion of the definition of 'trainee' implying that, unlike the old Record Book, this portfolio is intended also for the use of deck ratings.

In a series of guidance notes directed to the Master and deck officers, following instructions to the cadet/trainee, reinforcement is made of the importance of maintaining an accurate and complete record of the sea-going phases of their training between the nautical college placements. Considerable stress is placed on ascertaining that deck cadets are properly trained for their eventual promotion, in unequivocal no nonsense terms:

> "It is extremely important that the candidate is given adequate supervised bridge watchkeeping experience after the first sea phase. The MSA require that all deck candidates for a first certificate of competency must spend six of their final twelve months qualifying sea time engaged in bridge watchkeeping duties under the supervision of a deck officer. This requirement means that the balance of emphasis in terms of training during the latter sea phases should be directed towards watchkeeping."

The instruction means an end to the days when cadets spent most of their cadetship on little else than practical deckwork which was the practice for many youngsters in olden and not so olden times. The final paragraph to the guidance notes makes equally plain the aims involved in training:

"Masters and officers need to know that the standards expected of the candidate (when competence is reached) is that of a person about to take up the job for which the award is made. Cadets are expected at the end of their training to be competent to start to undertake the job of watchkeeping officer, but they will clearly be lacking in experience."

A similar mandatory monitoring system exists for the use of engineering officer cadets as well as for those fewer numbers who are recruited on dual certificate courses designed to offer them joint deck and engineer officer training.

The book, therefore, is designed to help the deck cadet assimilate much of the theory which supports the practical aspects of his work. It provides knowledge that is essential, not only for him/her to understand how the equipment or technique being used actually functions, but also so that a realisation may be gained of the limitations involved in its use, hence providing enhanced realistic expectations. The range of theory includes all of the tasks necessary for a first watchkeeping certificate such as areas of general seamanship, cargo and bridge work, including a planned programme for learning the vitally important Collision Regulations, port operations/duties, and aspects of construction and stability of the various ships upon which the cadet will serve during his cadetship.

The portfolio is far more comprehensive than the old Record Book with a range of very specific tasks planned around a number of units pitched at the different levels required for his NVQ. The tasks cover service aboard up to six ships, on a "pro rata" basis in case the trainee serves on less than this number, so that all of the duties can be authenticated as being completed to the satisfaction of the witnessing officer. Additionally, the Master has to review regularly the progress of the cadet trainee and to sign a declaration to this effect on a monthly basis. A number of entirely new duties, which never appeared in the Record Book, are included which take into consideration the changes reflected in the industry over the previous decades. These include ecology, SOLAS , MARPOL, stock control and a useful module on working relationships aboard ship. The portfolio appears to be an excellent training document. It has been thoroughly researched and thought through with the contents presented in a sensible seamanlike manner. Far superior to the old Record Book, it guides the cadet systematically and progressively through the theoretical and practical performance of the skills required from him eventually as a young officer. He is given the opportunity of learning about the deck and bridge duties expected of him as a member of the shipboard crew and then, towards the end of his cadetship, a chance to lead his team.

12.8.1. Deck Duties

During the early stages of his sea service aboard the requisite six ships, cadets invariably spend some of their time working on deck with the ratings learning what it actually means to be a member of the crew. Whatever painting preparation and maintenance duties are undertaken by the ratings, however cosmetic aboard a supertanker these may prove to be, the cadets also will do, thus fulfilling tasks under the Reference unit E18 of his portfolio. Occasionally on some ships, cadets are assigned specific areas of the funnel and superstructure of VLCC's. They rig their own stage (E19) and self-lower this by means of adjustable slip knots on either side (Photo. 12.1). Roller brushes have been used for many years now and enable even a comparatively inexperienced boy or girl to paint a wide area without too much loss either in material, or self-esteem! The job looks professional upon completion with the absence of too many 'holidays' or patches of missed or unevenly applied coatings. It can be a satisfying job for a quiet period at sea.

12.1. On some VLCC's, deck cadets are still given responsibility for the maintenance and painting of the funnel area and superstructure.
(By kind permission of British Petroleum Shipping.)

This 'menial' deck work is essential, even today, in order to learn the practical skills of their chosen profession. It helps them to know how long any particular job would take, what kind and, equally important how much of, the materials are necessary in order to complete the task effectively and proficiently. All of this information becomes invaluable when, as Mates themselves, they are responsible for running their own deck department. It is important also from another aspect. Roger Green, an ex-VLCC cadet with a leading company who rose through the ranks to chief officer, summed up the situation quite adequately in his comment:

"My own motto when training, and the view of many others with whom I sailed, was that the competent officer should be able to do the job and demonstrate this safely and efficiently before ordering the crew or deck cadets to do this. This ensures that the duty is carried out safely and efficiently and avoids the trap of some smart ass saying 'Well, how do you do this?' The only really effective way of gaining this experience is by having carried out the tasks as a deck cadet."

My own experience with deck hands has more than justified the veracity of this comment as I recall an incident down aft, during a berthing stand-by as a very new third mate when my own deck man, who happened to be in the mooring crew, asked me what action would I take if the tug passed us a wire that was not spliced with an eye. Without being too 'heavy' in my response, I pointed out that there would probably be little point in expecting him to put one in but, in this highly unlikely event so gleefully forecast, the crew would take the unspliced wire round the bitts in the normal way and then tie it securely between the coils with small stuff to prevent it uncoiling. Certain skills, whilst perhaps varying in principle, alter very little in practice. Certainly, for the foreseeable future, moorings will always have to be put ashore and tug's wires taken on, and stoppers will continue to be used, regardless of the degree of sophistication new ideas might introduce. Tank ullages, temperatures and soundings will have to be taken and routine maintenance will always have to be effected, however limited in scope this may become in the future. Even the mundane monotony of chipping the paint needs to be carried out with some concentration. I recall the deck cadet who in advertently put his chipping hammer through the saloon window, literally minutes before the captain and senior officers entered for first lunch sitting, causing great consternation to the stewards! Whilst they were bustling around cleaning up the mess, in walked the 'great man', who surveyed the scene as well as the somewhat subdued cadet at a glance, and uttered the memorable words: "*Glass does not corrode, Beavington. There is, therefore, no need for you to chip it!*". P&O Lines, he was one of yours! The anecdotes any deck officer can recall of his own direct or shared experiences as a cadet are legion, and could well fill an entire book, let alone being cited casually as illustrations providing a little light relief in a chapter which is, after all, about youth. In the magazine of ships and the sea, '*Sea Breezes*', articles occasionally appear offering a flavour of cadet life, including one, in the April 1990 edition, entitled 'Blackout in Sydney Harbour', and the February 1994 copy – 'There's no Smoke Without Fire ... . Or is There?' from the present author, relating some of the misfortune's invariably self-inflicted that can fall the way of unsuspecting deck cadets.

Amongst other traditional work associated with the sailoring skills still useful for cadets to learn, although this is now no longer required under the new training syllabus, is included canvas sewing. This is best learnt by watching, and then doing, under the supervision of a reliable deckhand, usually one of the able seamen (AB's) or even the Bosun, the senior deck rating in charge directly under the Mate for the deck (or General Purpose – GP) crew. (Photo. 12.2). Canvas, where this is still supplied, is in raw bolt-form so that when this is used it has first to be measured then chalk patterned, cut accurately to size and finally sewn. The kind of palm and needle stitching on canvas necessary on Supertankers today is very limited, but is still found occasionally and used for making and/or repairing lifeboat covers, in the open type of boat to be found at sea, or making covers for ventilators and even still, on some ships, anchor shapes however academic an exercise these may prove. Splicing skills are still regarded as necessary, under E18 and 19, but not for use so much on mooring ropes. Due to their complexity and sheer sizes involved the splicing of these is virtually impossible to perform at sea and would have to be dealt with ashore. Any splicing performed on a VLCC would be of "small stuff" used as, for examples, connecting lights to lifebuoys and, perhaps, for attaching bailers and other small items to the inside of life boats. (Photo. 12.3).

Additional cadet assigned practical deck duties may include, under the supervision of the third mate, refilling extinguishers that have been discharged during routine drills. Deck cadets are prepared, under Unit M9, and then assigned responsibility for the care and maintenance of the lifeboats (Photo. 12.4). Work here entails varnishing of oars used for pushing off the boat from the ship's side. All lifeboats on tankers, including VLCC's are always, to IMO regulations, powered by engines so the need for rowing has been largely superseded, except perhaps in the event of engine failure. Cadets would do the usual modest painting jobs, cleaning and refilling fresh water tanks, changing tinned milk and other stores and performing all of the essential work necessary to keep the boats on top-line so that they may be used with confidence in an emergency. These collective tasks provide excellent training when the newly promoted cadet accepts his first job as third officer. Regardless of maritime considerations, such skills invariably become useful 'around the house' when the accumulation of jobs to be done, whilst the husband has been at sea, becomes the reality once he is on leave! Generally, whilst the Master has become increasingly involved with all aspects of the cadet's training, the chief officer and the other mates 'keep an eye' on the cadet's progress with his bridge and deck work. This has probably become very important with some deck ratings who may lack the essential skills to be able to teach the cadet and, regrettably, if they possess the latter may not be able to have sufficient command of English to be able to impart their knowledge. It was directly as a result of my involvement

12.2. *Traditional skills such as canvas sewing are no longer required as part of the MSA examinations of competency, but are still useful to learn ...*
(By kind permission of British Petroleum Shipping.)

12.3. *... rope splicing, on the other hand, remains within the syllabus.*
(By kind permission of British Petroleum Shipping.)

12.4. *Routine cleaning and maintenance of lifeboat equipment continues to be assigned as a responsible duty to deck cadets.*
(By kind permission of British Petroleum Shipping.)

12.5. *Chief officer and deck cadet plotting position lines from star sight calculations.*
(By kind permission of British Petroleum Shipping.)

12.6. *Second officer instructing cadet in the use of azimuth mirror.*
(By kind permission of British Petroleum Shipping.)

12.7. *Navigating cadet studying in his cabin. The green covered 'Nicholson's Concise Guide' remains today a 'nautical Bible' for MSA examinations.*
(By kind permission of British Petroleum Shipping.)

with cadet, and some deck rating, training that my interest in becoming a schoolmaster ashore was undoubtedly initiated.

12.8.2. Bridge Duties

Deck cadets throughout their training, whilst continuing their deck learning, are now required to follow regular periods initially observing, and later understudying the officer of the watch on the Bridge. There they are taught navigational skills and put into practice the theory which they have learnt at nautical college and from their shipboard study. (Unit N4). Although electronic position-fixing has been available on board now for many years, and long-range methods such as satellite navigation, LORAN C and Global Position Fixing System (GPS) provide highly accurate and virtually instantaneous read-outs, it remains an examination requirement that cadets and navigating officers should still be able to fix the position of the ship by celestial means. It has already been seen, in Chapter 10, that cadets work out their own celestial 'sights' and they are encouraged also to become proficient at the plotting of these onto special astronomical plotting charts which have the large-scale necessary for accuracy. (Photo. 12.5). It is only on very rare occasions that the in-use chart is sufficiently large, to allow direct plotting of position lines obtained from sun/star sights, and then the ship would invariably be in sight of land. At the other end of the gyro-compass, as it were, he would have to learn the art of taking a number of bearings with confident fluency and regard for accuracy. (Photo. 12.6). If he were to spend say one minute taking his three shore bearings, whilst the ship was steaming at the optimum speed for many VLCC's of fifteen and a half to sixteen knots, then the ship would have sailed a distance of about 0.3 of a mile which would make the plotting of his position lines onto the chart impossible. The cadet would end up with a very large 'cocked hat' that could even possibly extend off the chart, instead of arriving at a neat single point fix.

A young cadet (Photo. 12.7) is shown working at the bureau in his cabin on some such relevant navigational exercise almost certainly covered by Unit N3. The green-covered book is Nicholls' Concise Guide, volume one. This is the 'nautical Bible' which continues to be used as a basic textbook for any aspiring navigating officer and, together with volume two, covers the requirements for all grades of certificate up to class one – Master Mariner. The book will be immediately recognised – and doubtless bring back memories – for generations of mates. The progress of the cadet is monitored closely not only by the second mate, but also by a series of supervised assessments designed to test his understanding of the material covered in any particularly lesson. These tests are sat, under close examination conditions, invigilated usually in the Mate's study or office. (Photo. 12.8).

Other Bridge skills learnt as a cadet include use of the 'Rule of the Road for the Prevention of Collision Regulations' kit that continues to be carried aboard many ships. (Photo. 12.9). In this shot, the second officer tests the cadet's knowledge of these important Regulations, which feature as an essential part of the oral examination for all grades of certificates. The mate is doubtless hoping that the cadet would have learnt the prescribed lights to be exhibited by a tug and tow and is expecting to receive the answer that, in this instance, the length of the tow exceeds 200 metres and that his own vessel, indicated by the arrow on the bottom LHS, would alter course to starboard for such a hampered vessel, slow her speed or even stop engines and perhaps go astern. Another very important facility, especially in the days of early-pre VHF ships, was the necessity of using accurately the Aldis – or long range signalling lamp – in order to 'speak' to other ships. (Photo. 12.10).

12.8. *Navigating cadet sitting on examination with the chief officer invigilating. Public examinations can be sat by arrangement with the Marine Society, London.*
(By kind permission of British Petroleum Shipping.)

12.9. *Instruction in ship's distinctive lights using Rule of Road kit. The Collision Regulations remain a vital part of the navigating officer's examinations.*
(By kind permission of British Petroleum Shipping.)

Supertankers Anatomy and Operation

12.10. *Cadet 'speaking to' passing ship with long range signalling lamp - an exercise he has to practise at least twice during his cadetship (By kind permission of British Petroleum Shipping.)*

12.11. *The carriage of an Aldis Lamp remains a legal requirement under SOLAS Rules. (By kind permission of Kelvin Hughes.)*

12.9. SIGNALLING, COMMUNICATIONS

Use of Morse Code has long since vanished in many areas ashore. This represents a serious loss, some would say regretfully, due to the extensively useful applications to which the code could be put. I am delighted to see that it has been retained, (M1), from the old Record Book, and that it remains an examination requirement that the cadet should be able to send and receive by lamp at a speed of six words per minute and, in Unit M1f, to actually practice this with at least two other ships during the course of his training. The carriage of the Aldis lamp is still very much obligatory under the Merchant Shipping Act and it serves, not only as a useful source of intense concentrated illumination, which it certainly is, but reinforces the 360° all-round signalling lamp carried on top of the main mast above the wheelhouse. I have used the Aldis for this purpose on board a Supertanker, in order to direct a specific beam into the wheelhouse of a crossing ship, when we were the hampered or 'stand-on' vessel, in order to send the 'waking up' signal of a series of at least five short flashes because the 'giving way' ship failed to do so, and we could not raise her on the VHF radio. It worked, for shortly afterwards she altered course to pass astern of our ship. A modern Aldis lamp, the 'KELAMP' manufactured by Kelvin Hughes Ltd, is shown in the accompanying photograph (Photo. 12.11) and plan. (Plan 12.1).

Although not specifically stipulated by the training syllabus, I believe that there is always justification for advising and widening the experience of deck cadets by "turning them to" with officers in other departments. Not the least to do so helps the trainee's awareness of problems likely to be encountered by colleagues and to see what goes on in other parts of the ship whilst the deck officers are engaged in an apparently 'deck oriented' operation. (Photo. 12.12).

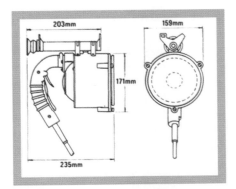

OVERALL DIMENSIONS:
KELAMP
IN CARRYING BOX: 33cm x 23 cm x 27cm
13in x 9in x 10⅝in (Imperial)
Weight 6kg/13lb
KELBAT 12V 23cm x 18cm x 21cm
9in x 7in x 8¼in (Imperial)
Weight 10.5kg/23lb
KELBAT 24V 30cm x 23cm x 21.5cm
11¾in x 9in x 8½in (Imperial)
Weight 19kg/43lb

Light Reduction Screen
Lamp Carrying Box
Battery Unit 12V
Battery Unit 24V
Red Filter
Green Filter
Bulbs 12V
Bulbs 24V
Universal Mains Transformer
Battery Charger—Standard
Battery Charger—Tropicalised

Plan 12.1 Plan showing dimensions of KELVIN-HUGHES "KELAMP" Aldis.
(By kind permission of Kelvin Hughes.)

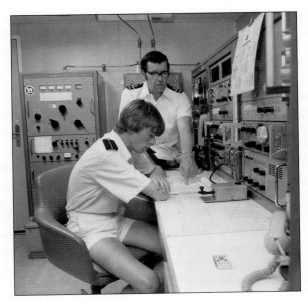

12.12. Turning the deck cadet to with the radio officer enhances the young man's knowledge of work in other shipboard departments.
(By kind permission of British Petroleum Shipping.)

In this photograph taken aboard one of the few pre-GMDSS supertankers which continue still to employ a 'Sparks', the cadet is receiving an insight into the radio officer's duties (M1h). He could be receiving advice on an aspect of the latter's work relevant to the deck department, such as the standard Marine Navigation Vocabulary for his Radio-Telephony Certificate that is a part of his communications unit, or in the use of an automatic distress alarm, or even the keeping of a radio message log. Even more useful is the practice of sending the cadet into the engine-room during a 'stand by', which could be justified under E10 dealing with the operating of marine plant, in order to observe what happens when the bridge telegraph instructions are received at the control panel below, or even during a normal ballasting or cargo transfer operation. It is quite customary to reciprocate the favour by taking onto the bridge an engineer cadet to enhance his own professional knowledge and experience.

12.10. CAREER OPPORTUNITIES ASHORE, DIFFICULTIES ENCOUNTERED FILLING POSITIONS

There would appear few difficulties concerning the quality of the young deck officer upon completion of his training and his first appointment as third officer. Some adverse remarks heard recently might perhaps need to be taken in the light of the 'generation gap' as not so much criticisms of competency, which I believe would have been examined most thoroughly when the cadet was 'up for his ticket', as more of temperament. The 'new officer' will have a different outlook on life in general which inevitably must also have an effect on his attitude towards responsibility. It does not necessarily follow that he or she will be such a totally different or more reprehensible person because of this. Possibly, the requirement is for a more tolerant adjustment in the liaison between each party.

For the cadet who completes his 'time' and goes on to gain his Class 1 (Master Mariner) or, occasionally a lower grade, certificate there remain hitherto unsurpassed opportunities ashore if they no longer wish for various reasons to remain at sea. A surprising and varied number of associated professions exist which regard the marine experience and unique qualifications of an ex-seafarer as essential prerequisites within their own particular specialisation. The decline in all cadet recruiting has had a profound knock-on effect in most of these areas because the required personnel are simply not available. A comprehensive and authoritative questionnaire-based survey was commissioned jointly by the Department of Transport, Chamber of Shipping and Marine Society that was undertaken by Bernard Gardner, an ex-seagoing Master Mariner, and Dr S.J. Pettit, at the University of Wales, Cardiff Institute which was intended, in the words of the executive summary of the final research document:

"to identify the demand for seafaring skills and experience in shore-based jobs; to assess the likely supply of ex-seafarers; and to indicate how any shortfall might be addressed".

The Survey examined *"a number of land-based, marine related industries who habitually use employees with seafaring experience"* and cited some 27 such organisations whose effective operations rely upon maritime expertise, very often up to the knowledge and experience necessary for the possession of a Class One certificate of competency. The following tables speak more than adequately for themselves: (Tables 12.2 and 12.3). Amongst the miscellaneous category are included the RNLI, Coastguard Service and Trinity House each of whom are employers not only of officers, but also a considerable number of deck and engine-room ratings. The summary continued by stressing the staggering number of:

"Approximately 17,000 jobs were identified which employers would prefer to fill with ex-seafarers. For 70% of these, seafaring experience was considered 'essential'; most such jobs (87.5%) were filled by former officers with Class 1 certificates.

Estimated number of jobs which employers consider it is essential to fill with people with seafaring experience

Business category	Type of Employee — UK Seafarer					Other type of Employee			Total
	Deck Officer ex MN	Engineer Officer ex MN	Officer ex RN	Engineer Officer ex RN	Other MN	Other RN	Foreign ex-seafarer	Non-seafarer	
Ports	1336	46	23	1	240	28	0	56	1730
Port services	223	0	0	0	35	2	10	0	270
Towage/Salvage	218	95	7	7	40	0	14	0	381
Dredging	18	24	0	15	12	0	6	0	75
Offshore	148	168	0	0	0	0	10	0	326
Pollution control	51	2	2	2	59	0	0	0	116
Surveyors/Inspection	294	175	4	19	0	0	56	4	552
Cargo Surveyors	55	5	0	0	0	0	5	3	68
Classification	19	1000	2	100	0	0	850	0	1971
Insurance	0	14	0	0	0	0	0	0	14
Banking	0	0	0	0	0	0	0	0	0
Ship/cargo Broking	88	26	0	0	0	0	0	0	114
Ship chartering	6	0	0	0	0	0	3	0	9
P and I	48	10	0	0	0	0	0	0	58
Loss adjusters	0	0	0	0	0	0	0	0	0
Legal	75	0	14	0	0	0	0	0	89
Consultants	918	326	10	10	30	276	59	0	1629
Marine equipment	79	316	30	79	49	0	0	0	553
Shipbuilders/repair	0	20	0	0	20	0	0	0	40
Federated Ship Co's	332	375	0	0	67	0	88	0	862
Non-Fed'd Ship Co's	216	276	0	0	50	0	40	0	582
Ship management	246	322	28	0	0	0	14	0	610
Crew management	21	18	0	0	0	0	0	9	48
Ships agents	12	4	0	0	0	0	0	0	16
Charitable Installations	3	0	0	0	0	0	0	0	3
Education & Training	710	212	25	21	29	16	0	0	1013
Publishing	7	0	0	0	0	0	0	1	8
Miscellaneous	87	106	6	2	204	236	0	0	641
Total	5210	3540	151	256	835	558	1155	73	11778

Table 12.2 The findings of research undertaken by the University of Wales, Cardiff Institute showing research findings on the number of shore employers who regard employment of ex-seafarers as essential ... (By kind permission of Captain Bernard Gardner, Dr S.J. Pettit and the Cardiff Institute.)

Estimated number of jobs which employers consider it is essential to fill with people with seafaring experience

Business category	Type of Employee — UK Seafarer					Other type of Employee			Total
	Deck Officer ex MN	Engineer Officer ex MN	Officer ex RN	Engineer Officer ex RN	Other MN	Other RN	Foreign ex-seafarer	Non-seafarer	
Ports	170	10	8	1	358	56	50	184	837
Port services	8	0	0	0	5	0	0	0	13
Towage/Salvage	74	27	0	0	20	0	7	0	128
Dredging	0	0	9	0	0	0	6	0	15
Offshore	59	0	0	0	30	0	0	0	89
Pollution control	28	21	2	0	10	4	2	0	67
Surveyors/Inspection	39	12	0	2	0	0	2	34	89
Cargo Surveyors	35	0	0	0	0	0	0	0	35
Classification	2	80	0	0	0	0	4	0	86
Insurance	27	0	20	0	0	0	7	0	54
Banking	0	0	0	0	0	0	0	0	0
Ship/cargo Broking	88	0	0	0	7	0	34	27	156
Ship chartering	0	0	0	0	0	0	9	0	9
P and I	104	5	8	0	3	0	3	35	158
Loss adjusters	21	4	0	0	0	0	0	0	25
Legal	115	0	40	0	0	0	7	7	169
Consultants	100	118	0	0	0	10	39	0	267
Marine equipment	20	493	0	69	69	10	0	227	888
Shipbuilders/repair	0	89	0	295	99	20	10	0	513
Federated Ship Co's	106	35	0	0	75	0	18	0	234
Non-Fed'd Ship Co's	138	0	0	0	0	0	20	0	158
Ship management	95	6	14	0	7	7	41	0	170
Crew management	15	0	0	0	0	0	0	15	30
Ships agents	34	0	4	0	4	0	0	21	63
Charitable Installations	0	0	0	0	0	0	0	0	0
Education & Training	4	0	0	0	46	29	8	0	87
Publishing	0	1	0	0	0	0	0	0	1
Miscellaneous	53	2	3	2	476	21	0	149	706
Total	1335	903	108	369	1209	157	267	699	5047

Table 12.3 ... or as an advantage.

(By kind permission of Captain Bernard Gardner, Dr S.J. Pettit and the Cardiff Institute.)

> *Demand for ex-seafarers ashore is estimated at between 640 and 740 per annum over the next nine years. (i.e.: to 2005). Demand for those filling posts where seafaring experience was considered 'essential' is estimated at between 450 and 580 per year.*
>
> *Recruitment by shipping companies of approximately 1,200 cadets a year is required to maintain the present status quo at sea and ashore, once the effects of under recruitment since 1981/1982 have worked themselves out. Current intake (late 2000) is about 400 per year".*

Inevitably, a widespread alarm has swept over employing industries and services, particularly that roughly one third of requirements only, at the initial recruiting stage, continue to be met. In Chapter 5, the authors considered the consequences of the shortfall and suggested some options for remedial action. Their forecast for the short-term consequences makes dismal reading:

> "It is clear from the analysis ... that it is inevitable that a shortfall in the supply of UK seafarers to fill job vacancies in the shore based maritime industries for which seafaring experience is considered essential will occur between now and the year 2004/5. Since the pool of seafarers to fill these positions was determined by intake levels over the past eight to twenty eight years the die was cast long ago and there are only a few measures that can be taken to increase the pool in the short term. Such measures include providing additional training for those seafarers who come ashore between the ages of 20 and 24 years and who have already been absorbed into the shore based industries in lower positions than those we have so far been considering, that is, seafarers who completed their cadetship and hold at least Class 3 or 4 certificates."

The article, 'Bitten by the Officer Gap' already cited, expressed direct concern following an examination into the repercussions of ex-officer shortages in management and allied industries ashore:

> "But perhaps even more worrying, in the longer term, for those managers based in Europe is that the impact of the so-called 'lost generation' of qualified European seastaff of recent years is now beginning to make itself felt ashore. David Rodger of Acomarit said that his company is already experiencing a shortage of suitably qualified shore staff for its UK management office. As the shortage of officers continues to grow, the lack of European managers and superintendents with necessary marine experience must only get worse and could pose a serious problem for European-based ship managers in years to come."

Regarding the long-term consequences, the report was equally unequivocal:

> "Lloyd's Register, which is by far the largest employer of former seafarers in this essential category, has given some thought to dealing with the long-term consequences of the projected shortfall. In fact, it is already unable to recruit sufficient suitably qualified former merchant navy engineer officers to meet its needs. In addition to employing over 800 suitably qualified foreign ex-seafarers in posts overseas, it is training graduates to fill positions which it filled previously with former UK seafarers with chartered engineer status, people whom, incidentally, it regards as a core business resource."

From all sections within the maritime industry and its associates the same cry is echoed. Roger Green may be cited again, speaking from a senior operations management position ashore:

> "The huge decline in training since 1980 is now having reverberations throughout the industry. We are finding it extremely difficult to recruit the right calibre persons for operational, safety and superintendents' positions. Also in twenty years' time, when my generation of cadets is retiring, it begs the question from where our replacements are going to be found."

A range of possibilities were examined by the Cardiff Institute report including raising retirement age limits, enhanced financial incentives, and the recruitment of ex-Royal Naval officers. Certainly, my own experience with the latter in terms of seagoing deck officer recruitment, has been far from inspiring. I have sailed with two ex-Naval officers at various times, whilst serving aboard VLCC's and, on each occasion, the Master dismissed them at the earliest opportunity as being unsuitable for various reasons. One practised an unfortunate approach to discipline. Whether or not this be good or bad is a matter of opinion, but it remains a fact that 'Jolly Jack' in the Merchant Navy will most emphatically not 'jump onto the shovel' once the officer says 'dirt'. Both gentlemen, not surprisingly, possessed a disastrously dangerous lack of cargo handling knowledge with one, additionally, 'knowing it all' and thus simply not prepared to learn. The report suggests that the recruitment of ex-Royal Naval engineering officers is rather more successful with shore employers, as well as at sea. A far more positive report appeared in the June 1997 edition of the 'Telegraph', which is the monthly newspaper of the Officers' Association for the Merchant Navy, NUMAST, reporting on an ex-Royal Naval seaman officer of twenty one years service who, as part of his resettlement course, studied at Warsash Maritime Centre. The Telegraph reported:

> "A former Royal Navy officer is carving out a new career in the merchant navy after becoming one of the first candidates to secure a Level 3 award in the new Merchant Navy Operations National Vocational Qualification ... this month he takes up a position as second officer ... the quality and standard of his portfolio was commended both by the Marine Safety Agency and the external NVQ verifier ..."

The success of this gentleman in settling into 'that other service' from the Royal Navy may well be attributable not only to an attitude of mind on the part of various parties, but also to the 'conversion' course which he was prepared to follow, at Warsash.

Numerous alternative schemes are then being examined to off-set the lack of ex-seafarers, but it seems apparent that the foreseeable employment future for qualified mariners wishing to come ashore remains, and is likely to remain, extremely favourable. Certainly I, and an increasing number of responsible shipping companies and agencies, remain strong advocates of deck cadet training and support the necessity for governmental assistance to encourage a more positive and enlightened approach to cadet recruiting. My own cadet days were really quite happy ones due undoubtedly, in light of the experiences of some less fortunate cadets, to the excellent quality and amiable dispositions of the officers who supervised my 'time' and, indeed, to the admirable fellow cadets with whom I have served.

Perhaps, rightly, the last word should be left to a young third officer, Grant Holmes of Stena Marine, who wrote about the reasons why he rejected the idea of working towards command of a passenger liner:

> "For me, large tankers are the greatest challenge. This is where I want to work and develop ... . The time I spend on board is extremely rewarding. Here I get to spend a lot of time on the bridge and am entrusted with carrying out responsible tasks. And there are plenty of skilled and experienced role models. Being able to stand beside and learn directly from them is extremely rewarding. That is why I feel that the cadet system is so good. It has to be, since I myself am a product of the system ... .'

As Stena commented: *"If and when Grant Holmes returns as third mate to the "STENA KING" (the eighth largest ULCC in the world), the environment and the bridge will be familiar to him. The routines and the safety philosophy will not be new to him. They are already part of him."*

Chapter Thirteen

LIFE ABOARD SUPERTANKERS

13.1. LENGTH OF VOYAGE, LEAVE

13.2. ACCOMMODATION, STEWARDS' STORES

13.3. BONDED STORE: ALCOHOL AND DRUGS POLICIES

13.4. CATERING

13.5. OFFICER AND CREW QUARTERS, MEDICAL CARE

13.6. RECREATION

Chapter Thirteen

LIFE ABOARD SUPERTANKERS

(INTRODUCTION – Voyages and Leave: ACCOMMODATION: Layouts of Different Ships – Plans – Officers and Crew: LIVING STANDARDS: Feeding – Laundry – Stewards – Bond Accounts – Cabins: Duty Mess – Crew Recreations – Ratings – Petty Officers – Officers – Differences According to Rank – Various Interiors – OFF DUTY TIME – Marine Society – Hobbies – Films/Video's – Darts etc. – Bar and Smoke room – Table Tennis – Swimming Pool – Practical insights into the life at sea as recorded in Articles by the Author in various copies of the magazine 'Sea Breezes')

13.1. LENGTH OF VOYAGE, LEAVE

Questions are often encountered expressing interest in the way in which officers and crew live on board VLCC's, and enquiring how they manage to pass the time when not actually working. The tour generally expected on Supertankers varies considerably between individual companies. To cite a few examples: some employers ask officers for two round UK/Arabian or Iranian Gulf trips which total about four months. On the Gulf/US voyages this might be slightly extended. Other companies have different arrangements for senior officers of captain, chief engineer, chief officer and second engineer ranks than those required from their junior officers so that the combination of patterns is considerable. With one of the companies with whom I served on VLCC's, the mates and engineers worked in teams such that a captain and his officers would sail together, go on leave simultaneously, and return together for subsequent voyages. Obviously, the pattern would change if an officer was to go for a higher ticket or to leave the ship permanently for other reasons. Very often, due to unpredictability in the final stages of a Supertanker's movements, this particular company would call us to meet initially in a hotel at a nearby port and spend a couple of days there until being driven by taxi or minibus to the ship. I always found this very useful because it gave all ranks the chance to acclimatise psychologically to the end of leave, thus diminishing partings, and to gel again into a team and become acquainted with any new members. I served continuously with the same Master and Mate, saw a complete change in other mates, especially as I gained promotion, and stayed with many of the engineers, including the same chief and second, over the subsequent voyages made on one VLCC with one company. My other employer changed European officers as they completed four months or so at sea, and the ratings every nine months. The longest spell for which I served aboard a Supertanker, without leave in England, was seven months which included two months or so standing by during a new building in Denmark. Certainly, I had found myself completing much longer trips on ships belonging to some dry cargo companies. For the voyage part of the VLCC period, so far as I was concerned, there was no question of shore-leave, distances always presented difficulties even if cargo watches made it possible to go ashore in the quick turn-round time. 'Life on board VLCC's' very often meant precisely what it said. Generally, once aboard, you remained there for the duration of the contract until relieved.

Supertankers Anatomy and Operation

13.2. ACCOMMODATION, STEWARDS' STORES

Doubtless, in compensation, the accommodation provided for all officers and ratings aboard reputable Supertankers was, and indeed, as I visit a range of VLCC's, remains extremely high. Whilst this varies considerably between ships, of course, generally the owners give a lot of thoughtful consideration to the planning and provision of cabin layouts, furniture and facilities. The ship's plans for a 'standard' early-generation VLCC (Plan 13.1) indicate how the spaces contained within the accommodation block of the superstructure on that ship were arranged. These make an interesting comparison with the accommodation layouts shown in the general arrangement plans of the traditional and modern Supertankers inserted in the end cover. The main, or upper deck, was devoted largely to deck and catering departments wet and dry store-rooms. In the photograph, (Photo. 13.1) the second steward is seen issuing stores to the catering boy, according to the day's menus. These were arranged, on early ships, following discussion between the chief steward and the cook. The adequately stocked shelves, had all tins side-stowed, resting against storm bars, thus taking into account any extreme rolling of the ship, as well as facilitating stocktaking. Very often, the stores would be stacked from deck level and stowed on gratings which allowed air to circulate around bags of flour, rice and similar foodstuffs requiring adequate ventilation. Every other voyage or so the storerooms were painted, either by a cadet or deckhand provided by the benevolent Mate, or one of the catering staff. Domestic fridges were generally separated into fish, meat and vegetable rooms, with access served by a mutual handling room. Most of the dry stores, frozen fish and meats were loaded in Europe, with sufficient quantities to cover the entire voyage, but fresh vegetables and sundries were taken on board whilst on passage, often by helicopter off Capetown during each of the round voyages, when the Supertanker was engaged on this route. (Photo. 13.2). This also gave the opportunity to provide seasonal foodstuffs common to the country of storing thus helping sustain variety in menu planning. The crew, moving in readiness to accept the stores being winched to the deck, emphasise the height of the cargo manifolds and indicate that part of the stores includes oranges from South Africa. The wet deck has clearly resulted from one of the many sudden and cold rain squalls which often occur in this region — giving lie to the 'Sunny South Africa' of so many holiday brochures. It was not uncommon for officers to be 'in blues', instead of white tropical uniforms, whilst the ship was trading in southern African waters.

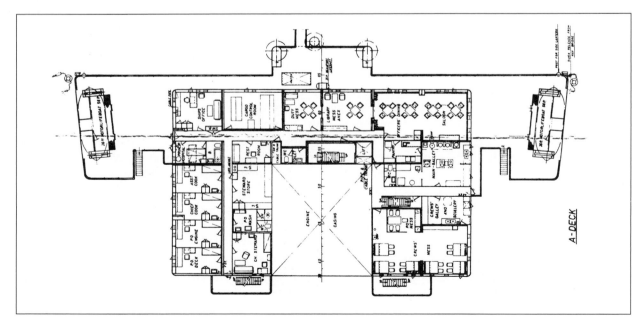

Plan 13.1 "A" Deck Accommodation layout situated on the Boat Deck.
(By kind permission of Builder's Plan of the VLCC 'RANIA CHANDRIS'.)

Life Aboard Supertankers

13.1. Catering Staff collecting items from the dry provision store for meal preparation.
(By kind permission of British Petroleum Shipping.)

13.2. Taking Stores on board off Capetown.
(By kind permission of British Petroleum Shipping.)

13.3. The Officers' Laundry on a First Generation VLCC.
(By kind permission of Shell International Trading and Shipping.)

The ship's laundry (Photo. 13.3) was adequately equipped with a number of Elto heavy-duty washing machines and spin dryers, along with electric irons and a generous company supply of powder. Some of the bedding and galley/catering cloths were sent ashore, as convenient, but much was handled on board with, of course, the crew and officers having facilities for their own personal laundry. Frequently, the Asian stewards took officers gear immediately after use, even whilst he was using the shower and returned the clean, neatly pressed uniforms within a few hours and rarely later than a day or so. Invariably, for this excellent service, the steward was financially rewarded with the traditional "bucksheesh" at the end of the duty tour. Companies also provide stewards to look after and change weekly, officers' bedding (Photo. 13.4) and to dust, vacuum and generally keep clean the cabins. (Photo. 13.5). Additionally, they would make up the bridge snack boxes with sandwiches and hot drinks for the night watches, as well as visiting the bridge at intervals throughout the day with tea/coffee trays and biscuits. Traditionally, the Captain shared the services of a steward with the radio officer, whilst other stewards were assigned to the chief and second engineers, and each of the deck and engine departments. They also waited in the saloon on their officers. (Photo. 13.6). With the reductions

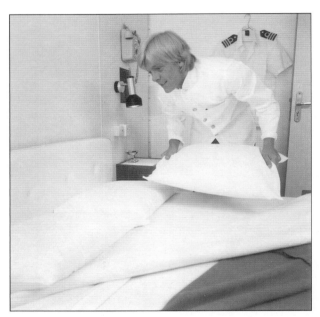

13.4. Catering staff duties include the upkeep and maintenance of Accommodation.
(By kind permission of British Petroleum Shipping.)

13.5. ... as above ...
(By kind permission of British Petroleum Shipping.)

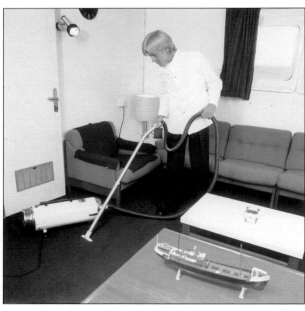

*13.6. Evening Meal in Officer's Dining Room.
(By kind permission of British Petroleum Shipping.)*

in manning prevalent today, the catering department is virtually non-existent. A chief cook/steward controls the department together with a second cook/steward and, with some companies, an assistant steward and boy trainees. Menu's are discussed, and stores purchased, by agreement between the Captain and the chief cook. One steward each continues to look after the totals in each of the deck and engineer officers, including cabin and table service.

13.3. BONDED STORE: ALCOHOL AND DRUGS POLICIES

The bonded store contains all wine, beer and tobacco products, as well as 'desirable' luxuries such as confectionery and toiletries. Following the "EXXON VALDEZ" Supertanker disaster, the availability of alcohol on board ship has become an issue of prime importance. A standard alcohol and drug policy operates with all reputable VLCC companies, the signing of, and adherence to which, is mandatory for officers and crew alike. Specific regulations state maximum amounts of alcohol which can be consumed and give time intervals, before watches are commenced, after which no alcohol may be consumed. Many ships are now totally dry whilst others have only wine and beer in their bond. Restrictions apply also to any visitors, and even shore employees, to VLCC's whilst the ship is in port, with unannounced testing occurring twice a year on board most vessels. On the United States coast, random checks are likely to be made by the shore authorities at any time when spot breath tests may be taken, and the option being reserved of full medical inspections in cases of suspicion. So serious is the problem of excessive drinking at sea being taken that an "Exxon Drug and Alcohol Clause" is incorporated into the Charter Party Agreement that is signed between the owners of the VLCC, or any other ship for that matter, and those who hire her for the conveyance of their cargoes. An extract from such a clause follows:

> *"Owner warrants that it has a policy on drug and alcohol abuse applicable to the vessel which meets or exceeds the standards in the Oil Companies International Marine Forum, 'Guidelines for the Control of Drugs and Alcohol on Board Ship'. Under the policy, alcohol impairment shall be defined as a blood alcohol content of 40 mg/100 ml or greater; the appropriate seafarers to be tested shall be all vessel officers and the drug/alcohol testing and screening shall include random or unannounced testing in addition to routine medical examinations. An objective of the policy should be that the frequency of the random/unannounced testing be adequate to act as an effective deterrent, and that all officers be tested at least once a year through a combined program of random/unannounced testing and routine medical examinations ...."*

Certainly, this action has resulted in a considerable focus of concentration regarding the implications of excessive drinking at sea by all Captains and their officers, an action that really had to be taken in the light of what was a very serious problem some years ago.

13.4. CATERING

On many ships, a 'chit' system operates so that officers sign for their drinks or slop chest purchases at the time of ordering and these are then lodged with the chief steward and settled monthly via the officers' salary system. Very often all of the bond would be purchased from the Botlek Stores in Rotterdam (or wherever) at exceptionally favourable prices. The ladder, in the accommodation plan, has been thoughtfully designed to lead conveniently to the galley and food preparation rooms on "A" deck. (Plan 13.1) The crews' mess room is comfortably and sensibly furnished with either a separate or an off-set mess for the PO's (Photo. 13.7). The refrigerator, on the top left-hand side of the shot, behind the door, enabled this room to be used out of meal hours for occasional snacks and drinks. The providing of a 'crew' galley indicates that this particular Supertanker was designed initially for a combined European officer, and Asian-rating crew which, in fact occurred, following completion of the maiden voyage when we changed the European ratings for Indians and Pakistanis.

The standard of catering on most VLCC's was, and it seems from recent visits remains, very high, with a varied menu provided by the catering department that would not disgrace a good hotel ashore. Many coursed breakfasts and lunches remain the norm (Draw. 13.1) whilst the excellent dinner menu encourages a tendency to over eat, unless sensible restraint is exercised. The following photograph shows the catering boy helping the chief cook with the preparation of food in the galley. (Photo. 13.8). Usually the boys, in companies where they are still carried, serve until they reach the age of eighteen years when they are promoted to assistant cook/steward. The very light and pleasant officers' dining saloon invariably looks out onto the main deck with large open windows. On the starboard side of this particular ship, there were separate tables for the various ranks. The chairs are cloth-backed and frequently fixed to the deck by storm chains necessary to encounter adverse weather. Bars are also fitted to the sides of tables which would be raised to form a storm-edge, thus preventing crockery, cutlery and food being swept onto the floor in the event of any exceptional motion caused by occasional rough seas. I do not recall them being used very often across the number of voyages on which I sailed in VLCC's.

13.7. The Crew's Mess Room is comfortably and sensibly furnished.
(By kind permission of Shell International Trading and Shipping.)

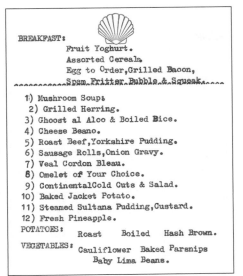

Draw. 13.1 Combined Breakfast and Luncheon Menu in Officer's Dining Room.
(By kind permission of Shell International Trading and Shipping.)

13.8. Part of the food preparation area of the galley on board.
(By kind permission of British Petroleum Shipping.)

13.5. OFFICER AND CREW QUARTERS, MEDICAL CARE

The accommodation at "A" deck, on this particular Supertanker, was for petty officers on the port side, with ship's working offices running thwartships forward. The duty mess is provided so that no deck or engineer officer whilst on boiler-suit duty would have to change into uniform from his working gear in order to have luncheon or dinner. "B" deck, (Plan 13.2), contains, on the port side, the crew recreation room and, in the starboard alleyway, the drying room for gear wet from deck/engine-room work. The recreation room is well furnished and provided with darts board, card-tables as well as a television, in order to receive programmes when in port or in-shore waters. (Photo. 13.9). There exist also facilities for video and/or film shows whilst at sea. Throughout the entire accommodation, extensive use is made of vinyl, laminated plastics and stainless steel, not only because they can be produced in cheerful colours and are durable, but also because the amount of cleaning and maintenance is kept to a minimum. All ratings are housed on this deck and provided with single cabins, although traditionally double-berthed cabins continue to be provided for deck and catering boys when,

Supertankers Anatomy and Operation

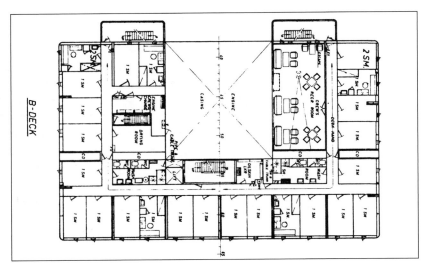

Plan 13.2 Accommodation on "B" Deck is for Ratings.
(By kind permission of Builder's Plan.)

13.9. The Crew's recreation Room is situated on "B" Deck on many VLCC's.
(By kind permission of Shell International Trading and Shipping.)

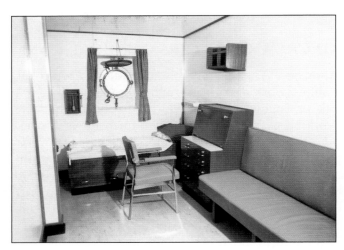

13.10. All Ratings are berthed in their own cabins (apart from boys) on "B" Deck.
(By kind permission of Shell International Trading and Shipping.)

rarely today, two of each are carried. Each cabin looks out either across the main deck or ship's sides. (Photo. 13.10).

"C" deck (Plan 13.3) indicates on this VLCC the start of officer accommodation with cabins for the ship's engineers. The brightly lit alleyways are practically furnished (Photo. 13.11) and clean, with touches such as outside carpets and door curtains to at least give the impression of homeliness and remove something of the spartan conditions associated with what is, after all, a working ship. Invariably, the chief engineer is treated to the same standard in his suite of rooms as is afforded the Master and, although practices vary according to individual companies, the chief is often housed on the same deck and next door to the captain. The photograph

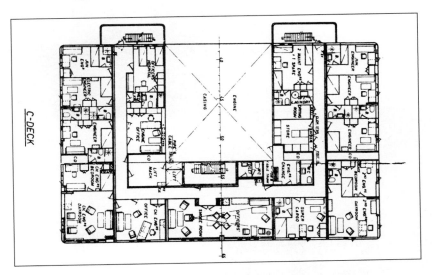

Plan 13.3 "C" Deck marks the start of Officer's Accommodation.
(By kind permission of Builder's Plan.)

13.11. An Alleyway on "C" Deck aboard a very large tanker.
(By kind permission of Mobil Ship Management.)

Supertankers Anatomy and Operation

13.12. The Chief Engineer's Bedroom aboard a modern VLCC.
(By kind permission of Shell International Trading and Shipping.)

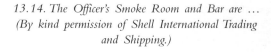

13.13. The Second Engineer Officer's Day Room is very well furnished.
(By kind permission of Shell International Trading and Shipping.)

13.14. The Officer's Smoke Room and Bar are ...
(By kind permission of Shell International Trading and Shipping.)

13.15. ... designed conducive to relaxation.
(By kind permission of Shell International Trading and Shipping.)

(Photo. 13.12) indicate the palatial accommodation typically enjoyed by these very senior officers serving aboard VLCC's. The chief engineer's bedroom, for example, with its bedside lockers, lights and thick fitted carpeting throughout could easily grace any reputable hotel ashore. The grill in the panel above the bunk is part of the circuit of air-conditioning provided in the entire superstructure block. It is worth mentioning that married officers sailing deep-sea are often allowed to bring their wives and every bedroom is provided with double-bunks for this purpose. This is generally a sound arrangement that can contribute much to the social life of the ship, especially when only one or perhaps two ladies sail at any time, as occasionally friction had been known to develop if three or more were on board. The second engineering officer's dayroom (Photo. 13.13) is on par with that of the chief officer, his equivalent rank in the deck department, and has again, sensibly practical furnishings, but in a modern style, which is comfortable and good for relaxation. The combined officers' smoking room and bar (Photos. 13.14 and 13.15) remains the focal point of the social life for, in its comfort, not only relaxation between all ranks can be established, but it aids also the bonding between deck and engine-room officers which is essential to the efficient working of a Supertanker, justifiably, I believe, and even more necessary than in other classes of ship. The officer's hospital and dispensary are situated on the after-part starboard of this deck, for quietness as much as anything else. All deck officers are qualified in first aid and each man, officer and rating, has to receive a thorough medical inspection ashore before being appointed to a VLCC, or any ship for that matter. The second mate includes running a 'sick parade' amongst his duties and certainly I became very experienced in the fields of 'disinclination to work', and a range of unsociable diseases, as a result of my time as 'ship's doctor'. I never recall loosing any patients however for, in the event of serious emergency, we always have recourse to medical advice from a shore station or warship and, as a last resort, can invariably summon a helicopter to lift an injured or sick man and land him ashore to a hospital in any country, or even call a nearby port for an authority launch to take the man ashore. I can recall no incidents of this nature during my time for, by and large, seafarers are quite a healthy lot. Any such incidents are more likely to be the result of an accident on deck or in the engine-room, but these again are mercifully rare.

The deck officers' cabins on "D" deck (Plan 13.4) continue the policy of grading according to rank. The third mate, for example, has a cabin that is equivalent to the junior and fourth engineers. (Photo. 13.16) It is perfectly comfortable and practical, but lacks the finesse afforded to those of higher rank. One of the things that always impressed me regarding the living quarters on all VLCC's, compared to dry cargo ships and smaller tankers, was the

Supertankers Anatomy and Operation

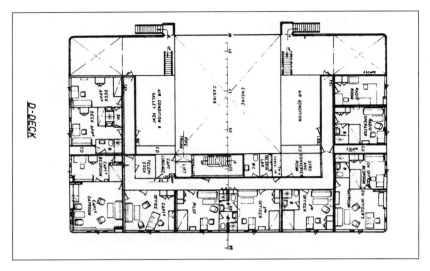

Plan 13.4 On this ship "D" Deck was assigned to the Captain and Deck Officers and any Apprentices or Cadets carried. (By kind permission of Builder's Plan.)

13.16. All Junior Officer's cabins are similarly fitted out.
(By kind permission of Shell International Trading and Shipping.)

13.17 The Chief officer's Bedroom would be virtually identical in its fittings to that of the 2nd Engineer Officer.
(By kind permission of Shell International Trading and Shipping.)

13.18. The Captain's Day Room is often used to entertain visitors and conduct Weekly Management Meetings. (By kind permission of Shell International Trading and Shipping.)

13.19. Master, Chief Officer and Cadet playing darts in the Officer's Smoke Room. (By kind permission of British Petroleum Shipping.)

almost excessive amount of room everywhere. The spacious accommodation experienced on these very large ships was certainly an 'eye-opener' when first I joined them from dry-cargo vessels, with their comparatively cramped spacing. The telephone is an important line of communication on VLCC and saves an enormous amount of lost energy. It is interesting to note the chief officer's bedroom (Photo. 13.17) for comparison with that provided with the Master and junior officers. Quite subtle variations can be seen: the more 'cramped' lay-out, for instance, still roomy, but not quite so spacious, whilst the Captain's quarters, had just that 'extra' refinement of plusher arm-chairs, with pictures on the bulkhead and a thicker carpet, rightfully befitting the man responsible totally for all departments within the ship and the safety of a multi-million pound VLCC, her valuable cargo and the lives of all on board. The Captain's day room is often used for the weekly officer meetings in preference to his office. (Photo. 13.18). Invariably, after the official ship's business between the senior officers has been conducted, time is made available for a period of social chat and a pre-luncheon drink. Generally, the 'Old Man' has to keep rather aloof for purposes of discipline. It can lead to a very lonely life if he so desires, especially if he is married and his wife is not accompanying him on the voyage. Although the crew meet the Master on only formal occasions, the Captain mixes socially to a limited extent with his officers. He often pops-down for a film show, for instance, or to play chess or darts (Photo. 13.19) and, on my ship, it became a regular practice for him to visit during my Watch on the Bridge for a 'cup of tea or cocoa and a chat'.

13.6. RECREATION

The Marine Society from its headquarters near the Imperial War Museum in London, performs quite unparalleled work for seafarers. They provide an excellent range of services covering distance learning courses in literally just about any subject the seafarer may desire. These may lead to academic and professional qualifications ranging from GCSE to university post-graduate degrees. Often, the Society arranges for public and other external examinations to be taken on board, with correctly supervised invigilation, usually in the Captain's or Mate's day room. They provide ship's libraries and often arrange for these to be exchanged by sending replacements out with the stores via the helicopter off Capetown – or wherever the helicopter service is used. Nothing is too much trouble for them to organise. If they do not provide the course or subject required then they will make enquiries on the sailor's behalf and contact an organisation who can help. Very few seafarers – whose Company headquarters have registered the ship with the Society and paid the modest fee requested – have not benefited from their services.

Apart from (unashamedly) using the Marine Society, I used to listen to my favourite music by taking away my recorder and a large collection of cassettes, even to the extent of sacrificing nautical books if I flew out to join a ship – relying on the fact that the books required would be on the supertanker when I reported on board. I was fortunate that I was never let down and shudder still to think what would have happened if they had 'become removed' before I joined. Reading is always an excellent past-time, along with creative writing. Some officers work on art courses, others study law, word processing, modern languages, history, both maritime and social – the list is endless.

Model making (Photo. 13.20) remains a very popular past-time and, in this shot, a cadet is putting the finishing touches to a remote-controlled racing car. The radio on the table-top would probably be seen only on the generally stable platform of a Supertanker. On most other classes of ship it would not have stayed there very long as, once the ship hit any kind of sea, it would have been thrown across the cabin on a unique sea voyage of its own.

*13.20. A Cadet puts the finishing touches to a much laboured over model racing car.
(By kind permission of British Petroleum Shipping.)*

Life Aboard Supertankers

*13.21. Junior Officer playing table tennis in the Games Room.
(By kind permission of British Petroleum Shipping.)*

*13.22. Officers preparing cinema projector in Smoke Room.
(By kind permission of British Petroleum Shipping.)*

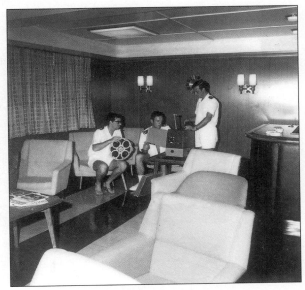

The recreational pursuits examined so far have been of the intellectually demanding kind, requiring various levels of concentration, clearly something which could be counter-productive if not balanced by a variety of alternatives. Casual games of darts, chess or draughts and playing cards can be effectively used for variety. Along with the radio officer and the cadet, on one occasion, we invented a card game which was a combination of whist and rummy. The problem was that the rules became so increasingly complex, as we added variations, that it was virtually impossible to introduce new blood into the game when other officers joined the ship. Table tennis (Photo. 13.21) is as popular at sea as it is ashore and many an intense competition occurs with a round of beer being the penalty for losing. Video's and films on documentary and safety topics are generally provided by the Marine Society, but entertainment films are sent to ships of all flags internationally by the Walport Video Company of London. These services are very much appreciated by all seafarers. Although today videos are prevalent, the film projector continues to be carried on some ships and, in this shot, (Photo. 13.22) the equipment shown is being prepared in the smoke room by officers of deck and engineering departments. Although the swimming pool on our Supertankers was situated on the main deck, it on many ships is aft of "A" deck. Any ship's pool remains a firm favourite. Most are spacious enough for a

Supertankers Anatomy and Operation

13.23. *Relaxing in the Swimming Pool.*
(By kind permission of British Petroleum Shipping.)

13.24. *Seaman cycling on the forward deck.*
(By kind permission of British Petroleum Shipping.)

13.25. *When duties permit, competitions are arranged such as this Officers v Crew cricket match on deck.*
(By kind permission of British Petroleum Shipping.)

comfortable dip even when the VLCC finds herself rolling in the notorious Cape rollers — mentioned in Chapter 6 — off the East coast of Africa. A uniquely invigorating swashing effect is experienced that would never be found in pools or beaches ashore. Certainly, the officers (and ratings on all-European crewed ships) (Photo. 13.23) enjoy pre-lunch drinks and the dip along with a social chat in between periods of duty. For the more energetic, there is always a walk — or even a bicycle ride (Photo. 13.24) around the main-deck, with two or so round trips equalling about one mile. Owing to the number of accidents caused by the conflicting motions of bicycle and Supertanker, however, these have now been withdrawn. A game of cricket (Photo. 13.25) between officers and a European crew can usually be fitted in between the pipes on the broad expanse of main deck until, that is, the ball disappears over the side. Deck golf is an equally popular past time, played with a wooden puck and clubs on a court inscribed by painted circles on the main deck.

Officer service aboard a Supertanker possesses unique advantages that compensate adequately for the restricted life-style. As hinted in the Preface, conversations with people ashore indicate that genuine interest and curiosity exist concerning the way of life at sea. Possibly, not the least of the reasons arises from the unsettled nature of ex-seamen in the eyes of their families, friends, neighbours and particularly employers, and the genuine efforts which many are seen to make in order to achieve successful adjustment to shore careers. It is extremely difficult to impart a true flavour of what is meant to be a seafarer because the levels of seagoing and non-seagoing experience differ so vastly. The obvious popularity of anecdotal-cum-factual books and articles, written by ex-seamen of all ranks, are indicative of the efforts made to build a bridge of mutual understanding.

To have served on VLCC's is different again. If anything, an even greater vacuum of incomprehension prevails. 'Life aboard Supertankers' seems to carry with it ashore an almost academically sociological fascination giving an esteem similar to that afforded to the airline pilot. It certainly attracts equivalent interest as a conversational topic. By using the pages of 'Sea Breezes' magazine, mentioned in Chapter 10, in the March and April 1991 editions under the titles, 'Memories of a Supertanker' and 'Capers on the Coast', as well as the April and May 1995 copies, entitled 'Tanker Officers' Turmoil on a Dry Cargo Ship', the author offers contrasting serious and humorous personal glimpses, attempting to fill some of the gaps regarding what it means 'to have served at sea' aboard a range of dry cargo ships, coasters and Supertankers.

PART 4: The 'Second' Generation VLCC

Chapter Fourteen

DOUBLE HULL CONSTRUCTION

14.1.   INTRODUCTION

14.2.   TORREY CANYON, AMOCO CADIZ, EXXON VALDEZ

14.3.   OPA90, MARPOL IMPLICATIONS

14.4.   CONSTRUCTION, COMPUTER AIDED TECHNOLOGY, SECTIONAL COMPLETION OF THE SHIP

14.5.   DEADWEIGHT CALCULATIONS, STABILITY INFORMATION

Chapter Fourteen

DOUBLE HULL CONSTRUCTION

(Introductory Remarks — Incidents to single hulled tankers leading to the stranding of VLCC. "EXXON VALDEZ" in 1990 — Repercussions in form of United States' legislation in 'OPA90' Regulations — Quote by Mr. A. G. Gavin: Principal Lloyd's Surveyor — IMO Recommendations in Amendments 13F and 13G to MARPOL 73/78 — Double Hulled Basic Construction — Nomenclature used to identify Midship Transversal Sections — The First Double-Hulled VLCC's: "AROSA" from Japanese Hitachi Zosen Yard and "ELEO MAERSK" from Lindo Yard, Odense in Denmark — Development of the Building of the VLCC "AROSA" Traced from First Plates to Completed Ship — Similarities and Differences Between Single and Double Hulled VLCC's — Double Hulled VLCC Stability and Loadline Information.)

NB: It may prove useful to read this chapter in conjunction with **Chapter Two** of **Part Two**, on the construction of single hulled VLCC's.

14.1. INTRODUCTION

The concept of a computer designed and manufactured double bottomed and double sided ship is not new. For many years now, bulk/ore carriers and gas tankers have been built with these features, together they constitute a 'double hull', in order to provide ships of these classes with enhanced strength and for additional safety factors. The idea, although it had certainly been carefully considered for many years in the past, is of comparative novelty in its application to Supertankers. There had been a 'first' in double-bottomed VLCC's which was a 280,000 sdwt, belonging to a major tanker company, built as long ago as 1969. The delay in implementation was caused by financial, rather than more practical considerations, and it was to be another two decades before the double sides were added, thus offering a protective skin around the crude oil cargo carrying capacity.

14.2 TORREY CANYON, AMOCO CADIX, EXXON VALDEZ

It was the wrecks of the m.t. "TORREY CANYON", which had been enlarged in 1964 to 120,000 sdwt, (although this ship was *not* a VLCC, even by 1967 standards when her grounding occurred), plus the 1973-built VLCC "AMOCO CADIZ" — of approximately 230,000 sdwt, which deposited around a quarter of a million tonnes of crude oil off the Brittany coast in 1977, that focused public and professional attention to the potentially alarming damage which could occur when these very large tankers suffer damage. A number of additional serious spills brought supertankers again into the World's limelight, but it was undoubtedly the grounding of the VLCC "EXXON VALDEZ", (Photo. 14.1), — (211,470 sdwt, built during 1986) — in Prince William Sound, Alaska, on 24th March 1989, that motivated positive action

Supertankers Anatomy and Operation

*14.1. The VLCC "EXXON VALDEZ" under the name "EXXON MEDITERRANEAN".
(By kind permission of World Ship Society Library.)*

and led directly to firmer governmental intervention and control than had previously been taken. Some fifty thousand tonnes, or around a quarter of a million barrels, of oil escaped from the ruptured number one centre and starboard wing tanks and, even though this represented less than one fifth, of the total cargo carried, the resulting slick was some forty miles in length and about half-a-mile wide. The oil contaminated around 10,000 square miles of sea and coastline and caused not only considerable and devastating ecological damage to the environment, but repercussions within the tanker industry that continue to reverberate.

14.3. OPA90, MARPOL IMPLICATIONS

Tough and compulsory legislation was rightly introduced by the United States government which resulted, in August 1990, (an amazingly prompt and decisive reaction of less than nine months from the incident), in the now famous or, perhaps to some, infamous Oil Pollution Act of 1990, known throughout the shipping industry as 'OPA90'. One can legitimately use the term "infamous" because of the non discriminatory nature of this legislation, as Mr A. G. Gavin, Principal Surveyor with Lloyd's Register and an internationally respected authority on matters pertaining to VLCC's, pointed out in a technical paper published during 1994/5:

"The new legislation for double hulled tankers does not take account of the tanker operator with a high safety and maintenance record. All owners/operators will be bracketed together and forced to comply if their ships trade to the U.S."

Certainly, the brief advocated in my Preface does not give me the task of becoming immersed in either the intricacies of the legal regulations, which abound in OPA90, or with the implications which might arise from implementing these rules. I lack both the knowledge and experience in these areas. Suffice to say, for my purposes, that the essence of Rules 9.3 and 9.4 made it compulsory that all tankers over 5,000 sdwt, which of course included all VLCC's, who wished in future to trade to United States ports must be constructed with double bottoms and sides. It seems to me, however, that this legislation virtually spells the death knell for the single hulled Supertanker. OPA90 imposed threats of severe financial penalties for any tanker which grounded and spilt oil in US waters and was taken so seriously by shipowners that some companies ceased trading temporarily to American oil ports, with a few threatening to discontinue permanently. Rigidly applied restrictions were applied to single-hulled tankers, which hinted at making compulsory the attendance of a stand-by tug for the whole of the time that ships of this class were carrying oil in American waters. Working groups were set up shortly afterwards by IMO, which included shipowners of all incorporated maritime nations, Oil Companies International Marine Forum (OCIMF), The International Association of Independent Tanker Owner (INTERTANKO), the International Chamber of Shipping (ICS), the International Association of Classification Societies (IACS), (See *Appendix* for brief summary of functions concerning the former organisations), and representatives of leading environmental groups. Their collective findings largely supported OPA90 and, in March 1992, new amendments, 13F, (and 13G, considered briefly in Chapter 19), were made to Annex 1, of MARPOL 73/78, the International Regulations preventing oil pollution. As part of regulation 13F to the amendments stated:

> "(1). *This regulation shall apply to oil tankers of 600 tons deadweight and above:*
>
> *(a) for which the building contract is placed on or after 6 July 1993,*
>
> or *(b) in the absence of a building contract, the keels of which are laid or which are at a similar stage of construction on or after 6 January 1994,*
> or
>
> *(c) the delivery of which is on or after 6 July 1996, or*
>
> *(d) which have undergone a major conversion:*
> *(i) for which the contract is placed after 6 July 1993; or*
> *(ii) in the absence of a contract, the construction work of which is begun after 6 January 1994; or*
> *(iii) which is completed after 6 July 1996."*

The initial factor which impresses is that the minimum sdwt tonnage stipulated by the regulations included many coastal tankers, virtually down to the capacity levels of large river and canal barges, and was extremely specific concerning the dates to which the rules should be applicable. The regulations continued with the all-important Clause 3, the preamble of which stated categorically:

> "(3). *The entire cargo tank length shall be protected by ballast tanks or spaces other than cargo and fuel oil tanks as follows:"*

The sections of rule three continued by advocating clearly defined distance measurements for wing tanks/spaces and double-bottom tanks/bilge's, together with capacity rulings for ballast tanks, suction wells in cargo tanks, and ballast and cargo piping arrangements. Quoting Section (a), governing the wing tanks or spaces:

Supertankers Anatomy and Operation

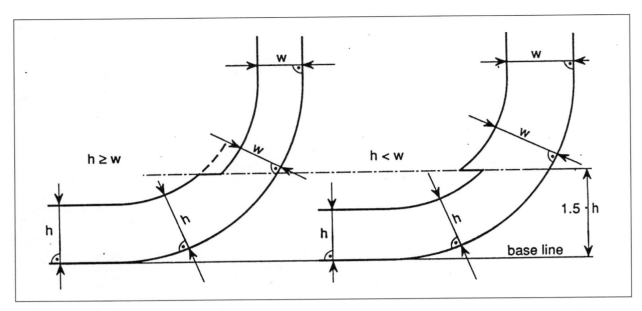

Diag. 14.1 Extract from Annex 1 of MARPOL 73/78 Regulation 13F concerning the Construction of Double-Hulled tankers. (By kind permission of International Maritime Organization.)

"Wing tanks or spaces shall extend either for the full depth of the ship's side or from the top of the double bottom to the uppermost deck, disregarding a rounded gunwale were fitted. They shall be arranged such that the cargo tanks are located inboard of the moulded line of the side shellplating, nowhere less than the distance h which, as shown in figure 1, is measured at right angles to the bottom shell plating as shown in figure 1 is not less than specified below:

$h = b/15$ (m) or
$h = 2.0$ m, whichever is the lesser.
The minimum value of $h = 1.0$ m."

(NB: Figure 1 is copied from the Regulations, (as Diag. 14.1), with acknowledgement to IMO.)

Under Regulation 13E, which remains applicable, the side water ballast tanks must be a minimum distance of two metres and, extend over the entire length of the cargo space.

14.4. CONSTRUCTION, COMPUTER AIDED TECHNOLOGY, SECTIONAL COMPLETION OF THE SHIP

Reference to the general arrangement plans for single and double hulled VLCC's (inside back cover pocket) shows that the basic arrangement of the cargo carrying capacity within the hull has not altered fundamentally. The traditional two longitudinal bulkheads, together with the transversal bulkheads, that separate the cargo carrying area into sets of either five or more centre tanks with port and starboard wings have been retained. The regulations permit the cargo tanks themselves to be narrower than previously although, in compensation, they may also be longer. The following diagram, (Diag. 14.2), is reproduced from the Tanker Structure Co-operative Forum's, (See *Appendix*), 'Guidelines for the Inspection and Maintenance of Double Hull Tanker Structures'. It shows their Table 2.1, and offers definitions of the nomenclature used in identifying the major parts of a 'typical midship section' of a Double Hulled VLCC, as well as, in their Figure: 2.2, the terms used when referring to a 'transverse bulkhead'. (Diag. 14.3). Most vertical stiffeners, cross ties and end brackets have

Double Hull Construction

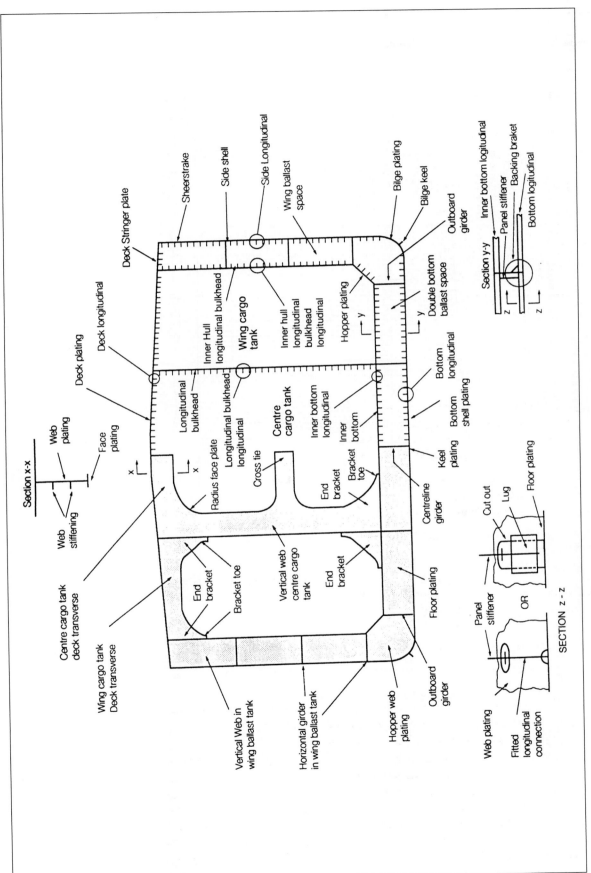

Diag. 14.2 Nomenclature for the Typical Midship Section of a Double-Hulled Tanker. (By kind permission of Tanker Structure Co-operative Forum.)

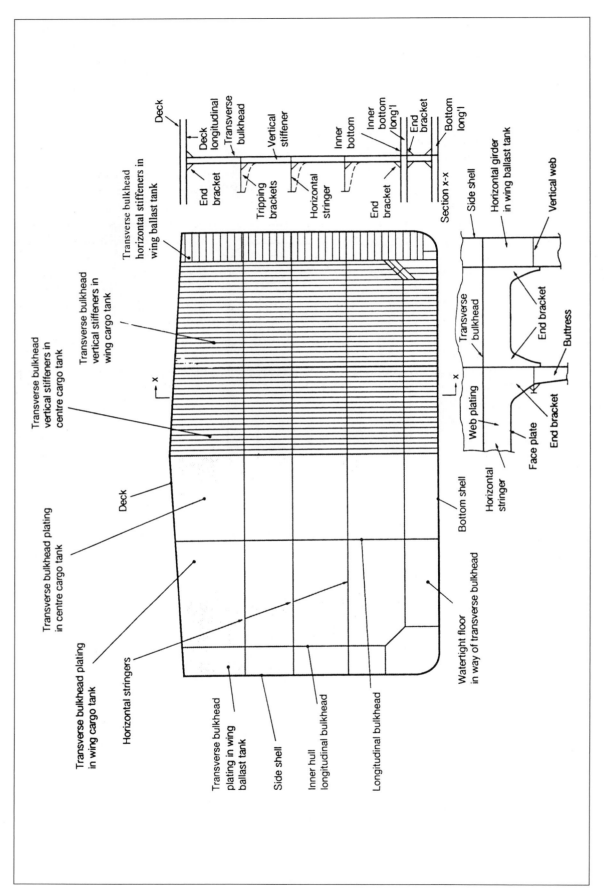

Diag. 14.3 Terms used when referring to a Typical Transversal Bulkhead on a Double-Hulled Tanker. (By kind permission of Tanker Structure Co-operative Forum.)

been retained, but many of the additional stringers and stiffeners, that were essential for additional strength in single hulled supertankers, have been removed. A relatively uncluttered tank remains which has eliminated most, but not all, of the difficult shadow areas which, as demonstrated in Chapter 8, could present challenges during tank cleaning operations in the single hulled VLCC. Although it is not the case with all double hulled construction most designs have cargo and/or stripping suctions recessed into the double bottom.

1993 saw the first of a line in double hull Supertankers from a Japanese Yard. This was the VLCC "AROSA", that was ordered by Lykiardopolu of London, on behalf of overseas clients, a company who are very much leaders and innovators in the field of VLCC technology. The same year saw the first new European construction, the VLCC "ELEO MAERSK", which was built by the Lindo Yard in Odense, Denmark for the Danish Maersk Line. These two Supertankers, because they represented both unique and prototype ships, attracted considerable international maritime and media attention. Since that year, however, a considerable number of double hulled VLCC's continue to be delivered by numerous yards throughout the world which are gradually replacing the "early" single hulled generation of very large ships. This fundamental question concerning rates of change is faced briefly in Chapter 19. The "AROSA" started life as Hull 4855 in the Ariake Works of the Hitachi Zosen Corporation's Yard at Nagasu, Japan. (Photo. 14.2). Hitachi Zosen have had considerable experience constructing Supertankers — amongst, of course, a range of other ship types. Proof of the quality in this Yard's workmanship can be seen in the ULCC's, "ESSO ATLANTIC", (Photo. 14.3), and "ESSO PACIFIC", (Photo. 14.4), each in excess of 516,000 sdwt, that Hitachi Zosen built for the Exxon Group tanker fleet in 1977. Under the names "KAPETAN GLANNIS" and "KAPETAN MICHALIS", respectively, they continue trading as the third and fourth largest ships afloat in the world.

In keeping with conventions of building for all very large tankers, that were examined in Chapter 2, the first of the main construction blocks of the "AROSA" to be placed in the dry dock, comprised the parallel body section. This occurred during on the 24th June 1992, shortly after the cutting of the first plates in March of that year. It is of interest to note that a total of 109 blocks was used to make up the entire Supertanker and that the final one was completed, set and welded into position during early October of that year, less than four months later. The naming ceremony took place in January 1993. Some 37,000 tonnes of steel was required and the total length of welding came to about 805,000 metres, or some 502 miles. The entire construction and fitting out took about 520,000 man hours.

14.2. Hitachi-Zosen's ARIAKE Works in the Kumamoto Prefecture, Nagaso. Japan.
(By kind permission of Hitachi-Zosen.)

Supertankers Anatomy and Operation

14.3. ULCC "ESSO ATLANTIC". Built by Hitachi-Zosen in 1977 continues trading as the ULCC "KAPETAN GLANNIS".
(By kind permission of Joachem Pein.)

14.4. ULCC "ESSO PACIFIC". Laid-up in Norway in 1983 alongside the ULCC "PRAIRIAL" of 555,051 sdwt built in 1979 at St. Nazaire.
(By kind permission of Joachem Pein.)

Double Hull Construction

14.5. *The Initial Plates of the VLCC "AROSA" following Shot Blasting were Protected by a Primer Coat.*
(By kind permission of David Manning.)

14.6. *Computer Controlled PLASMA Arc Cutting Machines ...*
(By kind permission of David Manning.)

14.7. ... *Used in the Cutting of Steel Sheet.*
(By kind permission of David Manning.)

14.8. *The First Plates of the "AROSA" can be seen lying on the floor ...*
(By kind permission of David Manning.)

Double Hull Construction

14.9. ... of the Cutting Shop.
(By kind permission of David Manning.)

14.10. The Forepart of the "AROSA's Bows under Development.
(By kind permission of David Manning.)

Supertankers Anatomy and Operation

*14.11. The Lower Part of the Forward Section being assembled in the Dock bottom.
(By kind permission of David Manning.)*

Hitachi Zosen's were one of the first builders to introduce computer controlled technology into the field of ship construction, by using numerically designed robots for the cutting of ship's plates, as long ago as 1966. They devised also the world's first three dimensional, totally computerised, design, production and manufacturing information system which has been applied to most stages and sub stages of double hulled construction in the various assembly shops. Such totally inclusive automation reduced considerably the number of hours necessary for welding and indeed produced an improved finish. As with single-hulled construction, examined in Chapter 3 of Part 2 all plates were shot blasted and then treated as soon as possible against initial stages of corrosion by the application of a primer coat. (Photo. 14.5). The steel plate was cut by a computer controlled plasma arc cutting machine, (Photos. 14.6 and 14.7) that works directly from stored disc data and is about five times as fast as traditional oxy-acetylene cutting. The range of photographs show various stages in the construction of the "AROSA". They commence with an interesting 'historic' shot of the ship's very first plates, (Photos. 14.8 and 14.9), which are seen lying on supports in the cutting shop after leaving the machine that had worked the steel sheet into shapes of the required dimensions. Once treated and cut, they were then fabricated into the appropriate elements necessary for the completion of any specific block. As with single hulled construction, a number of sections were worked simultaneously under covered sheds within the Yard. The fabrication of the plates comprising the upper forepart of the bows is shown under development (Photo. 14.10) and, whilst this was taking place, the lower part of the forward area was already under assembly in the dock bottom. (Photo. 14.11). Ariake Yard possesses two x 700 tonne swl gantry cranes, each with a rail span of 115 metres capable of lifting to a height of seventy one metres from ground level. They are ideal cranes to transfer the various blocks as these are constructed.

14.12. The Various Stages ...
(By kind permission of David Manning.)

14.13. ... in the Building of ...
(By kind permission of David Manning.)

*14.14. the Cargo Tanks.
(By kind permission of David Manning.)*

The following series of photographs shows the sequences involved in the building of the cargo tanks. (Photos. 14.12, 14.13 and 14.14). The arrangement of the hull blocks on supports in the dock bottom is the same as that used in single-hulled construction. The way in which the double sides and bottom are constructed is plainly visible. A further glance at the general arrangement plans shows that, compared to a double hulled VLCC of another leading tanker company, the "AROSA" has a distribution in her permanent ballast tanks that is a feature totally unique to this ship. Whilst the five centre cargo tanks extend conventionally over the entire length of the cargo area, the segregated ballast tanks, situated in way of number three cargo tanks, have been enlarged to include part of the wing cargo tank area. This, together with a high segregated ballast tank capacity, permits a wider distribution of ballast, thus greatly reducing maximum bending moments when the ship is in partly loaded or in ballast conditions. It also virtually reduces the necessity of ballasting with sea water some of the cargo tanks on those occasions when extreme weather conditions are unexpectedly encountered whilst on passage. The method of easing ship stresses by ballasting cargo tanks was discussed briefly in Chapter 6 and, although certainly not environmentally desirable, simply has to be used when necessary, on the single hulled VLCC, as an essential safety feature. Clearly, the double hulled tanker with its extra 'skin' has, in effect, four longitudinal bulkheads and therefore quite considerably enhanced strength.

The cargo tanks on the "AROSA" are fitted with suction wells that ensure a highly efficient stripping system which results in negligible 'remains on board' on completion of discharge. The significance of this comment can be appreciated by referring to the loading plan of the ULCC, considered in Chapter 5. This plan indicated that in excess of 2,300 cubic metres of previous, different grade, cargo oil in her slop tanks had to be segregated as remains on board and hence excluded from the loading totals.

Cargo loading/discharging and tank cleaning procedures for double hulled VLCC's using COW/IGS methods, are similar to those described in Chapters 7 and 8.

Double Hull Construction

14.15. The Build-up of ...
(By kind permission of David Manning.)

14.16. ... the Hull Develops ...
(By kind permission of David Manning.)

14.17. ... with Successive Blocks.
(By kind permission of David Manning.)

14.18. The Accommodation Block is Lowered into Position.
(By kind permission of David Manning.)

14.19. *A View of the Inside of the Wheelhouse Rarely Seen by Ships' Officers.*
(By kind permission of David Manning.)

14.20. *The Block Completing the Main Deck Area is Added to a VLCC.*
(By kind permission of Shell International Trading and Shipping.)

*14.21. The View along the Main Deck shows a VLCC nearing Completion.
(By kind permission of Shell International Trading and Shipping.)*

The blocks comprising the transverse and longitudinal bulkheads were each of some 400 tonnes, whilst the double bottom blocks averaged around 640 tonnes each. These weights contrast significantly with the much smaller approximately 200 tonne blocks of a number of the earlier single skinned Supertankers and are derived not only from the additional weight arising with the extra plating involved, but also the larger bulk of the individual blocks. The placing of each individual section and the gradual build-up of the hull, as each block was placed and welded, can be followed in the sequence of the next shots which indicate various stages of the building. (Photos. 14.15, 14.16 and 14.17). The lowering of the 680 tonne accommodation block makes an impressively dramatic photograph (Photo. 14.18). It indicates also the state of completion to which each unit was advanced before being fitted finally into position. The internal view of the Wheelhouse, (Photo. 14.19), shows one of the two radars in position with suitable protection, along with the pedestals for the steering units and other navigational aids yet to be fitted. The conglomeration of wires and support brackets, that are later to be covered, show to the deck officer who operates from the wheelhouse, a rarely seen view of the way in which cables are run in the spaces behind bulkheads and deckhead.

The fitting and welding of the main deck of a double hulled VLCC is worthy of specific mention. This would also have been constructed in block form in a separate workshop and, upon completion, lowered into place. The shot of the 650 tonne main-deck block being shipped into position, on a VLCC different to the "AROSA", indicates how the deck area would be finished. (Photo. 14.20). This particular supertanker was one in a series of five VLCC's, each of around 300,000 sdwt, completed between 1995 and 1997 by the South Korean Daewoo yard. It is worth noticing that the cargo manifolds were already fitted onto the main deck module thus ensuring further saving of time. The view along the main deck (Photo. 14.21) shows this particular ship nearing its final stages of completion.

Double Hull Construction

14.22. *The Main Cargo and Stripping Suctions within a tank showing the Bell Mouth or Elephant's Foot. (By kind permission of Shell International Trading and Shipping.)*

14.23. *Forward Slop Tank Bulkhead showing Ladders, Heating Coils and Anodes. (By kind permission of Shell International Trading and Shipping.)*

14.24. Lowering the Middle Part of the Pump Room on the "AROSA".
(By kind permission of David Manning.)

14.25. The Pump Room Bulkhead is Fitted.
(By kind permission of David Manning.)

14.26. The Main Engine of the "AROSA" on its Test Bed Ashore prior to Fitting on the VLCC. (By kind permission of David Manning.)

The 'elephant's foot' or 'bell mouth' of the main cargo and stripping lines, inside one of the cargo tanks, (Photo. 14.22), shows the 17 mm or so distance of the mouth from the plating which allows maximum suction. The vertical stiffeners, reinforcing the transversal bulkhead are easily identifiable. The forward slop tank bulkhead (Photo. 14.23) shows the access ladder arrangements usual within cargo and ballast tanks. The inclined ladder reaches down to a number of permanent 'rest' platforms, conforming to IMO and Classification Society regulations. The heating coils and protective anodes in the slop tank stand out quite distinctly from the tank bulkhead.

The heavy scantling blocks of the pumproom on the "AROSA" are shown in the next sequence of photographs. (Photos. 14.24 and 14.25). The shots show the middle part containing the pumps, and mid section bulkhead, being lowered into position. The main cargo and stripping pumps can be seen in their protective covers of blue reinforced plastic, as well also as the bulkhead pipes that will be connected to those of adjacent blocks as these are fitted. On either side of the pumproom are the port and starboard fuel oil tanks whilst, abaft, lies the ship's engine and machinery spaces. The single 980 tonne engine of the "AROSA" is shown on its test bed ashore prior to being fitted. (Photo. 14.26), Consistent with most VLCC's, be they single or double hulled, the engine is within the range of 32,000 shp, thus offering a fully loaded service speed of between 14 and 15 knots. This particular engine was made by Hitachi. During normal operational speeds, the fuel consumption is to the order of 227/282 litres per mile dependant, of course, upon the loading condition of the ship, and about 72 tonnes of fuel oil per day is used when the ship is fully loaded. The generators on the "AROSA" have been designed to burn either fuel oil or diesel or a blending.

Supertankers Anatomy and Operation

14.27. The Propeller Duct in its Final Stages on Completion.
(By kind permission of David Manning.)

14.28. The After End Block with the Stern Frame.
(By kind permission of David Manning.)

Double Hull Construction

14.29. The Rudder is Positioned for Fitting.
(By kind permission of David Manning.)

14.30. The Rudder and Propeller are placed on the
"AROSA".
(By kind permission of David Manning.)

Supertankers Anatomy and Operation

A series of shots, (Photos. 14.27, 14.28, 14.29 and 14.30), of the "AROSA" show the propeller duct in its latter stages of completion and also in its final fitting into the after end block, along with the stern frame for the rudder. It is another of the special features of this Lykiardopulo supertanker that the ship has been fitted with superstream propeller duct for enhanced propulsive efficiency. In the third photograph, the rudder is shown in the stage of being fitted. This is a conventional type similar to that examined in earlier chapters, but included because it shows perfectly clearly the cradle, pins and fittings essential for hoisting the rudder onto the stock. The final shot shows both the rudder and propeller in place ready for sea-trials. The propeller here was made by the Nakashima Propeller Company and is of the four bladed solid type with airfoil sections. It is made of nickel-aluminium bronze whilst the cone is made from magnese bronze. It is slightly larger than the 9.1m propeller, found on the traditionally built VLCC, being ten metres or about thirty three feet diameter and weighing around fifty tonnes. The figure below the propeller gives a strong visual comparison highlighting the dimensions involved.

The final photographs, (Photos. 14.31, 14.32 and 14.33), show the completed supertanker "AROSA". The first is of the VLCC in dry dock immediately prior to flooding and being towed to the fitting out berth. Barely visible, below the bulbous bow of the stem in the centre line of the ship, the figure of the ship's first Master can just be recognised dwarfed by the sheer size of the hull. The two other shots show the ship undergoing sea trials prior to handing over and acceptance into the Company's fleet, so completing the recording of a process which commenced with the cutting of the ship's first plates. The VLCC has subsequently completed a number of voyages between Arabian Gulf ports, the Far East, USA and Europe.

14.31. The Completed Hull is ready for Launching, dwarfing the figure of the AROSA's first captain.
(By kind permission of David Manning.)

Double Hull Construction

14.32. *From the "AROSA's First Plates to the Completed VLCC ...*
(By kind permission of Hitachi-Zosen.)

14.33. *... as the ship successfully undergoes her Sea Trials.*
(By kind permission of Hitachi-Zosen.)

Further study of the general arrangement plans, for VLCC's that have been constructed in different yards, offers some interesting comparisons. There exist a considerable number of factors that are common not only to each of the double hulled VLCC's we have considered in this chapter, but also to many 300,000 sdwt single hulled ships The principle particulars, for instance, are very similar in terms of overall deck layout and particulars, including dimensions of length, deadweight and gross tonnage's, complement of officers and crew, basic deck and anchoring machinery, including leads to winches and bollards, and the twenty tonne hose handling cranes situated by the manifolds on each of the port and starboard sides. The deck cranes are a significant feature on all new buildings, replacing the set of derricks, but which continue to facilitate immediate identification between Supertankers and bulk carriers. Cargo capacities, and those of fuel, diesel and lubricating oils are also similar between most Supertankers.

Apart from the unique layout of ballast tanks in the "AROSA", the major differences, between single and double hulled VLCC's, are in the beam and depth caused, of course, by the additional 'skin' around the cargo carrying capacity. From Chapter 1, and glancing at general arrangement/capacity plans, the differences are quite significant. The single hulled VLCC, for example, has an average 50 metre midships beam and 22 metre midships moulded depth, whilst the double hulled very large tanker averages 58 metre in the beam, and 31.5 metres depth. The scantlings also may differ considerably. Plate thickness, for instance, at the bottom plating on a single hull VLCC has been seen, in Chapter 2, to be around 27/29 mm which varies, considering the length of the ship, from the 18/20 mm on the inner to 20/22 mm on the outer bottoms of the newer buildings.

14.34. The double-hulled VLCC "EAGLE", (301,686 sdwt), was built for a major tanker company in 1993. (By kind permission of Mobil Ship Management.)

14.5. DEADWEIGHT CALCULATIONS, STABILITY INFORMATION

Using the plans provided inside the back cover, a glance at the deadweight tables and deadweight scales for double-hulled VLCC's should be considered with those examined previously for the single hulled ULCC "JAHRE VIKING". The double hulled VLCC "EAGLE", (301,686 sdwt), was built for a major tanker company in 1993 by Sumitomo Heavy Industries in Japan. (Photo. 14.34). The deadweight table and scale for this ship, along with stability information for normal ballast and fully loaded conditions are shown in the following series of tables and drawings. (Tables 14.1 and 14.2 – Drawings 14.1 and 14.2). Some interesting comparisons may be made with the tables for similar VLCC's as well as the stability diagrams for the single hulled supertankers that were considered in Chapter 6.

DEADWEIGHT TABLE				
ITEMS		FREEBOARD (M)	DRAFT (M)	DEADWEIGHT (M.T.)
Tropical Fresh Water	TF	8.455	22.994	309,315
Fresh Water	F	8.913	22.536	301,712
Tropical	F	8.961	22.488	309,471
Summer	S	9.419	22.030	301,686
Winter	W	9.877	21.572	293,925

Table 14.1 The Deadweight Table of the VLCC "EAGLE"... and ...
(By kind permission of Mobil Ship Management.)

Supertankers Anatomy and Operation

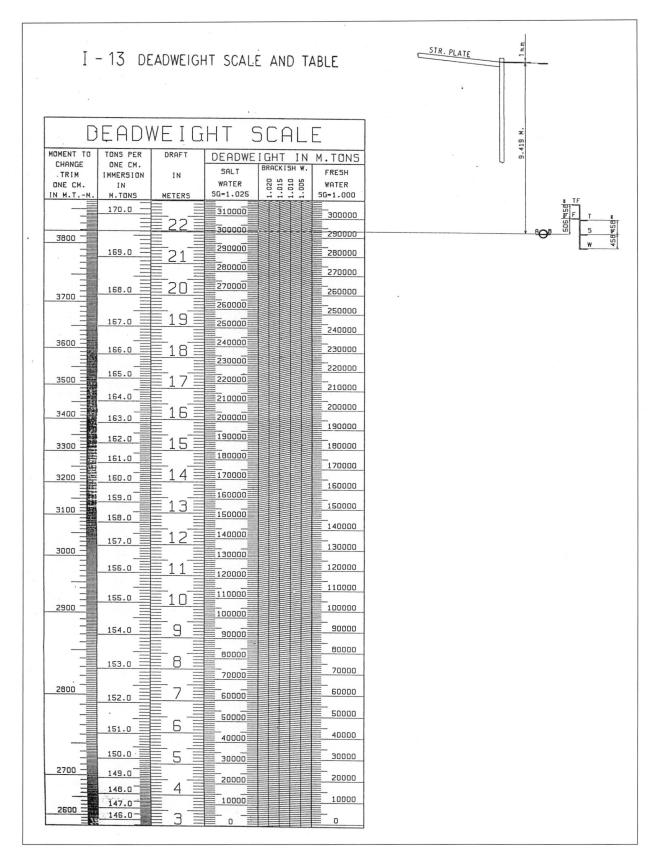

Table 14.2 ... the Deadweight Scale for the same ship.
(By kind permission of Mobil Ship Management.)

Double Hull Construction

```
S.NO.1185   L O A D I N G   C A L
COND.NO.22   LONG VOYAGE  NORMAL BALLAST (ARR.)(
```

ITEM	%	WEIGHT (MT)	LCG (M)	MOMENT (MT-M)	K G (M)	MOMENT (MT-M)	I MOMENT
D.W.CONSTANT	100	410	107.76	44182	19.93	8171	0
PROVISION	10	1	123.75	124	33.70	34	0
HEAVY OIL							
FORE D.F.O.T. (P)	0	0	110.70	0	0.0	0	0
FORE D.F.O.T. (S)	0	0	110.54	0	0.0	0	0
AFT D.F.O.T. (P)	10	129	118.02	15225	12.33	1591	728
AFT D.F.O.T. (S)	11	128	118.09	15116	12.31	1576	728
F.O.SETT/SERV.T.	96	331	115.15	38115	19.42	6428	108
SUM		588		68456		9595	1564
DIESEL OIL							
DIESEL O.T. (S)	5	26	123.70	3216	13.75	358	215
D.O.SETT/SERV.T.	96	23	114.45	2632	21.85	503	7
SUM		49		5848		861	222
FRESH WATER							
DRINK.W.T. (P)	10	31	146.05	4528	23.37	724	193
FRESH W.T. (S)	10	31	146.05	4528	23.37	724	193
SUM		62		9056		1448	386
BALLAST WATER							
FORE PEAK TANK	0	0	-152.12	0	0.0	0	0
NO.1 W.B.T. (P&S)	45	8000	-119.72	-957760	1.99	15920	6052
NO.2 W.B.T. (P&S)	100	19782	-72.79	-1439931	10.26	202963	0
NO.3 W.B.T. (C)	100	19796	-24.80	-490941	10.25	202909	0
NO.4 W.B.T. (P&S)	100	19490	22.96	447490	10.39	202501	0
NO.5 W.B.T. (P&S)	41	8000	74.86	598880	2.07	16560	8710
NO.6 W.B.T. (P)	100	2137	130.47	278814	24.40	52143	0
NO.6 W.B.T. (S)	100	1263	131.70	166337	23.63	29845	0
AFT PEAK TANK	0	0	151.85	0	0.0	0	0
SUM		78468		-1397111		722841	14762
CARGO							
NO.1 C.O.T. (C)	0	0	-119.35	0	0.0	0	0
NO.2 C.O.T. (C)	0	0	-72.80	0	0.0	0	0
NO.3 C.O.T. (C)	0	0	-24.80	0	0.0	0	0
NO.4 C.O.T. (C)	0	0	23.20	0	0.0	0	0
NO.5 C.O.T. (C)	0	0	78.07	0	0.0	0	0
NO.1 C.O.T. (P&S)	0	0	-119.13	0	0.0	0	0
NO.2 C.O.T. (P&S)	0	0	-72.80	0	0.0	0	0
NO.3 C.O.T. (P&S)	0	0	-24.80	0	0.0	0	0
NO.4 C.O.T. (P&S)	0	0	23.20	0	0.0	0	0
NO.5 C.O.T. (P&S)	0	0	67.65	0	0.0	0	0
SLOP TANK (P&S)	0	0	96.92	0	0.0	0	0
SUM		0					
DEADWEIGHT		79578	-15.95	-1269445	9.34	742950	16934
LIGHTWEIGHT		41287	7.58	312955	15.43	637058	0
DISPLACEMENT		120865	-7.91	-956490	11.42	1380008	16934

DRAFT AND TRIM

DRAFT AT LCF	M	8.31	T.P.C.	T	153.3
DRAFT FORE	M	6.14	M.T.C.	T-M	2839.1
DRAFT AFT	M	11.04	L.C.G.	M	-7.91
DRAFT MEAN	M	8.59	L.C.B.	M	-19.43
TRIM	M	4.90	H.B.G.	M	11.52
PROPELLER I.	%	54.0	L.C.F.	M	-17.95

STABILITY

T K M	M	35.56
K G	M	11.42
G M	M	24.14
GGO	M	0.14
GOM	M	24.00
KGO	M	11.56

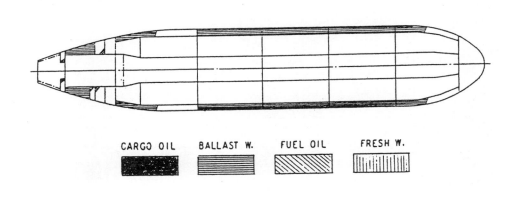

Draw. 14.1 (a) Loading Calculations and …
(By kind permission of Mobil Ship Management.)

Supertankers Anatomy and Operation

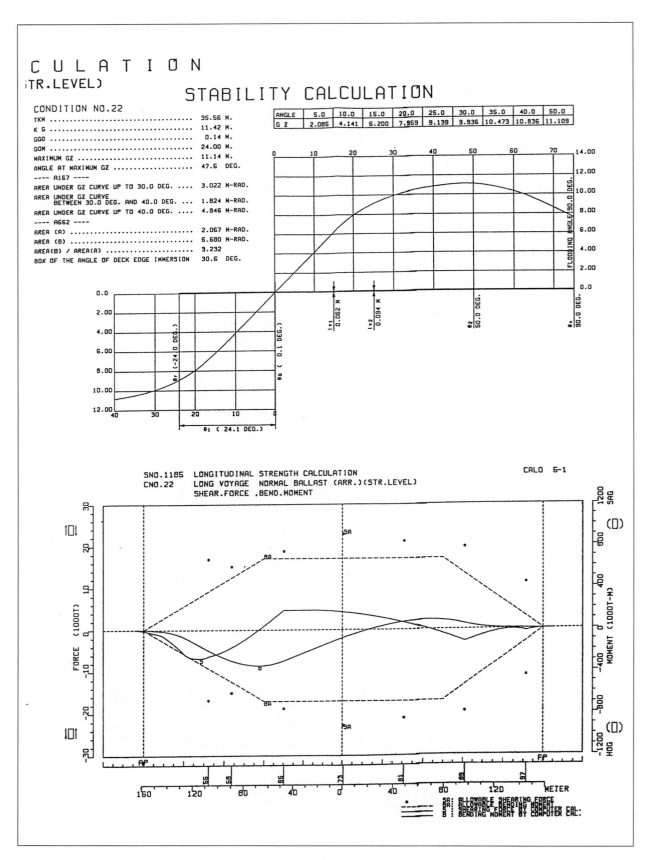

Draw. 14.1 (b) ...Stability Curves for the same ship whilst in Normal Ballast Condition ...
(By kind permission of Mobil Ship Management.)

S.NO.1185 LOADING CAL
COND.NO.2 LONG VOYAGE FULL LOAD (DESIGN DRAFT)

ITEM	%	WEIGHT (MT)	LCG (M)	MOMENT (MT-M)	K G (M)	MOMENT (MT-M)	I MOMENT
D.W.CONSTANT	100	410	107.76	44182	19.93	8171	0
PROVISION	10	1	123.75	124	33.70	34	0
HEAVY OIL							
FORE D.F.O.T. (P)	0	0	110.70	0	0.0	0	0
FORE D.F.O.T. (S)	0	0	110.54	0	0.0	0	0
AFT D.F.O.T. (P)	10	129	118.02	15225	12.33	1591	728
AFT D.F.O.T. (S)	11	128	118.09	15116	12.31	1576	728
F.O.SETT/SERV.T.	96	331	115.15	38115	19.42	6428	108
SUM		588		68456		9595	1564
DIESEL OIL							
DIESEL O.T. (S)	5	26	123.70	3216	13.75	358	215
D.O.SETT/SERV.T.	96	23	114.45	2632	21.85	503	7
SUM		49		5848		861	222
FRESH WATER							
DRINK.W.T. (P)	10	31	146.05	4528	23.37	724	193
FRESH W.T. (S)	10	31	146.05	4528	23.37	724	193
SUM		62		9056		1448	386
BALLAST WATER							
FORE PEAK TANK	0	0	-152.12	0	0.0	0	0
NO.1 W.B.T. (P&S)	0	0	-119.72	0	0.0	0	0
NO.2 W.B.T. (P&S)	0	0	-72.79	0	0.0	0	0
NO.3 W.B.T. (C)	0	0	-24.80	0	0.0	0	0
NO.4 W.B.T. (P&S)	0	0	22.96	0	0.0	0	0
NO.5 W.B.T. (P&S)	0	0	74.86	0	0.0	0	0
NO.6 W.B.T. (P)	0	0	130.47	0	0.0	0	0
NO.6 W.B.T. (S)	0	0	131.70	0	0.0	0	0
AFT PEAK TANK	0	0	151.85	0	0.0	0	0
SUM		0		0			
CARGO							
NO.1 C.O.T. (C)	98	17512	-119.35	-2090057	17.81	311889	13612
NO.2 C.O.T. (C)	98	18919	-72.80	-1377302	17.80	336758	16407
NO.3 C.O.T. (C)	98	18919	-24.80	-469191	17.80	336758	16407
NO.4 C.O.T. (C)	98	18919	23.20	438921	17.80	336758	16407
NO.5 C.O.T. (C)	98	24703	78.07	1928663	18.15	448359	27272
NO.1 C.O.T. (P&S)	98	31355	-119.13	-3735320	17.74	556237	24896
NO.2 C.O.T. (P&S)	98	35043	-72.80	-2551129	17.59	616406	28134
NO.3 C.O.T. (P&S)	98	35043	-24.80	-869066	17.59	616406	28134
NO.4 C.O.T. (P&S)	98	35043	23.20	812997	17.59	616406	28134
NO.5 C.O.T. (P&S)	98	29160	67.65	1972673	18.14	528962	24617
SLOP TANK (P&S)	98	7673	96.92	743667	20.32	155915	4960
SUM		272289		-5195244		4860854	228980
DEADWEIGHT		273399	-18.54	-5067578	17.85	4880963	231152
LIGHTWEIGHT		41287	7.58	312955	15.43	637058	0
DISPLACEMENT		314686	-15.11	-4754623	17.54	5518021	231152

DRAFT AND TRIM | STABILITY

DRAFT AT LCF	M	20.36	T.P.C.	T	168.3	T K M	M	23.85
DRAFT FORE	M	20.69	M.T.C.	T-M	3733.6	K G	M	17.54
DRAFT AFT	M	20.02	L.C.G.	M	-15.11	G M	M	6.31
DRAFT MEAN	M	20.35	L.C.B.	M	-14.31	GG0	M	0.73
TRIM	M	-0.67	H.B.G.	M	-0.80	G0M	M	5.58
PROPELLER I.	%	142.1	L.C.F.	M	-4.18	KG0	M	18.27

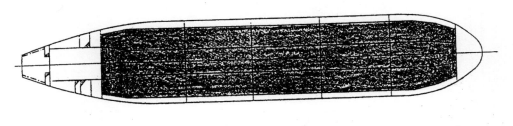

CARGO OIL   BALLAST W.   FUEL OIL   FRESH W.

Draw. 14.2 (a).
(By kind permission of Mobil Ship Management.)

Supertankers Anatomy and Operation

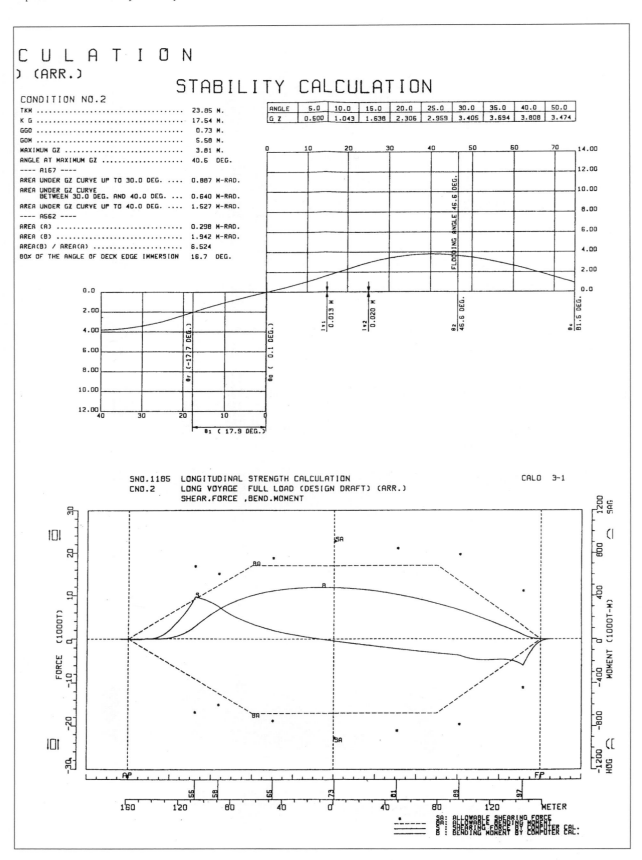

Draw. 14.2 (b) ... and Fully Loaded.
(By kind permission of Mobil Ship Management.)

Chapter Fifteen

INVESTIGATING COMPUTERS AT SEA

15.1. THE EARLY DEVELOPMENT OF A NEW PHILOSOPHY

15.2. NEW SHIP CLASSES, DIFFERENT PROBLEMS, INADEQUATE TECHNOLOGY

15.3. BASIC SEAMANSHIP AND NEW SYSTEMS

15.4. APPLICATIONS IN THE ENGINE ROOM

15.5. SHIPBOARD ADMINISTRATION

15.6. SATELLITE NAVIGATION, GPS

Chapter Fifteen

INVESTIGATING COMPUTERS AT SEA

(Innovations from the 1960's – City of London School of Navigation postgraduate courses – Emergence of different trading patterns and problems handling new classes of ship – overcoming opposition from some 'traditional senior officers' – "BRITISH AVIATOR/ CRYSTAL JEWEL" 'radar assisted' collision-Sperry Marine Systems early contribution – computers in wider use within all shipboard departments – 'Transit' Satellite system on "QUEEN ELIZABETH II" – refinements leading to Global Position Systems (GPS) – extension to other ships – Lloyd's Register Staff Association Report – Comments regarding advances made in the 1980's – Necessity of technological advances to continue fulfilling navigational and other deck officer related objectives – Effects of Research upon the Author.)

15.1. THE EARLY DEVELOPMENT OF A NEW PHILOSOPHY

Rather like growth of Supertankers themselves, so shipboard word processing and the database applications have also developed imperceptibly to assist officers on all classes of ocean going ships for over thirty years. Unlike the VLCC, however, they have not yet reached their maximum saturation point and stabilised, but continue to develop. I recall following a highly interesting post graduate specialist Science and Technology of Navigation course at the School of Navigation at the City of London Polytechnic, now Guildhall University, in the early 1970's which was sandwiched between my professional DTp qualifications. The course included presentations and the study of academic papers on integrated navigation systems many of which were based on, what were then, very basic computerised techniques. Thus it was that, under the auspices of Captain, (now Professor), John Kemp, PhD, head of the school, I served my formal introduction to computerisation. In keeping with all navigating officers, I too had heard numerous rumours and had discussed with other officers whilst serving aboard ships at sea, 'this new development', but the general picture was really rather woolly, to put it mildly. Like my contemporaries, and indeed some of our lecturers, I do not feel that we fully appreciated in those days the exciting revolution that was unfolding before our eyes and in whose birth we were destined to become a part. Prior to these developments, we deck officers generally regarded ourselves more or less as 'modest scientists practising the art of navigation', but now a new concept was looming over the horizon. The new breed of technologists ashore, who supported the shipping industry, had recognised long before this time that certain areas of ship board life would lend themselves very readily to logical and mathematical software programmes. It was seen that such applications would increase not only officer efficiency, but also competency, as well as removing the tedium from a number of monotonous, but necessary routine tasks.

15.2. NEW SHIP CLASSES, DIFFERENT PROBLEMS, INADEQUATE TECHNOLOGY

It was undoubtedly the specific interest aroused by a new generation of trading patterns that led to the emergence ashore of different designs in hull forms, themselves computer generated, which encouraged initial research. Experience soon found that existing Bridge equipment, whilst continuing to perform admirably aboard the conventional cargo vessel and product tanker, possessed considerable limitations when applied to the generation of rapid tonnage growth in Supertankers, oil-bulk-ore (OBO's) and bulk carriers, and in the face of increasing sizes and speeds of container ships that often cruised in excess of thirty knots. There were, as Chapter 10 on bridge duties and ship handling discussed, additional problems encountered with the increased weight, length, beam and draft of VLCC's whose larger turning circles, and slower response times, created difficulties in bringing these ships safely alongside jetties within the parameters of critical limits. Voyage momentum, accrued with the steaming of a very large mass for perhaps a period of weeks without stopping, or even easing power from the engines, meant also longer stopping distances. This was especially aggravated when handling Supertankers through the increased traffic conditions in what were fast becoming more congested coastal waters. All of these developments occurred at a time when Traffic Separation Schemes at sea were being introduced, but whilst many were still going through the slow process of international agreement. The density of traffic, as demonstrated in Chapter 10, led to the need for more sophisticated digitised radars that could aid the prediction, accurately and virtually 'at a glance', of potential collision situations.

Increased cargo capacity, carrying with it a greater emphasis on speed, was aggravated by the knowledge that fuel consumption on VLCC's constituted the greatest single running cost. An awareness of market needs, not only in the oil industry but more universally, directed renewed interest on the whole issue of shipborne economics with particular emphasis being placed on voyage planning. This latter covered not only the immediate chartwork by the second mate, in consultation with the Master concerning the ship's immediate course line, but began to encompass a wider view that took a more concentrated look at other aspects related to ship routing.

15.3. BASIC SEAMANSHIP AND NEW SYSTEMS

New technological advances were therefore introduced both ashore and on board VLCC's and other large ships. These were seen not only to perform the task for which they had been adopted, but whose obvious advantages were passed to craft of other classes for their benefit. An example of this was shown in Chapter 2, on 'first' generation VLCC construction, concerning the effectiveness of the bulbous bow that enabled these very broad beamed ships to move more easily through the water. The success achieved with Supertankers soon led to modified forms of bulbs becoming an almost standard fitting on most smaller ships with, albeit a marginal, but nevertheless definite increase in their speed.

The road to technology however had to be taken cautiously, not the least of the reasons being the need to overcome entrenched ideas and, indeed, cynicism experienced from some senior officers. There existed a potential for danger in a manner not too dissimilar, I suppose, from that which led to the 'radar assisted collisions' of the fifties when some mariners at sea failed to appreciate fully the capabilities and, more importantly the limitations, of their modern sets. I still recall, even in the mid sixties, serving on the bridge with a Master, of the 'old school', who was not prepared to alter or modify his views and would have nothing to do with 'this modern contraption' the term he used to dismiss contemptuously the ship's radar set. He was alas not alone. Indeed, collisions occurred when, had the ships not been fitted with radar, they would have passed each other in complete safety. The sad case of the "BRITISH AVIATOR" and "CRYSTAL JEWEL" collision, in September 1961, serves as just one of the numerous cases which appeared before the courts that arose, in this case, partially from both ships failing to appreciate information provided by their radars and, apparently, with one taking action on incomplete radar information. The fear that computerised information could be similarly treated needed to be allayed, along also with the potential threat that the traditional responsibilities of a navigating officer would not be underplayed by machines to the extent that he could find himself under utilised and therefore bored.

Mr J. B. Carr, of Sperry Marine Systems, writing in a paper presented to us students at the London School of Navigation, clearly 'put his finger on' these aspects by emphasising the importance of not losing sight of basic seamanship objectives or even, at the other extreme, of regarding the computer as infallible. In memorable advice, at least that was how I regarded it then, ... and continue to do so ..., he emphasised the need for maintaining the whole question of what I term, 'realistic computerisation perspective', when considering the implications of using technology on the bridge:

"Systems must remain as aids to the ship's officer, the servant but never the master. Computers only know what is told to them in their software programme. The human brain, on the other hand, contains experience gained from years of knowledge of the sea and ships, has data inputs such as sight, hearing, smell and feel which are denied to the computer and has the gift of imagination which allows the human brain to re-write its own software programme to suit unexpected or entirely new situations."

Until the day arrives when innate human intelligence, exercising value-judgements following structured professional teaching and possessing bridge experience, becomes superseded by advances in computer applications, advanced technology must remain a subordinate role model to the marine navigating officer. Certainly at the present time, and in the immediate future, possession of computerised technology can neither undermine the duties of a bridge watchkeeping officer nor negate the need to glance with binoculars out of the wheelhouse window at frequent intervals.

15.4. APPLICATIONS IN THE ENGINE ROOM

It was not initially on the bridge, but in the engine room, that on-line applications of the computer were put to greatest effect. Programmes were devised which controlled the main turbine, boiler and auxiliary machinery, as well as monitoring valves, steam pressure and water levels, to name just a few. It was then only a matter of time before software was applied that led to fully automated bridge control. This meant that whilst the VLCC was on deep sea passage only, hence not in the close confines of pilotage waters, the navigating officer was able to bypass a duty engineering officer and control directly from the bridge the speed and manoeuvrability of the engine.

15.5. SHIPBOARD ADMINISTRATION

Soon, the question of shipboard administration was addressed. Domestic applications were devised that included the more mundane clerical tasks, essential to the running of any ship, that have to be undertaken by most officers on ships, where pursers are not carried, as "optional extras" in the ordinary course of their duties. The ubiquitous crew lists, port health papers, immigration and custom forms along with lists covering store indents, spare parts and maintenance programmes for deck, engine room and catering departments were put onto disc. Perhaps the most welcomed innovation was when the dreaded portage bill, for those many VLCC's where this was not done in the office ashore, was discovered by the computer. Almost immediately, crew's wages, along with the whole paraphernalia of income tax and national insurance contributions, calculations for leave due, with Sundays at sea, subs abroad and on board ship, along with currency conversions, the slop chest account and numerous 'etceteras', became stored in memory banks. Certainly, I can recall being stood down on one occasion by the Master from stand by, at the late stage of the ship being manoeuvred by tugs alongside the berth at Finnart in Scotland. I was ordered frantically to 'have my go also', at sorting out the narrow column muddle of a sheet of paper, measuring at least one metre in length by a half and covered in virtually indecipherable figures, in order to 'find the twenty pence deficiency' before the European ratings could be paid off.

15.6. SATELLITE NAVIGATION, GPS

As early as 1968, the first Aerospace commercial shipborne satellite navigation system was installed, on the liner "QUEEN ELIZABETH II", using a PDP8 computer and a teleprinter. Very soon afterwards, log-speed and gyro compass course data were fed into an improved memory bank which led, once initial 'teething troubles' had been resolved, to a series of position fixes that were updated every few hours. Almost concurrently, this development led to the inclusion of data which served the automatic steering pilot and off course alarm systems. The use of the doppler shift in the United States Navy's 1964 Satellite Navigation System, known as 'TRANSIT', was adopted with great effect to develop the doppler radar equipment that was of such major assistance in the berthing of VLCC's, as it was seen in Chapter 11.

An even more important universal maritime development from transit is the Global Position System (GPS) that has, over the years, developed into a sophisticated, accurate and continuous position fixing service. In excess of twenty satellites orbit the earth in twelve hour sequences so that any ship board receiver is able, at any time, to receive the on line transmissions of at least four satellites. This means that at least four position lines may be obtained whose intersection indicates the location of the ship. A comparatively recent refinement enables GPS receivers to re-locate around localised radio beacons and thus obtain what is called a 'Differential GPS' signal which offers a higher degree of accuracy.

Later computational software consisted of a data base that contained Elements from the HMSO's Abridged Nautical Almanac and Reed's Tabulated Sight Reduction tables, as well as the contents from those 'Navigator's Bibles' — Norie's or Burton's Nautical Tables. Once these were combined with readings from sextant observations, fed in manually, sight reductions of celestial bodies could be immediately and readily obtained which helped fix the ship's position when out of sight of land. Later additions solved calculations that led to estimated positions of the ship, from her dead reckoning position, that took into account set and drift of the current from tidal stream data. An extract from a Lloyd's Register Staff Association Report, the forerunner to their Technical Association, by one of their officers, Mr J. Hancock, published as long ago as 1970, indicates something of the progress that was being made even in those early days. On page six of his 'Application of On Line Process Computers in Ships', Mr Hancock stated:

> "The computer can also be used to shorten the calculation time with great circle navigation. When the destination latitude and longitude positions are fed into the computer a print-out is obtained on the bridge typewriter of calculated course, distance to destination, time of passage and average speed of the ship. This information can, of course, be used for the start of the program for steering by computer."

The solving in seconds of this arithmetical marathon certainly proved of considerable benefit to navigators and, equally as important, helped gain credibility in their eyes for computerisation as a package.

Captain Geoff Cowap, the Extra Master Mariner and Director of Energy Marine International Limited, whose comments were included in Chapter 6, remarking on the development of computerisation specifically to cargo loading instruments, makes a pertinent comment regarding the more general use of computerisation at sea:

> *"In the 1980's, IBM changed the computing world by introducing and flooding the market with the low cost personal desktop computer. This low cost technology now made it possible to offer a software package incorporating all the calculations for shear force and bending moments. Initially, it was thought that the software could be supplied to be operated on the ON BOARD computer. The ON BOARD computer, however, did not happen for, even today, there are relatively few ships with more than one or two PC's on board."*

The gradual appearance of computer technology over a period of years has offered assistance to the ship's officer as new and advanced software has been introduced. Concurrently with advances in the refinement of discs, improvements have occurred in the hardware aboard Supertankers that has grown from the large, slow, unwieldy, but quite accurate, solid state computer, into the more refined and compact 'user friendly' microprocessor. The old 'typewriter' has given way to the modern keyboard along, of course, with far more sophisticated printers. There have, indeed, been some positive moves and a gradual increase continues in the actual number of stations located in various strategic sites aboard the ship which are now becoming linked to produce a system approaching the common ON BOARD unit of which Captain Cowap has remarked. It remains essential, however, that nobody should lose sight of the objectives which the deck officer might reasonably expect from more intensive computerisation. These objectives vary, I believe, according to specific functions to which the computer is to be applied. Aboard a VLCC, a far more comprehensive computerisation, for example, can probably be applied to cargo control and handling, than would be relevant at present to navigation.

Certainly, in the next four chapters, it is intended to examine, albeit briefly, many of the applications of computerisation in the practice of navigation, seamanship and cargo work. There may be many surprises in store for the professional lay person. Indeed, undertaking research for this part of the book, including conversations with professional practitioners, has 'brought me up with quite a jolt' as the realisation has dawned of the extent to which computers have been and will undoubtedly continue to be used at sea.

Chapter Sixteen

USE OF COMPUTERS IN NAVIGATION

16.1. INTRODUCTION

16.2. RADAR

16.3. COMPUTERISED CHARTS
RASTER CHART SYSTEM – VECTOR CHART SYSTEM

16.4. ADVANTAGES OF ELECTRONIC CHARTS

16.5. INTEGRATED NAVIGATION SYSTEM

16.6. INTER-LINKED NAVIGATION AND VOYAGE MANAGEMENT

16.7. A PASSAGE PLAN EXAMINED

16.8. FUTURE TRENDS

Chapter Sixteen

USE OF COMPUTERS IN NAVIGATION

(Introductory Remarks – Daylight Viewing Computer and Video Based Radar for Collision Avoidance and Navigation – Developments in, and Implications of, Electronic Charting Systems, Including RASTER and VECTOR Data Bases and Developments – ECDIS (Electronic Chart Display and Information Service) – Working Towards a Form Acceptable to SOLAS – Voyage of the mv. "KATRINE MAERSK" – Need for IHO or IMO to Resolve Current Difficulties and Ratify ENC – Use of Electronic Charts and ARPA for Collision Avoidance – Integrated Navigation Systems – Navigation and Voyage Management Systems – Inevitable Advantages of ECS – Practical Experiences of Author Running a Past VLCC Passage Plan Electronically – Future Trends.)

16.1. INTRODUCTION

It is true to say that a technological revolution has occurred in the field of navigation which affects directly the work of watchkeeping officers aboard all ships, not only Supertankers, as current research unfolds innovative ideas for the future. The advantages in shipboard computerisation are exciting, far reaching and profound and, as demonstrated in Chapter 15, have been extended gradually often over a period of one, and up to even three, decades. Advances continue even as I write ... . The process of introducing both hardware and software has been concurrent, in the sense that improvements in a particular item of gear did not occur in isolation until the 'ultimate piece' was produced, whilst developments in other items stood still. There was, therefore, no particular sequence which arose affecting directly the advance of any specific piece of equipment. The sudden appearance of the weather facsimile machine is one example that I immediately recall which was not in the wheelhouse when I went on leave during the coastal continental voyage, but was fully operational upon my return six or so weeks later before signing on for the next deep sea voyage. It was some considerable time, to continue expanding this example, before the next development in forecasting appeared. Only very rarely was any kind of formal instruction offered, and correct use of 'whatever it was' had to be learnt 'on the job' and from manufacturers' manuals, until the gear became integrated within the next nautical college course. Inevitably, however, the pace of implementing developments progressed faster in some areas than others. To offer another example in clarification: new chart technology almost certainly developed following improvements in radar PPI presentation and the introduction of electronic collision avoidance (ARPA) packages, but each of these, in turn, was preceded by research undertaken in other areas of electronic, gyroscopic navigation and satellite systems. This need not present my readership, however, with any particular problems when discussing computerisation in its wider aspects as this has affected, and continues to affect, bridge and wheelhouse work, and no effort, therefore, will be made to consider any chronological implications of improvements to equipment.

16.2. RADAR

Since the Second World War, major developments in radar technology and practices, both as an aid to collision avoidance and navigation, have been of paramount importance. They have affected directly virtually all navigational technology which has followed. Radar thinking and practice have been revolutionised by the introduction of multi-coloured computerised video graphics to raw radar information. There has been the provision of a day light viewing screen, and the sophistication of push button interface controls, which have incorporated keyboard inputs as well as the basic set and performance controls. To my mind, it is the latter which has been significant in reinforcing the psychological fact that permanent changes *have* arrived and that modern, and future, bridge watchkeeping operations will never again be the same. The physical fact that no longer working a knob or switch, but pressing a button which lights up and emits a sound when activated and yet does the same job, but with an extra ingredient, indicates the impetus of 'something different'.

There are a number of commercial organisations which have been involved with research and development of the radar equipment that has had such a profound effect upon Merchant Navy thinking. One such international company is Consilium Marine, whose range of 'SELESMAR' radars, plus other, 'cargo handling' aids, some of which will be examined in Chapter 18, are amongst the leaders in the field. Day light viewing radars, (Photo. 16.1), have the considerable advantage whereby watchkeeping officers are able to discuss simultaneously with the Master, and others, any collision avoidance targets or points of coastal interest on the PPI, without having to take turns in looking through the radar hood and the obvious lack of continuity which that necessitated. Of course, the devilish minded young deck cadet will no doubt feel deprived from no longer being able to smear boot polish around the edge of the viewing piece, so that the Master, hopefully, would look up from his concentrated study with an imprint of the hood outline around his face. Still, I suppose, even at sea every silver lining has to have its cloud!

Consilium's MM950 radar offers a choice of three different sized screens, the maximum of which is a 28" (400 mm) CRT that projects the PPI radar picture complete with computer graphics. When the radar is console mounted it stands in isolation and thus may be viewed totally independently of other equipment. (Photo. 16.2). The radar in this photograph shows, incidentally, three of my navigating cadets on the starboard side of the wheelhouse of this very large crude carrier being addressed informally by the ship's Master. The advantage of placing

16.1 Most Daylight Viewing radars have a multi-coloured PPI.
(By kind permission of Consilium Marine.)

Use of Computers in Navigation

16.2. *A Pedestal Mounted Daylight Viewing radar on board a VLCC.*
(The Author)

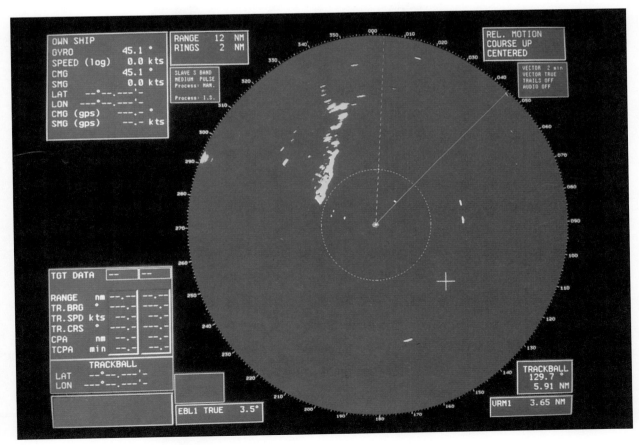

16.3. *The Information contained in the Windows, and a Closer View of the PPI.*
(By kind permission of Consilium Marine.)

Supertankers Anatomy and Operation

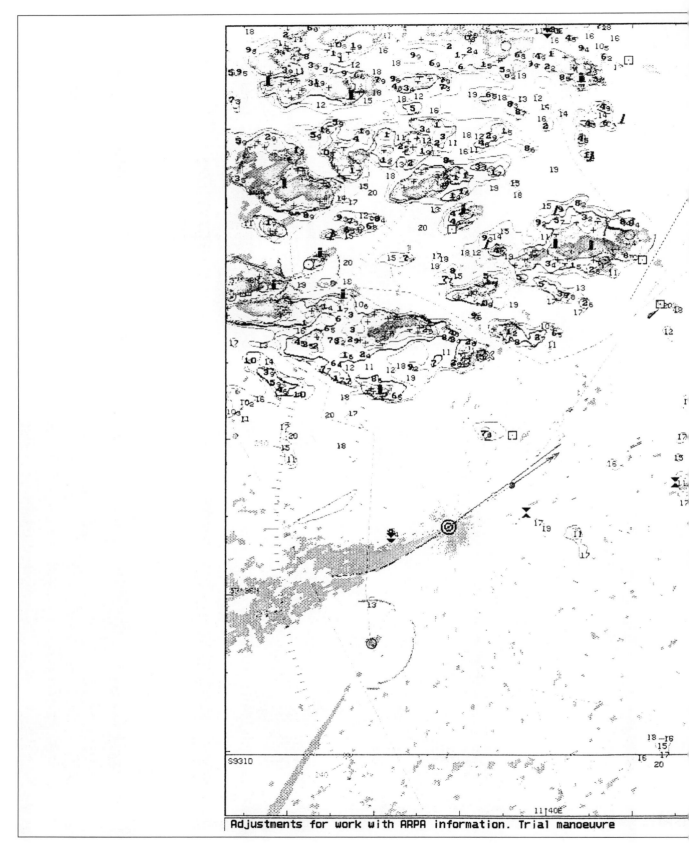

Drawing 16.1 Electronic Charts may be used also for ...
(By kind permission of Transas Marine.)

Use of Computers in Navigation

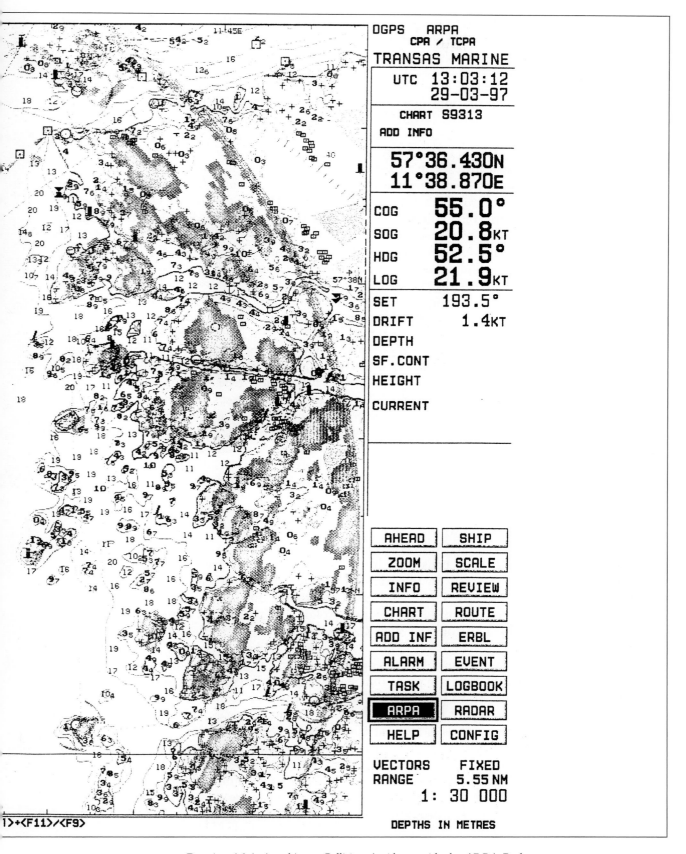

Drawing 16.1. (contd.) ... Collision Avoidance with the ARPA Radar.
(By kind permission of Transas Marine.)

the set in this position is largely to assist collision avoidance, rather than for navigational usage, so that visual viewing of a potential situation on the PPI may readily be reinforced with binoculars in order to detect any very small targets.

In the collision avoidance mode, a number of windows can be superimposed onto the PPI which continue to offer the information expected from a traditional ARPA digital hooded computer. A close up of the details (arranged for demonstration purposes), for example on the MM950, (Photo. 16.3), include a window on the top RHS that advises a Relative Motion display, with the ship's course at 045.1° (T), and zero speed. The continuous line represents the heading marker, with a centre screen instead of an offset display. The window, on the lower RHS, gives details of the current situation regarding the trackerball, namely that it is located at a position 129°.7 (T), at a distance of 5.91 miles from the PPI centre. The longer lowest window, together with the adjacent one on the LHS, shows the condition of the electronic variable range marker, and electronic bearing indicator respectively. This is the 'pecked' line on the PPI which reads 003°.5 (T). Other LHS windows give data that is pertinent to the target and an observer's own ship. There is also a window that would allow exploration of any trial and/or forecast manoeuvres which the officer of the watch may wish to explore. The two mile range rings on this particular display have been temporarily suppressed, an action that is frequently taken at sea for reasons of clarity, thus avoiding cluttering the PPI with unnecessary detail. This particular radar set allows the tracking of up to forty targets which may be individually selected by means of the tracker ball. The other usual computerised collision functions are available, including guard zones and target identification numbers. There is also a system failure as well as dangerous and lost target alarms, and a particularly useful relative motion true trail mode which blends both traditional presentations, (see Chapter 10), and enables a target's position to be indicated together with a forecast one/sixty minute vector. Although it is not shown, this facility is aided by a digital 'after glow' of targets, thus offering an immediate indication of the possible speed of other observed vessels. For the plotting of fast moving targets there is also a dual tracking facility which has proved particularly useful. One channel is available, that offers a wider overview of the target in relation to its situation, plus a second more intensive channel which gives a closer thus more accurate plot.

The radar can be used also in conjunction with an electronic chart so that the positions of both own ship and any target vessels may be seen in relation to the surrounding land or underwater obstacles such as sand banks etc. In the chart extract portrayed, (Draw. 16.1.), part of the radar screen may be observed on the RHS whilst the position of the plotting or observer's ship is indicated by the concentric circle, and proposed course line, mid-lower LHS of the chart. The target under review is shown as an enlarged solid dot, with a six-minute vector, and is bearing approximately 330° (T) from the West Cardinal buoy, fine to starboard of observer's ship heading marker. The recommended course line is seen as a pecked line set to pass safely along the channel indicated by the cardinal buoys to starboard and the buoyage 'squares' to port. The obstructions may be clearly seen to port and starboard. Any avoiding action that might be considered, if it were necessary to overtake the target for instance, could be assessed beforehand in relation to the actual position of the ship and the surrounding navigational dangers. The broken circle with the flash, off own vessels' starboard quarter, indicates a RACON beacon – discussed in Chapter 10. Additional information is given in the window on the RHS of the display. The ship's position is plainly indicated, whilst below is shown Course of 052°.5 (T) and her speed of 21.9 knots. The Course over Ground (COG) and Speed over Ground (SOG), of 055° (T) and 20.8 knots, indicate the Course and Speed of the ship, in terms of the set and drift of the Current actually experienced. The current is therefore acting adversely at just under one knot. The Set and Drift, recorded in the window below, of 193°.5 (T) and 1.4 knots, indicates the theoretically calculated current from the system. These are based on tidal predictions of the kind that would appear on the paper chart as tidal diamond information, and require manual calculation and geometrical application.

Use of Computers in Navigation

16.4. As a Desk Mounted display, Daylight Viewing Radars may be integrated with other Equipment. (By kind permission of Consilium Marine.)

16.5. Windows are fitted around the PPI offering additional information to assist the Navigating Officer. (By kind permission of Consilium Marine.)

For navigation purposes, the MM950 relies on video processed information that presents land masses, in an immediately identified colour, obtained from a digital sweep that is adapted to the pulse length and beam width of a specific set. Continued use can be made of either the traditional north or ship's head up presentation, on Relative or True Motion display, offering the PPI own ship position that is centred, or off centred with automatic re-set, according to preference. The display can also be integrated with its keyboard as part of a system with wider applications. (Photo. 16.4). In this instance the two apparently identical radars, on each side of the central control unit, would be the 10cm and 3cm sets. A zoom display, on the far RHS, would be used for close quarter work in collision avoidance situations, or for more accurate position fixing in coastal navigation.

The information offered by the additional windows which surround the PPI is of direct assistance to the navigating officer. (Photo. 16.5). The lower LHS windows, with the red panel, indicate continuously displayed ARPA functions with, on this occasion, the second ship flashing a red warning of too close a proximity to a guard limit. Details of his own ship are displayed, including gyro course and speed, together with the course and speed made good that again can

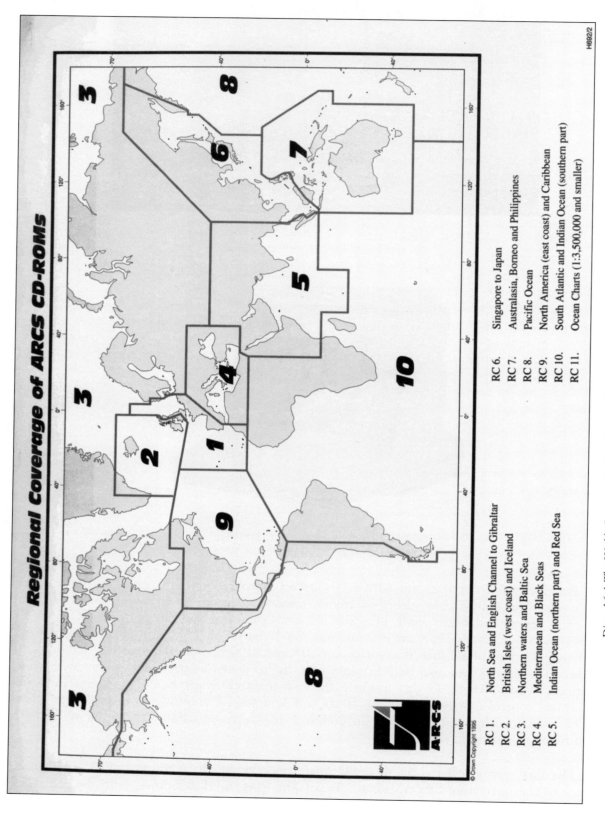

Diag. 16.1 *The World Coverage of Admiralty RASTER Charts is in a total of Eleven Regions, including a Small Scale 'Ocean' Area. (By kind permission of Admiralty Hydrographic Office.)*

also take into consideration manually fed in calculations covering the set and drift of the current. Thus the true course, according to the chart, might be 060° (T) but, with say, an algebraically applied 3° for S&D, the gyro compass course actually being steered would be 063° (G) or 057° (G) depending upon whether the ship was being set to the north or south of the course-line respectively by the tide, and thus needed to be brought back onto course. The updated position is shown, in terms of latitude and longitude, together with course and speed from the Global Position System (GPS).

16.3. COMPUTERISED CHARTS
RASTER CHART SYSTEM – VECTOR CHART SYSTEM

The idea of placing computerised charts on board ships, as part of a wider electronic navigational service, has been pursued by the United Kingdom Hydrographic Office, and interested commercial organisations, since the mid-1980's with the investment of considerable sums of money, time and other resources. As a result, a range of systems have been running for some years now with a considerable degree of success although, inevitably, a range of "teething troubles" continue to be experienced.

It is perhaps unfortunate, and a sign of the times, that the introduction of new technology has brought with it a plethora of acronyms and jargon. In fact, without being too cynical, it appears that the use of specialised terms seems to have increased more or less in proportion to advances with the systems. In order to simplify and cut a clearer path, through what could potentially be a maze, only a selection is offered of some terms in use. CDU is the term applied to a Chart Display Unit, which is part of an Electronic Chart System (ECS), whose function is to project electronic charts onto a video screen. The data, constituting the charts themselves, is obtained by computer scanning information portrayed on ordinary paper based, internationally used, British Admiralty navigating charts, as well as the variety printed by other nations including those on the Continent, the United States and Russia.

There are basically two different methods involved in portraying digitally the chart information. The first step is the scanning of a conventional paper chart, or its printing masters, into facsimiles by means of a multi series of coloured lines and dots, or 'pixels'. The name given to the information obtained in this way is 'Raster data' and it has been pioneered by the British Admiralty, and developed under the acronym RCDS, or Raster Chart Display System. These are termed 'official' charts because they have been created and issued by an authoritative Hydrographic office, in this case, the Admiralty at Taunton in Somerset. The diagram, (Diag. 16.1), shows the world wide coverage of the Admiralty Raster Chart System (ARCS). As indicated,

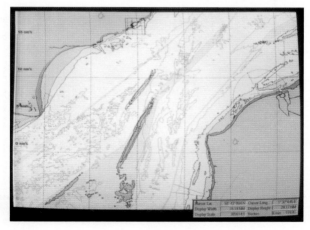

16.6. Screen Shot of Racal-Decca's CHARTMASTER Electronic Chart.
(By kind permission of Racal-Decca.)

Supertankers Anatomy and Operation

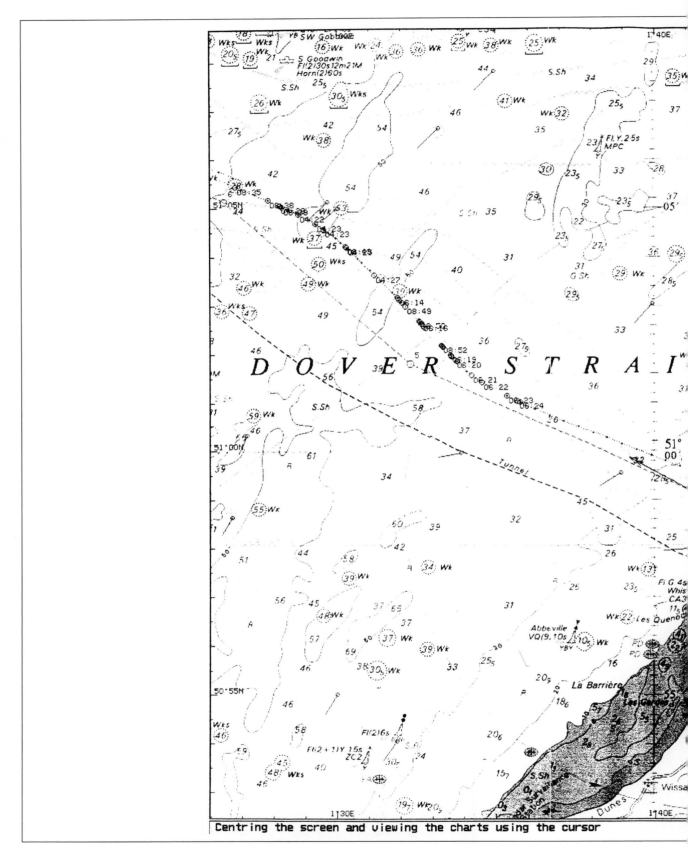

Drawing 16.2. A Larger Scale Chart may be used on the Computer in exactly the same way that a larger scale ...
(By kind permission of Transas Marine.)

Use of Computers in Navigation

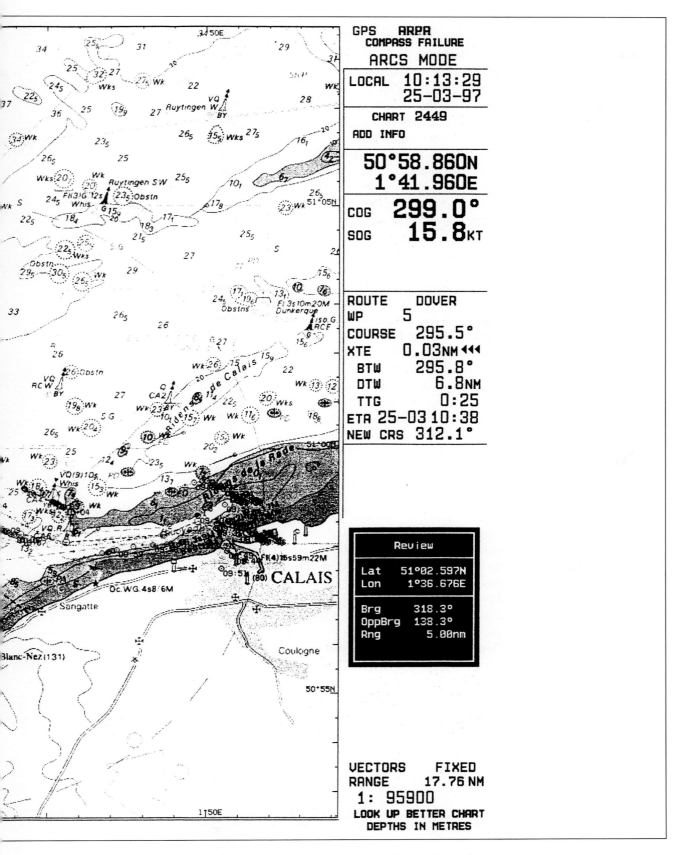

Drawing 16.2. (contd.) … Paper Chart could be chosen.
(By kind permission of Transas Marine.)

Supertankers Anatomy and Operation

the entire world's issue of charts has been computerised into just ten regions, each containing around 350 charts, to make an eventual total of around 3,500 charts that are digitised into just *one* compact disc per region. There is a separate CD for the smaller scale ocean charts. The scanning and creation alone of all Admiralty charts is a quite breathtaking feat. In view of the fact that traditional charts have been scanned, or copied as they stand, the electronic charts convey precisely the same information as the originals with, of course, the same degree of accuracy. RCDS, therefore, because they are facsimiles cannot be ratified for exemption, under the SOLAS V20 Regulation, for the carrying of up-dated paper charts aboard SOLAS registered ships. The Admiralty produce only the charts themselves, and not the equipment upon which they are used but, and this is an important consideration, *no specific hardware* is necessary in order to use RCDS. The CD's containing them can be used on any 486 plus personal computer which has a minimum 8 Mb RAM and a 256 colour SVGA screen in Windows, DOS, Apple Macintosh and other formats. It is for this reason that they are becoming increasingly popular amongst the "other end of the shipping scale", as it were, and are finding considerable favour with both coastal and deep sea yachtsmen.

The second method of producing electronic charts is by means of 'Vectors' which use co-ordinates based on geographical positions of charted objects which are then stored in a data base. The information is obtained from the paper charts of Hydrographic authorities, but are not sanctioned by the authority whose charts have been used. The companies would pay the authority an appropriate copyright fee. This information may be presented on the screen in 'layers' such that items, soundings or lights, may be temporarily removed, and then reinstated, at the direction of the navigating officer thus, it is contended, avoiding cluttering the screen with too much information. This can have an advantage on those occasions when intense navigation concentration needs to be directed to specific areas of interest at the momentary exclusion of other information. An example of such use could be the coming alongside a jetty when adjacent, buoyage in excess of say one half of a mile would, at that moment, be of only secondary concern. These charts are produced by commercial companies, and are called 'non official'. Again, Vector charts may be used on PC's that possess similar specifications for RCDS.

Selection of a required chart for use on board is made possible by using 'review' and 'zoom' facilities. When the system is operated initially, a map showing the entire world appears. The scale is changed by using the Review function to select small scale charts relevant to the passage being planned. Large scale charts can then be chosen. The process is followed in exactly the same way that a second mate would operate in order to choose his chart folios from the drawers beneath the chart room table, which was examined in Chapter 10. On the PC, it is best perhaps not to fit too much of a large scale chart into one window, but to break the area into a series of smaller sections, as will be seen in the discussion accompanying photographs 16.6 and Drawing 16.2 below. Similarly, if charts are shown that would be relevant for navigating in the area, but are ones not required for the specific voyage planned, then these can be ignored in the same way that the unrequired paper charts in the folio would also be rejected. To facilitate the process even more efficiently, a video plotter may be used to file a route plan. On some systems this may already be stored on disc for future reference, hence a series of passage plans may be constructed and saved for future amendment and use. In the same way that Admiralty charts (et.seq.) are numbered, so also may the electronic charts be catalogued and selected by use of this function.

'Typical' electronic charts are shown in the following photographs. (Photo. 16.6). Racal-Decca's CHARTMASTER screen shot indicates the Dover Strait and shows quite clearly the Varne and adjacent sand banks as well as other areas of shallow water whose decreasing depth is indicated by the darker shades of blue. The land and drying areas continue to be depicted in the now familiar yellow and green. The absence of reference indicators, to the grid superimposed over the screen, should be noted and, combined with the suppressed layers indicating details of depths, buoyage and other navigational information, indicates that this is a small scale vector chart. The idea of suppressing information by reducing layers was seen in the ARPA collision avoidance chart, (Draw. 16.1), where it might have been noticed, that all names relevant to the buoyage, and many of the metered depths, had been removed in the immediate vicinity so that attention could be focused totally on the collision aspects. Quite clearly, these layers could be re-introduced at a seconds' notice if it were considered expedient to do so.

A different chart of part of the same area of the Dover Strait, taken on another ship, indicates the versatility of the electronic system. (Draw.16.2.). Own ship is just visible in the now familiar concentric circle, on a Course 299° (T) and a speed of 15.8 knots. The 'layer' of buoy names is activated so that own ship may be seen just departing the CA4 and CA5 channel cardinal buoys slightly north east of Sangatte. Numerous targets may be observed with their appropriate vectors, indicating their respective course and speeds. The windows on the RHS indicate Observing Ship's position, course and speed. The computer has indicated also that problems exist with the gyro compass upon which the ARPA is based-thus advising remedial action – and, equally as important, recommendations have been made, in the LLHS, that a change to a larger scale chart is required – in exactly the same way that the Master on the Bridge would have changed to a larger scale paper chart by now. The windows indicate also a "cross track error" of 0.3 nm advising that the ship has drifted from her course line by this amount and needs to be brought back. The voyage plan, with the ETA Dover, is updated automatically as the ship proceeds on her passage.

The debate concerning the use of either system of Raster or Vector produced charts is being hotly pursued at the time of writing. Certainly, the International Hydrographical Office (IHO) of IMO are acting very cautiously. In the summer of 1997 they failed to ratify RCDS. IMO will not allow electronic charts to be used unless they are backed up by traditional, fully corrected, paper charts. The entire issue, at the moment, is a potential minefield. It appears to revolve around acceptance of responsibility, and accountability, in the event of a navigational accident to a ship that has occurred directly through the use of information conveyed on an electronic chart. It is one which is difficult fully to understand, as complex issues are presented in complex ways – depending on who is doing the presenting.

There is no reason why the two chart systems cannot be operated simultaneously. A demonstration was made recently in the South China Sea when a container ship, the mv. "KATRINE MAERSK" owned by A.P. Moller, used both chart styles very effectively. In a press release from Taunton, shortly after the event, the United Kingdom Hydrographic Office announced:

"The demonstration uses official ENC (Vector) data in the critical harbour areas of Singapore and Hong Kong and their approaches, with Admiralty (Raster) charts linking the two across the China sea. The combined use of official Vector and official Raster charts is a major step forward in marine safety."

The container ship is fitted with the Sperry VMS and uses a worldwide folio of electronic charts covering the entire passage from North Europe to the Far East. The UK HO release confirmed the views of the Master of the container ship:

"Captain Lyse of the 'Katrine Maersk' believes the combined use of Vector and Raster data is a safe and effective solution to the ship's navigational requirement. He said that the Sperry VMS system using ENC and Raster charts had great benefits for vessel safety, as the navigating officer is able to ascertain the vessel's position in relation to navigational hazards and during the passage through restricted water, as well as making fast and easy voyage plans."

Transas Marine's NAVI SAILOR, also permits the display of both Vector and Raster charts. It seems that further investigations by the International Hydrographic Office and IMO must inevitably – **and before too long** – find a way of overcoming the obstacles and ratify fully the use of electronic charts.

16.4. ADVANTAGES OF ELECTRONIC CHARTS

Having seen systems in operation, I believe that the very concept of electronic charts is extremely exciting and undoubtedly, to my mind, the system of the future. It can be only a matter of time before the controversial and legal issues concerning their use will be resolved. I should like therefore to consider some of the considerable advantages gained by the use of electronic charts.

Not the least of these, is the accuracy obtainable by the integration of satellite based information, as well as other authoritative maritime sources, including Admiralty list of lights and tidal atlases. Another major advance concerns the updating of charts instantaneously by means of a CD-ROM containing the weekly Notices to Mariners. By this means will be rendered obsolete not only the carriage of numerous folios of paper charts, but also that time consuming, often messy, but always painstaking task of 'chart correcting', much beloved of deep sea and coastal second mates, mentioned in Chapter 10 when discussing bridge duties.

The 'zoom' facility of Racal-Decca has an advantage other than to select passage charts. It allows extremely accurate close up views of the immediate area concerning the navigator. (Photo. 16.7). This can be either by a window at the top of the screen, or wherever, or can fill the entire screen so that buoyage, bridges and berths are clearly shown. Other tiled windows enable route tracking and planning facilities to be shown and include contour checking and warnings. As seen, the position of an observer's ship is readily superimposed on the electronic chart, so that the ship's passage may be monitored along her course line in direct relationship to the navigational land masses and noted hazards. A point watching mode, whereby the continuous calculations of bearing and distance from specified points of navigational interest, may be inserted. (Photo. 16.8). For many years now, with the advent of Consilium Marine's GPS systems, it has been possible for all ships to fit 'way points' into their passage planning. Generally, these represent an 'object of navigational interest' and usually indicate places along the course line where major alterations of course are to be taken. Way points are easily keyed into the electronic chart, (Photo. 16.9), so that alterations can be seen at a glance and the distance off scanned immediately and accurately. By using Racal-Decca's Differential GPS, that is GPS associated with a station in the near vicinity of the ship, it is possible actually to 'see' the position of the ship superimposed on the screen. This can be seen to move not only as it manoeuvres in closely restricted waters, but also when it is actually alongside, or being brought alongside, a jetty. (Photo. 16.10) At the moment, this part of the system experiences some early problems that appear to be caused by 'wow and flutter' in the atmosphere which affects the electronics. This has the effect, under certain conditions, of making the outline of the ship hover around, and even upon, the jetty. The advantages of the facility, however, are very impressive and I believe that, *"once these gremlins have been put to rest"*, it could well prove possible to dock a VLCC in restricted visibility by using the close up version of the electronic chart together with an integrated doppler.

At the other extreme, electronic charts may be shown at the maximum sensible working range. (Drawing 16.3). Here, using the systems of Transas Marine, the entire working area may be seen at a glance, the 46.28 mile range in this instance is actioned, so that course vectors may be inserted and the passage planned some considerable distance and even days ahead of the current position. The ship is at Lat. 33° 57'0 N. Long. 132° 43'0 E whilst the area under planning is around that of Lat. 21° 00'0 N. Long. 157° 00'0 W. The actual position of the ship would be inserted nearer the time in the light of adjustments necessary due to external influences caused by set and drift of the current etc.

Information so presented in the form of Electronic Navigation Charts (ENC), complies with performance standards and specifications under a SOLAS ratified system known as ECDIS, 'Electronic Chart Display and Information Service'. Both Raster and Vector induced data is compounded into the single electronic chart system, but ECDIS also continues to be subjected

Use of Computers in Navigation

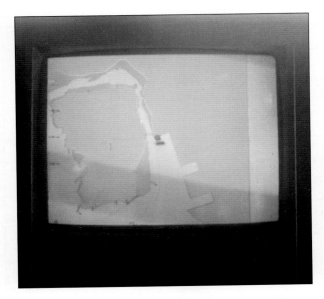

16.7. *The Zoom Facility allows closer inspection of the Ship's track and Adjacent Marks.*
(By kind permission of Racal-Decca.)

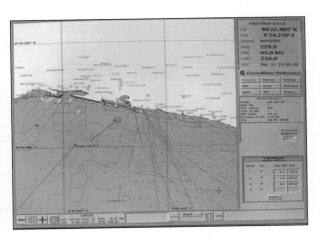

16.8. *The 'Point Watching' Mode allows continuous tracking of the ship from a navigational point of interest.*
(By kind permission of Consilium Marine.)

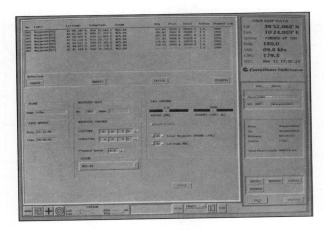

16.9. *Way Points along a Ship's Track may be keyed in (LHS Diagram) and then shown clearly on the chart.*
(By kind permission of Consilium Marine.)

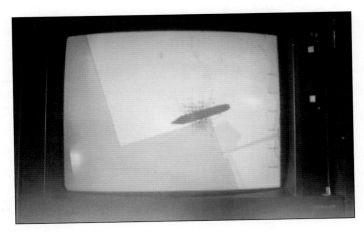

16.10. *Close-up Definition allows the Ship to be seen Approaching and alongside her Berth.*
(By kind permission of Racal-Decca.)

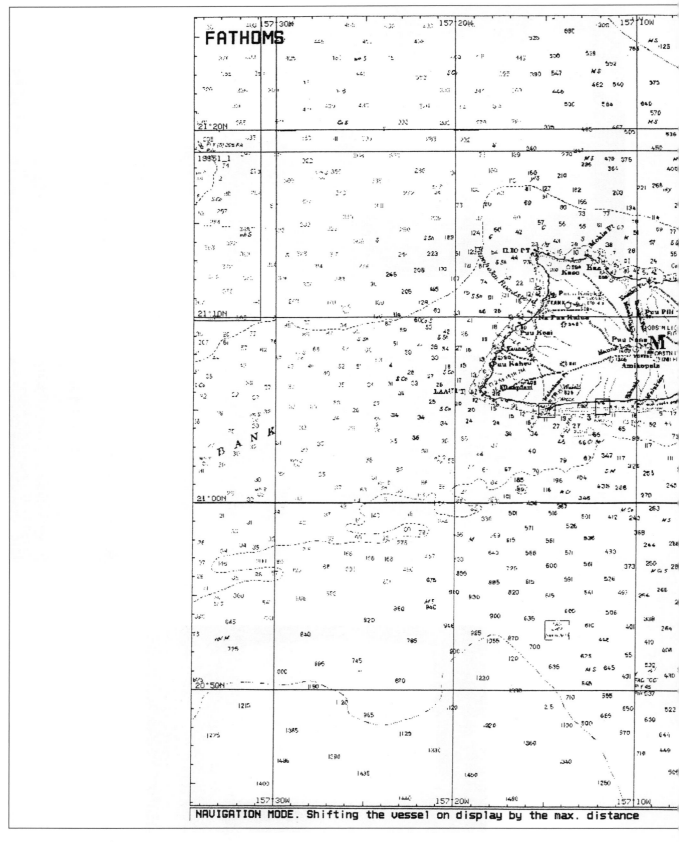

Drawing 16.3. An Entirely New Passage may be planned far from the position currently held by the Observer. (By kind permission of Transas Marine.)

Use of Computers in Navigation

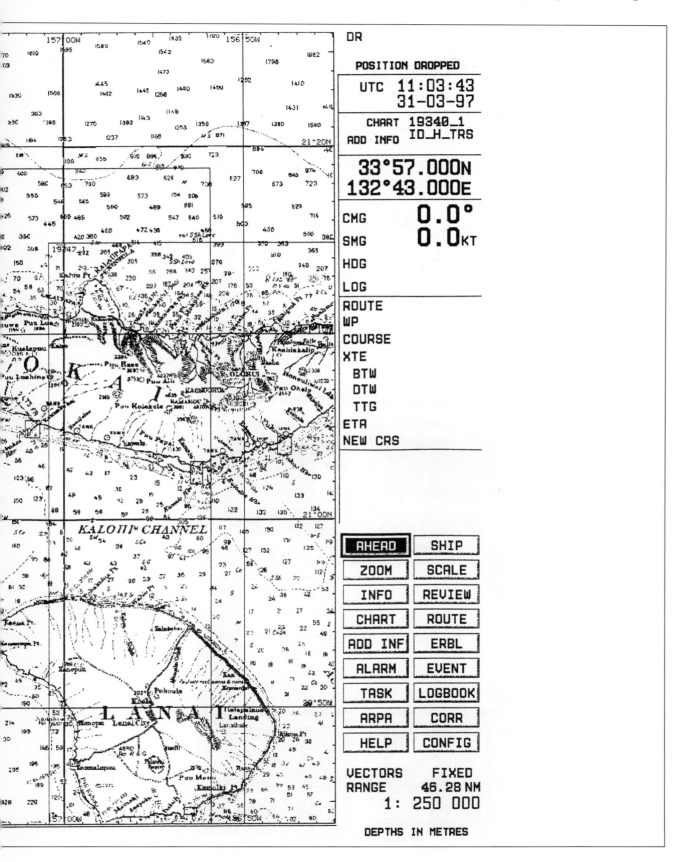

Drawing 16.3. (contd.)
(By kind permission of Transas Marine.)

Supertankers Anatomy and Operation

to performance standards, arising largely from safety implications concerning internationally accepted standards in chart production, even though the adoption of ECDIS was made by IHO/IMO in November 1995. As discussed, the new technologies have created considerable re-thinking and the need for precise definition on the part of the International Association of Classification Societies and IMO. Wherever legalities are concerned, authoritative sources have to be quite certain of all implications likely to arise from their definitive pronouncements. Naturally enough such deliberations, and their international maritime ratification, take time to be considered by all parties and agreement reached that could lead to mutual international acceptance. IMO approval, through their International Electrotechnical Commission (IEC), for example, of ECDIS hardware equipment is not expected immediately.

16.5. INTEGRATED NAVIGATION SYSTEM

The use of an 'Integrated Navigation System' (INS) was a logical stage of development in conjunction with ECDIS. This entails an interface with all other equipment externally utilised in navigating the ship much of which, as will be examined briefly in Chapter 17, has itself been computerised. In the Consilium Marine's 'INS 970' System, inputs to a Data Concentrator Unit (DCU) and a Data Processing Unit (DPU) can be fed into the system from external sensors, including both the magnetic, gyro compasses and autopilot, speed logs, echosounders, the 10cm and 3cm radar displays and position sensors. (Photo. 16.11). Extended memory banks offer

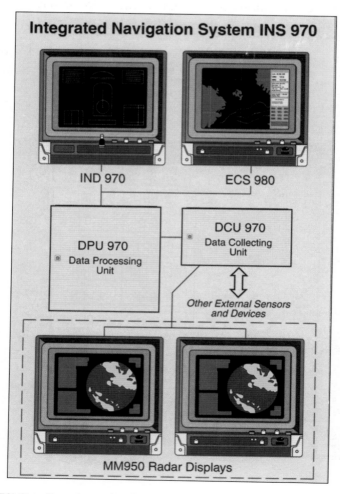

16.11. *'SELESMAR INS970' Data Processing and Collecting System with Charts, Route Planning, Steering and Radar systems. (By kind permission of Consilium Marine.)*

historical information concerning the ship's position, such that points of wheel over for alteration of course etc may be recalled, as well as full assessments made regarding the tracking of targets. Overall voyage and immediate route planning may be made directly onto the screen by use of a mapping facility, and permanently stored for subsequent use via 'floppy discs'. Thus, the previously used parallel indexing techniques, that covered distances off coastlines and safety margin track plotting, have been developed into a more sophisticated route steering system integrating the ship's course consoles. This means that navigation and also collision avoidance techniques in narrow waters, by means of an Automatic Navigation and Track-keeping System (ANTS), that has been invariably a time of intense concentration to the officer of the watch, is made considerably less stressful. This is particularly important when these situations are confronted in conditions of restricted visibility. An indication of the way in which the integration of the units occurs, in the Sperry Marine Integrated Navigation System, can be seen in the accompanying functional block diagram. (Diag. 16.2). The X-Band and S-Band radars are in direct link with the ECDIS Navigation Station and connected to the planning and coming displays. Outputs are made to bridge Repeaters on the wings, internal bearing pelorus and steering compass. The doppler speed log is fed from its transducers into the radars and an analogue display in the wheelhouse. The master gyro compass and rate of turn input in the steering pedestal are supported by a print out giving records of all course alterations and rudder movements.

Complete voyage management systems that have been fitted to VLCC's include the MIRANS 5000, or Modular Integrated Radar and Navigation System, that is produced by Racal-Decca who have been amongst the leaders in the field of navigation and radar for over fifty years and have brought this wide spread experience to fruition in their computerised equipment. The 1996 built ULCC "RAMLAH", 300,361 sdwt, (Photo. 16.12), carries this system, integrating the radars on board with electronic chartwork, weather and speed inputs, and presenting all critical information on Racal-Decca's, Live Situation Report (LSR). This display

16.12. A VLCC fitted with the MIRANS 5000 INS.
(By kind permission of Racal-Decca.)

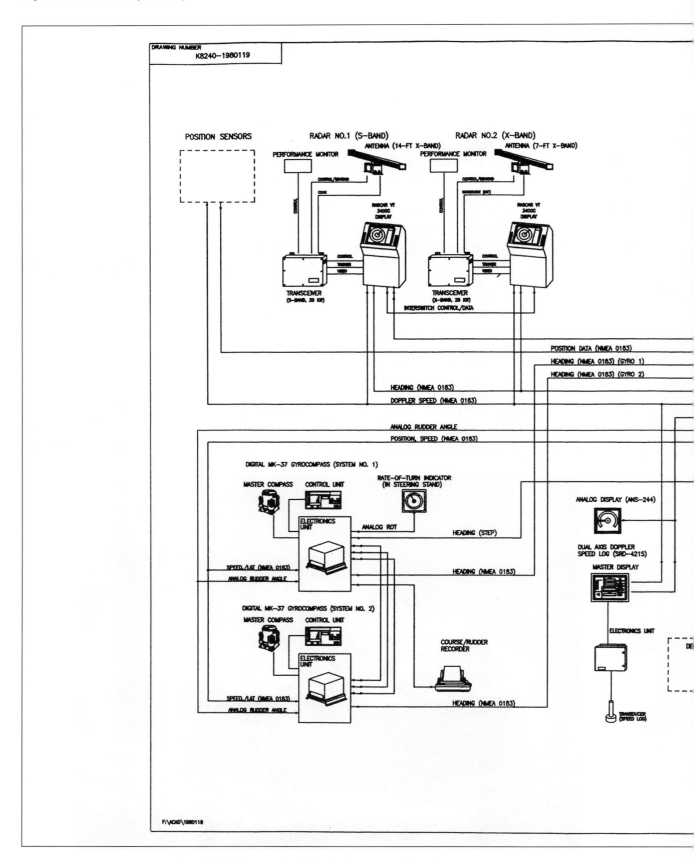

Diag. 16.2 Schematic Diagram for the Integrated Bridge System.
(By kind permission of Sperry Marine.)

Use of Computers in Navigation

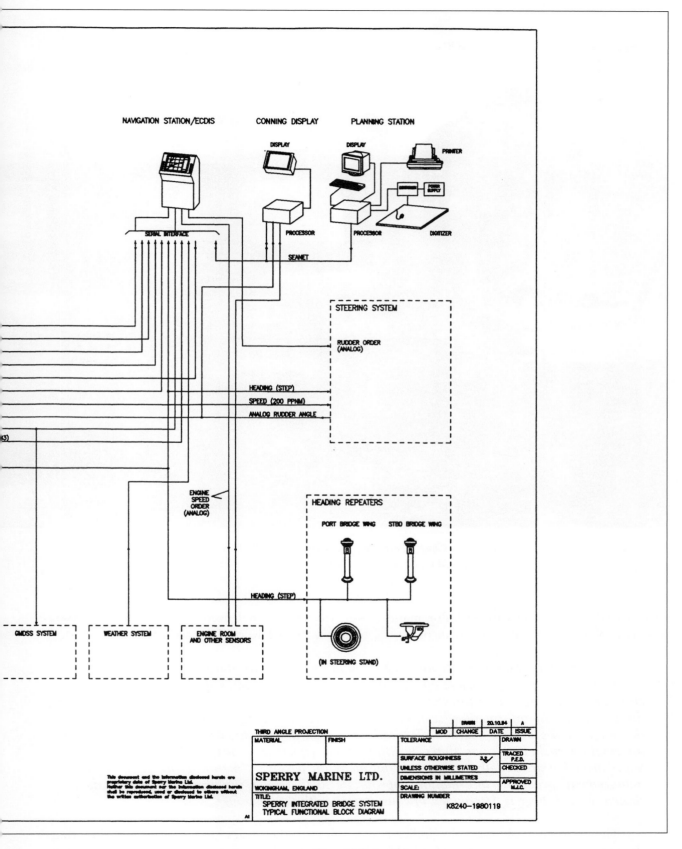

Diag. 16.2 (contd.)
(By kind permission of Sperry Marine.)

Supertankers Anatomy and Operation

16.13. A Chief Officer using the MIRANS 5000 INS.
(By kind permission of Racal-Decca.)

of essential information can be tailor made to suit the operational needs of each ship and uses a system of pages in windows to incorporate, in the words of Racal-Decca's manual:

> "The Live Situation Report (LSR) module brings together all the information essential to the safe and efficient operation of a vessel. This centralised display unit presents clear and unambiguous information which is easily read from a distance. The module displays pages of information, selected by operators according to their current needs or the stage of the voyage reached. Typically a standard page would display navigational information including own ship's position, tidal effects and way point data, ... The pages available are usually designated 'full away on passage', 'manoeuvring' and 'voyage management' providing information appropriate to the ship's situation. The precise data shown on each page is determined in consultation with the customer."

The following photographs indicate examples of the pages and the information that can be conveyed. (Photos. 16.13, 16.14, 16.15 and 16.16). A further facility enables data from the electronic chart display to be overlaid on the paper chart so that two different views may be

Use of Computers in Navigation

16.13 (contd.)
(By kind permission of Racal-Decca.)

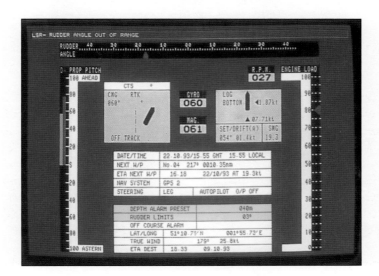

16.14. Live Situation Report (LSR) Full Away on Passage Page ...
(By kind permission of Racal-Decca.)

Supertankers Anatomy and Operation

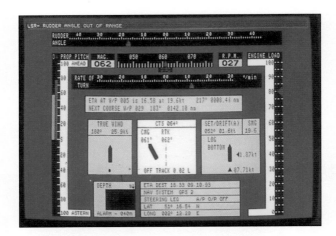

16.15. ... LSR A Typical Manoeuvring Page ...
(By kind permission of Racal-Decca.)

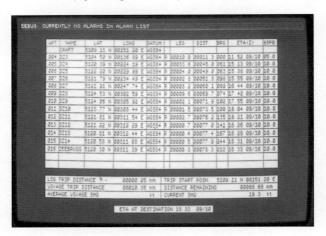

16.16. ... LSR Voyage Management Page.
(By kind permission of Racal-Decca.)

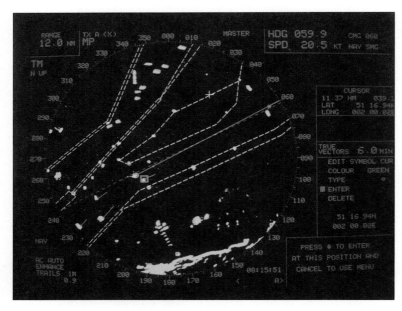

16.17. A Fully Computerised Collision Avoidance and Navigation Radar Operational aboard a Ship at Sea.
(By kind permission of Racal-Decca.)

16.18. The MIRANS System is seen fully operational aboard a Large Ship.
(By kind permission of Racal-Decca.)

obtained instantaneously of the ship's passage. The Chartmaster CM500, for example, could be kept at an optimum navigating scale chart is able to provide a wider overall view of the entire passage within that chart's geographical limits.

In the accompanying PPI, (Photo. 16.17), the area under consideration is in the vicinity of the Sandettie Bank and is, coincidentally enough, only just astern of the same position where I experienced the tide rip that was related in Chapter 10. The Ruytingen Bank is to starboard and the Sandettie Bank on the port side. The radar is on the 12 n/m range and the ship's course is 059.9° (T) at a speed of 20.5 knots. The sand banks in this notorious area and their attendant buoyage are clearly marked, along with the RACON 'flash' of the Falls Light Vessel, which is bearing approximately 340° (T), and is identifiable by the close triple echo. There is a considerable build up of shipping in the southwards heading Traffic Separation Zone of the Noord Hinder Channel. The Sandettie Light Vessel can be identified at the intersection of the zone markers astern of own vessel. For obvious reasons, only the targets appropriate to own ship are being tracked. The one with the "acquired" box and vector, fine on the starboard beam of own ship, is being overtaken. This is indicated by the length of the six minute vector attached to the "box" indicating her true course and speed. The heading marker of our 'own ship' indicates a track that is deviating slightly outside the extreme limit of the channel: this is probably because 'we' have opened up to port slightly for the ship being overtaken in order to give her more sea room and thus, perhaps, avoid interaction effects, also mentioned in Chapter 10. Three other targets astern have been acquired previously, as can be seen by their vectors. There is a further target, near the heading marker, at about eleven miles distance on the starboard side. The different colours assist considerably identification of own heading marker, targets and land masses to the south off Dunkerque. The radars are seen here, (Photo. 16.18), as part of the complete MIRANS system in fully operational mode aboard a large ship.

16.6. INTER-LINKED NAVIGATION AND VOYAGE MANAGEMENT

Sperry Marine's "VISION 2100", fully interlinked navigation and voyage management bridge console has also been designed specifically for the larger ship and numerous complete modules have been fitted on ships of the Supertanker class. Additional to all of the information

Supertankers Anatomy and Operation

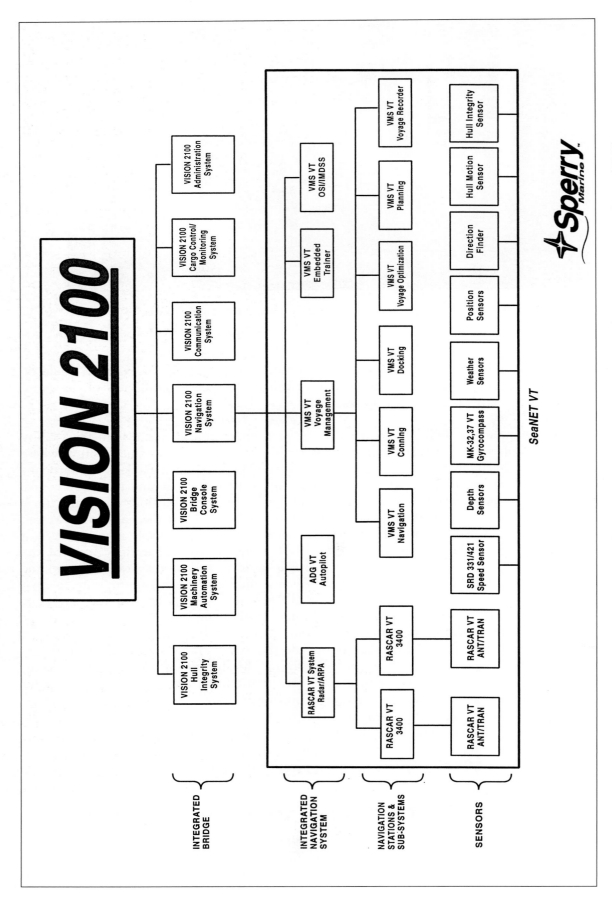

Diag. 16.3 The VISION 2100 Schematic Plan Showing inputs and sensors contributing to the Voyage Management System (VMS). (By kind permission of Sperry Marine.)

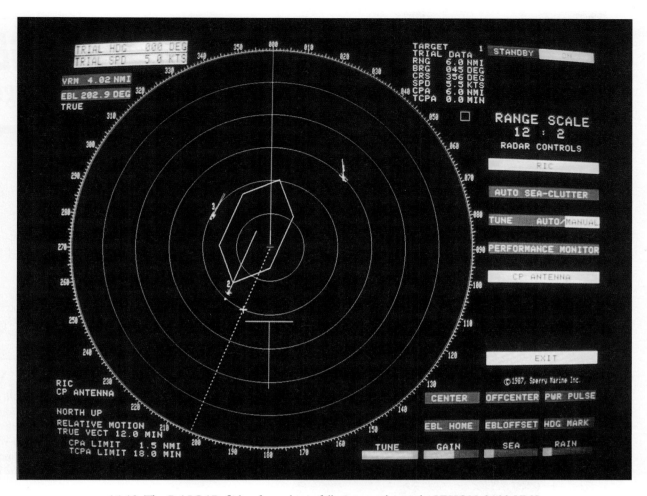

16.19. *The RASCAR flicker free radar is fully integrated into the VISION 2100 VMS.*
(By kind permission of Sperry Marine.)

already absorbed in INS, which is of course incorporated into VISION 2100, are sensors which monitor additionally the weather including wind and current data, hull stress motion and hull integrity, a trainer, engine room machinery, cargo loading and control as well as ship administration and Global Maritime Distress and Safety System. (GMDSS). There is very little which takes place on a VLCC voyage which cannot be fed in, and integrated, into this comprehensive package, including log book records and the Master's night orders to his navigating officers. The following flow diagram indicates the way in which the various units interlink. (Diag. 16.3). Each of the modules can be directed onto the screen by means of a number of windows and, as with other systems, it is possible for more than one window at any time to be projected. The Sperry VMS uses their 'RASCAR VT' (Vision technology) colour radar display, (Photo. 16.19), which is entirely devoid of any flicker due to the implementation of a high pixel count with latest radar technology. The Sperry VISION 2100 VMS fully integrates autopilot and electronic charts into the navigation management station and their planning system uses an electronic plotting table which offers interface with paper charts. (Photo. 16.20). The following extract from Sperry Marine's VISION 2100 brochure offers a clear illustrated profile of the elements involved. (Plan 16.1).

Supertankers Anatomy and Operation

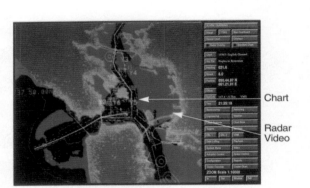

Plan 16.1 The VISION 2100 VMS is illustrated to show the Range of Equipment and Sensor Inputs Available. (By kind permission of Sperry Marine.)

Use of Computers in Navigation

SYSTEM – VMS VT™
gh our Windows...

Waypoint No. 2
Route Plan
ARPA Targets

tion Window

rack Keeping System (ANTS)
ce and commercial
ncluding British Admiralty ARCS

Window in Window

VMS VT Planning and Navigation Management Station

- Independent workstation typically located at main chart table
- Optimized route for "on time arrivals", efficient fuel oil consumption, reduced risk of cargo damage and loss, and heavy weather damage prevention.
- Electronic chart creation and updating
- Electronic chart data base translation to compatible format

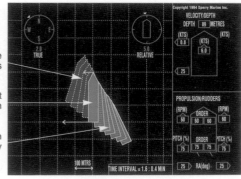

Future Positions
Present Position
Position History

VMS VT Docking (Optional) Window*

- Mimic presentation of ship's velocity against a scaled grid
- Selectable future and history interval
- Multi-sensor configuration to include doppler, rate gyro and GPS
- Bridge wing displays suitable for enclosed and outside wing stations
- Video Window in Window display capability is available as an option, e.g. bow camera.

a successful heritage

e continued evolution of Sperry
eneration system is user friendly and
nd Microsoft's Windows NT®

Plan 16.1 (contd.)
(By kind permission of Sperry Marine.)

16.20. The Complete VISION 2100 VMS.
(By kind permission of Sperry Marine.)

It is very easy to discuss the developments of computerisation and to speak, dare I say almost glibly, of comprehensive voyage management systems and the integration of oceanography and meteorological data, GMDSS, and a range of electronic and technological 'inputs'. The implications of radar computerisation on collision avoidance have been demonstrated practically, but perhaps an insight into the use of the computerised VMS package to navigation might be more readily reinforced by considering a practical example. This would enable the quite outstanding contribution which has been made by computers to the practice of deep sea and coastal navigation to be fully appreciated.

16.7. A PASSAGE PLAN EXAMINED

A passage plan for the voyage of a VLCC, say from Coryton to Mena al Ahmadi, is briefly examined. By 'conventional' means to plan and plot such a run would be quite a time consuming business for the second mate. The various small scale charts have to be selected and the overall run from the River Thames, through the Suez Canal, (in ballast remember), to the Arabian Gulf carefully planned. A range of reference books, including various oceanographic publications, would have to be consulted, along with ocean and in-shore tidal predictions, local and oceanic weather forecasts and route charts. The next step would be to select the appropriate large scale charts, from a number of folio folders, and then undertake the plotting of courses onto the series of charts for immediate use once the ship left port. These would then have to be placed in the 'ready use' locker immediately below the chart table working surface. At all stages, of course, the various legs of the trip would have to be discussed with the Master and the process for an entire four week or so voyage could well take a number of days spread out of course over the trip. Following completion of the various legs of the voyage, the charts would have to be erased and returned to the various folio folders and the next ones prepared. On the computer at Transas Marine's, Southampton office, I 'planned' the entire voyage outlined above in under forty minutes. By zooming in on the large scale charts, this time factor included

all of the coastal work necessary, including distances off the coast throughout the passage. Taking into account the electronic inputs which could obviate the necessity of sextant work and written records, (see Chapter 10), the actual passage work would be simplified equally as effectively.

The entire ship's operating and administrative systems are linked comprehensively in a way that could never be envisaged even a decade or so. Each office may have monitors that show any aspect when it is required and a constant monitoring of the ship's position at any time, plus all relevant navigational data, is available for immediate viewing by the Captain, a very sobering thought for all watchkeeping officers. A sobering thought for the Captain, just to demonstrate the adage *"fleas having smaller fleas that bit'em"*, is the monitoring of the ship's position by a similar system fitted in the company's head office so that the exact geographical situation regarding the ships' whereabouts is available on sight by the company's directors and superintendents.

16.8. FUTURE TRENDS

Perhaps an even more sobering thought with bridge voyage management systems might concern the potential problem of which input actually controls other inputs, or has some effect on their performance capability. With one system that was examined practically at sea, for example, the efficiency of one part is in fact controlled by the performance of the 10cm radar. This, in turn has not the same discrimination for detecting small targets under certain conditions as does the 3cm radar. The latter, on the particular system being discussed, has also an effect on the accuracy of the differential GPS, thus if not causing, at least contributing to the electronic 'wow and flutter' referred to earlier in this chapter. As I write, these specific problems are being confronted. Remedial steps, for instance, have paid close attention to the so-called 'roll over' system of GPS, the counter that stores the number of weeks upon which the data base is stored, which was given its testing in August 1999. Moves are being made to interswitch between the 3cm and 10cm radars, as well as to eliminating sea clutter without leading to the loss of weaker targets. In other areas, advances are examining how Bridge Management Systems might be developed into voyage data recorders (VDR's) and already hints are circulating regarding fitting the 'Black Box' into merchant ships that has proved so highly successful in aircraft. The way in which technology is progressing makes, not only to my mind, the distinct feasibility of an eventual crewless ship for deep-sea passages, with personnel being put on board only for river/harbour pilotage and berthing duties. With regard to ships leaving their berths indeed, as will be seen in Chapter 17, automation is already becoming a distinct reality.

Chapter Seventeen

OTHER BRIDGE COMPUTERISATION

17.1. INTRODUCTION
17.2. GYRO COMPASS
    17.2.1. Steering Console
17.3. SPEED LOG
17.4. FIRE DETECTION
17.5. GENERAL ALARM SYSTEMS
17.6. GMDSS
17.7. INMARSAT
17.8. TIDAL PREDICTION, BAROMETRIC PRESSURE
17.9. MEASURING HULL STRESS
    17.9.1. Bending Moments
    17.9.2. Pitching, Slamming
17.10. DECK CRANES – MOORING HOOKS
17.11. SEA-KEEPING PREDICTION SYSTEM, DATA COLLECTING UNIT (DCU)
    17.11.1. Harbour Prediction Mode
    17.11.2. Operational Advisory Service
    17.11.3. Cruising Prediction Mode of Ship Motions
17.12. OCEAN PASSAGE PLANNING – WEATHER ROUTING
17.13. USE OF COMPUTERS ASHORE – SIMULATOR TRAINING
    17.13.1. Nautical Colleges
    17.13.2. Transas Marine "Navi-Trainer"
    17.13.3. GMDSS
17.14. "VIRTUAL REALITY" TRAINING
17.15. CONCLUDING REMARKS

Inmarsat system comparisons: The following table provides a detailed comparison of the physical and technical characteristics of each of the inmarsat systems.

FEATURES	INMARSAT-A	INMARSAT-B	INMARSAT-C	INMARSAT-M
Worlds Coverage [1]	Global	Global	Global	Global
Overall Weight	Average 100kg	Average 100kg	Average 4kg	Average 25kg
Size of Antenna (diameter & height)	Approx 0.9m - 1.2m	Approx. 0.9m	Approx 0.3m	Approx 0.5m
Antenna type & means of tracking	Parabolic antenna mechanically steered & gyro stabilised, against vessel motion	Parabolic antenna mechanically steered & gyro stabilised, against vessel motion	Small omni-directional antenna, with no moving parts, does not need to be steered or stabilised	Parabolic antenna mechanically steered & gyro stabilised, against vessel motion
Communications Type	Real time (Immediate)	Real time (Immediate)	Store forward	Real time (Immediate)
SERVICES				
Voice	Yes	Yes [7]	No	Yes [7]
Telex	Yes	Yes [7]	Yes	No
Group 3 fax rates	To 9,600 bps	To 9,600 bps [7]	No	To 2,400 bps [7]
Data rates [2]	To 9,600 bps	To 9,600 bps [7,8]	600 bps	To 2,400 bps [7,8]
X-25 (Dedicated data channel)	Yes	Yes [7,8]	Yes	Yes [7,8]
X-400 (Electronic Mail	Yes	Yes (Enhancement) [7,8]	Yes	Yes [7,8]
High Speed data	56/64 kbps	56/64 kbps	No	No
Full motion "store-&-forward" video	Yes	Yes [7,8]	No	No
Short Data Position	No	No	Yes	No
GROUP CALL [3]	Yes	Yes (Enhancement) [7,8]	Yes	Yes (Enhancement) [7,8]
SafetyNET™ [4]	Yes, if Inmarsat-C/EGC Receiver installed	Yes, if Inmarsat-C/EGC Receiver installed [7]	Yes	Yes, if Inmarsat-C/EGC Receiver installed [7]
FleetNET™ [4]	Yes, if Inmarsat-C/EGC Receiver installed	Yes, if Inmarsat-C/EGC Receiver installed [7]	Yes	Yes, if Inmarsat-C/EGC Receiver installed [7]
DISCREET & SAFETY				
GMDSS Compliant	Yes, if properly installed (See Imarsat Design & Installation Guidelines)	Yes, if properly installed (See Imarsat Design & Installation Guidelines)	Yes, if properly installed (See Imarsat Design & Installation Guidelines)	No
Distress Button	Yes	Yes	Yes	Yes

Note   [1] **World coverage:** worldwide availability except at popular latitudes (above 76N and below76S). - [2] **Data Rates:** Higher throughput may be achieved with data compression techniques. - [3] **Group Cells:** Simultaneous broadcast to selected groups of users or geographic areas. - [4] Services broadcast include distress & Safety information, weather & navigational information.- [5] for Fleet management, subscription services such as news and other commercial applications. - [6] Design and Installation Guidelines (DIGs) fot the GMDSS are available from Inmarsat.- [7] For the new systems, services will come on-line at staggered intervals dependent on the LES operators. - [8] Expected service availability during 1996.

Chapter Seventeen

OTHER BRIDGE COMPUTERISATION

*(Introductory Remarks. 1. **SOME COMPUTERISED APPLICATIONS TO OPERATIONAL EQUIPMENT:** Digital Gyro Compasses – Gyro Pilot for Steering – Speed Log Equipment – Docking Logs for VLCC's – Fire Alarm Systems – General Shipboard Alarm Systems – GMDSS Communications and the Radio Officer – IMO/SOLAS Creation of GMDSS – Brief Explanation – INMARSAT description – Extensions of INMARSAT. 2. **COMPUTERS AIDING PREDICTIVE EQUIPMENT:** Tidal and Barometric Predictions – Sea-keeping/ Loading Prediction Systems for Stress Reduction on the Hull – Ocean Passage Planning and Weather Forecasting Data Management. 3. **USE OF COMPUTERISATION IN THE BRIDGE TRAINING OF NAVIGATING OFFICERS:** Comprehensively Equipped Software Packages for 'own Ship' enabling Navigation, Radar and Bridge Management Techniques to be exercised, practised and demonstrated from Cadet to Master levels – Independent and Integrated Bridge GMDSS Simulator Training – 'Virtual Reality' Computer Training for Emergency Reaction Assessment – Concluding Comments.)*

17.1. INTRODUCTION

It was suggested in Chapter 16 that the quickness, accuracy and reliability of computerisation had made indispensable its application to equipment utilised into an integrated navigation, bridge and voyage planning system, under whatever name this may be marketed. In fact, it is true to say that any sub-system which attempted to operate without computerised components could undoubtedly work, but would have its effectiveness considerably weakened. Non-computerised inputs, such as extensive voyage records or the Master's night orders for instance, are readily assimilated and become digitised merely by being integrated into the wider bridge system. But a non-computerised input, from say the ship's speed log, would affect detrimentally the efficacy of the entire system, even though the radar, compass and other elements would undoubtedly continue to work. In this chapter, the aim is to comment briefly on an exclusive range of developments in modern technology which work as directly operational components to integrated bridge navigation and voyage planning systems, and then to consider some of the latest digitised developments in predictive equipment.

17.2. GYRO COMPASS

Notwithstanding computerisation the compass, as discussed in Chapter 5 on Section Analysis, is still the focal point of reference in terms of a ship's direction and charted course line. Magnetic compasses remain a compulsory fitting aboard all IMO ships and, apart from being able to input the reading from this into the wider planning system and, perhaps, 'modernising' the appearance of the casing, there is little that can be done without altering the fundamental

Supertankers Anatomy and Operation

17.1. The SPERRY MK37 VT Digital Gyroscope with Inset showing settings for Latitude and Speed as well as 'read at a glance' Ship's Course.
(By kind permission of Sperry Marine.)

workings of the earth's magnetic field, something which, so far, man has not been able to do. Technology has been applied to the ship's master gyro compass, however, with considerable advantage. The Sperry Mk 37 Digital Compass, (Photo. 17.1), for instance, has automatic compensation for latitude, as a result of GPS input, whilst retaining the important option of manual feed in. Recalling to mind a coastal ship upon which I served, during a long university vacation, that was fitted with a gyro compass with the latitude set for the Equator, manual correction reduced the yaw around the course line quite substantially. Automatic amendment of the difference of 50° Latitude would have prevented the poor steering condition in the first place. Automatically applied speed adjustment is available by link up with the ship's log, assuming that this mode is preferred to manual input, and the compass takes less than three hours to work through its 'starting, wobble and settling' procedures. It is also fitted with both power and compass alarms in the event that any drifting occurs beyond stipulated limits. Should the ship's head on the automatic pilot, for instance, deviate more than say ten degrees from the course set then an audible alarm would sound in the housing pedestal. The heading data is presented in a separate and easy to read format so that the all important course may be seen at a glance by the officer of the watch.

The compactness of this gyro-compass may be compared with the veritable monster previously fitted on all classes of ship which was examined in Chapter 5. Even as I write, plans are being implemented, by Radio Holland Marine, for the production of a 'fibre-optic' gyro compass that will measure automatically the Coriolis Force, determining the earth's rotation, using sensors to determine True North. The compass will integrate light and fibre, which will negate the necessity for fluids and moving parts, and it will have a settling period of less than twenty minutes.

Other Bridge Computerisation

17.2. *Console Display of the Sperry Marine MK37 Digital Gyro Steering Compass.*
(By kind permission of Sperry Marine.)

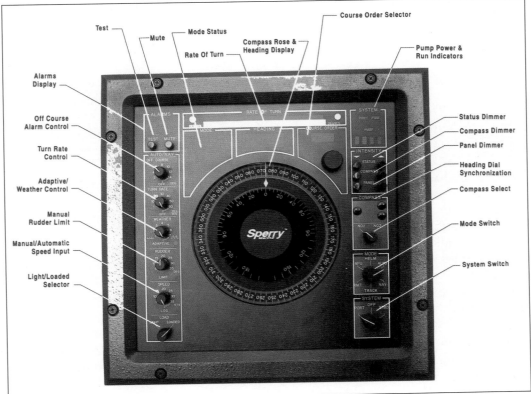

17.3. *The Face of the ADG Steering Control System retains many of the traditional features …*
(By kind permission of Sperry Marine.)

17.4. *… but possesses additional LED Interfaces assisting a greater control over manual and automatic steering.*
(By kind permission of Sperry Marine.)

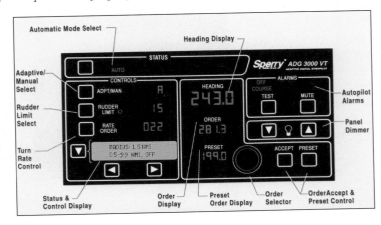

481

Supertankers Anatomy and Operation

17.2.1. Steering Console

Computerised steering consoles reduce to a minimum even the most minor of fluctuations in rudder movement with a considerable reaction in fuel and savings in maintenance costs of the steering gear. Sperry Marine's Adaptive Digital Gyropilot (ADG) series of steering control systems can also be linked directly to the voyage management system operating aboard the ship. The steering equipment can be integrated into the management console or fitted as a separate stand. (Photo. 17.2). There is much to be said for the latter because during docking operations or close quarter work it often remains desirable, for greater control over the ship, to have a hand at the wheel manually steering a VLCC. If the ship is fitted with a pedestal then the other controls are more readily available for the pilot, ship's Master or officers to have a greater freedom of movement around a system which has, after all, been specifically arranged for one man operation. The face of the ADG 6000 steering panel retains much of the traditional features, seen in Chapter 5, but with the addition of more LED interfaces for greater ease in consulting the heading references. (Photos. 17.3 and 17.4).

17.3. SPEED LOG

The speed log for measuring the ship's relative speed through the water, without taking into account the set and drift of the tidal streams, remains an essential item of equipment in the handling of ships. The SAL Rl Log, produced by Consilium Marine, acts on the acoustic principle by transmitting a beam through a transducer fitted to the hull of the ship in the foremost area of the vessel. (Photo. 17.5). The SAL Rl transducer is flush mounted to the hull, thus avoiding the necessity of a pitot tube extending out of the hull with the risk of damage being sustained in shallow water. It is essential that the water flow below the transducer must not be turbulent or affected by skew water flows. Selecting the correct transducer is always done in co-operation between the Shipowner and Consilium Marine. The following diagram, (Diag. 17.1), indicates how the system is fitted to a single hulled ship, with indications of the read out facilities available. The log is micro-processor controlled, for greater sensitivity, and works accurately to within ±1%, at speeds of greater than ten knots, and within 0.1 knot at speeds of less than ten knots. Analogue speed indicators, as additional read outs, remain available above each of the bridge wing entrances and as digital inputs into the wider management system, as required.

It has been shown, in Chapter 11, that bringing a VLCC alongside, without demolishing the jetty, requires the assessment of ship movement which is beyond the realm of human senses and that a variety of doppler radar and sonor equipment is used specifically to assist the pilot and Master in this task. Consilium's SAL 860 Log uses, again, acoustic vertical sound beams via a very small diameter transducer that is fitted through a sea valve and transducer in the ship's hull to measure both the ground and water speeds of the ship in the final stages of docking.

17.5. The Complete SAL RI Speed Log Equipment with Pitot Tube.
(By kind permission of Consilium Marine.)

Other Bridge Computerisation

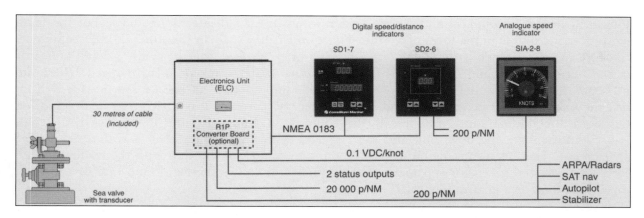

*Diag. 17.1 The Fitting and Wheelhouse display interfaces of a Speed Log.
(By kind permission of Consilium Marine.)*

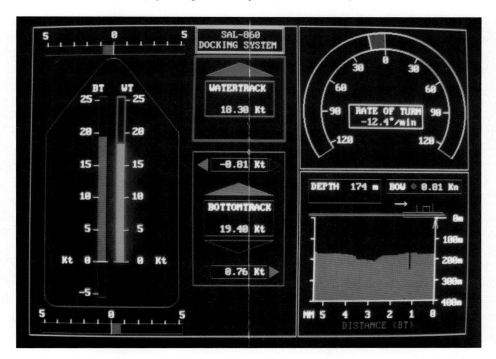

*17.6. The Multi-Functional Display (MFD) of a Docking Log showing the Docking Situation at a glance.
(By kind permission of Consilium Marine.)*

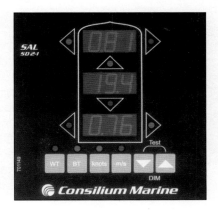

*17.7. The SD2 Docking Log indicates the Water Depth and Rate of Approach of the Ship Fore and Aft.
(By kind permission of Consilium Marine.)*

Interfaces in the wheelhouse and on the bridge wings again indicate the readings, but the main log processing unit is in the form of a Multi-Function Display (MFD). (Photo. 17.6). BT represents the bottom track, which is the longitudinal ground or true speed, together with a depth indicator in the bottom right hand side. WT is the longitudinal water track or relative speed. There are no separate sensors at the forepart or stern of the ship so that all speeds are measured from just the one sensor. Transversal ground speed is also measured as it is a two axis dual speed log (ie: both longitudinal and transversal). In addition, a Rate of Turn Input (NMEA signal) is used to calculate the bow, mid and stern speeds, with an accuracy that is within 0.5 degree at 0.1 knot. There is also a digital display of the rate of turn of the ship with the traditional red for port turns and green for starboard. The SD Indicator shows the depth and rate of approach forward and aft at a glance. (Photo. 17.7).

17.4. FIRE DETECTION

As mentioned in Chapter 10, the possibility of fire at sea is an ever present risk and Consilium's SALWICO Model CS3000 Fire Alarm System provides passenger trades in the shipping industry, and services ashore in hospitals etc, with a highly sophisticated computerised analogue system of smoke and heat detection. The CS3004 is a smaller version more suitable for ships that do not require the wider cover of so many points or stations. A series of up to four detector loops, covering different fire zones, is fed into an independently powered central control unit. Each system loop, (Diag. 17.2), interfaces with a wider display management system, as well as the normal fire patrol, and can monitor and control external fire door and extinguishing systems. The system is able to sense parameters of heat, smoke and flames, as collective or individual selections, by use of SMART/TM detectors that are each equipped with a microprocessor. (Photo. 17.8). The detectors measure, on a totally individual basis, pre-programmed levels and feed these into a central unit which constantly collects measurement values and, upon receipt of an apparent anomaly, interrogates a number of samples from the detector concerned to determine whether or not limits have been exceeded. The central unit itself may initiate action by questioning the detectors and obtaining their current readings. Consilium Marine's analogue addressible system has a triple security in the alarm process. The first of these is an auto alarm by which the detectors themselves will trigger an alarm when the limits have been exceeded. This is done quite automatically as the detectors do not have to wait to be addressed by the central unit. The central unit verifies the alarm by asking for the measured value to be repeated several times and by comparing this each time with its own pre-programmed alarm level.

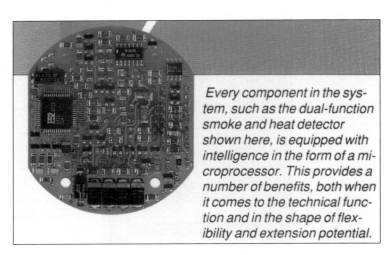

17.8. SMART/TM fire detectors.
(By kind permission of Consilium Marine.)

Other Bridge Computerisation

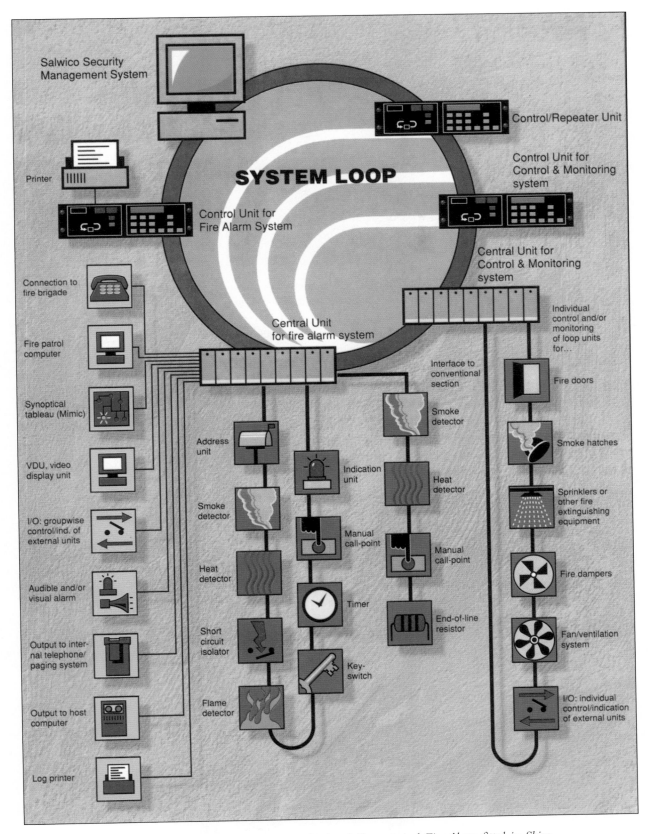

Diag. 17.2 The 'System Loop' for a Sophisticated Computerised Fire Alarm fitted in Ships.
(By kind permission of Consilium Marine).

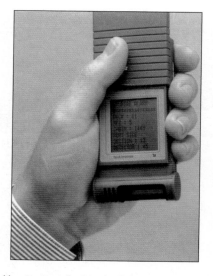

17.9. *Modern Technology enables Paging of selected officers and crew members discreetly and efficiently. (By kind permission of Consilium Marine).*

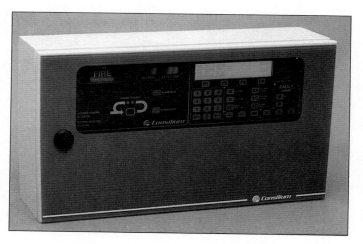

17.10. *The Local Buzzer (Seen in Photo. 17/8) activates an Alarm situated in the Wheelhouse. (By kind permission of Consilium Marine.)*

17.11. *Use of a Graphic 'Split Screen Display' enables a range of Fire Zones to be monitored continuously. (By kind permission of Consilium Marine.)*

*17.12. The Safety Management System (SMS) on a large ship incorporates Fire Detection into a wider range of monitoring by Ship's Officers.
(By kind permission of Consilium Marine.)*

The second security device is known as 'polling' by which the central unit interrogates the value of each detector triggering an alarm when the limits have been exceeded. Finally, there is a basic back-up alarm which is triggered by a detector if the latter does not receive the appropriate response from the central unit.

In the event that danger levels are reached, a series of different actions may be selected to activate. Immediately situated fire doors may be made to close, ventilation fans regulated to minimise the air circulation, sprinklers and fire dampers become activated, and alarm information can be paged to essential crew members, (Photo. 17.9), discretely without causing a wider initial alarm, whilst prepared automatic vocal alarm transmits messages to selected loudspeakers. Additional information is displayed via a LED screen, (Photo. 17.10). indicating the source of the alarm. The system is fitted with a back-up device which becomes operational in the event of the central unit ceasing to function partially or fully. The SALWICO CS3000 can be integrated into a wider Safety Management System (SMS) enabling fire detection to be monitored continuously using a graphic split screen presentation screen, (Photo. 17.11), which can be readily assimilated into overall bridge control and monitoring of all alarms. (Photo. 17.12).

17.5. GENERAL ALARM SYSTEMS

The computerisation of a ship's wider general alarm system covering machinery, especially in the engine room, but also elsewhere on the ship, as well also as tank contents and calculations was, as demonstrated in Chapter 15, one of the earliest areas to be processed. Amongst the various types of equipment available to the marine industry, the comprehensive coverage offered by Racal-Decca's Integrated Ship Instrument System, 'ISIS 250', works by a highly sophisticated series of individually controlled alarm units which are situated adjacent to the machinery to be monitored. Annunciators, (Photo. 17.13), consisting of bulkhead or panel mounted units with a 2-line LCD that show the channels actually in alarm on the upper line, with the duty engineer status/call system on the lower line. Once activated and acknowledged, the alarm may be silenced by depressing a button fitted to the annunciator below the LCD panel. A data recording device enables logs of the alarms to be retained for about one month on a single standard 3.5 inch disc. The alarm units are called either 'Local Scanning Units (LSU)

Supertankers Anatomy and Operation

17.13. *A Cabin or Public Room Annunciator shows the Channel in Alarm and gives the Status of the Duty Engineer. Here the Alarm/Status is 'Zone 002/Second Engineer'.*
(By kind permission of Racal-Decca.)

17.14. *Pressing any of the Highlighted Keys enables an Officer who has newly joined the Ship to familiarise himself with the entire system of related displays and Coverage.*
(By kind permission of Racal-Decca.)

and local scanning and control units' (LSCU) depending upon the extent of their function, which can be paged onto a twenty inch colour SVGA screen, (Photo. 17.14), in the form of a number of constantly updated displays.

A typical system configuration is shown indicating the versatility and extent of the cover available. (Diag. 17.3). Separate loops, each with its own data capture unit (DCU), govern the areas under surveillance. An additional sequence control unit (SCU) monitors the operation of the DCU's and provides a series of specialised functions including running hours of the machine concerned. Other functions are incorporated, such as blackout recovery, tank contents calculations and tasks. A centralised control unit is also available which has extended software applied to the existing distributed system that provides the required dimension of control. A 64 channel alarm unit (LSCU's) collects data via the loop concerned and sends this to the workstation involved. The control units are equipped usually with local interrogation panels allowing both local and central control functions to be operated.

Other Bridge Computerisation

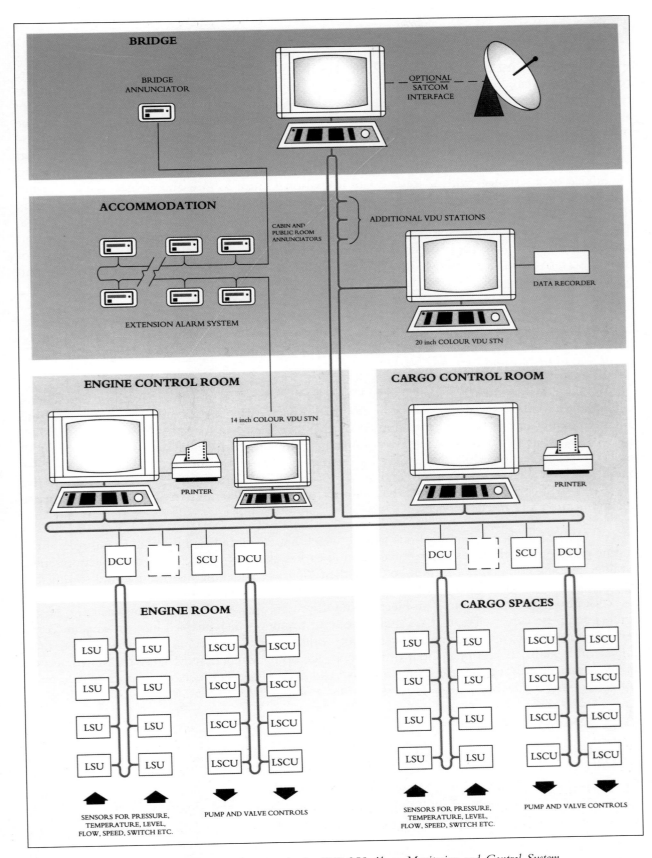

Diag. 17.3 A Typical Configuration in the ISIS 250 Alarm Monitoring and Control System.
(By kind permission of Racal-Decca.)

17.6. GMDSS

Considerable progress has been made since the satellite system mentioned in Chapter 15 was placed on board the "QUEEN ELIZABETH II". In the 1970's, IMO commenced a complete review of the situation regarding the SOLAS'74 Convention which had regulated ship's communications at sea. Advances in technology led to a need for improving the quality and effectiveness of ship to ship, and ship to shore radio traffic. The range of transmissions, for example, had to be increased and the existing quality of Morse signals was being affected considerably by atmospheric interference. IMO created the GMDSS, Global Maritime Distress and Safety System, so that all ships, whatever their location, should have the ability to communicate easily and clearly, for advice or assistance, with a Rescue Co-ordination Centre ashore and, if necessary, transmit a distress message. The United Kingdom based International Maritime Satellite Organisation, (INMARSAT), was created in 1979 basically to provide the satellite element supporting GMDSS. The title then became The First Mobile Satellite Organisation, as they extended coverage to land and aeronautical units. Their function and authority is best perhaps described in the words of their descriptive brochure, 'INMARSAT Maritime Operations', published in October 1996, and quoted with permission:

"Inmarsat is an internationally-owned co-operative which provides mobile communications worldwide ... (and) ... has since evolved to become the only provider of global mobile satellite communications for commercial and distress and safety applications, at sea, in the air and on land."

Since 15th April 1999, Inmarsat has been a plc, having become the first Intergovernmental Organisation to make successfully the transition to a private national company. The new distress and safety communications system was adopted by IMO, in the form of Amendments to SOLAS'74, establishing what has become a largely automated computerised based system that is fitted to all cargo ships, exceeding 300 grt, and all ocean going passenger liners. GMDSS became fully operational on 1st February 1999 and now covers all Satellite GMDSS requirements. It has expanded to incorporate a considerable number of essential computerised and other maritime communication and rescue services.

GMDSS is the communications feature which has revolutionised the traditional role of the marine radio officer, that was commented upon in earlier chapters, to the point of extinction. Some leading shipping companies continue to carry their 'Sparks', widening his training and so utilising the expertise of his knowledge, particularly of equipment maintenance and servicing, by integrating his work into a wider function of an Electronics Officer. To my knowledge, at least one radio officer made a complete career change at sea and underwent training which enabled him to transfer to the deck department as a navigating officer. Many others, of course, came ashore and used their considerable skills in marine related professions. As Brian Mullan, INMARSAT's Maritime and Safety Manager advised me, they sometimes spread their shore-side nets even more widely, with *"one becoming a Paedeatric Surgeon, and another a Methodist Minister"*.

17.7. INMARSAT

Using the experience of twenty years successful mobile satellite operations, the extent of INMARSAT operations is quite considerable and it continues to expand its range of markets. The services cover direct-link telephone, telex, facsimile, electronic mail and PC data connections, as well as the transmission of high definition still photographs and slow scan video material. It is used for all maritime two way communications, including international coverage during disasters.

Inmarsat delivers maritime communications and other services using Inmarsat-2 and Inmarsat-3 satellites that are partially self-owned and partially leased from major commercial organisations. They cover the entire globe apart from the direct area of the poles. The satellites are powered by solar panels, orbiting the earth at an altitude around 35,700 km., and revolving at the speed of the earth's rotation thus remaining in the same position relative to the Equator. Inmarsat-2 consists of four satellites launched between 1990 and 1992. The more powerful Inmarsat-3 has a constellation of five satellites, with the last one launched in February 1998 to act as an in-orbit spare. Additionally, through its majority shareholding in ICO, plans are ahead that will be operational by 2000, for a whole series of dual-mode, hand-held telephone service links operating with cellular systems where these are available and satellite systems elsewhere. These satellites will orbit the earth, at an altitude of around 10,000 Km, arranged in two intermediate circular orbital planes inclined at 45° to the equator.

A series of land earth stations (LES) or, as they are sometimes referred to, coast earth stations (CES), in the maritime, and ground earth stations (GES) in the aeronautical environments, link Inmarsat's satellites with national and international communications networks. There are about forty LES's distributed around the world, with at least one in every continent. The following diagram, (Diag. 17/4), indicates the geographical placing of these stations which cover the INMARSAT-B/M and A/C systems, as the network was organised early-1999. A break down drawing, (Diag. 17/5), indicates more clearly how world wide coverage is achieved. The four satellites each cover its own region. IOR, for instance, services the area covered by the operational Inmarsat-3 satellite for the Indian Ocean region, with Inmarsat-2 acting as a spare. Each area has its own Network Co-ordination Station (NCS) which monitors the entire process of transmission and reception of all signals and ensures the quality and correct functioning of all traffic.

INMARSAT-A was the first system devised and introduced commercially in 1982. The terminals are small self contained earth stations comprising a small parabolic light weight antenna, with its own power supplies and electronic units, and direct dial telephone and telex facilities, with modems for computers and facsimiles etc. This is the system used by around 18,000 ships of all classes, ranging from VLCC's to fishing vessels. The ship carries an antenna in an enclosed glass reinforced radome, a photograph of which appears in Chapter 5, that is called Above Deck Unit Equipment (ADE), the average weight of which is about 100 Kg. The electronics unit and associated equipment is carried below decks (on the Bridge or sometimes in what was the old radio room on older ships) and is called the BDE — or, below deck unit equipment. The terminal consists of direct dial telephone along with telex, fax, PC and other connections. The service provides completely the distress alert and general communications requirements of GMDSS. The antenna is kept pointing in the direction of the satellite by means of the ship's gyro system and it automatically corrects itself as the ship alters or deviates from her course line. INMARSAT-B was introduced in 1993 as an advance on INMARSAT-A which it will ultimately succeed — although the two systems will continue to co-exist well into the next century. INMARSAT-B, being digital, makes improved use of satellite resources. There are well over 5,000 terminals utilising the latest digital transmission technology requiring less satellite capacity and power. It has lower operating costs which are passed on to users. A high speed data service (HSD), linked to INMARSAT-B, offers an improved range of sophisticated services to consumers who have large amounts of data to transmit. This includes video-conferencing, local area network communications, and a telepresence facility that brings expert assistance aiding, for example, the repair of ships' engines whilst at sea, together with medical and educational services as required.

INMARSAT-C is used for distress alerting, marine safety information and general communications. It is a highly versatile, refined system that will allow anything which can be coded into data bits to be transmitted. It is particularly useful for sending navigational information, especially position reporting, and as was seen in Chapter 16 is readily integrated into wider Voyage Management Data control systems. INMARSAT-C is used by the Admiralty Hydrographic Office to send Notices to Mariners, by satellite, facilitating chart corrections both to paper and electronic charts. Currently there are over 40,000 users of the system, a number which is increasing rapidly. INMARSAT-E provides global alerting services enabling distress messages to

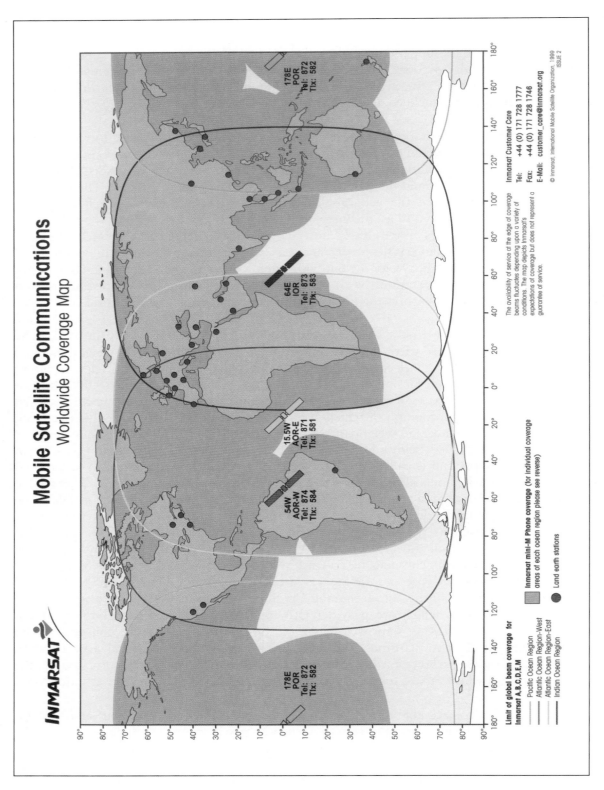

*Diag. 17.4 Worldwide Placings of Inmarsat.
(By kind permission of Inmarsat.)*

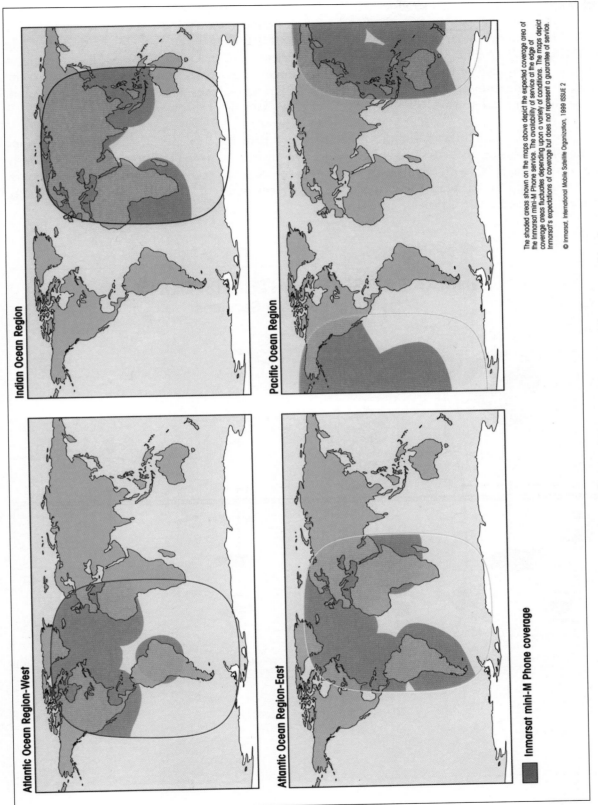

Diag. 17.5 A More Detailed Look at How their Placings are achieved. (By kind permission of Inmarsat.)

Supertankers Anatomy and Operation

be transmitted from Emergency Position Indicating Beacons (EPIRB's) carried, for instance, aboard ship's lifeboats. They are automatically activated, either from the Bridge, or when released into the sea by means of a hydrostatic switch. They offer a response time between 2 and 5 minutes with around 200m accuracy and can be fitted with an optional search and rescue radar transponder (SART). To facilitate the approaching rescue vessel, EPIRB's are fitted with a low duty flashing light. INMARSAT-M and Min-M have no direct part to play in GMDSS, other than offering a supporting role under certain circumstances. The dimensions of the satellites and the way the equipment is synchronised aboard ship is shown in Drawing 17.1. A table comparing the main physical and technical features of INMARSAT systems can be seen on page 478.

From the point of view of the practical user, there have been voiced guarded reactions by some officers using the system on the Bridge. The distress alarms become activated frequently by far distant signals, with a corresponding loss in concentration at inconvenient moments. There would appear to be aspects concerning the sensitivity of GMDSS which may require assessment and some correction.

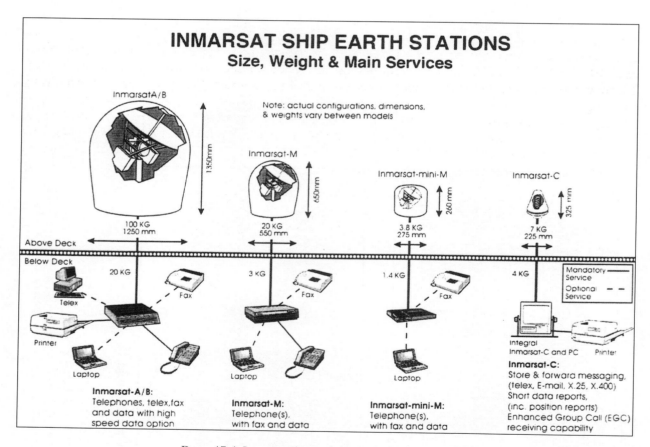

Draw. 17.1 Inmarsat Ship and Earth Station characteristics.
(By kind permission of Inmarsat.)

Other Bridge Computerisation

17.8. TIDAL PREDICTION, BAROMETRIC PRESSURE

The range of computerised equipment, manufactured by Prosser Scientific Instruments, of Ipswich-England, provides some extremely powerful aids to navigation. Amongst a range of tidal clocks are their 'TIDECLOCK 3' (Photo. 17.15) and 'TIDESTREAM 3'. The former offers a tidal prediction and passage planning facility which gives continual tide data predictions for over 600 European ports, ranging from Norway to the Azores, or over 200 in the United States, up to the year 2100. The only input necessary is an initial date and time, and there exists a facility for the user to insert ports of his own choice. It will be, no doubt, only a matter of time before tidal data covering all ports of the world are incorporated into this kind of predictor system. To the practising deep sea navigator, the advantages are many. To the coastal man, or the competent professional yachtsman, the advantages are profound.

Predictions for use with the Standard ports, and to determine set and drift for off-shore and ocean passages, are of course used frequently on VLCC's and it will certainly prove extremely advantageous to be able to insert ports into the clock and see at a glance all of the relevant tidal data.

'TIDESTREAM 3' can be integrated into a voyage planning and bridge navigation system, of the kind examined in Chapter 16, and thus be linked to all other navigational aids. Whilst the intricacy of tidal calculations for Secondary ports, for example, are rarely used by the VLCC mate, he will certainly be interested in set and drift calculations, say off Torquay Bay, for the lightening operations discussed in Chapter 11. Based on the Admiralty sources, tide and current data for a complete voyage may be determined including, at a glance, routes, waypoints and an offering of a best time of departure. The conventional, and up until recently, 'irreplaceable' annual tide tables and associated tidal atlases, diamonds, together with port and vector calculations have in fact been replaced by a machine that is the size of a small laptop. Additional equipment manufactured by Prosser Scientific includes merchant ship

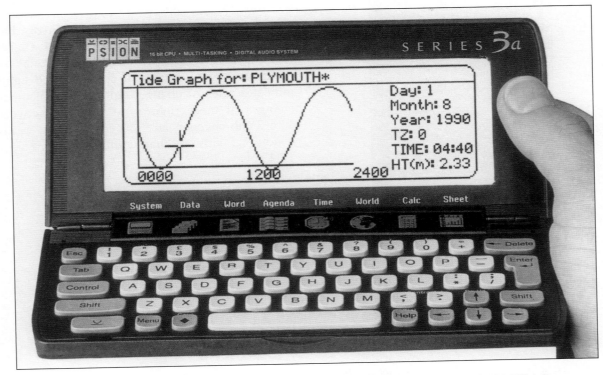

17.15. The TIDECLOCK 3 Calculates the Height of Tide for over 600 European and 200 USA Ports.
(By kind permission of Prosser-Scientific.)

Supertankers Anatomy and Operation

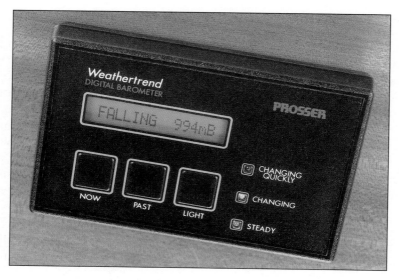

17.16. *The WEATHER TREND Digital Barometer monitors air pressure, currently, and historically, over the previous 24 hours. (By kind permission of Prosser-Scientific.)*

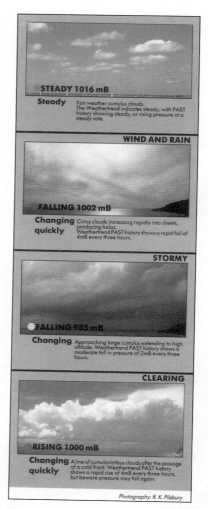

calculations, based on draft surveys, oil tanker cargo calculations and merchant ship stability, draft and trim devices. Each item is offered at an exceptionally reasonable cost and in convenient easy to use format. Their 'WEATHERTREND' digital barometer, (Photo. 17.16), for example, offers not just a continuous read-out of the current atmospheric pressure, but includes also rates of change over the previous hour, or twenty-four hour period, so that air pressure patterns may be determined. The information is presented in the form of a large LCD, with a clear read out for either day or night viewing. A system of colour coded lights indicates rates of change: red advising of quick changes of more than 3 mb; yellow, between 1 to 3 mb changes, and green, indicating steady conditions with less than 1 mb changes. (Photo. 17.17).

17.17. *The System of Coded Lights on the WEATHERTREND Digital Barometer indicates Rates of Pressure Changes. (By kind permission of R. K. Pilsbury & Prosser Scientific.)*

17.9. MEASURING HULL STRESS

In Chapter 6, Part 2 of this book, a very elementary explanation was offered of the extremely complex subject concerning stresses to which VLCC's are subjected when they are in conditions of ballast and, more particularly, when fully loaded. Strainstall Engineering Services and Consilium Marine, are two amongst other leading manufacturers, who have devised systems concerned with strain and load measuring problems for VLCC's and other classes of ship. Strainstall have devised a measuring system, whilst that of Consilium is predominantly for predictive purposes.

17.9.1. Bending Moments

Strainstall's 'STRESSALERT 11' stress monitoring system for VLCC's, and indeed other classes of ship, gives indications of longitudinal bending moments not only in a seaway, but also during loading or discharging operations, and compares findings with pre-determined levels approved by the relevant Classification Society. The system consists of four deck mounted strain gauges fitted along the main deck, plus an optional bow pressure transducer, together with appropriate display monitoring units and independent power supply. The following diagram, (Draw. 17.2), indicates the placement of the units, aboard a large bulk carrier, in a hull distribution similar to that fitted by the company aboard product tankers and a number of VLCC's. The long baseline strain gauges are fitted three on the port side of the maindeck, approximately every quarter length between perpendiculars, with one on the starboard side amidships. They are each temperature compensated and consist of a two metre rod which is fixed at one end and measures a nominal two metre base. (Photos. 17.18 and 17.19).

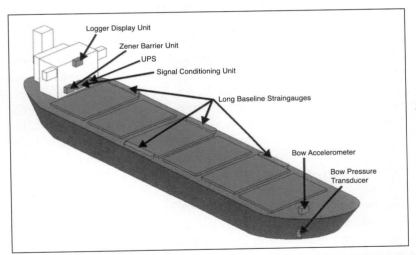

Draw. 17.2 The Displacement of Hull Stress Monitoring Units aboard a Bulk Carrier. (By kind permission of Strainstall Engineering Company.)

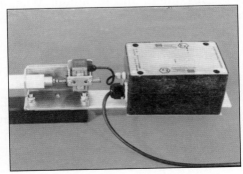

17.18. External Fitting of a stress Monitor on the Main Deck of a Large Ship. (By kind permission of Strainstall Engineering Company.)

Supertankers Anatomy and Operation

17.19. External Construction of a Deck Stress Monitor. (By kind permission of Strainstall Engineering Company.)

The operating arm is so adjusted that it remains free to move within its bearings from a setting that can be zero positioned. The gauge has a range of ±4mm, is highly accurate and can be fitted with perfect safety in hazardous areas. The casing is waterproof and fitted with a reinforced GRP cover. The bow slam accelerometer is a strain gauge beam type that is scaled for a deflection of ±2g. Like the pressure transducer, with its gate valve assembly that measures the sea water pressure at the bow, the slam accelerometer is connected to, what is termed, a Zener barrier unit that, situated in the non-hazardous area of the ship, provides the safety interface for the entire system between the hazardous and non-hazardous areas of the deck. The display unit, like most maritime computerisation, runs from an IBM compatible PC, using a Pentium unit with large capacity hard disk and floppy disk drive, with a 15" SVGA colour monitor. The 'STRESSALERT 11' displays a number of screens including a mimic diagram, trend display, vessel profile, and slam warning amongst other categories.

The mimic display on the screen offers a gull's eye view of the ship giving a bar graph for each of the sensor positions. (Draw. 17.3). The format of the bar graph varies depending upon which of three modes have been selected from either harbour, short or long sea voyages. The harbour mode is illustrated and demonstrates the maximum still water bending moment (SWBM). The red coloured, mid port sensor, for example, has indicated stresses of 16.2% and 8.5%, whilst the yellow, forward port sensor, reads 31.0% and 11.8% of the maximum values. These have been induced by theoretical loading or discharging of cargo. These SWBM allowable maximum and minimum values are fed in via a set-up screen with alarm level inputs that can be inserted in five degree steps. There is also the facility to obtain statistical information that covers the previous five minutes. (Draw. 17.4). Using the mouse, the bar graph of any specific sensor may be selected. Although set on the same day, the illustrated diagrams differ from each other by a few hours and should not therefore be regarded as directly comparative although, of course, there remain common features. Also available to be shown are the relative maximum and minimum values set for the sea modes. These values are less than the maximum SWBM of the harbour mode in order to allow for wave action. In either of the voyage modes, the bars change to dual graphs showing the maximum and minimum dynamic stress values of the SWBM, with the addition of wave motion effects, expressed as a percentile of any maximum values fed into the computer.

The selection of the trend display indicates for any selected sensor, or for all sensors collectively, (Draw. 17.5), the graphical percentile of the maximum and minimum total bending moment values experienced over a four hour period. It is a particularly valuable facility because, by seeing collectively the stress tendencies experienced, a wider overall view can be taken of, literally, the trend to which the ship is being subjected. The ability to isolate the readings from any specific sensor leads to greater clarity and possesses the obvious advantage afforded by an opportunity to study more closely individual stress areas. In a VLCC, as mentioned in Chapter 6, stresses experienced when the ship is fully loaded are often in excess of 95%. This is a fact of life that has to be lived with and regarded as a norm, without cause for concern; a situation

Other Bridge Computerisation

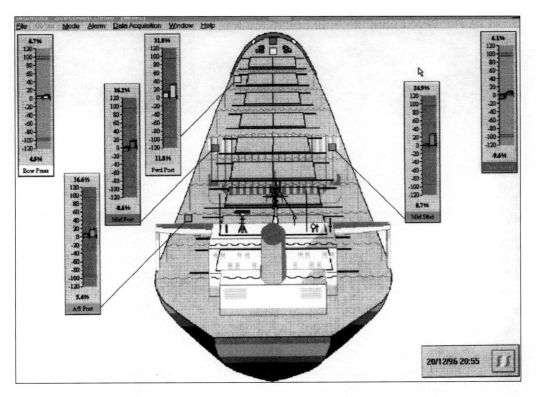

*Draw. 17.3 Mimic Display on a Computer Screen gives a Bar Graph Showing Readings on Each of the Hull Sensors.
(By kind permission of Strainstall Engineering Company.)*

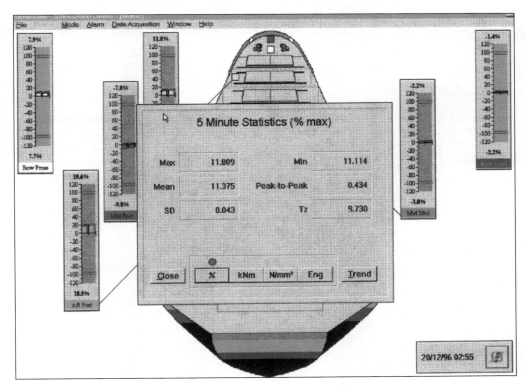

*Draw. 17.4 A "Window" can be Superimposed on the Mimic drawing which gives Historical Stress,
experienced over the preceding five minutes.
(By kind permission of Strainstall Engineering Company.)*

Supertankers Anatomy and Operation

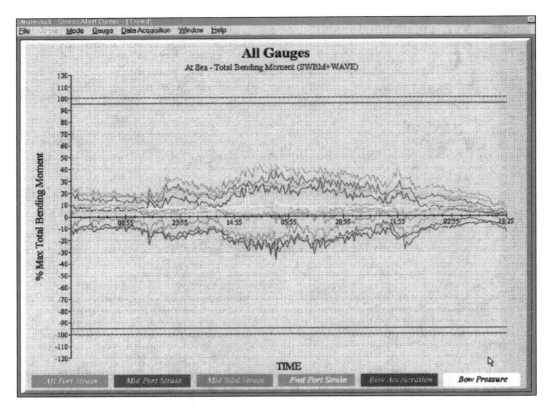

*Draw. 17.5 Trend Displays may be shown for all Sensors Collectively.
(By kind permission of Strainstall Engineering Company.)*

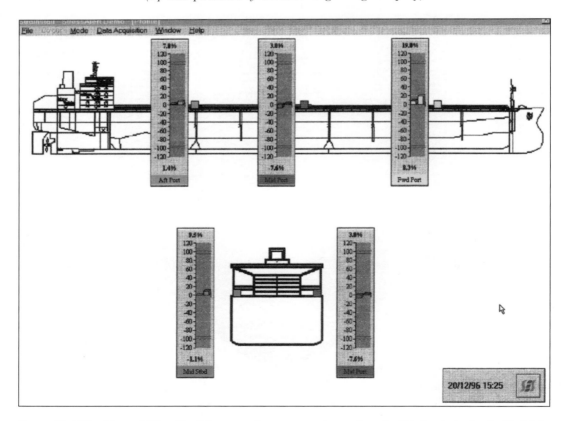

*Draw. 17.6 The List and Trim Conditions may be seen at a glance when the Display is in the Profile Mode.
(By kind permission of Strainstall Engineering Company.)*

Other Bridge Computerisation

that would not be tolerable with other classes of ship. The profile display enables the ship to be seen from a beam and also end-on indicating the longitudinal positions of the sensors as well as those mid-port and mid starboard. (Draw. 17.6). Again, alarms can be activated when particular limits are approached that have been fed in from the alarm menu.

17.9.2. Pitching, Slamming

The 'slam warning', I would regard of considerable importance upon VLCC's. As mentioned in previous chapters, it is almost impossible for the human senses to discriminate at a glance the amount of slamming experienced at the stem and bows from a position that is anything between and quarter and third of a mile astern. It is very easy to pound the Supertanker to a quite damaging extent and, prior to the days of sensors, a shrewd 'guestimate', based largely upon a combination of intuition and experience, determined the extent, and at which stage, the speed of the ship ought to be decreased. On occasions still, in instances when the wave frequency is deduced to be similar in phase to the natural pitching frequency of the VLCC, an *increase* in speed might be the more appropriate alternative. In some cases, even a minor alteration of course could be equally as effective in reducing considerably the slamming stress. Certainly, the old fashioned kind of 'box and cox' arrangement seemed to work quite effectively but it is, of course, no substitute for the peace of mind that is provided by a far more accurate, scientifically applied, determination. A realistic example was demonstrated on pp 180/1.

The bow pressure transducer displays the measured draught forward, (Draw. 17.7), tracked over a period of time, thus indicating the surge forward of the ship as she makes way through the water. In the diagram indicated, the bow of the particular ship has always been submerged by at least nine metres and only on one occasion has the bow come closer to the surface than one metre. There has obviously, over the parts of the time period cited, been

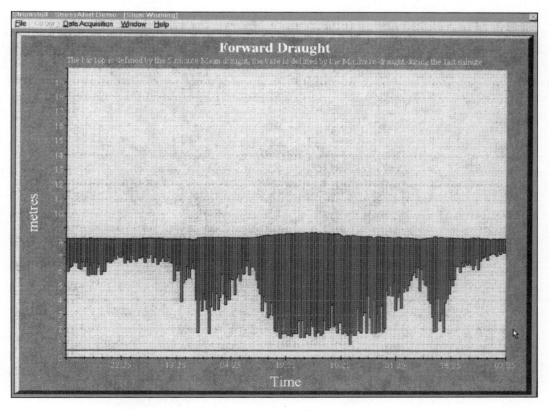

Draw., 17.7 The Slam Warning is of Specific Importance to VLCC's as it gives an Indication of Bow Pressure.
(By kind permission of Strainstall Engineering Company.)

Supertankers Anatomy and Operation

considerable pitching experienced which has shown itself in quite a distinctively strong head wave or swell pattern. Again, alarm limits may be fed into the system so that the attention of the officer of the watch may be directed to potentially dangerous slamming situations.

17.10. DECK CRANES – MOORING HOOKS

It is worth citing that Strainstall Engineering have devised computerised monitoring for the deck cranes which have overtaken the midships derricks fitted on the early generation of VLCC's. These enable safe working loads to be assessed at a glance as well as offering payload and capacity monitoring. The Company make also mooring hook transducers for the VLCC (et al) trade that are generally fitted on the mooring jetty, but which give precise information regarding the load sharing of mooring lines along the length of the ship. The quick release mooring hook, which has gained an extremely wide application, eliminates the use of mooring crews when leaving port, thus cutting considerably running costs of the ship, as the moorings can be cast-off by direct computer control activated from the jetty operations room.

17.11. SEA-KEEPING PREDICTION SYSTEM, DATA COLLECTING UNIT (DCU)

Consilium Marine's SAL SPS Seakeeping Prediction System also assists mariners by providing information regarding wave induced dynamic effects on the ship. SAL SPS monitors bow slamming, the effect of green water shipped onto the main deck, the vertical bending moment of the hull, and effective heel angles and acceleration levels, the majority of which were examined briefly in Chapter 6. The system works on a series of windows by using a combination of theoretical inputs, rather than a range of physical sensors fitted to the hull of the ship. The inputs are connected with data from existing navigation instruments, including the ship's course and speed, these are fed into a central Data Collecting Unit (DCU) and linked to a VDU screen. (Draw. 17.8). The screen, like so many computerised aids, may be deck or desk mounted and can be integrated into the wider voyage management system. The following flow diagram indicates the relationship between the main and advisory displays as these are fed by inputs to provide output displays. (Photo. 17.20). The main display is designed to enable monitoring of all vital data in one single display. It is organised into several groups of data, and functionality. (Photo. 17.21). As required individual windows may be keyed to provide a closer view of the monitored information. SAL SPS offers two modes of working: a cruise mode and

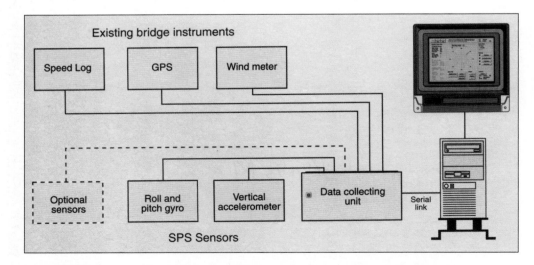

Draw. 17.8 The Findings of the Seakeeping Prediction System (SPS) on a VLCC relies on inputs fed into a Data Collecting Unit (DCU) which are then displayed on a VDU Screen.
(By kind permission of Consilium Marine.)

Other Bridge Computerisation

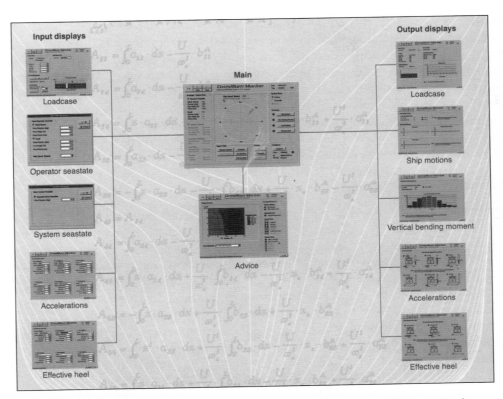

17.20. *A Flow Diagram showing the relationship between the Main and Advisory Displays as they are fed by inputs to provide output displays.*
(By kind permission of Consilium Marine.)

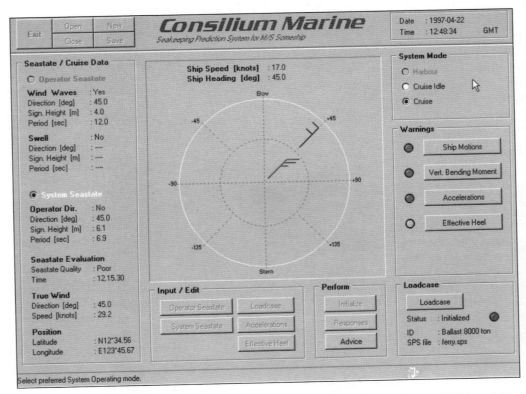

17.21. *The MAIN DISPLAY monitors inputs of Sea and Meteorological Conditions with Manual Input against Output Displays to give an Overall view of the Seakeeping Qualities.*
(By kind permission of Consilium Marine.)

503

harbour mode. Seakeeping predictions in the cruise mode are based on a combination of measurements and theoretical calculations. It is a fully automated default 'at sea' operational mode which evaluates continuously the ship motion measurements experienced on passage. Stresses such as the rate of rolling, pitching, yawing, as well as heeling angle experienced, are fed into the main display by sensors. As an essential feature, the computer holds formulae relating to the geometry of the hull and combines all values from the sensors presenting findings in the form of outputs. The only information which needs to be inserted by the navigating officer's keyboard is the loadcase/ballast condition, as well as the observed sea state, in terms of wave height and velocity, condition of swell, wind direction and force.

17.11.1. Harbour Prediction Mode

The harbour mode may be used at any time to predict wave induced effects on the ship for arbitrary combinations of loadcase and seastate. All of these predictions will be carried out based on theoretical models of the ship's seakeeping characteristics. This makes it possible to explore the effects of changes to loadcase and/or sea state data, and to preview the ship's behaviour whilst still in harbour. The SAL SPS warning function is continuously active. The warning sections on the main display hold collective warning LEDs for all groups of responses. They represent the worst case from the detailed output displays. (ie: if any defined warning level on a particular detailed display is exceeded, the corresponding LED on the main display will alert the operator). By applying current weather forecasts, and those received immediately prior to sailing, the operator may select a range of potential sea states, and experiment in complete safety, by making an assessment of the longitudinal and transversal stresses to which the ship may be subjected relevant to her loaded or ballast condition. The potential for safety with this equipment for VLCC's is really quite profound, notwithstanding its values for dry cargo ships and certain bulk carriers. It is well to be reminded that one large deep sea ship is lost without trace on average every five or six weeks, probably due to cargo shifting and/or structural failure in heavy weather.

17.11.2. Operational Advisory Service

A main feature of the SAL SPS is the readily available advisory service situated just below the main display called the Advice Display, (Photo. 17.22). This is an operational guidance tool designed to help determine correct actions in critical situations. The advice display may present detailed information for a selected response component. The graph indicates how forecast changes to the ship's course and speed would be likely to affect the response level. The response levels in the graph are drawn as relative to the corresponding warning criteria, with the red section representing areas where the criteria is exceeded and a dangerous stress

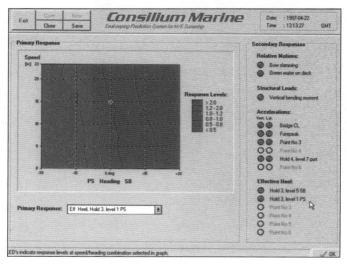

17.22. The ADVICE DISPLAY indicates that whilst the relationship between the Ship's Course and Speed do not exceed Warning Criteria the (Yellow Coloured) Cross Hair image remains in the Blue area.
(By kind permission of Consilium Marine.)

situation is likely to occur. The cross hair image marks the actual course and speed combination and is moved by a mouse or tracker ball. By moving the cross, the corresponding effect on other warnings such as bow slamming, green water shipping, accelerations, bending moments and accelerations and heeling error, would be simultaneously shown to the right of the display.

17.11.3. Cruising Prediction Mode of Ship Motions

The ship motions cruise mode display, (Photo. 17.23), holds two output groups called relative motions and measured ship motions. The relative motions group presents the most probable number of bow slams and occurrences of green water on deck during the next hour of service. The Measured Motions group presents readings from the SPS motion sensors for the last two minutes. This information is provided as a means to ensure that all sensors are working.

The vertical bending moment display, (Photo. 17.24), is in the form of a graph that shows the predicted wave induced Bending Moment along the hull, given as maximum responses, and again predicted over the next hour. The horizontal bar indicates the longitudinal distribution in terms of the ship's frames with the vertical bar being the percentile stress limits experienced which are measured against a maximum, or critical bending factor. The red coloured part of the graph indicates the section of the hull that is under maximum threat as well as the specific frames, or range of frames, directly involved. As it has been seen, with a VLCC in a loaded condition, that sector would consist of most of the hull. It is quite safe if to the officer new to Supertankers rather disconcerting. An alarm alerts the navigating officer when any maximum limit selected is exceeded.

The accelerations and effective heel displays, (Photo. 17.25), forecast the maximum vertical and lateral accelerations at any of six selected positions in the ship. Points for evaluation are defined in the accelerations input display, and can be given labels for easy identification. On a VLCC they could, for instance, be vulnerable areas such as the vicinity of the bridge and forepeak. The effective heel output display is similar to the acceleration output display, but shows the real time combined effect of gravitation and accelerations in different directions at a certain position.

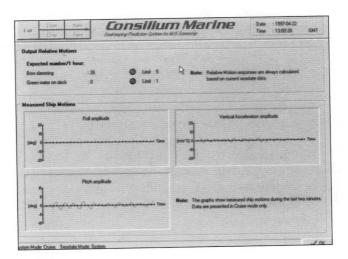

17.23. The SHIP MOTIONS OUTPUT DISPLAY offers historical information confirming SPS Sensor readings over the previous two minutes, and predicts Bow Stress for the next hour.
(By kind permission of Consilium Marine.)

Supertankers Anatomy and Operation

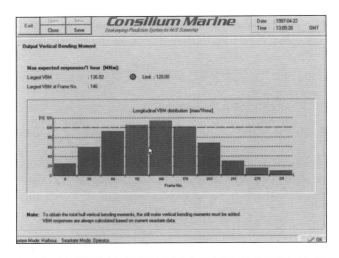

*17.24. The VERTICAL BENDING MOMENT DISPLAY (VBM)
Indicates the Largest Forecast VBM over the next hour as well as the Ship's frame where this might be expected.
(By kind permission of Consilium Marine.)*

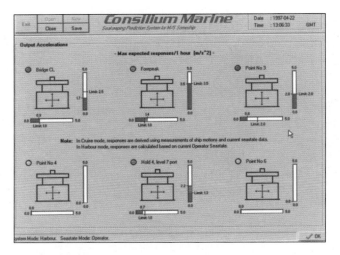

*17.25. The ACCELERATIONS OUTPUT DISPLAY predicts from six points on the Hull the Maximum Vertical
and Lateral Acceleration experienced.
(By kind permission of Consilium Marine.)*

17.12. OCEAN PASSAGE PLANNING-WEATHER ROUTING

Ocean passage planning is the first task undertaken by the deep sea navigating officer. Prior to selecting the chart folios, he will discuss in detail with the Captain all aspects of the forthcoming voyage and, in the past, would consult the necessary oceanographic publications in order to gain an idea of ocean tidal drifts and weather patterns. His concerns at this stage are with the direction and behaviour of ocean currents and seasonal meteorological influences likely to lie in the track vicinity of the voyage.

For some years now, a number of commercial organisations have been engaged internationally in providing a service to the mariner in these fields and it was inevitable, therefore, that the weather facsimile map of the 1970's would soon lend itself to a more instant solution by computers carried aboard ship. From the range of software widely in use is the package provided by Oceanroutes, part of the WNI (Weathernews) Group, who have developed their 'ORION On Board Guidance System', based on over forty years' experience of advising weather routing services. An extract from the manual offers relevant details concerning the system:

> *"Orion for Windows has been developed ... to improve the efficiency of weather routing by providing the master with all available marine weather information for up to 10 days ahead together with the tools to simulate the progress of the vessel along alternative routes, even before he sails. The master can therefore select the best option with an increased level of confidence. He can also review the effects of updated forecasts received during the voyage and take decisions on altering course or speed to minimise the effects of changing storm tracks."*

The advantages in passage planning are invaluable. Better passage times, reduced bunkers which, to a VLCC, is vital and decreased damage to vessel and cargo. There is also the psychological factor of reduced stress to the officers and crew on board that comes with knowing, with a greater degree of certainty, the weather patterns ahead of the ship.

The latter 'psychological' factor is all too often regarded merely as a 'manufacturers' blurb' included largely as an aid to helping sell equipment. There is far more involved for, as a mariner who very nearly lost his life some years ago when the dry cargo ship in which I served caught the tail end of a hurricane off the American coast which had backed unexpectedly along its track, I can speak with some heartfelt authority on the importance of being able to forecast severe weather conditions that might lie ahead. The officers and crew of the mv. 'MATRA', of Brocklebank Lines, a Company now, alas, lost to the Industry forever, came pretty close to joining the ranks of deep sea ships which, in those days, amounted to an average of one ship every five weeks throughout the year being lost with all hands and without trace. The ships, often of large ocean going size, left port A to proceed to port B and were simply never heard of again until they were reported 'overdue'. An air/sea search would then be automatically organised, but frequently no trace was ever found of the ship. The largest ship lost to date is the mv. 'DERBYSHIRE', of 91,655 grt which set out from Les Sept Isles in Canada for Japan, in September 1980, but failed to arrive. The investigation into the loss of this ship, and all of the people on board, (along with the pain of surviving relatives and friends) is approaching its definitive conclusions. All too often, the causes have been attributable to suspected cargo shifting, or structural failure, that were both assessed to have occurred during severe weather conditions. Certainly to me, the sight of a wave nine metres higher than a bridge, that itself was fifteen metres above sea level, was very frightening and has provided me with a healthy respect for the power of the sea. As we would say today, it has subjected me also to a 'traumatic experience' with lasting memories.

For readers who like factual, but authentic, details the forward derrick on the port side was thrown out of alignment, an accommodation ladder leading from the lower bridge to the main deck was bent; the wheelhouse windows on the port side were smashed, with minute fragments of glass so deeply embedded into the after wooden bulkhead/chartroom screen that we were unable to dig them out later with a pen knife. The overside lights and life boats, on that side, were almost carried away and we shipped about sixty tons of water in the after accommodation. The Asian crew living there thought, like us who witnessed and knew what was going on, that their last hour had arrived. The ingress of water had the effect of bringing up the ship's head so that we faced the following wave sequence with lifting rather than submerging bows, hence saving the ship.

In a published paper, 'Weather Routing Using Onboard Guidance', Mike Slavin of Oceanroutes Ltd, commented:

Supertankers Anatomy and Operation

"Several years ago the company recognised both the need for an onboard guidance system and the possibilities opened up by low priced powerful computers, working together with the Inmarsat system for data communication. Thus began the development of orion. Orion has been under development for nearly seven years, the science being extensively tested and proven in field trials with a two screen system in the first phase. Since 1992 the development has been with Windows as the standard interface."

ORION takes into consideration the characteristics of an individual ship and the weather, winds and current likely to be experienced over the trip. The Company uniquely go one step beyond the implications of this bland comment. They have issued a 'mission statement' in which is expressed their clear commitment to merging people and the latest developments in weather technology:

"Meteorology is the prince of physical sciences' – one of the most difficult and demanding of all scientific disciplines whose results are nevertheless vital to the safety and well-being of people everywhere. At the same time our ecosystem is fragile and we have a duty to protect it for ourselves and future generations."

The weather data base offered by Oceanroutes is world-wide and, through GPS and INMARSAT, can be integrated into a voyage planning system on board ships. Details of observed weather data in selected areas, for a preceding period, and forecasts for up to ten days are offered. The meteorological data includes contours of surface pressure and 500 Mb heights, cold, warm, occluded and stationary fronts, wave heights and speed, text forecast for designated areas and, of great importance, advance warnings and descriptions of severe weather conditions. Oceanographic data includes wave heights and directions, sea surface temperatures, ocean and coastal tides and currents, and ice information. The information can be presented visually and/or numerically. (Draw. 17.9). Here, the screen shows surface pressure isobars on a chart of the

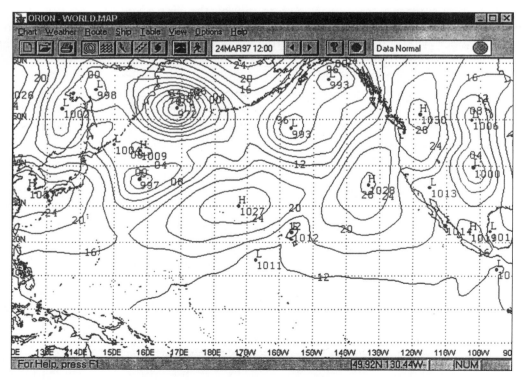

Draw. 17.9 Surface Pressure Isobars on a World Map in the North Pacific Region.
(By kind permission of Oceanroutes (UK) Ltd.)

North Pacific. By using the mouse, the displayed weather can be moved forward or backward in time thus allowing the changes in weather patterns to be clearly seen. The weather data is colour coded making it that much easier to distinguish at a glance the patterns presented.

As already stressed, tracking and forecasting the path of Tropical Revolving Storms (TRS) is of vital concern to all ships' masters and navigating officers. TRS usually include winds of Force Ten or more that, briefly, are known as typhoons in Far Eastern waters, cyclones and monsoons in the Indian Ocean and hurricanes off the West Indies, parts of the North and South Pacific and American Seaboard. Whilst the visual portrayal of a selected storm track is useful, the numerical representation indicates clearly the position on successive dates and times, and offers other information relevant to decision making. The current methods for TRS tracking appear to have shown little improvement in accuracy for the twenty years up to 1992. A new TRS model has recently been developed by the Global Fluid Dynamics Laboratory, (GFDL), and this was introduced in that year. The initial results are very promising. In this diagram, (Draw. 17.10), the regular plotting of a storm's movements over a period of three days shows a clear pattern of direction thus enabling voyage data to be entered such that an alteration of ship's course may be profiled, and suggestions made regarding forecast tracks and alternative routing. Storage facilities by means of templates in a directory are possible which enable a navigating officer to build up a series of route profiles for future use and, indeed, for these to be circulated within a flcet for the dissemination of such knowledge.

When using ORION initially, the weather data has to be downloaded from WNI Oceanroutes, via INMARSAT, and then processed so that it can be used in the ORION programme. Downloading generally takes place daily from new data, published at 1730 hours and 0330 hours UTC, generally for the local area of the voyage undertaken. Facilities are available for a larger area to be incorporated thus taking into account lengthy ocean passages.

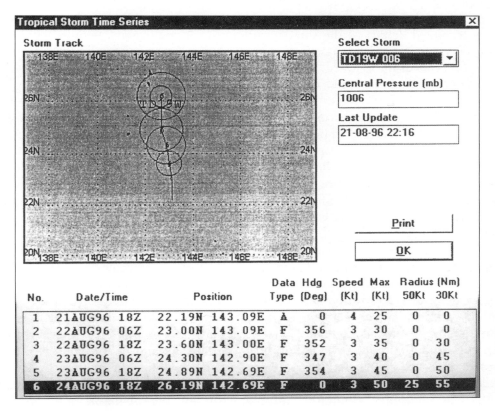

Draw. 17.10 The Computer indicates the Plotting of a Tropical Revolving Storm (TRS).
(By kind permission of Oceanroutes (UK) Ltd.)

Supertankers Anatomy and Operation

Draw. 17.11 Considerable Shipboard Information may be fed into the Computer to Assist the Accuracy of the Forecasts. (By kind permission of Oceanroutes (UK) Ltd.)

The refreshing of databases on a regular basis is recommended ensuring that only the most recent information is being processed.

The information is used via a number of menus that can be viewed independently, but worked collectively, to provide a complete voyage profile. The Ship menu, for example, contains information relevant to the specific ship which is stored in the ship design file. This contains general particulars of the ship such as her name and call sign, details of length, tonnage, speed and engine characteristics, etc. There is also an advanced ship entry form that has more specialised information germane to the vessel. (Draw. 17.11). The information stored here contains technical details of the ship, including block coefficients, detailed characteristics of the bulbous bow etc, and is designed to ensure maximum accuracy in the profiling of the eventual voyage. The draft forward and aft for the particular voyage is also entered into this menu. The information is then used to calculate the ship performance curves.

Amongst other options are included the Chart menu. This allows the creation of charts for any area in the world, based upon any of the normally used projections, including Mercator, Gnomonic, and even Stereo. These can, of course, be worked on as the 'in use' version and printed and/or saved. The Weather menu allows the display of several types of weather information on the selected chart, including animating of the proposed voyage to gain an overall view. The menus work collectively as well as individually thus allowing the proposed route alternatives to be simulated in advance so that the voyage may be accurately defined, and then modified as appropriate.

Voyages are entered in the following way. First the correct chart must be displayed in ORION. Then the ship and advanced ship entries are completed, and the weather is downloaded in the manner described previously. The voyage details are then entered. (Draw. 17.12). In this demonstration example, a voyage from San Francisco to Yokohama has been planned, with the relevant details inserted. A Great Circle track has been selected, as a waypoint, and the information concerning initial course, speed, fuel consumption and ETA established. Weather and oceanographic data are then used by the programme to calculate the optimum route that may be followed. It is at this stage that alternative routes might be compared. (Draws. 17.13 and 17.14). The computer simulates automatically all the available alternatives so that the master may then review and, if necessary, modify the calculated optimum voyage. Clearly, the same departure and arrival positions must be used, and the stored data will help determine

Other Bridge Computerisation

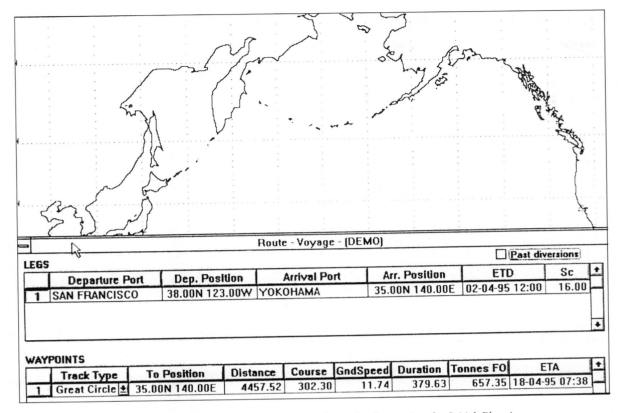

Draw. 17.12 A Map of the Overall Passage to be undertaken assists the Initial Planning ...
(By kind permission of Oceanroutes (UK) Ltd.)

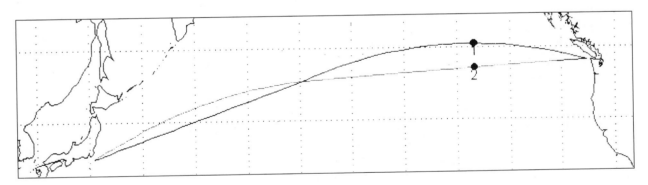

Draw. 17.13 ... and enables Alternative Routes to be considered ...
(By kind permission of Oceanroutes (UK) Ltd.)

Draw. 17.14 ... in the light of most recent Meteorological Information.
(By kind permission of Oceanroutes (UK) Ltd.)

which one will prove the most effective. The comparison of routes can be utilised at any stage of the voyage and is particularly useful when the voyage extends beyond a ten day period.

17.13. USE OF COMPUTERS ASHORE – SIMULATOR TRAINING

17.13.1. Nautical Colleges

All UK nautical colleges have computerised training aids and equipment and it would be invidious to cite by name any one in preference to advances made by others. The modern techniques enable Courses to be offered that cover all aspects of deck and engineering officer training. Some colleges have additionally had manned models for ship handling for some years now, but all offer 'state of the art' equipment regularly up-dated and conforming closely to STCW95 regulations. Many courses are modular offering tailor made tuition for individual officers and their employing shipowners. Innovative teaching enables frequent amendments to be made as advances in ship-board technology leave the manufacturers. It is a far cry from the 'onboard ship training', via a manual and a colleague who knew only slightly more than the learner, outlined in chapter 16.

Although not directly used on the bridge of a deep sea or coastal ship, it is considered highly relevant to comment briefly on the way in which computerisation is being used ashore for the training of navigating officers, at all stages in their development, from cadet to master. For years now, computerised simulation has been a part of officer training. Radar simulators have been used for forty years and, from the late 1970's/1980's, integrated ARPA and bridge trainers have also been readily available and have been very much a part of an officers' preparation in the period prior to 'going for their ticket', or refresher and up date courses between voyages. The courses have always been renowned for the intensity of pressure that has been created by the atmosphere of realism that they have generated. The control of any selected ship and her handling capacity, for example, evokes a reaction which is in real time in terms of speed adjustment and the turning circle of the ship in response to answering her helm. If a ship takes ten minutes to reduce from full ahead, through stop engines, to full astern in real life, then so the same time scale will operate on the simulator. This can sometimes be very disconcerting even to the more senior student. I recall handling a very fast container ship in the approaches to Vancouver, during a simulator exercise at Plymouth Nautical College. Having come directly from a fifteen/sixteen knot cumbersome VLCC onto the course, I failed to take into consideration the more ready manoeuvrability of this fine ballerina and so placed her 'firmly on the putty'. Alarm bells were activated in the simulator and other "supportive" students, and the lecturers, came running to see who and what etc. The incident proved not only most embarrassing for me, as well as providing also great entertainment, but ended up by my having to buy a round of drinks at lunchtime. I am happy to recount, however, that by the end of the course, I too had been suitably treated on more than one occasion.

17.13.2. Transas Marine's 'Navi-Trainer'

Transas Marine's Navigation and GMDSS training facilities have helped to revolutionise the sophistication of teaching available for navigators to a 'state of the art' due to the totally comprehensive applications of the company's navigation and bridge 'NAVI-TRAINER'. These include bridge watchkeeping procedures, all aspects of passage planning and execution, Rule of the Road training and testing, ARPA and general radar familiarisation including blind pilotage, feasibility studies for very large ships using existing channels and berths, and development of specialised ship handling skills. Each of these elements enable scenarios and passage planning to be devised, run, modified and experienced in perfect safety. Ship accident and near miss situations may also be simulated enabling precision ship handling, within the intent and application of the Collision Rules, to be practised. There are provisions also offering oil spill situations and search/rescue management exercises.

Other Bridge Computerisation

Amongst other programmes, offered by 'NAVI-TRAINER' are those designed specifically for port authority and Vessel Traffic Management Systems (VTMS). These include dedicated pilot training, port development and dredging operations as well as highly specialised applications of patrol and surveillance, and tug efficiency. There is an additional facility offered to the ship owner by this versatile company. Using INMARSAT-C, Transas Marine's 'NAVI-MANAGER' enables the courses and positions of ships in a companies fleet, to be monitored continuously at the head office. This is done by means of configurations taken from up to 3500 electronic charts on board the vessel. It will be recalled that this system was mentioned briefly towards the end of the last chapter.

'NAVI-TRAINER' can be set up to consist of an instructor's console in an ante-room, rather similar in principle to those fitted to supervise instruction in the old radar simulators, that enables the programming and monitoring of a number of work stations. In the privacy of these self-contained 'own ship' bridge cubicles, the students or trainees, use all of the equipment normally to be found on the bridge of a ship that has been completely modernised and can make their own command decisions. They are able to do this without the stress of "looking over their shoulder" to see if the instructor approves or otherwise of their manoeuvres. A briefing room may also be integrated into the teaching complex in which students receive prior and post 'wash up' instruction relevant to the exercise practised. The scope and range of equipment used would be dependant upon their level and skill, bearing in mind that the object of all exercises is always to train and encourage, thus building confidence. The potential danger of overwhelming a student, likely to happen with trainees at any level, is thus avoided.

The student can stand, or sit in a comfortable chair, facing a number of VDU screens, which can be multi-purpose representing electronic chart display and information system, (Photo. 17.26), or ARPA/radar with chart module only, depending upon the level of training required. The range and complexity of equipment is very impressive. Hardware represents

17.26. Equipment placed in a Training Simulator Cubicle.
(By kind permission of Transas Marine.)

Supertankers Anatomy and Operation

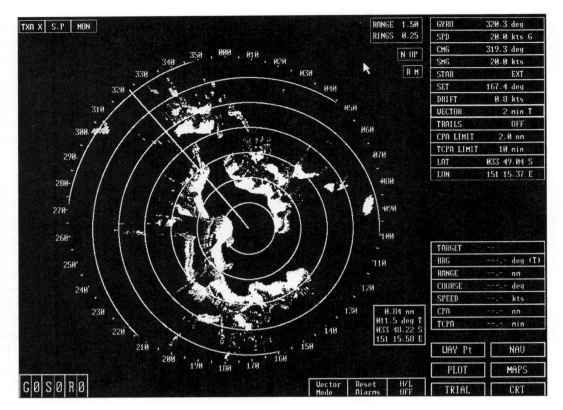

17.27. *A Radar Simulator PPI on the 1.5 nm Range Scale ...*
(By kind permission of Transas Marine.)

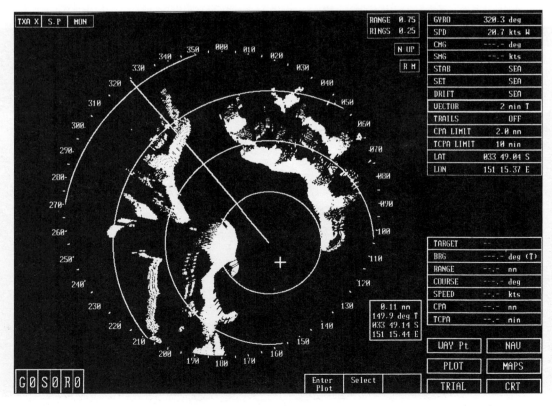

17.28. *... and the same area on the 0.75 nm Range.*
(By kind permission of Transas Marine.)

Other Bridge Computerisation

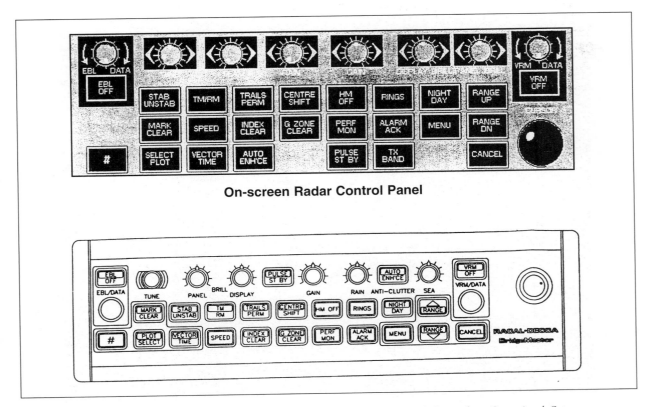

Draw. 17.15 The Differences between Simulated Radar Controls and those of an Operational Set. (By kind permission of Transas Marine.)

replicas of all actual equipment found on the bridge, whilst the software programs can be made to simulate virtually every conceivable situation that is likely to be encountered at sea. Transas Marine have used their total Company experience, that has provided navigation systems in varying degrees to around 1500 ships, to create programs capable of simulating all classes of ship, including a small sdwt VLCC. The student navigator has his collision avoidance ARPA/radar in front of him. The equipment in the simulator is authentic and performs all of the functions normally found in an operational set on board ship, as examined in Chapter 10. The following comparison shows the simulated and the real radar control panels, (Draw. 17.15), and is really quite self-explanatory. The original set is a fully computerised, modern, Racal-Decca 'BRIDGEMASTER'. There is perfect clarity of the simulator PPI displays. (Photos. 17.27 and 17.28). The initial photograph shows a set on the 1.5 nm range scale with north up display and the ship's course speed. For me personally, I do not find the computerised figures distinguishing the numbers zero and eight consistently easy to read as quickly as I should like. The bar across the zero causes confusion at first glance and requires, for me, a very close inspection before I am completely satisfied which figure is actually being displayed. This irritation, I hasten to add, is not the fault of Transas Marine, as I have experienced this with other computerised bridge equipment. The second companion photograph indicates the clarity obtained with the same display on the 0.75 nm range. The 'cross' at around 140° (T) is the present position of the cursor. Either 10 cm or 3 cm radar sets may be integrated. The guided layout of the PPI display screen, (Draw. 17.16), indicates the various components available to the officer of the watch and offers a breakdown of the functions involved in each part. The use of many of these have been examined elsewhere.

The current electronic chart, covering the area under practice, is also displayed, together with the control panel. The ECDIS module also is exactly the same as that found at sea and incorporates the DGPS which, as demonstrated in Chapter 16, can work with a

515

Supertankers Anatomy and Operation

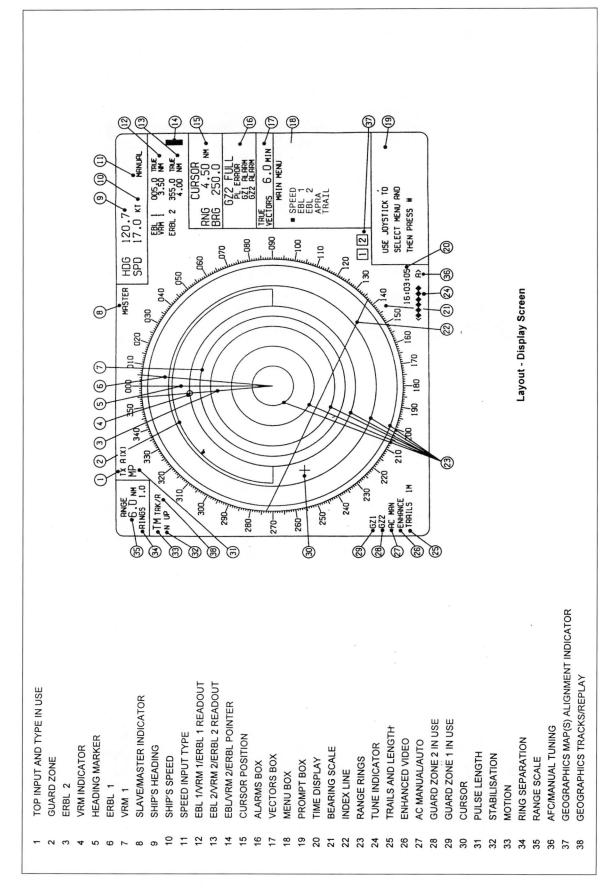

1 TOP INPUT AND TYPE IN USE
2 GUARD ZONE
3 ERBL 2
4 VRM INDICATOR
5 HEADING MARKER
6 ERBL 1
7 VRM 1
8 SLAVE/MASTER INDICATOR
9 SHIP'S HEADING
10 SHIP'S SPEED
11 SPEED INPUT TYPE
12 EBL 1/VRM 1/ERBL 1 READOUT
13 EBL 2/VRM 2/ERBL 2 READOUT
14 EBL/VRM 2/ERBL POINTER
15 CURSOR POSITION
16 ALARMS BOX
17 VECTORS BOX
18 MENU BOX
19 PROMPT BOX
20 TIME DISPLAY
21 BEARING SCALE
22 INDEX LINE
23 RANGE RINGS
24 TUNE INDICATOR
25 TRAILS AND LENGTH
26 ENHANCED VIDEO
27 AC MANUAL/AUTO
28 GUARD ZONE 2 IN USE
29 GUARD ZONE 1 IN USE
30 CURSOR
31 PULSE LENGTH
32 STABILISATION
33 MOTION
34 RING SEPARATION
35 RANGE SCALE
36 AFC/MANUAL TUNING
37 GEOGRAPHICS MAP(S) ALIGNMENT INDICATOR
38 GEOGRAPHICS TRACKS/REPLAY

Draw. 17.16 The Layout of the Radar Display Screen in a Simulator. (By kind permission of Transas Marine.)

Other Bridge Computerisation

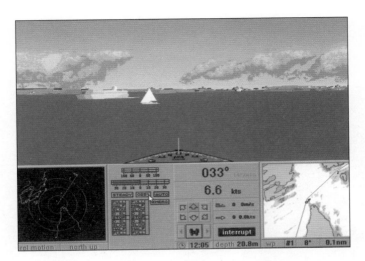

17.29. *Close-up of the Simulator Screen showing a view from the 'Wheelhouse Window'.*
(By kind permission of Transas Marine.)

17.30. *The Simulator Radar seen in the Daylight Viewing Mode ...*
(By kind permission of Transas Marine.)

Supertankers Anatomy and Operation

17.31. ... and the same display shown as at Night Time.
(By kind permission of Transas Marine.)

computerised Consilum Marine SAL 860 Log to assist in the berthing of all classes of ships. 'On screen presentations' enable a 360° view to be obtained, as this would appear from the windows of the bridge, were the scenario to be encountered actually at sea, thus contributing to the reality of the exercises practised. A conning display can be zoomed that shows a replica tip of the fo'c'sle jackstay, thus offering a point of reference on the screen, otherwise indicating merely the observed view from the wheelhouse. (Photo. 17.29). In this simulation, own ship is shown under way with both a yacht and ferry off the harbour approaches, with the harbour itself merged into the perspective of distant land. Both ships have crossed the student's own ship's heading and are, currently, proceeding safely. The targets each appear on the anti-collision radar PPI, on the lower left-hand screen, and the course off the headland can be seen clearly from both chart, screen and radar. Training is made more valuable by offering daylight or night time viewing facilities. (Photos. 17.30 and 17.31). In the daylight simulation of a ship entering Sydney Harbour, the view of the channel below the famous bridge is authentically displayed, whilst the density of shipping depicted is certainly equally as probable in this congested region. The immediate viewing of chart and radar show, again in these examples, the navigation and shipping situation at a glance. Judicial use of engine and steering controls would enable any appropriate collision avoidance action to be assessed and taken. The night time view is equally realistic. The stern lights

Other Bridge Computerisation

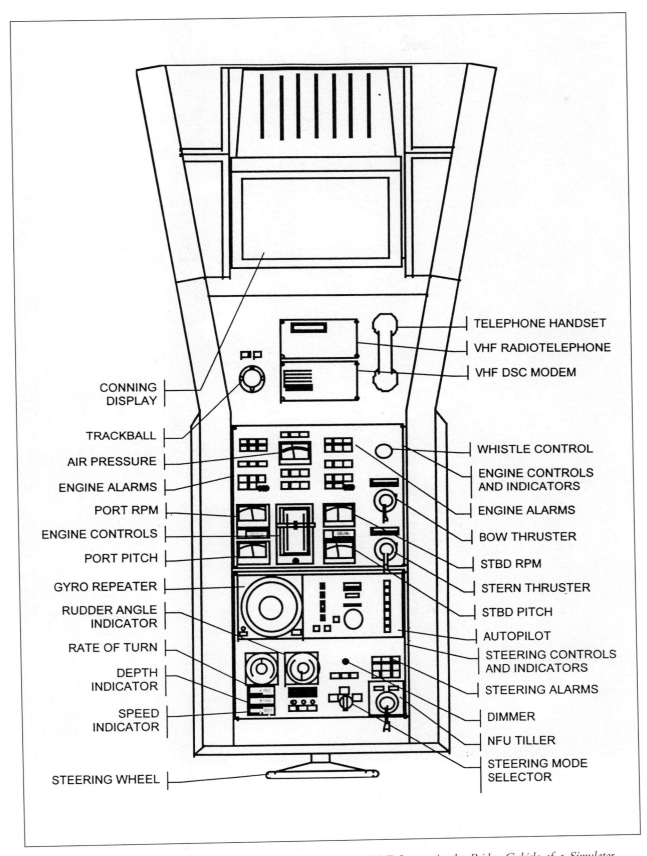

Draw. 17.17 Drawing showing the MANOEUVRING CONSOLE Layout in the Bridge Cubicle of a Simulator. (By kind permission of Transas Marine.)

Supertankers Anatomy and Operation

17.32. The Simulator Control Panel has all of the gear likely to be found on a real ship.
(By kind permission of Transas Marine.)

Other Bridge Computerisation

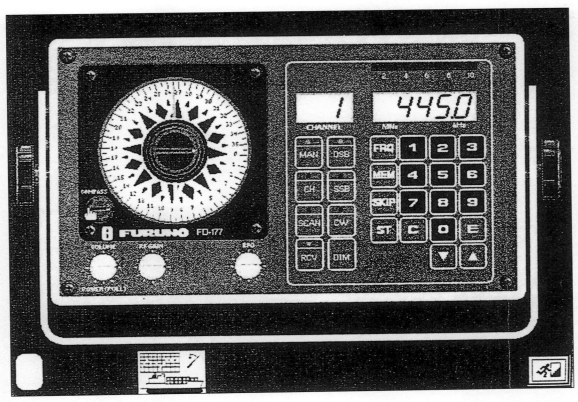

17.33. *The FURUNO FD-177 Direction Finder in the Bridge Simulator Cubicle.*
(By kind permission of Transas Marine and Furuno Electronic Company.)

of the ships ahead are clearly visible, along with the starboard and two masthead steaming lights of the dry cargo ship leaving port, a couple of points off our starboard bow. The open distance between the forward and after masthead lights, with the closer inspection necessary to distinguish the former from the near proximity of the shore lights, indicates realistically the present safe passing of this ship, eventually to clear our own starboard side. It is probably worth directing attention to the difference in colours between the day and night shots of both radar and chart displays. Initially, the different shading of the chart meets the eye as an alien colour scheme, but adjustment takes only a few moments following which an easy familiarity occurs. The equipment, on the lower left-hand side, is the TGS-2000 basic version of GMDSS, discussed later in this chapter.

The controls on the manoeuvring console of the simulator consist of the conning display unit and those likely to be found on a real ship. (Draw. 17.17) and (Photo. 17.32) All are conveniently to hand, and much of the equipment shown here has been examined previously. Gyro compass and engine room telegraph along with navigation light and signal indicator switches are available, together with sound and fog signal controls. There is also a facility for using a pelorus, or 'dumb' compass, aligned manually to the ship's head, for taking bearings, and a binocular viewer enables similar zooming facility as would be found with conventional glasses used at sea. All other ship's instruments are simulated, as the following example of the Furuno FD-177 'Direction Finder' indicates. (Photo. 17.33). Faults and errors may be entered by the instructor thus allowing the same type of differences found between measurements in time and reception of signals in real life to be equally successfully simulated. It was demonstrated in Chapter 10 that the ship's log book was kept up-to date at the end of every four hour watch, with relevant facts concerning course, speed followed/alterations, major points of land sighted with range off and bearing, compass error, weather report etc.etc. Voyage documentation on the simulator is recorded every ten seconds, thus providing an accurate

Supertankers Anatomy and Operation

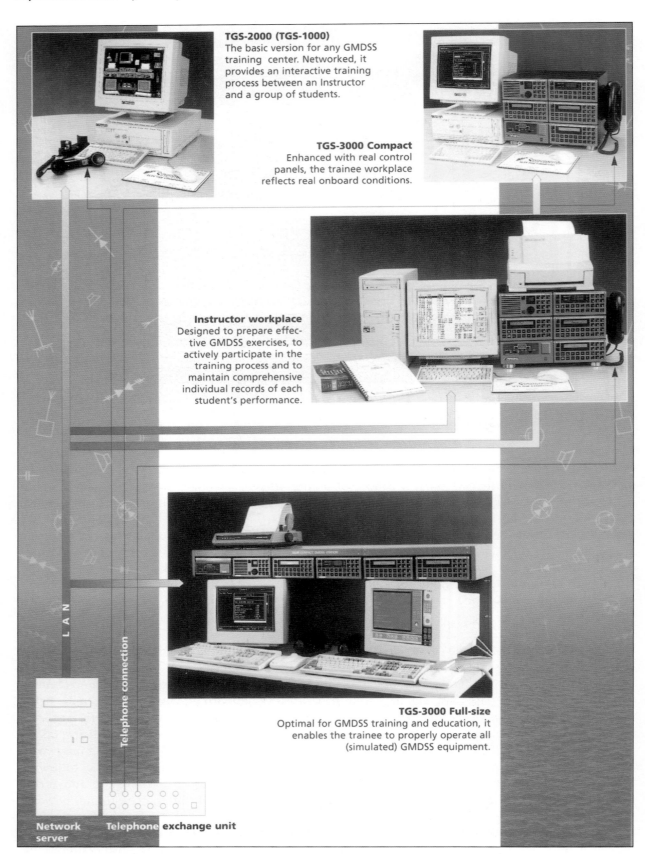

Diag. 17.6 GMDSS Training and Education Simulator offering practical work that reinforces technical knowledge. (By kind permission of Transas Marine.)

Other Bridge Computerisation

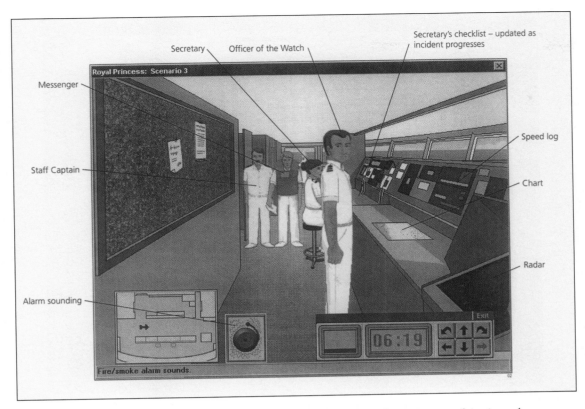

*17.34. The Captain Communicating with other crew members to obtain reports and issuing orders.
(By kind permission of Sanderson CBT.)*

retrospective indication of the ship's navigational history. Paradoxically perhaps, but probably inevitably, real life will eventually imitate simulation and records undertaken by computer will eventually follow as standard practice legally on board ship.

17.13.3. GMDSS

It has been seen that GMDSS is paramount now in terms of communications at sea and all operators require either a general operator or a restricted certificate qualification in order to fulfil both IMO resolutions and STCW requirements. Transas Marine has used practical marine shipboard equipment to develop a training facility which consists of four major components. (Diag. 17.6). Students of all ability levels can be trained realistically in the correct procedures to imitate communications between ships, ship and shore stations using both satellite and land lines. The GMDSS simulator may also be integrated with the navigational bridge simulator, including radars, to provide a more authentic approach to the inter linking between radio and chart co-operation essential in successful Sea Air Rescue (SAR) work that is an important part of any course. Additional to the instructor based assistance, there is a self-help facility provided at every stage of training so that students can 'feel their own way' out of a problem without causing any real damage, but gaining considerably in confidence. The trainer permits access to a database of coast radio stations, the entire Inmarsat network, and Navtex stations. A print out is obtainable which adds verisimilitude to the practicality of exercises.

17.14. 'VIRTUAL REALITY' TRAINING

Glancing through the officers' union NUMAST magazine, 'Telegraph', for October 1997, my interest was attracted by an article which appears to have taken computerisation a step forward from the almost 'traditional simulation' discussed previously. A company called Sanderson CBT, of Sheffield, has worked in conjunction with shipowners, Princess Cruises in the USA,

Supertankers Anatomy and Operation

17.35. *The Captain views the status of the Sprinkler and Fire Alarms.*
(By kind permission of Sanderson CBT.)

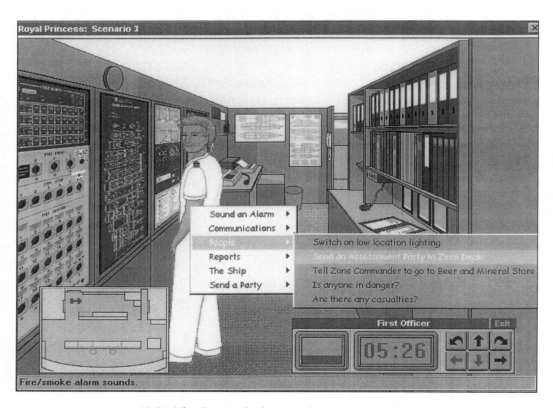

17.36. *The Captain decides to send an Assessment Party.*
(By kind permission of Sanderson CBT.)

a sister company of P&O Cruises (UK), to develop a multimedia simulation, incorporating sound, graphics and animation.

The training is conducted using an onboard computer, located in a dedicated training area aboard a cruise liner, to which open access is made available so that the program may be used by crew members during periods whilst standing-down from normal shipboard duties. A ship's bridge, complete with officers, is simulated in precise detail. (Photo. 17.34).

There are two modes, assessment and practice, which represent a major advance in command and control, as well as emergency response training. The objectives of the program, called 'Capter', are explained in the company's literature:

"First, to assess the performance of ships' Captains during an emergency by judging responses he makes to a number of realistic situations. Secondly, enabling them to experience certain types of shipboard emergencies without the pressure of assessment so that they may learn by experience thus gaining confidence in their ability to make judgements and decisions. Both modes can be used as a training aid to captains and officers. People can actually experience 'what if' scenarios and explore the consequences of a variety of actions without pulling themselves, others, or equipment at risk."

The programs work in real time and are completely interactive so that questions may be asked and answered, orders issued and responded to, controls manipulated and the 'role play' of a complete scenario enacted. It is important to stress that all relevant control panels and status indicators are 'live' so that complete verisimilitude is added to the exercise practised. The Captain, who can be the Master himself or any officer, and his bridge team are confronted with a number of multi-faceted scenarios, containing extremely complex issues. The action taken for a fire on board scenario, for example, can be seen in the following series of photographs. In the first shot, (Photo. 17.35), the mimic board on the bridge indicates the scene of the emergency to the Captain. The sensors for this purpose have been discussed earlier in this chapter. The incident is marked on the plans as being in the crew accommodation on zero deck, in the area by cabin 164. (Photo. 17.36). The Captain then alerts the assessment party by personal radio pagers or using the public address system. He then sends them to zero deck to investigate and take initial action. (Photo. 17.37). The fire is then confirmed and subsequent reports made by the on-scene commander to the bridge. These would advise either successful extinguishing of the blaze or request additional teams.

Other scenarios include an explosion below the waterline due to a bomb and an incident which includes manoeuvring the ship whilst she is under way and involves a major engine room fire. In the latter case, the Captain is able to alter the ship's heading, view the radar, which updates as the incident progresses, and use the engine control console to change the ship's speed. (Photo. 17.38).

A glance at a scenario offers a flavour of this advanced program. For example, on one scenario the Captain has the brief:

"The time is 0830 hours. The ship is cruising at twenty knots. There is a low swell with a light south westerly wind. The ship is in the Caribbean Sea, 40 miles, from Cartagena, on course for St. Maarten."

Having seen this, the Captain hears a tannoy announcement sending the assessment party to Deck 5. He arrives on the bridge to find alarms sounding in passenger cabins on that deck. The Captain then has to deal with the incident, which includes two passengers trapped in their cabin, as well as keeping passengers and the Company informed as the incident progresses. The outcome is determined, as it would be in an actual emergency, by the decisions the user makes.

Supertankers Anatomy and Operation

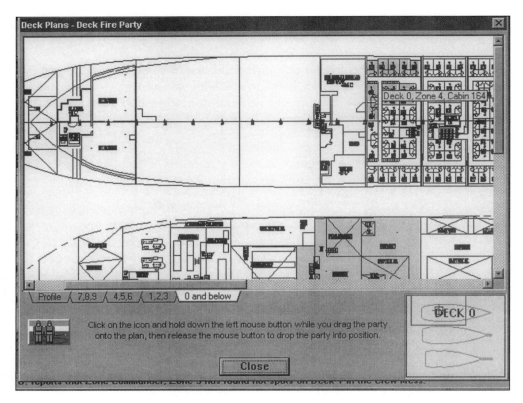

17.37. *The Captain Positions the Boundary Cooling Party at a chosen location on deck zero.*
(By kind permission of Sanderson CBT.)

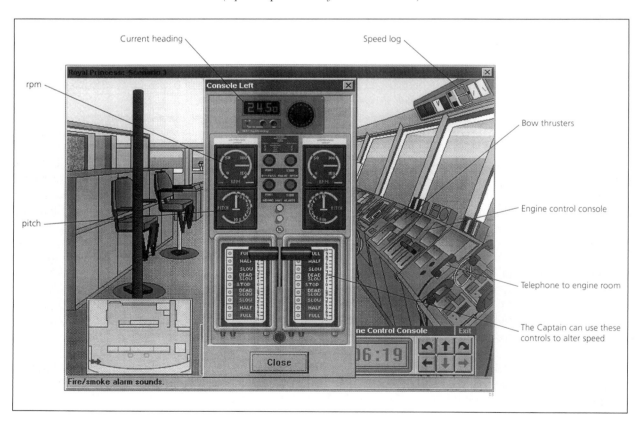

17.38. *Console Selected and Display Enlarged.*
(By kind permission of Sanderson CBT.)

As Keith Shaw, Sanderson CBT's Sales Director explained:

> *"At the end of each scenario, the Captain is given a feedback of what he did, other things he could have done and suggestions for further actions he did not appear to consider. This enables the Captain to recognise his own training needs and begin to address them through further practice."*

The principle behind this type of computer based training, although the nature of the scenarios differ, is not too far removed from the weekly emergency drills currently carried out on VLCC's that were examined in Chapter 10.

In some respects, it seems almost strange to report that computerisation across so many areas has been adopted in such a natural manner. As NUMAST commented on the company's findings of the virtual reality" program operating with Princess Cruises:

> *"Those who have tried it are saying: 'this is wonderful – stuff-please can we have more of it?' and it has already made a valuable contribution to training among Princess Cruises personnel."*

As I read the various scenarios, echoes were awakened in my mind of the "pre-imagined situation and my possible response" which, without becoming paranoid, runs occasionally through the mind of virtually all ships' officers. I am reminded also of the Orals for Masters' and Mates' where the candidate is confronted by a battery of instant situations requiring similar responses. The added bonus of 'pre-examination' time, as it were, in which to consider situations and possible responses would make the programs extremely valuable training aids in nautical colleges for the whole range of various grades in professional maritime examinations.

There exists also an enhanced factor, in which Sanderson CBT's 'Capter' holds for future development related specifically to classes of ships other than cruise liners including, of course, drills specifically germane to Supertankers.

17.15. CONCLUDING REMARKS

Undoubtedly, there appears to be an easy familiarity by seafarers with all kinds of computerised equipment and its use, but I do sense a few concerns raised in the minds of some users of PC's, aboard VLCC's specifically, that raise a question concerning reliability in some aspects. Reports, for instance, come my way that speak of problems with hard discs and difficulties experienced in obtaining spare parts. Some argue, although not totally convincingly to my mind, that the equipment is too 'complicated' and is not 'marine equipment' at all. In the Gulf, extremes of heat encountered seem to affect radar by producing spurious echoes on the heading market on the PPI, whilst the almost inevitable and, dare I suggest, traditional VLCC vibration problems appear to cause its own batch of difficulties to much Bridge equipment. One Captain commented that the computer controlling the ECDIS on his Supertanker was operational for only ten days in just over three months. These views, however unpalatable they may be to industry and technology, are those experienced by the users of the equipment and, as such, need to be aired so that they may be confronted and remedied. Personally, I think that such difficulties can be, and indeed will be, sorted out so that in cases where confidence has been hit, this may soon be recaptured and retained.

Far more worrying is the trend I find reinforced continuously in articles of authoritative magazines expressing concern regarding the stress levels to which officers on board all ships are being subjected. It is well known today that, if any person has employment, especially in management either ashore or at sea, then stress has become regarded as part of the 'career development' package. On VLCC's this could be particularly lethal. The problems here can be identified specifically as too much information on the Bridge, acting as a distraction particularly to the 'look out' duties of the officer-of-the-watch, and the commercial need for 'quick turn-rounds' in port in order to enhance payloads. The increase in opportunity for human error becomes immense with potentially far reaching consequences. Currently, there does not seem to be equally as specific an answer.

Chapter Eighteen

COMPUTERISED CARGO CONTROL

    18.1.   CARGO MONITORING

    18.2.   TAPE SENSORS
            18.2.1.   Ullaging
            18.2.2.   Gauging
            18.2.3.   Safety

    18.3.   BEAM MEASURING

    18.4.   TANKRADAR WORK STATION
            18.4.1.   Transmitters, Safety

    18.5.   TOTAL CARGO HANDLING
            18.5.1.   Data-Link Sensors

    18.6.   CARGO PLANNING SYSTEMS

Chapter Eighteen

COMPUTERISED CARGO CONTROL

(Introductory Remarks: Monitoring Systems: – Centralised Tank Level Gauging – Sensitised Tape Measuring – Gas Sampling System – Radar Beam Gauging for Ullaging and Temperature Measuring – Complete Cargo Monitoring and Cargo Control Systems – Concluding Comments).

18.1. CARGO MONITORING

As examined in the previous chapters of Part Four, computerisation has had a profound effect upon all navigation and bridge control equipment aboard all types of ships, not only VLCC's. Even though the engine room has not been examined in this book, for reasons stated in the Preface, computers have had an even more profound effect upon the monitoring of machinery and equipment in this department. The trend is continued, with an equally marked influence, as computerised applications have been applied in every area of tank gauging with these being extended to incorporate all aspects of cargo handling.

There are a number of cargo monitoring systems offered to the VLCC industry, each of which has made a considerable contribution to safety and efficiency. From amongst those suppliers responding, a sample has been offered that considers different approaches to current practices in measuring of ullages, temperatures and volumes, with the addition of alarm devices that become activated when input limits are approached or reached. Many of the systems incorporate monitoring also of ship's fuel and diesel oil bunker levels and that of water ballast tanks and draught gauging.

18.2. TAPE SENSORS

Consilium Metritape Inc, a wholly owned subsidiary of Consilium Marine AB of Sweden, offer such a system using their SENTRY II unique resistance tape type sensor. Designed to operate in the harsh marine environment, the resistance tape sensor is constructed by using a stainless steel base strip, wound round the base strip, but prevented from touching it by insulation. Wrapped around both edges is a nichrome wire. This wire is wound round the base strip in a spiral (HELIX), each being precisely 5 mm apart. The next stage is to apply a corrosion resistance jacket over the total length of the sensor. This is then heat sealed and the sensor ready for use.

An additional advantage of the Consilium Metritape sensor is that a temperature detector can be installed to the back of the level sensor, thereby reducing the number of tank penetrations to one only. Sensors are suspended from the tank top sensor housing and installed into a stilling tube which gives protection to the sensor and also allows free movement of the liquid being measured.

The principle of sensor operation is really quite simple. As liquid being measured is introduced into the tank, its hydrostatic pressure collapses the sealed outer envelope and shorts out the helical winding. The wound helix above the liquid level remains unshorted. Its resistance and its length can, therefore, be measured precisely so that accurate measurement and repeatability are given.

The element itself is extremely narrow, (Photo. 18.1), and can extend for a depth of thirty metres, which is sufficient for inclusion in all modern double hulled VLCC tanks. It can be retro fitted to older Supertankers. There is also a programmed overfill alarm offered which is fitted independently of, but adjacent to, the ullage sensor and within the same deck mounting, thus avoiding the inconvenience and expense of a separate installation. (Photo. 18.2). The alarm allows not only the 98% ullage approach to be covered, but also other fault monitoring such as system deficiencies. Alarms are sounded by means of an audible horn that has acknowledgement functions once activated by the officer of the watch. Both systems operate within the various Classification Society regulations and fulfil the requirements of SOLAS, OPA90 and MARPOL.

18.1. *The Actual Width of "SENTRY 11" Sensor Gauging Tape.*
(By kind permission of Consilium Metritape Inc.)

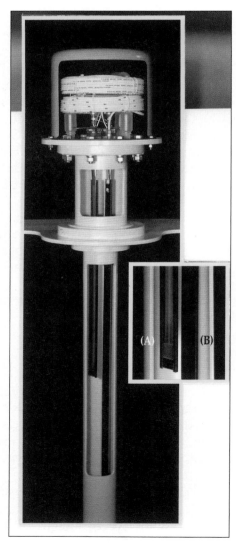

18.2. *An Independent Hi-Hi Alarm Sensor (A) fitted next to the Ullage Sensor (B).*
(By kind permission of Consilium Metritape Inc.)

Computerised Cargo Control

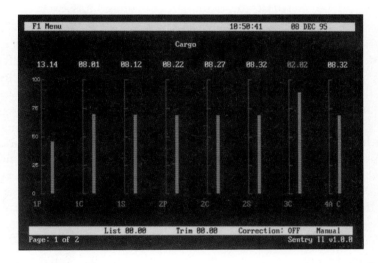

18.3. *The Tabular Display of the 'SENTRY 11' Gauging System allows the loading or discharging State of the Cargo to be Seen at a Glance. (By kind permission of Consilium Metritape Inc.)*

18.4. *The Bar Graph Display permits Tanks Approaching Completion to be Immediately Recognised. (By kind permission of Consilium Metritape Inc.)*

18.5. *The Ullage Display and Tank Filling Condition on a large tanker. (By kind permission of Consilium Metritape Inc.)*

18.2.1. Ullaging

From the sensor, which is the heart of the level gauging system, Consilium Metritape have developed various types of display systems for the presentation of data received from the sensor. The most common presentation is the straightforward individual or shared digital display. The VGA offer of 'Sentry 11' can be console or desk top mounted and windows information in either tabular or graphical format. The tabular display, (Photo. 18.3), is in digital format and covers all tanks simultaneously. These may be selected also on their own page for closer observation. Ullage, volume, temperature and, in the case of ballast tanks, soundings are shown along with list and trim information and the situation concerning the alarm. In the photograph, this is portrayed in red as a 'Hi' alarm showing that the ullage of number three centre tank is reading 2.02 metres, with a sounding of 27.97 metres, thus indicating that the tank on this particular Supertanker is approaching completion. Information can be shown also as a bar graph display. (Photo. 18.4). The value of this output is that the situation regarding all tanks on the ship may be seen at a glance. An analogue bar, ranging from zero to full percentile, enables the loading or, of course, discharging state to be displayed. Here, the ullage in three centre, referred to above, can be immediately seen. It is quite easy for the window showing the overfill protection condition of each tank to be observed independently. (Photo. 18.5). Here, the overfill state is indicated on a number of tanks where the demonstrated ullage has exceeded the limits required.

18.2.2. Gauging

The latest development by Consilium Metritape is the VANGUARD system. This is designed to operate with the current onboard trend of integrating all instrumentation systems. Vanguard collects all tank level and temperature data, converts level into volume and, by use of an RS232 interface, transfers all data (volume, level and temperature) into the shipboard control system. Built into the front cover, of the Vanguard signal conditioning (data acquisition) unit, is a panel with keypad which will display level and temperature for any tank selected. This is used also for sensor calibration when starting up a new system. The Vanguard system provides through its direct serial data communication a cost advantage over traditional analogue systems. Consilium Metritape's level gauging systems are frequently interfaced with the ship's loading computer offering the operator real time levels of cargo and ballast tanks.

18.2.3. Safety

Consilium also produce a gas sampling unit for the tanker industry. Their new 'SALWICO SW2010' detects the presence of explosive and toxic gases, as well as oxygen gases,

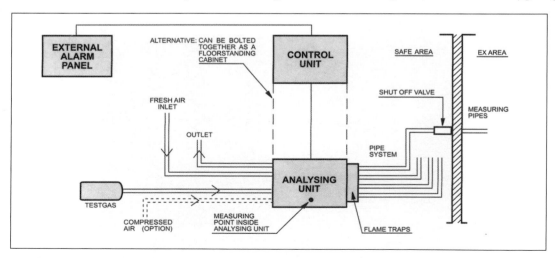

Diag. 18.1 Diagram Showing the Main Elements in the 'SALWICO SW2010' Gas Sampling System. (By kind permission of Consilium Marine AB.)

in pumprooms, cofferdams, pipe tunnels, double bottoms and all types of cargo, ballast and slop tanks. It is a permanently installed system that consists of the following elements. (Diag. 18.1). The control unit is usually located in the cargo control room and contains all of the functions necessary to operate the system, as well as showing the results of systematic samples on an alphanumerical display. It is the analysing unit which is at the 'operational end' of the equipment for it is here that samples of the gas obtained from the source pipes are situated. A 'shut off' valve isolates the measuring pipes at the hazardous area on deck and, for additional safety, flame traps are fitted preventing sparks etc, from entering the pipe system. A timewire individually programmable presuction function draws the sample through a filter, thus preventing contamination by dust, dirt, salt and moisture. Automatic purging of the entire pipe system with clean air, between the sampling phases, ensures that continuous cleanliness is maintained. An external alarm panel maintains constant monitoring of potential faults within the system and operates acoustic and optical alarms on the bridge and cargo control room.

18.3. BEAM MEASURING

An alternative to the physical strip inserted into a tank, is the use of radar based systems to measure ullage and temperature etc. Saab Marine Electronics have been involved in using radar gauges in tank gauging for over twenty years and virtually half of the tankers at sea today use their equipment. The Saab systems not only measure ullages and temperatures etc. by the Saab 'TANKRADAR G3' but, working in conjunction with the Saab 'TANKRADAR MaC', offer also a complete cargo monitoring and control system.

18.4. TANKRADAR WORK STATION

Saab 'Tankradar G3' consists of four major parts. (Diag. 18.2). The work station is used by the officer on duty to read tank ullages, temperatures and the other data previously

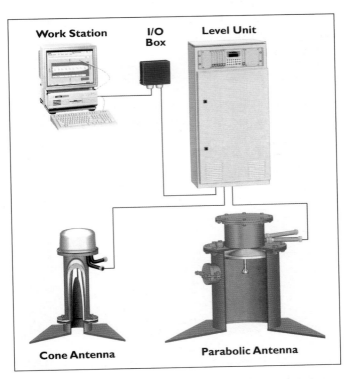

Diag. 18.2 Diagram Showing the Three Main Features of the 'SAAB TANKRADAR G3', together with a Choice of Two Tank Transmitters.
(By kind permission of Saab Marine Electronics.)

examined in this chapter. The system offers also an option which enables inert gas pressures to be measured. The software is used with a type approved PC, which has a VGA coloured screen, and can be placed on the bridge or wherever is considered most convenient.

The main functions of the work station are to calculate, log, store and display measured values and it is equipped with an on-line help function that offers direct access to on-screen help texts and extracts from the company's operating manual. The workstation is delivered with a light pen as standard, but it can be operated also with a mouse or tracker ball. The light pen is pointed directly at a command box on the screen in order to activate any selected function. (Photo. 18.6). The work station handles also a variety of alarms and fail safe devices of the sort already examined. It may be integrated also into a wider network so that several operator consoles may be fitted into various offices on any ship.

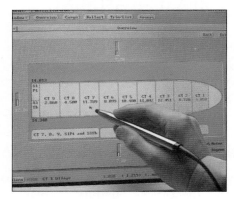

18.6. The Work Station of a Tank radar Aboard a Modern VLCC is Operated by Means of a Light Pen. (By kind permission of Saab Marine Electronics.)

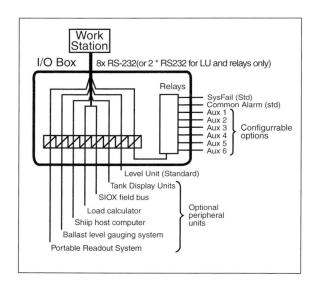

Diag. 18.3 The Input/Output, I/O, Box is the Unit connecting the Work Station and the Level Unit.
(By kind permission of Saab Marine Electronics.)

If the work station is the control centre of the system, then the Input/Output, or I/O, box is the unit into which a number of communication interfaces and sensors are connected between the level unit and the work station enabling intelligence to be added to the system. The possible connections are shown in the following diagram. (Diag. 18.3). The work station receives information from a range of other sensors and systems. The system failure relay control is also situated in the I/O box along with other alarms covering both output as well as input signals.

Computerised Cargo Control

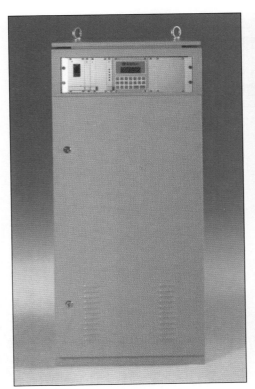
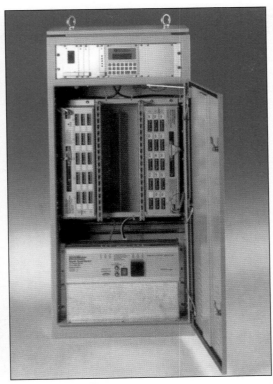

*18.7. The Level Unit is an Integral Part of the Cargo Control System.
(By kind permission of Saab Marine Electronics.)*

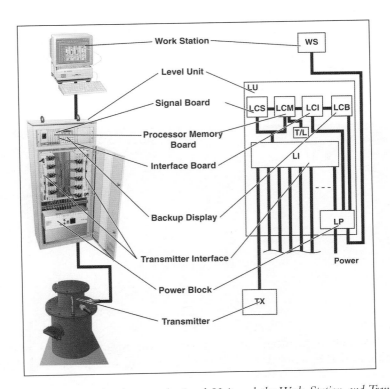

*Diag. 18.4 Information Flow between the Level Unit and the Work Station and Transmitters.
(By kind permission of Saab Marine Electronics.)*

Supertankers Anatomy and Operation

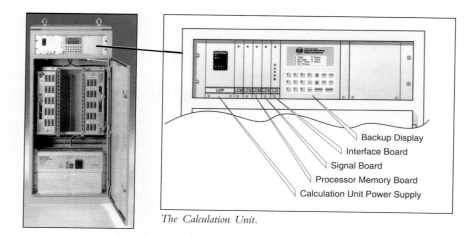

The Calculation Unit.

18.8. The Components of the Calculation Unit, CU.
(By kind permission of Saab Marine Electronics.)

The Level Unit (LU), (Photo. 18.7), is a cabinet containing the main communication modules and the system power and transmitter interfaces. (Diag. 18.4). A major component in the LU is the calculation unit (CU), (Photo. 18.8). This consists of a series of printed circuit boards containing the LCS-LCM-LCI-LCB and T/L, which are respectively the Signal Board, Processor Memory Board, Interface Board, Back-up Display and Trim/List Unit. The signal board, in essence, gives intelligence to the signals received from the transmitters situated on the individual tanks. The LCM processes the various readings received from the signal board and contains various memory databases. The trim/list unit measures the trim and list angles which are used to support the echo detection process and to correct ullages when the ship is inclined.

18.4.1. Transmitters, Safety

The transmitters are the work force of the entire system. As the latter operate through the potentially hazardous deck area, a Zener barrier board, with intrinsically safe connections, acts as an interface between the electronics in the calculation unit and the transmitters. Complete safety is therefore assured as no contact is made that can cause sparking to occur, nor can

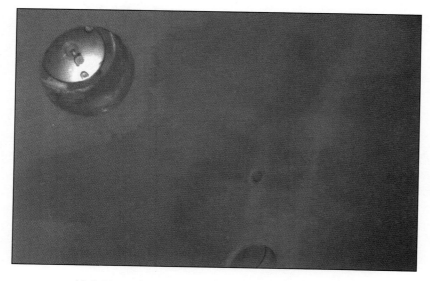

18.9. View of the Parabolic Antenna from inside the Tank.
(By kind permission of Saab Marine Electronics.)

Computerised Cargo Control

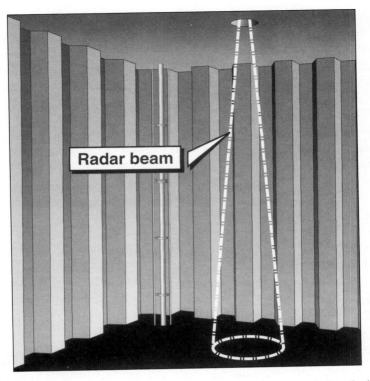

*Draw. 18.1 The Narrow Concentrated Radar Beam is Highly Accurate and Completely Harmless.
(By kind permission of Saab Marine Electronics.)*

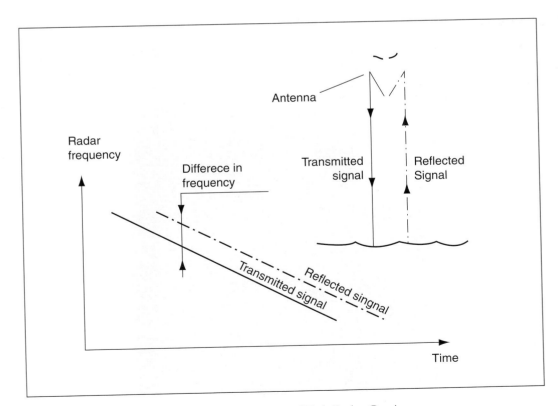

*Draw. 18.2 The Principle of Tank Radar Gauging.
(By kind permission of Saab Marine Electronics.)*

heat be generated to an extent that causes a flame or explosion. There are two different types of transmitter available depending upon the depth of the tank involved. The cone antenna is designed for use with shallow tanks of the type found, for example, aboard product and small crude oil carriers. The other type, the parabolic antenna, is of specific interest for VLCC's owing to its diversity for use with all types of tanks, particularly those with the complex internal structures found on all Supertankers that were examined in Chapter 2, especially in wing tanks.

The radar beam of the parabolic antenna inside the tank is narrow and concentrated and completely accurate and reliable. It emits radar waves from an internal microwave module, (Photo. 18.9 and Draw. 18.1), that are totally harmless to humans should they be encountered while working in the tanks, with the system continuing to operate, however close to the antenna one may approach. The Saab Manual explains perfectly clearly the principle upon which the system operates (See also Draw. 18.2):

> *"The frequency of the transmitted signal decreases over a time period. The incoming signal is compared with the outgoing signal. The difference between these two signals is a low frequency signal. Its frequency is directly proportional to the distance from the transmitter to the surface of the product. This is called the Frequency Modulated Continuous Wave (FMCW)."*

The parabolic reflector and antenna feeder A/V made of stainless steel are the only parts of the transmitter which are situated in the tank. The remainder of the transmitter is housed in a pedestal type socket that is fixed onto the main deck. (Photo. 18.10). The following drawing, (Draw. 18.3), indicates the working parts of the equipment showing clearly the dimensions and internal workings. The direction of the antenna beam may be fixed to within ±2°. A cleaning hatch is provided so that the equipment may be cleaned whilst the system is in operation using a brush that is entered by a check valve. An ullage plug is provided for hand dipping and sampling and there is a separate inert gas pressure sensor. Temperature sensors may also be fitted via an optional protective hose connection. Both a portable and local read-out display are available, additional to the work station read-outs and a separate tank display unit (TDU) for each tank may also be provided that shows ullages in bar graph and

18.10. The Parabolic Antenna Transmitter on the Main Deck of a Large Tanker.
(By kind permission of Saab Marine Electronics.)

Computerised Cargo Control

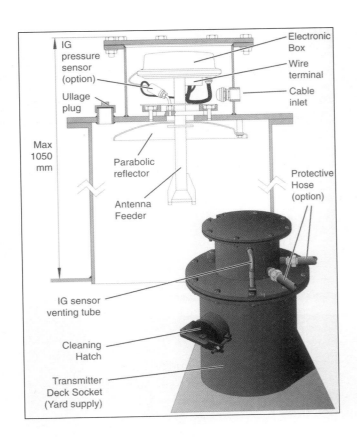

Draw. 18.3 The Working Components of the Saab 'TANKRADAR G3' Transmitter.
(By kind permission of Saab Marine Electronics.)

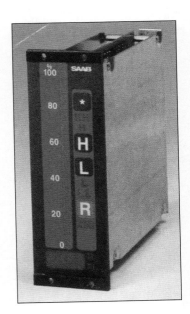

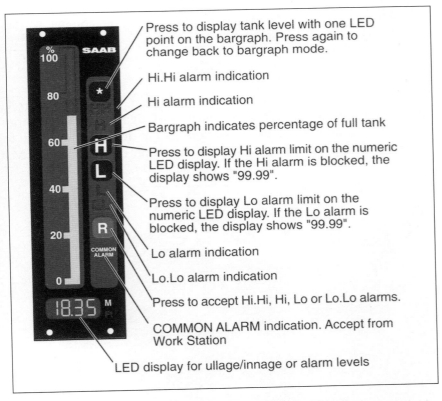

18.11. Tank Display Unit, TDU, for Ullage Presentation is Fitted with a Range of Safety Alarms.
(By kind permission of Saab Marine Electronics.)

Supertankers Anatomy and Operation

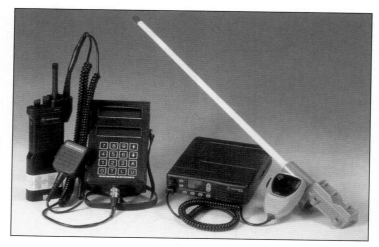

18.12. *The Portable Readout Equipment is Totally self-contained and Enables Constant Monitoring by the Duty Officer.*
(By kind permission of Saab Marine Electronics.)

Diag. 18.5 The Portable Readout Equipment is Connected to the Work Station by the I/O Box.
(By kind permission of Saab Marine Electronics.)

digits as well as an indication of high and low alarms. (Photo. 18.11). The portable display is totally self-contained, (Photo. 18.12), enabling the officer of the deck to monitor ullages constantly and, by means of a base radio, constant contact is maintained through the I/O box to the work stations. (Diag. 18.5).

18.5. TOTAL CARGO HANDLING

Additional to the monitoring equipment Saab 'Tankradar G3', the system may be fitted with a Saab 'Radartank MaC' which is designed for total cargo handling. This is achieved by use of an additional work station and it may be installed together with a Kockumation 'LOADMASTER', as a load calculator. Based on the earlier Saab MaC/501, which continues to be

18.13. *A Saab TANKRADAR MAC Work Station Situated in the Wheelhouse of a Large Tanker.*
(By kind permission of Saab Marine Electronics.)

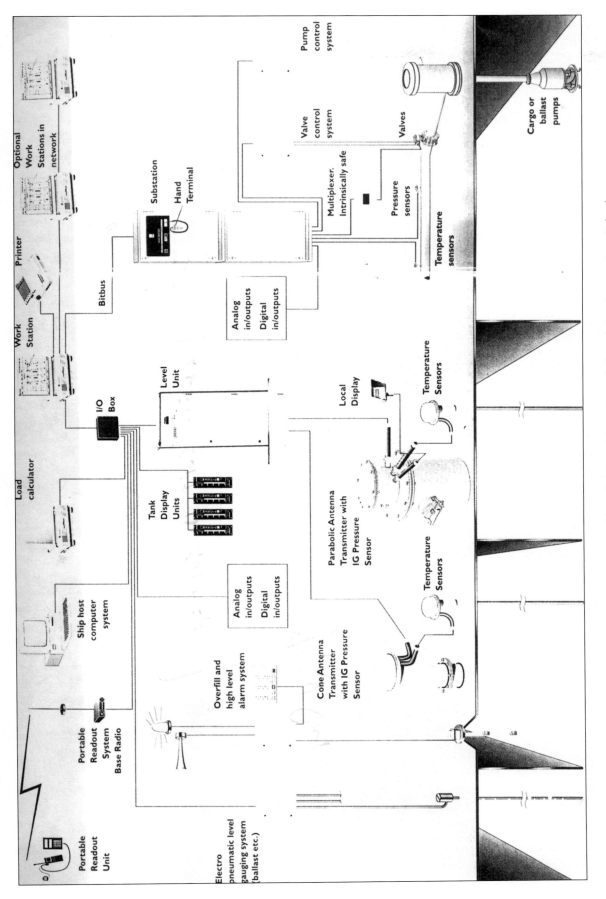

Diag. 18.6 Flow Diagram Indicating how the Cargo Monitoring and Control Functions are Inter-Related. (By kind permission of Saab Marine Electronics.)

Supertankers Anatomy and Operation

used on many tankers, the duty officer on the ship is offered complete overview and control of all aspects of cargo operations. The second work station is generally placed in the cargo control room or on the bridge in the wheelhouse next to the work station of the Saab 'Tankradar G3' its companion equipment. (Photo. 18.13). The following diagram indicates how the control and monitoring functions are inter related for a complete cargo handling package. (Diag. 18.6).

The Saab 'TANKRADAR MaC' uses a larger 534mm screen for greater clarity and is operated again by means of a light pen which has proved to be the most efficient method for interacting with the system. Control windows in the PC display a mimic diagram which outlines the tanks under consideration, be they cargo, ballast or fuel, and shows the positions of all valves, pumps and sensors in the form of graphical symbols imposed upon the layouts. Individual tanks may be activated as well as the entire tank system thus allowing either a closer or an overall ship view to be taken. Additional control stations may be implemented and an optional redundancy box that, in the event of a failure with one of the master stations, becomes activated within one minute so that no control is lost of the operation. An explanation of the system is offered by Saab:

> "Pumps, valves and other equipment are controlled by pressing the symbol using the light pen. An indication shows that the symbol is selected. A small control window is opened close to the symbol, allowing the operator to give commands such as open/close, stop or percentage settings. Control commands can be issued from any work station in a network."

As the system is networked the possibility of one operator countermanding the actions of a second are removed because each of the displays will show the current situation of the valves or pumps simultaneously on each set. The current state of operation may be seen at a glance. Extensive alarm devices and test capabilities are fed into the system so that any abnormalities may be detected instantaneously, thus allowing the entire system to be used with confidence.

18.5.1. Data-Link Sensors

Two essential additions are required for the system to work. The first is a series of sensors that are attached to each valve, pump and any other desired input which are connected to the second important element, the sub station, by means of high-speed Bitbus data links. As a hazardous area of main deck is involved, the by now conventional Zener barriers are integrated into the circuit. The sub-station cabinet, (Photo. 18.14), contains the intelligence necessary to convert the signal into a format that enables the information to be activated and

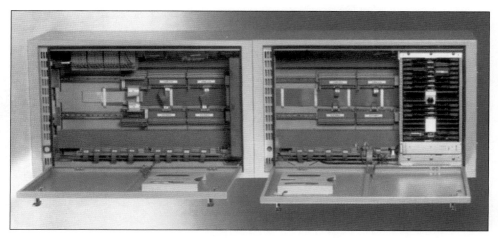

18.14. The Substation Contains the Intelligence Necessary to Convert the Signal into a Format Facilitating Information Via the Screen.
(By kind permission of Saab Marine Electronics.)

Computerised Cargo Control

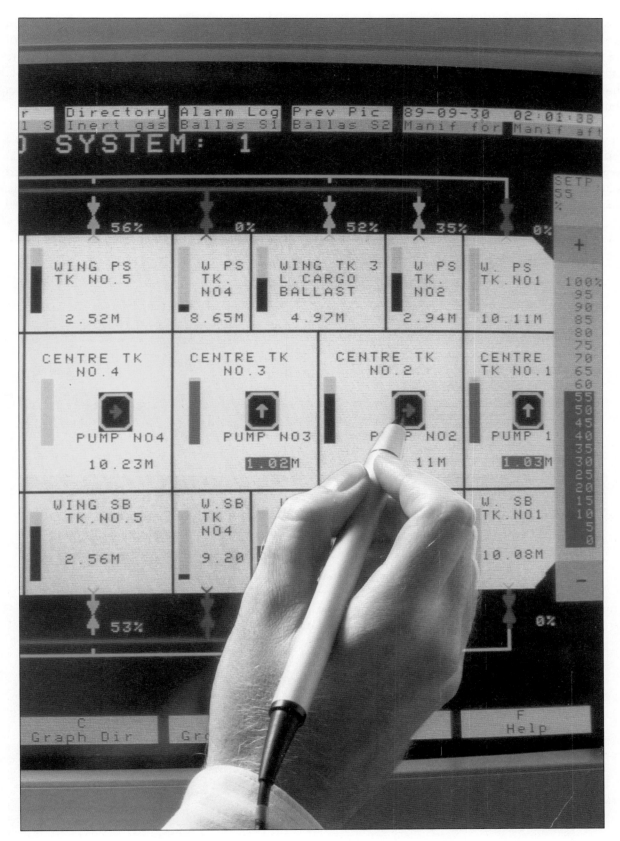

18.15. *The Rate of the Pump serving Number Two Cargo tank can be Regulated by the Touch of a Light Pen on the Screen. (By kind permission of Saab Marine Electronics.)*

Supertankers Anatomy and Operation

pro-activated via the screen. In many respects, although with obvious differences, the sub station serves a similar function to the level unit of the monitoring equipment. By touching the symbol of a valve with the light pen, (Photo. 18.15), it may be opened or closed and its current state observed immediately. Additionally, the sub station may be worked manually through a portable hand held alphanumeric display and keyboard. This serves not only as a back up alternative, but is used also for testing purposes.

Whessoe Varec Limited, whose gauges as has been shown in previous chapters have provided the ullaging requirements for literally hundreds of 'early' generation Supertankers, continue to offer ship's deck officers their services in the form of a new and totally computerised package which monitors all liquid cargoes. Their Marine Whessmatic 550M System and Tank Control Inventory Management System (TCIM/M) is a versatile and accurate Supervisory control

18.16. An Approved PC Acts as a Centralised Database for the Monitoring of all Gauging Parameters.
(By kind permission of Whessoe Varec Ltd.)

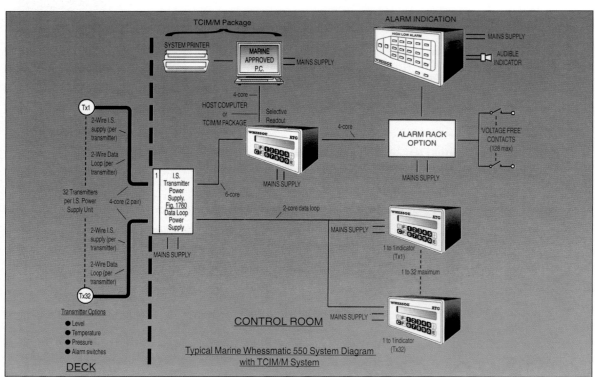

Diag. 18.7 Typical Marine Whessmatic 550 System Diagram with the TCIM/M System.
(By kind permission of Whessoe Varec Ltd.)

Computerised Cargo Control

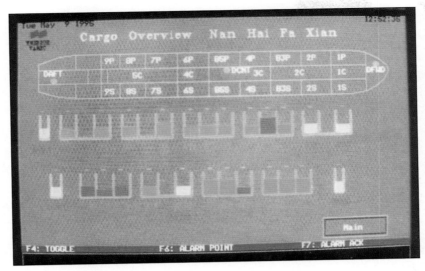

18.17. *The Cargo Overview on Whesso Varec's TCIM/M Offers an Instant View of the Cargo Control Situation.*
(By kind permission of Whessoe Varec Ltd.)

18.18. *Model 7600 Radar Level Gauge can be Used on Tankers of all Sizes, as well as Oil Farms etc. Ashore.*
(By kind permission of Whessoe Varec Ltd.)

and Data Acquisition System (SCADA) which has been fully approved for marine use and has been fitted to a very wide range of tankers. TCIM/M/SCADA also can be used on an approved bridge PC to offer a centralised database for the monitoring, logging and display of all tank gauging parameters. (Photo. 18.16). Whessoe Varec's latest 'FUELSMANAGER' Microsoft Windows NT operating system has been designed for terminal and refinery use (inter alia) and, at the time of writing, is currently under test for approval aboard sea going tankers.

TCIM/M interfaces with the Whessmatic 550M Marine System to offer a centralised, fast, reliable and accurate data presentation that is displayed in high quality graphics with data back up support and fully pro-active alarm systems. (Diag. 18.7). The alarm indicator possesses the ability to cover fifty alarm points that are aligned in chronological order showing occurrence dates and times together with a facility that displays a maintenance page for any alarm occurrence. The main tank inventory link can display level, pressure, temperature and gauging information for a total of thirty tanks, showing up to ten tanks per page, whilst the cargo overview, (Photo. 18.17), offers a view of the immediate cargo control situation at a glance.

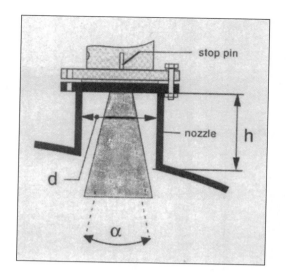

Diag. 18.8 The Mounting on Deck of the 7600 RLG.
(By kind permission of Whessoe Varec Ltd.)

Whessoe Varec's latest Radar Level Gauge, Model '7600 RLG', integrates with other sensors to provide all required measurements necessary for the rigid control of crude oil cargo loading and discharging operations. (Photo. 18.18). The operating principle is based on the proportional distance/time measurement of a microwave pulse that is transmitted from the RLG cone to the surface of the cargo and back to the horn. (Diag. 18.8).

18.6. CARGO PLANNING SYSTEMS

There are a number of excellent fully computerised cargo planning systems available to the VLCC market each of which offers totally accurate and reliable assistance to the deck officer. Kockum Sonics AB of Malmo, with their almost one century of experience for instance, have produced their software program 'LOADRITE', which is developed, licensed and supported by Transas Marine (UK)'s Loading Control System, LCS 97. The products of this latter versatile

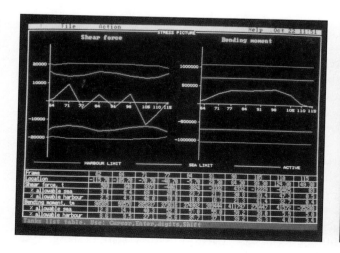 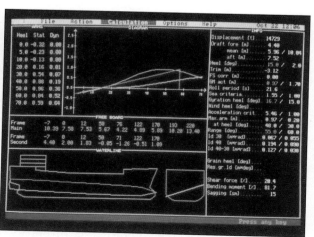

18.19. Actual Shear Forces and Bending Moments,
with Sea and Harbour Conditions and Static/Dynamic Stability, Including Water Lines.
(By kind permission of Kockum Sonics AB.)

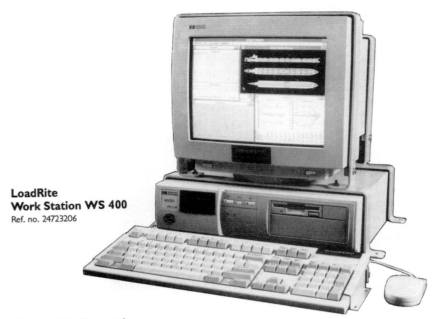

LoadRite® Work Station WS 400

LoadRite Work Station WS 400
Ref. no. 24723206

General Information

WORK STATION WS 400 is an IBM PC–AT Compatible Personal Computer intended as a central processing and display unit for LoadRite. The WS 400 is:

- made rugged to withstand the harsh conditions that may occur on board ships,
- type approved by all major classification societies,
- fitted, as standard, with a colour monitor, hard disk and floppy disk drive.

In the LoadRite system, WS 400 works with the MS–DOS® operating system and Windows™. This enables WS 400 to be used as a general PC running programmes such as word processing, calculation or data bases.

Technical Specifications

Central processing unit:
486 SX 33, DX2, 50/66 MHz
3 x16 bit ISA slots
1 x 32 bit VL/ISA slot
VESA connection, PCMCIA prepared, CD-ROM prepared, Keyboard lock, Low noise fan
Upgradeable to Pentium Overdrive

Primary memory:
8 Mbytes expandable to 96 Mbytes

Floppy disk drive:
1.44 Mbytes, 3.5 inch, 135 TPI high density

Hard disk drive:
210 Mbytes or more

Input/Output ports:
2 serial ports, 1 parallel printer port

Real time clock:
Battery backed–up clock calendar

Video output:
Ultra VGA + up to 1280 x 1024 at 256 colours

Ambient temperature:
Operational + 5 – + 40° C, storage –40 – + 70° C

Power:
100–240V AC
47–63 Hz, 100W max./230V
EPA certificate

Dimensions:
369 x 354 x 400 mm (H x W x D)

Weight:
Central unit: 10 kg
Keyboard: 3 kg
Display: 14 kg

Postal Address	Address	Telephone	Telefax	Telex
Kockum Sonics AB Box 1035 S-212 10 Malmö, Sweden	Industrigatan 39	Nat 040-671 88 00 Int + 46 40 671 88 00	Nat 040-21 65 13 Int + 46 40 21 65 13	33792

KSM 395E/9505
Replaces KSM 395E/9409

Diag. 18.9 The Kockun-Sonics LOADRITE WS400 Work Station. (By kind permission of Kockum Sonics AB.)

company for the navigation and training markets have been examined in previous chapters of Part Four. The 'Loadrite' kit is intended to calculate and control loading/discharging of liquid cargoes and water ballast through their Workstation WS400, a detailed description of which is offered in the accompanying leaflet. (Diag. 18.9). The equipment also estimates and corrects trim, stability and longitudinal strength data, (Photos. 18.19), including damage control, safe sailing speeds as well as the issuing of all documentation relevant to any particular cargo.

Kockum Sonics 'LEVELMASTER-CALM', or Computer Aided Level Measurement, (Photo. 18.20), uses the electro-pneumatic principle of operation, or so called 'bubble measuring'. This technique only demands clean compressed air, but keeps the installation as well as the maintenance costs down. To obtain higher cost efficiency, the company have designed the system in such a way that each pressure transmitter (needed for measuring the hydrostatic pressure) is engaged for a number of tanks. (Diag. 18.10). One pressure transmitter can handle up to six tanks. The system has features for automatic calibration of all measurements and permits purging by blowing air through the pipes at, typically, four hourly intervals in order to keep the piping serving the tanks free of any potential obstruction. A digital flow controller keeps the air flow constant by varying the main valve's opening time. The CALM system can transmit on-line data to an automatic system and/or integrate totally with Kockum Sonics work station to provide a fully comprehensive cargo handling and stability system. A glance at the following tank information photograph of their DU350, (Photo. 18.21), clearly justifies the company's claim that Levelmaster-Calm displays 'are easy to read'.

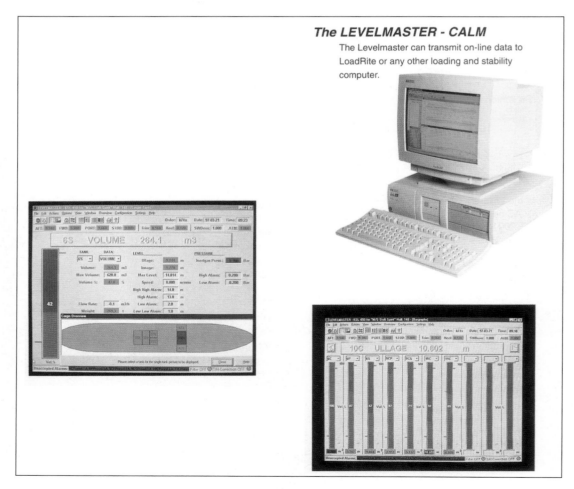

18.20. The 'LEVELMASTER-CALM' can transmit on-line data to 'LOADRITE' or any other Loading and Stability Computer. (By kind permission of Kockum Sonics AB.)

Computerised Cargo Control

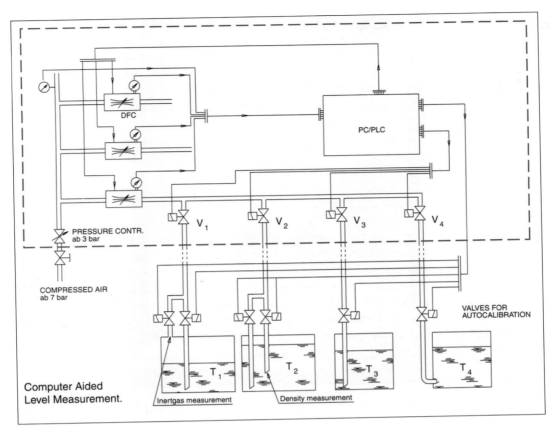

Diag. 18.10 The Operating Principle of 'LEVELMASTER-CALM'.
(By kind permission of Kockum Sonics AB.)

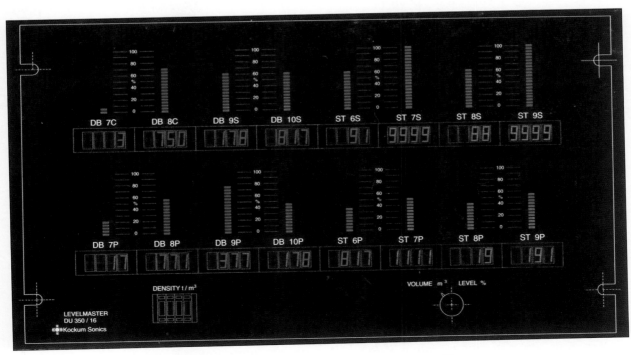

18.21. The Overall Tank Situation may be Seen At a Glance.
(By kind permission of Kockum Sonics AB.)

Supertankers Anatomy and Operation

*18.22. The 'LOADSTAR' Loading Computer is very much a total Ship Management Tool.
(By kind permission of Maersk Data AB.)*

Maersk Data AS of Copenhagen, a company within the A.P. Moller Group, have supplied their 'LOADSTAR' loading and management package computer to some of the world's leading shipping companies. 'Loadstar' is based on a total of twenty-six different modules which, in various combinations, are relevant to the operations of virtually all types of ocean going shipping and nearly half of which are directly applicable to crude oil carriers including the VLCC market. (Photo. 18.22). It is an advanced loading programme that is fast and easy to use which integrates into one single process all necessary functions.

There can be little doubt that computerisation has had more than a profound effect upon cargo handling. The first generation of VLCC's, it will be remembered from the earlier chapters of Part Three, relied totally upon a fully fitted cargo control room with all of the paraphernalia necessary to work a crude oil cargo. This has now been superseded fundamentally by two PC's placed in the cargo control room or wheelhouse. As Captain Geoff Cowap observes:

"The modern loading instrument software will calculate and display not only shear forces and bending moments, but all aspects of stability. This includes draft, trim, heel, metacentric height and also the curve of dynamic stability which is a measure of the righting force that will bring the ship back to the upright condition as she rolls in a seaway.

The program will also give warnings to the operator when any aspect of the ship's condition becomes unsafe or when pre-set limitations are exceeded such as when the draft exceeds the loadline, or when the GM is less than the statutory minimum. The program can also give warnings if local deck weight limitations are exceeded."

Computerised Cargo Control

18.23. An Example of LOADSTAR's Module Z
an example of a G-Z Stability Curve.
(By kind permission of Maersk Data AB.)

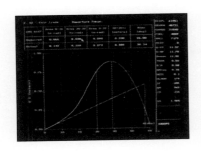

18.24. Module Y-An Example of
Damage Stability Calculations.
(By kind permission of Maersk Data AB.)

How the comments made by Captain Cowap can operate is demonstrated in Module Z, of the Maersk Data AB's Loadstar program, and its accompanying explanation. (Photo. 18.23). The GZ-Curve illustrated is a graphical presentation of the calculated stability curve that offers a clear picture of current stability figures. A table which compares the calculated stability with the relevant stability criteria gives a complete analysis of the stability of the ship in the present loading condition. A print out of the screen picture may be handed over to the authorities as appropriate. The Loadstar has also a Module Y, (Photo. 18.24), that performs calculations which verify whether the present loading condition complies with the relevant IMO damage stability codes. Due to a unique calculation method, these calculations are extremely fast compared to market standards.

The idea does not stop here. Captain Cowap, again, suggests an overview which links computerised loading with the GMDSS system examined in Chapter 17:

"Satellite communications can now be utilised to link the loading computer directly to a shore based station so that loading plans can be transferred between ship and shore. This information can be of considerable use both commercially and in the event of accidents when a record of the loaded state of the vessel is instantly available from a shoreside source."

Certainly, SHIPWRITE/TANKERLOAD, the programs produced by the company of which Geoff Cowap is a director, Energy Marine (International) Limited, are sufficiently comprehensive to function with all types of tankers, including gas and chemical carriers, using versions of MS Windows 95, 98 and MS DOS.

It is worth stressing, perhaps the obvious, by emphasising that all computerised cargo handling systems incorporate concurrent facilities that indicate the dynamic and static stresses, examined briefly in Chapter Six, which are involved during loading and discharging.

Although not investigated, complete automation of the ship's engines and machinery could also be combined into one system and monitored directly from the wheelhouse. This would not detract, of course, from the carrying on board of engineering officers with the expertise necessary to fault trace and remedy.

PART 5: Conclusion

Chapter Nineteen

SUPERTANKERS – THE FUTURE

19.1.   THE AGEING SINGLE HULL TANKER

19.2.   MARPOL REGULATIONS

19.3.   SHIPS' STRENGTH – CAUSE FOR CONCERN?

19.4.   REGULATIONS APPLIED TO THE OLDER SHIP

19.5.   HYDROSTATICALLY BALANCED LOADING (H.B.L.)

19.6.   DOUBLE HULLED NEWBUILDINGS

19.7.   TEMPERATURE AND ITS EFFECT ON CORROSION

19.8.   COATINGS, COLOUR AND THICKNESS

19.9.   INSPECTION AND VENTILATION IN DOUBLE HULLED TANKERS

19.10. THE DOUBLE HULL AND POTENTIAL OPERATIONAL DIFFICULTIES

19.11. POSSIBLE SOLUTIONS TO DAMAGE CONTROL, HULL STRESSES

19.12. CONCLUSION

Chapter Nineteen

SUPERTANKERS – THE FUTURE?

(1998-VLCC Statistics and some Implications of Single and Double Hulled Ships – General Integrity of many existing Single Hulls approaching Fourth and Fifth 'five year' Surveys – VLCC "STENA CONVOY" – VLCC "STENA CONGRESS" – Comments on Hydrostatically Balanced Loading – Future Prospects for Single and Double Hulled and Sided VLCC's – Some Potential Difficulties with the New Class of VLCC's – Some Solutions Suggested – Concluding Remarks.)

19.1. THE AGEING SINGLE HULL TANKER

Chapter 14, in a brief outline, confirmed that double hulled Supertankers will undoubtedly be the carriers of bulk crude oil in the future. MARPOL 73/78, SOLAS 74/78 Regulations, with Amendments, together with OPA90 and other legislation have made certain of that. A major, but somewhat rhetorical, question that might be posed however could ask: 'How far ahead *is* the future?'

The current situation (late-2000) indicates that there is still a fair way to go before the number of double-hulled VLCC's becomes a majority. In the Preface, it was stated that the world's fleet of Supertankers, in excess of 160,000 sdwt under AFRA definitions, totals around 450 ships. Figures provided by Mr Cliff Tyler, of Clarkson Research Studies in London, indicate that on the 1st July 2000, a total of 455 Supertankers were operational in the world – including 14 built between 1974 and 1980 (totally nearly 4 million sdwt) being used as long term storage vessels for periods in excess of sixty days. Of the remaining 441 ships, *134 only* are double-hulled (the earliest of which is a ship of 164,000 sdwt, built in 1986). The 307 single-hulled VLCC's include 7 built pre- and during 1973 that are due for scrapping within the near future. From the year 1974, 14 Supertankers are still in service, 1975–28, 1976–42, 1977–21, with 22 from the remaining years before 1980. From 1980 until 1990, 73 single-hulled VLCC's remain in service. From 1991 until 1996, 100 single-hulled vessels were constructed: the latter year representing that in which, under older regulations, the final single-hulled Supertanker was built.

I have selected 1980 as a 'watershed' year for two reasons. Research would indicate that this date seems to be the one under which to categorise VLCC's into conventional 'pre' and 'post' MARPOL ships. This is due to legislation that came into force around then which covered the construction of new oil tankers. Secondly, a number of ships that were built in the 1980's experienced problems of structural weaknesses, although this does not exclude *some* VLCC's from earlier years where defects occurred due to severely neglected maintenance. This led, amongst other things, to instances (cited elsewhere in the text) where entire bow units became detached during severe weather conditions, and complete holes appeared through forward ballast tanks. Inevitably, these tales captured the public's imagination, in more ways than one, and soon led to quite horrendous, but justified, criticism in the media. Evidence indicates that the 1980's was a period in which some owners created an era of ships that had severe economies in terms of the thickness and quality of steel used in many areas of construction, and 'cut as many corners as possible', including minimal tank coatings, to reduce building costs to a minimum. It is highly unfortunate that such criticism, in the eyes of the general public, has come to embrace all Supertankers as a class without regard to age or, more importantly, ownership.

19.2. MARPOL REGULATIONS

When discussing any aspect of MARPOL 73/78, concerning single or double hulled Supertankers, I find myself confronted with problems similar to those found when attempting to consider OPA90 and inevitably MARPOL, Regulation 13F, in Chapter 14. Bearing in mind that my background is that of a deck officer, not a barrister, I am in severe danger of becoming bogged down in legislation which is far too complex for a book of this nature. Keeping to the simple facts, crude oil carriers over 20,000 sdwt have to comply with Regulation 13G which requires a regular programme of inspections to be carried out, modified as necessary by Classification Society rules, and to carry on board a condition evaluation report. They must also, depending on whether or not they are pre or post MARPOL tankers, for a period varying between 25 and 30 years from their delivery date, comply with Regulation 13F for New Oil Tankers (i.e.: those post July 1993). It is stipulated also in 13F that 'other arrangements' are permissible for pre MARPOL tankers.

What this amounts to is that when a tanker reaches the age of twenty-five years, the owner has open to him a number of options. If the tanker is double hulled there is no problem and his ship can continue to operate. If the tanker is not double hulled then he must either scrap it or convert it to double hull. If the ship is post-MARPOL, he can continue to trade it until a thirty year period is reached, and if pre-MARPOL, can still trade it for thirty years providing that a proportion of the cargo wing tanks are converted to either ballast or void tanks, and/or ascertaining that he always loads the tanker so that the tanks are hydrostatically balanced with the sea level. No action is necessary if the tanker has a double bottom. The VLCC might also be turned into a non-trading storage or offshore vessel. To conform to the regulations by adding an extra hull around the cargo carrying capacity is, to most intents and purposes, a quite unrealistic proposition. This is not only on the grounds of costs to the ship owner, but also in terms of feasibility to the yards who would be approached to make available the facilities necessary to do the work. Another legal requirement, that became effective in July 1998, insisted that all VLCC's have to possess certificates issued under the International Safety Management Code (ISM) of IMO.

19.3. SHIPS' STRENGTH-CAUSE FOR CONCERN?

The question foremost in the mind, therefore, concerns the situation regarding the condition of the current 307 or so single hulled VLCC's trading worldwide that are involved and which have 'survived' until 2001 to continue trading. Many of these are by now approaching their fourth or fifth mandatory five year survey, thus falling into the twentieth-plus year category. The fears, however, that are publicly expressed on all too many occasions in the media regarding the age and serviceability of these VLCC's are, in the majority of cases, quite unfounded.

Captain Geoff Cowap, the Extra Master Mariner, marine consultant and company director, whose contributions have been quoted elsewhere in this book, has made the following comment regarding early VLCC construction:

> "Up until the 1960's, ships were built as extremely rigid structures, with shell thickness considerably greater than modern ships. The plating was, with only a few exceptions, all riveted and the overall longitudinal strength of the vessel was of a magnitude that could never give any cause for concern.
>
> During the 1960's, ship builders and Classification Societies became more adventurous and ship owners demanded lighter structures in order to carry more deadweight. So it was that scantlings were reduced, welding techniques were improved and emphasis was placed in building ships with little more than adequate strength. It was at this time that tanker operators, in particular, were concerned with the economic aspects of larger capacity vessels."

Certainly, Captain Cowap's remarks remain very true of many 1970 VLCC's. Indeed, as it has been mentioned in Chapters 6 and 17, it is the very 'elasticity' of many single hulled VLCC's, a direct result from using high tensile steel, which actually guarantees their resilience in rough seas, hence their safety. It is also true that not all shipping companies reduced scantlings immediately, with the result that many of the Supertankers which were built to the mid-1970's remain working ships, although not *so much* true of the later ones, are of a *very* solid construction indeed. Many were in their embryo stages at a time when ship building remained a matter of considerable pride. Most of the names traditionally associated with ship construction in British, United States and European shipyards, although competing for increasing world orders, worked to a high standard. The ships from these yards were solid, often continuing to be made largely from mild steel, with scantlings involving heavy longitudinals, transversals and solid hull components in terms of frames and girders. They were built to last, being capable of withstanding considerable wear and tear and there were few fears that such ships would bend under severe weather conditions or break in half.

Contrary to the belief in some, not surprisingly ill-informed areas, reputable owners *are* in a majority and continue to maintain correctly their single hulled VLCC's. They cover the costs of maintenance across a wide range of areas encompassing wear and tear. They continue to pay, for example, to have enhanced ballast tank coatings thus combating the corrosion which causes metal fatigue. The effect of this increases with a ship's age. In terms of steel replacement, this can amount up to and even exceed 1,000 tonnes in some cases, especially for metal fatigue in the area of longitudinals at the ship's side which is a particularly vulnerable region as VLCC's grow older. Pipes and valves continue to require modernisation and maintenance, and all of these ships have had to be retro-fitted as new IMO regulations have been introduced and new ideas implemented. This was not the least concerning IGS and COW and, as it was shown in Part Four, the innovations that have been achieved by computerisation. Also, as Rob Drysdale, a Superintendent with a leading London tanker company, pointed out to me:

> "Corrosion is not the only cause for material failure in ships — both old and new — for such failure can be dependant also on the thickness of the material. Fatigue in the form of cyclical bending of the vessel over many years can also lead to serious fractures especially in the thinner 'high tensile' steel locations and where the detail design of structure (changes in section) is less than adequate."

19.4. REGULATIONS APPLIED TO THE OLDER SHIP

One of the oldest ships to continue trading is the VLCC. "STENA CONVOY", (Photo. 19/1), a ship of 262,630 sdwt that is one of the two VLCC's built in 1972, to remain in service: even though both are due soon to be scrapped. As a direct result of considerable owner investment in the kind of maintenance and support mentioned above, this Supertanker, in 1996, was granted an American Bureau of Shipping 'Certificate of Condition Assessment',

Supertankers Anatomy and Operation

19.1. The VLCC "STENA CONVOY" which was built in 1972 and in 1996 was Awarded a Certificate of Condition Assessment by ABS Marine Services Inc. (By kind permission of Concordia Maritime AB.)

19.2. VLCC "STENA CONGRESS"-Another Example of a Highly Maintained Early Generation Supertanker. (By kind permission of Concordia Maritime AB.)

Project ID No. HO56420

Certificate No. MST-HO564206

Date: 22 March 1996

Office: New York

CERTIFICATE OF CONDITION ASSESSMENT

This is to Certify that a condition assessment has been made of the vessel "STENA CONVOY" ABS I.D. 7217624 at the request of Universe Tankships, Inc.

CONDITION ASSESSMENT:

This certification is based on the recent condition assessment survey of the hull and machinery, Report No. CZ7019, dated 17 December 1995 which was performed in accordance with the ABS "Guide for SafeHull Condition Assessment - Tankers," January 1994.

Additionally, the vessel structure was evaluated using the criteria in ABS "Guide for Dynamic Based Design and Evaluation of Tanker Structures," September 1993 and the ABS "Guide for Fatigue Strength Assessment of Tankers," September 1993 as identified in Report No. H054206, dated 11 March 1996.

Based on the condition assessment survey and structural evaluation the vessel has been assigned a Grade of 1.

James J. Gaughan
SHCA Coordinator

This certificate is granted or issued subject to the condition that it is understood and agreed that nothing contained herein shall be deemed to relieve any designer, manufacturer, seller, supplier, repairer or operator of any warranty, express or implied and MSInc. liability shall be limited to the acts and omissions of its employees, agents, and subcontractors. Under no circumstances whatsoever shall MSInc. be liable for any injury or damage to any person or property occurring by reason of negligent operation, misuse, or any defect in materials, machinery, equipment or other items other than defects in items actually inspected by MSInc. and ascertainable by normally accepted testing standards, or defects reflected in documents reviewed by MSInc. and ascertainable by normally accepted testing standards, or defects reflected in documents reviewed by MSInc. and which are covered by this certificate or report.

*Cert 19.1 The ABS Marine Services Incorporated "Certificate of Condition Assessment awarded to the VLCC "STENA CONVOY".
(By kind permission of Concordia Maritime AB.)*

(Cert. 19/1), with an accompanying report that awarded this ship a grade one designation covering her hull, machinery and structural members. As recently as 1998, this same ship was granted a Bureau Green Award Certificate. This is a solid achievement in standards for such an early generation VLCC. Incidentally, all Supertankers in the Concordia fleet, which includes ULCC's exceeding 457,000 sdwt, and a number of VLCC's, continue to be maintained to a very high standard and to pass statutory Classification Society fourth and fifth surveys.

The following comments were taken from a recent Concordia Annual Report and covered a tank inspection of the VLCC "STENA CONGRESS" (Photo. 19/2), of 273,00 sdwt built in 1974:-

> *"Down in No. 1 port wing tank, the Classification Society's inspectors comment that the structure is in excellent condition and that the tank is well cleaned. The first measurements show that the decrease in plate thickness is either non-existent, negligible or sometimes even negative compared with the new building's dimensions. The reason for the last mentioned is due to some over-dimensioning in the plate production process compared with the design.*
>
> *The tanks are inspected and measured one after the other. When all the empty tanks have been completed, they are partially filled with water at the same time as the water level in the previously filled tanks is lowered. This is carried out in stages so that four levels in all the tanks gradually become available: below deck, upper cross ties, lower cross ties, and bottom, and vice versa. In the tanks filled with water, the inspection and measurements are carried out with the help of inflatable rubber rafts ... What the inspectors are looking for are e.g. cracks caused by fatigue, a decrease in dimensions due to corrosion, or corrosion in the form of so called pittings. All ships suffer from these problems in some form. The difference lies in how the problems are handled and, as a consequence, how large they become."*

It should be stressed that the enhanced fourth, and any subsequent special survey inspection for older VLCC's is particularly stringent. Much of the examination is carried out in ballast condition at sea, which is then followed by as much as a twelve day dry docking. These surveys are therefore expensive. This is not only in terms of taking the VLCC out of service for the duration of the docking, and the preparatory labour involved, but also to cover the additional work that is inevitably necessary to bring a very large ship up to required standards. They are carried out by Classification Societies and include a very thorough examination of the ship's structural condition, with particular emphasis on plate thickness, measured from 15,000 points, which is virtually the entire ship. A moment's reflection indicates the enormity of this task. A close inspection would be made also of all machinery, equipment, systems and fittings.

Concordia Marine's standards are undoubtedly high, but it must be stated that high standards themselves are *not* unique. There is much evidence which confirms that *all* reputable owners strive to maintain to a high standard their single-skinned VLCC's. It is pleasing to report that many of the other not so well managed Supertankers are increasingly being identified by spot inspections, and the mandatory surveys just mentioned, which are leading frequently to their eventual scrapping. It is these sub-standard ships that can find themselves spending anything up to sixty days in dry dock thus incurring for their owners the enormous expenses involved. The world is left, therefore, with a fleet of single hulled Supertankers that is, by and large, structurally sound, well maintained and quite safe.

19.5. HYDROSTATICALLY BALANCED LOADING (H.B.L.)

Amongst alternatives considered by the shipping industry to Regulation 13G, under the umbrella of 'other arrangements' mentioned earlier, has been the possibility of Hydrostatically

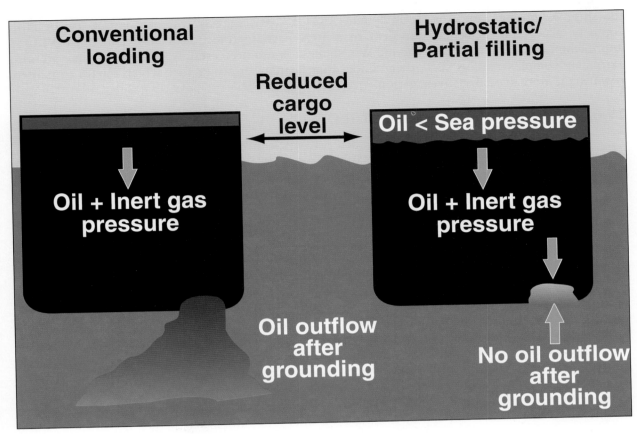

Diag. 19.1 The Principle of HBL.
(By kind permission of Paul Doughty and Colin Sowman of "The Motor Ship")

Balanced Loading, (HBL), that was approved by IMO in 1994. The success of HBL is based on the principle that when cargo tanks are partially loaded below the 98% ullage, to a level that is consistent with the capacity of the ship and the specific gravity of the oil, and this is combined with an increased pressure of inert gas then, in the event of a cargo tank being breached, the internal pressure would be less than the sea pressure outside. The sea water would tend to come into the tank rather than the crude oil escape into the sea. (Diagram 19/1). In rough seas, however, or if the VLCC happens to ground on an ebbing tide, there could still be an escape of crude oil.

The success of HBL, therefore, depends, in general terms, on the differences in density between the mixture of oil, itself lighter than water, of course and that of sea water, at the ship's draught. Clearly, once a tank or set of tanks has been designated for HBL use, then this condition would exist for all future cargo loadings. There are one or two arguments, however, which express concern regarding the efficacy of the system. The major one involves the obvious loss of between 5 to 10%, some authorities suggest up to 13 or even 15%, pay load to the shipowner. In practical terms, this could work out from around 14,000 to 42,000 tonnes on a 280,000 sdwt cargo at extreme assessments, a tab that realistically no one would be happy to pick up. It might be suggested however that, in the majority of cases, actual deadweight loss under HBL could still be much less than alternative options such as converting wing cargo tanks to SBT or to have them remaining void. Incidentally, the VLCC "STENA CONVOY", was one of the first Supertankers to be certificated for hydrostatically balanced loading towards the end of 1997.

Other controversial views, regarding HBL, consider points for and against potential 'sloshing' effects that such a reduced payload could have within the tanks upon the stability of the ship. Additional concerns, related to MARPOL 13G, suggest whether or not use of HBL could be acceptable as being equivalent to a 30% partial protection location, in those VLCC's approaching their 25 year plus survey. Whilst these issues are certainly interesting, they are also extremely complex and, as such, come under that heading in my Preface which relates to matters requiring a more specialist investigation.

19.6. DOUBLE HULLED NEWBUILDINGS

Certainly, it seems as if single hulled VLCC's can continue legal trading until the turn of the century and possibly, in the case of the post 1985 ships, until the year 2015. By then the world's fleet of trading Supertankers will have to be double hulled even if other proposals attract approval from the International Association of Classification Societies. New orders continue to be placed. These totalled 11 completed in 1997, 21 in 1998 with more due for delivery during 2000/1. VLCC orders are increasing already for later in the 21st Century. Trying to forecast in mid-2000 the demands of a freight market for 2000+ however, and the extent to which an extra 350 or so Supertankers will be required to necessitate a one-to-one replacement by 2015, has to be a matter of conjecture bordering on guess work.

One or two facts do emerge. Once the present class have been scrapped, it appears that the days of the 400,000 and 500,000 sdwt VLCC will be over. Since writing this in late 1999 however, this forcast has been firmly squashed by news, in February 2000, that two 450,000 sdwt VLCC's have been ordered from a Korean shipyard, with an option placed for a further two Supertankers of similar size. The tendency generally for new building is to order VLCC's that are of a maximum tonnage of around 300,000 sdwt. There may also have to be some stages of re-thinking in instances where the double hulled VLCC does not always match up with the forecast ideal. One or two weaknesses appear to have been discovered in the new buildings that have had to be confronted and remedied. Other failings may also occur as the newer ships come into service, but I have no doubt that these too will be resolved as experience confronts weakness. The nature of the defects seems to cover a wide field and mainly concerns constructional problems. Some operational difficulties have also come to light many of which, some would argue, could be the result of insufficiently developed regulations governing the construction of these newer very large ships.

19.7. TEMPERATURE AND ITS EFFECT ON CORROSION

There have certainly been problems experienced in the cargo tanks of some newer VLCC's concerning bacterial growth. In the single-skinned Supertanker, there was perhaps 17 mm or so thickness of the shell plating which was next to the cooler sea water, when the ship was on her loaded passage, thus cooling the cargo. This was used to advantage, of course, with pre-COW tankers when in ballast because it permitted the warmer waters south of the Canaries to facilitate tank cleaning. The waters paradoxically also had the effect of keeping down the temperature in the cargo tanks and maintaining this at a level which discouraged marine growths. The two-metre insulation of the cargo tank from the cooler sea has meant a quite considerable increase in temperature within the cargo tanks. The proliferation of marine growths (many of which were portrayed in Chapter 3) have found the warm moist internal tanks to be an ideal breeding ground, with an abundance of food in the form of steel, thus leading to corrosive effects. This required the necessity of having protective coatings, often in the form of modified epoxy, applied to the lower three metres or so of the cargo tanks, and has proved an unexpected expense. In some instances, the top metre or so of the cargo tank has also been coated in order to afford protection from detrimental condensation effects due to the 2% ullage.

19.8. COATINGS, COLOUR AND THICKNESS

It was shown, also in Chapter 3, that an application of paint acts as a defence mechanism for the steel structure of the ship. This means that not only cargo tanks, on the new VLCC's, but also ballast tank coatings continue to have the same significance as they had with single skinned ships. Again, correct initial preparation of the steel is extremely important so that a firm base is created before the coatings are applied. Although it is not currently mandatory under IMO Regulations, it is becoming a matter of industrial practice now, in many but not all shipyards, for the coal tar epoxy coatings to be replaced with lighter coloured epoxies. David Penny, a marine engineering superintendent with a large London based shipping company involved with new building double hulled tankers, emphasised how important it is for ship owners to insist on the application of lighter colours:

"The failure to insist on minimum coating specifications will be defended on the basis of not increasing owner's costs on a matter which is not, strictly speaking, safety related. I would argue that ease of inspecting these spaces is safety related and dark coatings (which make leakage's from small defects in welds/parent material, pinholes, hairline fractures, that much more difficult to locate) make inspections slower with that much more time spent in tanks. This is not a task most people at sea (or anywhere else) relish and, if the job is difficult, human nature usually means that it will not always be completed properly. The bottom line means that leaks of a small nature go undetected and uncorrected with the resultant potential danger that this brings."

There exists a consideration, therefore, regarding the number and thickness of coatings. It seems quite apparent that should there be increased numbers of coats, in ballast tank areas, or even a sizeable increase in the thickness of the initial coating, that a considerable decrease in corrosive effects will occur. Thus the probability of steel replacement as the newer ships continue in service could be cut considerably so that an initial increase in outlay might well lead to a substantial saving. This leads to a reinforcement of the plea, made in Chapter 3, that representatives of paint manufacturer's should be included as early as possible in the initial stages of a ship's construction. It is not known for certain, at this stage, the extent of steel replacement in these very large newer ships, but some forecasts seem to indicate that these could, in fact, be more excessive than has proved the case for single hulled VLCC's.

19.9. INSPECTION AND VENTILATION IN DOUBLE HULLED TANKERS

Consistent with the inspection of tank coatings is the importance of access to ballast tanks. A glance at the general capacity and arrangement plans of the double hulled Supertankers (inside back cover pocket) indicates that there is considerably more area to be inspected in the ballast space of a new VLCC than was even imagined with the single skinned ship. This implies that, from a safety point of view for the ship's officers and other inspectors, a ready and easy access is vital. The International Association of Classification Societies, (IACS) — (See Appendix) have drawn up a list of recommendations directly concerning the layout of ladders and catwalks within these spaces to enable inspections to be made 'in a safe and practical way'. David Penny has made the following comment:

"The recommendation is for a fore and aft catwalk along the lower part of the side tank, with (in the area of the 'lower hopper') inclined ladders leading down into each bay (between the transverse frames) in that part of the tank under the cargo tank. This, together with access ladders from the main deck at each end of the double hull ballast tank, gives good and easy (ie: fast) access to and from all parts of the tank."

The question of ventilation is also one of vital importance, due to the potential for gas accumulation. The fitting of a purge pipe is one legal requirement with the attachment of gas exhaust fans, perhaps of the type examined when considering tank cleaning procedures in Chapter 8. This would have the effect of circulating fresh air throughout the entire double bottom and allowing this to permeate the side ballast tank areas. Similarly, gas detection is vitally important. The possibility of a gas leak from a cargo tank into the ballast tanks is ever present. It has not happened to date, but that is not to confirm any belief that it will never occur and, as a result, it is essential that there are a number of gas sampling points, not just one, situated throughout the tank areas. There could be a case, from a safety point of view alone, for having the double hull ballast spaces filled with inert gas when they are not actually carrying ballast water.

19.10. THE DOUBLE HULL AND POTENTIAL OPERATIONAL DIFFICULTIES

There are a number of potential difficulties which could arise from a purely operational point of view with the new breed of Supertanker. Safety of life, collision avoidance and the prevention of stranding are the main objectives underwriting all duties of a navigating officer. Indeed, as was seen in Chapter 10 specifically, and has been stressed throughout every comment in this book, *all* officer training from initial cadet or trainee level to that of serving Master is designed purely with these aspects in mind. Externally, traffic separation schemes and restricted channels for all very large ships hampered in their ability to manoeuvre, with rigid VTS applied in confined harbour waters, are equally as important measures concerned directly with safety. The idea of the double hulled tanker was an extension to these measures, which arose directly from the "EXXON VALDEZ", as examined in Chapter 14. The intention is to minimise the leakage of crude oil in the event of tank broaching due to either a collision or stranding. HBL seems to be an efficiently working alternative on single hulled tankers.

The double hulled VLCC which runs aground will almost certainly have quite a marked reduction in the outflow of crude oil from her cargo tanks. This will depend, clearly, on the force and/or speed of impact and the depth to which the obstruction penetrates. It is highly probable however that some of the cargo will escape into the sea. This means that an element of environmental risk remains with its attendant pollution and the consequences. It also means that the Supertanker herself will continue to be surrounded by hydrocarbons with the dangers, of fire and explosion, and all that this implies to those on board and to those assisting. Clearly, the size of the cargo tanks would affect the amount of outspill. A reduction in length and beam of each wing and centre tank will mean that much less oil will escape depending, again, upon the point of impact in relation to the waterline. The fitting into double hulled VLCC's of even more cargo tanks, as seen in the seventeen sets comprising the deadweight capacity of the ULCC "JAHRE VIKING" for instance, although costly, would obviously assist in preventing less spillage.

The case concerning a collision, although equally as undesirable, may have wider and more important consequences. Dave Penny makes another shrewd comment that is worth taking on board:

"In a high energy impact, such as a collision, the double hull may well be a mixed blessing. A case in which a single hull tanker was rammed by another vessel at approximately ten knots resulted in a deep penetration of the moving vessel into the loaded tanker. A large volume of oil was released suddenly. However, the moving vessel penetrated the tanker easily and just as easily parted company and slid down the side of the tanker. The heat energy created as a result of the impact was to some extent dissipated into the large volume of oil released, and the initial impact and subsequent detachment was minimised due to the 'soft' nature of the construction of the single hull vessel. Had the tanker been a double hull vessel, with a 'harder' or stronger construction in way of the impact, then it is most likely that more heat would have been generated at the time of impact and that, given the force of contact, the inner hull or cargo tank full of oil, would still have been breached. However, the release of oil would most likely have been slower, allowing a more dangerous atmosphere to build. During the detachment of the moving vessel there would have been greater heat energy generated, due to the greater friction caused, because the moving vessel would not have become detached so easily."

The lesson is worth bearing in mind. One, of course, fervently hopes that such an incident will never happen, but the scenario so vividly painted by Mr. Penny, could well be maximised if it does occur, and it transpires that a double hulled VLCC is involved, with the potentially greater risk of explosion and pollution.

The computerised cargo loading software, examined in the previous chapter, clearly has relevant application to damage control assessment, but there would appear to exist two major difficulties that have been confronted regarding its use in practical terms actually working a ship involved as a casualty at sea. Captain Geoff Cowap succinctly points out the nature of the problems:

"Some consider that the instrument should be capable of remodelling the properties of the vessel in the damaged condition, others consider that the on board instrument can only be of use as a good estimate of how to counteract the effect of relatively minor damage.

The first concept is indeed technically possible. Given modern computing power, the software would be capable of calculating a new hull form from the ship's lines and removing any buoyancy from the hull form in the region of the damaged area or areas. The program could rebuild all of the hydrostatic information of the vessel from the new hull form and present results to the operator relating to the vessels draught, trim, heel, residual buoyancy and stability, and show the calculated results of shearing forces and bending moments along the length of the vessel. This sounds like a wonderful concept, giving a wealth of information to the operator who in theory could then make immediate decisions on the likelihood of the vessel remaining afloat, or otherwise. The operator would also be in a position to try alternative flooding actions to counteract the damage.

The arguments against this approach are numerous. The results of the new hull form depend on the accuracy of the input to the program in relating the extent of the damage. It is extremely difficult to assess the extent of the damage from within the ship. The limited numbers of personnel on board the vessel which has just suffered major damage would restrict the amount of time any one person could allocate to using a computer when the ship is sinking around him and the mental state of those on board would not be conducive to computer operation.

Such a program would be ideal in the comfort of a shore based office with a fully trained emergency response team at hand, but not on board the vessel.

The second concept is much more realistic for shipboard use. The software would simulate a minor damage such as a hull pierced, by automatically calculating the amount of water the damaged compartment would take and absorption into the cargo. The flooding would take place up to the water line and the program would assimilate the behaviour of the vessel until a point of equilibrium would be reached. The concept is based upon limited damage which can be quickly identified."

19.11. POSSIBLE SOLUTIONS TO DAMAGE CONTROL, HULL STRESSES

It might, perhaps, be tempting to look at Sanderson CBT's virtual reality program of Chapter 17 in order to suggest a possible solution to damage identification and control. This could be a particularly seductive thought, particularly as one of the scenarios considered actions that might be undertaken to deal with an underwater hull explosion. The program however, certainly in its present form, is designed as a solid training aid for personnel reaction. The kind of approach advocated by Captain Cowap would need to be in a diagnostic format more adaptable for practical damage identification in terms of hull stress and stability.

It was shown in Chapter 14 that, by definition, a double hulled VLCC has four longitudinal bulkheads. Undoubtedly, this gives a considerably stronger structure. It means also that the ship is that much 'stiffer' in terms of being more rigid and will not, therefore, display the same reliance which has been seen to be so much a strength of the single hulled VLCC, particularly in terms of handling under heavy weather conditions. As Chapter 6 highlighted, the navigating officer on the older type of Supertanker could literally both see and feel the deck moving under his feet and so gain an assessment of the pounding and slamming effect, hence he was able to reduce speed or even alter course moderately in order to reduce stress on the hull. On the double hulled VLCC, he would have to rely far more on using sensors of the type discussed in Chapter 17. This is not such a problem, but the fact that a 'soft' single hulled forepeak and forecastle head is attached to the 'rigid' double hulled body of the ship, could possibly lead to a weakening in structure when the ship is pounding into heavy seas. Again, there is insufficient evidence for much of a value-judgement to be made, but it may well prove to be an area which might require further investigation as these ships themselves start to age. At the stern of the ship, all engine rooms are already fitted with double bottoms for additional strength even if this does not extend throughout the entire stern section. As Dave Penny observes:

> *"Problems have been experienced in the pump room area because of flexing from loaded to ballast to loaded conditions. This has shown itself in the changing alignment of cargo pumps due to the degree of relative movement between turbine and pump when going from ballast to loaded condition and vice versa. On VLCC's, this has already been sufficient to require re-alignment of pumps."*

The situation would seem to require the extension of the double bottom across the entire stern section in order to provide a constant level of hull support in this highly vulnerable area.

19.12. CONCLUSION

There is not much indication at the moment of the likely life of the newer VLCC's, but I have heard figures mentioned around the twenty year mark. This figure was, of course, voiced about the single hulled Supertankers, some of which are well on the way to their twenty fifth or thirtieth year of service. It seems to be a matter, once again, of waiting to see how the ships stand up to expectations in the light of the conditions they experience. Certainly, all of the newer VLCC's have modern supporting "state of the art" computerised equipment, beyond the legal requirements, justifying safe operation. Most are indeed fitted with an adequate number of sensors for both gas detection in all tanks, and hull stress monitoring, particularly of slamming stresses when in ballast. Planned and comprehensive maintenance programmes are implemented and, with the ships of these highly reputable owners, there remain few fears. There is always, however, the possibility that some of the problems discussed in this chapter might develop, and perhaps also, a number of unforeseen difficulties that may appear as these new VLCC's age. This could be particularly true if and when they might be sold to 'other owners' at some later date in the 21st Century?

APPENDIX

ORGANISATIONS MENTIONED IN THE TEXT ASSOCIATED WITH SUPERTANKERS

The Honourable Company of Master Mariners.
The International Association of Classification Societies.
The International Chamber of Shipping
The International Maritime Organization.
The International Association of Independent Tanker Owners.
The International Tanker Owners Pollution Federation.
The Marine Society.
The Nautical Institute.
The National Union of Marine, Aviation and Shipping Transport Officers.
The Oil Companies International Marine Forum.
The Royal Institute of Navigation.
Tanker Structure Co-operative Forum.

THE HONOURABLE COMPANY OF MASTER MARINERS.

The Honourable Company of Master Mariners was founded in 1926 and they received their Royal Charter in 1930 and Grant of Livery in 1932. They were the first new City of London Livery Company to be created for more than 200 years. Their Headquarters are situated appropriately on board the ex-RN 1934-built sloop, "WELLLINGTON", that is moored at Temple Stairs on the River Thames. This ship was purchased from the Admiralty, in 1947, following her service on the China and New Zealand stations and as an escort ship on the North Atlantic convoys during the Second World War.

The Charter, recently amended, allows holders of current Class 1 Certificates of Competency, traditionally known as Master (Foreign Going) to join the Company, as well as Royal Naval Officers with appropriate Seaman Branch qualifications and Command experience. The Honourable Company's function is, as far as practicable, *"to promote and protect the interests of the British Merchant Navy and its seafarers"*. The Company is represented on the Parliamentary Maritime Group and other bodies which concern or affect the Merchant Navy, providing professional advice and counsel, including Nautical Assessors to the Court of Appeal.

Additionally, the Company administers various charitable funds, is represented on the boards of governors of various schools, and has an active Apprenticeship Scheme for those pursuing a career in the Merchant Navy.

THE INTERNATIONAL ASSOCIATION OF CLASSIFICATION SOCIETIES – IACS

IACS, with its eleven Members and two Associates, classifies over 90% of the world's merchant tonnage. The Association was formed in 1968 and has held consultative status with IMO since 1969. With its unique technical knowledge of the world's fleet, it is the only non-governmental organisation whose representatives participate as observers and technical advisers at Assembly, committee, sub-committee and working group meetings which is able to develop rules for ship structures and essential engineering and electrical systems. IACS was formed originally to promote the highest possible standards in ship safety and to prevent pollution and, in order to meet the considerable challenges involved in implementation, has since taken the initiative to develop standards which cover virtually every major advance in ship technology. Because the Association performs an integral role in IMO, its Members make also a vital contribution to every aspect of research and development.

"At the heart of ship safety, classification embodies the technical rules, regulations, standards, guidelines and associated surveys and inspections covering design, construction and through-life compliance of a ship's structure and essential engineering and electrical systems. Ships are built under survey in accordance with the technical rules, and thereafter through-life compliance is effected by further surveys on an annual and five yearly basis. The standards are extremely stringent, particularly those applied to the older ship. Failure to meet the relevant standards, or non-compliance with recommendations issued as a result of a classification survey may result in disclassing-which is the withdrawal of class certification for any specific ship". This penalty, as examined in Chapter Nineteen, spells the 'death knell' for such ships operated by offending owners. "Whilst these aspects are not addressed in detail in the various IMO international conventions the latter, nevertheless, require certification that these elements are in all respects satisfactory for the service for which the ship is intended." Compliance with a classification societies rules is accepted by the international community as the only recognised technical basis. This is recognised by SOLAS Regulation, 11-1/3, which entered into force on 1st July 1998. This states as follows:

'In addition to the requirements contained elsewhere in the regulations, ships shall be designed, constructed and maintained in compliance with the structural, mechanical and electrical requirements of a classification society which is recognised by the Administration in accordance with the provisions of regulation X1/1, or with applicable national standards of the Administration which provide an equivalent level of safety'.

In addition to developing and applying their own classification rules for ship structures and essential engineering and electrical systems, IACS Members are authorised by more than 100 IMO member states to undertake statutory international and national/regulation compliance surveys and to issue the necessary certification on their behalf."

INTERNATIONAL CHAMBER OF SHIPPING-ICS

The International Chamber of Shipping (ICS) is the international trade association for merchant ship operators. It represents the collective views of the international industry from different nations, sectors and trades and its membership comprises national shipowners' associations representing over half of the worlds' merchant fleet.

A major focus of the ICS is the International Maritime Organization and it is heavily involved in a wide variety of areas including any technical, legal and operational matters affecting merchant ships. It is a unique organisation in that it represents the global interest of all the different trades in the industry: operators of bulk carriers, tankers, passenger ships and container lines, including shipowners and third party ship managers. ICS has consultative status with a number of inter-governmental organisations which have had an impact on shipping. Its close ties with IMO stretch back to this bodies inception in 1958. Other partners include the World Customs Organisation, the International Telecommunications Union, The United Nations Conference on Trade and Development, and the World Meteorological Organisation. ICS enjoys also close relationships with industrial concerns representing different maritime interests such as shipping, ports, pilotage, the oil industry, insurance and classification societies responsible for the surveying of ships.

The Chamber is committed to the principle of maritime regulation being formulated at an international level. Shipping is by nature international: the regulations that apply to a ship when it sails from Buenos Aires must apply equally when it arrives at Brisbane. The alternative to an international system of shipping legislation would be chaotic web of local rules and regulations that would result in commercial distortions and mass economic deficiencies. The objective of ICS is the maintenance of a sound, well considered global regulatory environment in which well-run ships can operate safely and efficiently.

INTERNATIONAL MARITIME ORGANIZATION – IMO

The International Maritime Organization (IMO) was established originally in 1948 under the auspices of a United Nations Convention. It was then named the Inter-Governmental Maritime Consultative Organisation, (IMCO), and was inaugurated and held its first session, in 1959 as a specialised agency to deal with maritime affairs in order to improve safety at sea. It changed its name to IMO in 1982, by which time it had widened considerably its international function to incorporate a considerable number of 'matters maritime' probably the most important of which was legislation controlling the prevention of pollution, particularly from oil carried by the world's tankers. The function of IMO, in the words of Article 1(a) of the UN Convention is:

"to provide machinery for co-operation among Governments in the field of governmental regulation and practices relating to technical matters of all kinds affecting shipping engaged in international trade; to encourage and facilitate the general adoption of the highest practicable standards in matters concerning maritime safety, efficiency of navigation and prevention and control of marine pollution from ships."

The Organisation consists of an Assembly, a Council and four main Committees, assisted by a number of Sub-Committees.

IMO has concluded Agreements with a large number of Intergovernmental Organisations, including INMARSAT, as well as arranging consultative status with many non-governmental institutions, including IACS, ICS, INTERTANKO, ITOPF and OCIMF, each of which have been mentioned in this book and are examined briefly in this Appendix. They have extended their brief to include also representatives of Greenpeace and Friends of the Earth.

IMO achieves its objectives by promoting the adoption of Conventions and Protocols which have legal binding and hence implementation amongst the (current) 157 member States. *As seen, the term 'IMO Regulations' has appeared consistently throughout this book* and covers such recommended areas as Cargo, Technology, Environmental Issues, Navigation, Life Saving and SAR, Communications, Training and Certification. IMO have instituted also the World Maritime University in 1983 at Malmo and the International Maritime Law Institute in 1989.

THE INTERNATIONAL ASSOCIATION OF INDEPENDENT TANKER OWNERS – INTERTANKO

This Organisation was formed in 1971 and consists of over 270 member companies, in forty-plus maritime countries, operating around 155 million of deadweight tanker and combined tonnage, representing approximately 80% of the world's eligible tonnage, as well as over 220 associate members and subscribers. Full Membership is open to independent tankers owners, owning or managing crude oil, chemical or product tankers. Companies who are not specifically in this category, such as oil companies and state owned companies, shipbrokers and others with a commercial interest in tanker operations and related activities, may become associate members.

The prime goals of INTERTANKO are: *"to promote a free and competitive tanker market; to work for safety at sea and the protection of the environment."* All tankers owned by members must be classified by a classification society that is a full member of the IACS and they must be in 'good standing' with a P&I club.

INTERTANKO is a spokesman for the Independent Tanker Owners and represents the members *vis a vis* in international and governmental organisations. As such, it is represented on all major IMO Committees and Sub-Committees with representatives of member companies taking an active part in INTERTANKO's work in IMO. It acts also as a forum for discussion in tanker related matters. The Association provides, additionally, comprehensive information services and has produced a number of influential publications.

THE INTERNATIONAL TANKER OWNERS POLLUTION FEDERATION – ITOPF

The International Tanker Owners Pollution Federation is a non-profit making organisation, involved in all aspects of preparing for and responding to oil spills from tankers. Funded by the world's tanker owners, it was established in 1968, in the wake of the *TORREY CANYON* incident, to administer a voluntary oil spill compensation agreement, which has since been superseded by two international Conventions and associated Protocols – *the Civil Liability Convention* and *the Fund Convention* – developed under the auspices of the International Maritime Organisation. ITOPF now focuses exclusively on the provision of a wide range of technical services to and on behalf of its membership, their P&I insurers and others prominently involved in oil spill pollution. It regularly contributes to national and international discussion on matters relating to oil pollution and, since 1980, the organisation has had observer status at both IMO and the International Oil Pollution Compensation Funds.

Responding to marine spills is ITOPF's priority service. Its small technical team (consisting of marine biologists, chemists and engineers) is at constant readiness to respond anywhere in the world.

To date, over 350 spills in over 70 countries have required on-site attendance by a member of the ITOPF technical team. Whilst most of these spills were from tankers, the staff have frequently been called upon to respond – on a consultancy basis – to oil spills from other marine sources, such as dry cargo ships and offshore oil installations. Occasionally, advice is given also in relation to chemical spills.

ITOPF's other chief functions include damage assessment, claims analysis, contingency planning, advisory work and training. It is also a comprehensive information source and produce a wide range of technical publications and papers. It maintains various databases as well as a web site at *http://www.itopf.com*, where further information on the organisation may be found.

THE MARINE SOCIETY

The Society was founded in 1756 by Jonas Hanway and *"for over two centuries it has, with the practical and financial support of individuals and organisations, helped young people to go to sea and those at sea."* The Society continues to support mariners and potential seafarers in constructively practical ways, including provision of assistance by distance learning through voluntary tutors with professional and academic examinations – including the organisation and administration of exams at sea. They offer, also, practical advice and assistance with a wide range of leisure pursuits that incorporate language learning and the much appreciated provision of ship's libraries and videos which are changed at regular intervals – as indicated in Chapter 13. Along with occasional publications, they produce an illustrated three-monthly magazine called, *"The Seafarer"*, and arrange annual competitions for a range of seafarers including retired member, wives, and *all* ranks of both the Royal and Merchant navies and the fishing fleets. The Society is extremely active with a range of schools and, through their "Sea-Lines", arranges active partnerships between seafarers and education.

THE NAUTICAL INSTITUTE

1971 saw the Foundation of the Nautical Institute following a gradual feeling of disquiet caused by increasing fears that politically, commercially (and indeed socially), the nautical profession was becoming under valued. Very real concerns were being widely expressed on a number of important issues. The attitudes of some organisations towards investment in nautical training were hardening and becoming increasingly serious. Not the least of these doubts involved the professional education of seafarers, with particular reference to the training of deck and engineering cadets who, after all, represent the major source eventually of quality officers. *(Some results of policies exercised in this area were discussed briefly in Chapter 10)*.

There existed a genuine need for an independent Body which could act as a neutral forum offering a closer negotiating framework to intervene where appropriate regarding industrial disputes. There was a positive requirement for closer links to be forged between a range of organisations representing maritime interests concerned with ship operations: for those in command at sea, and ex-master mariners serving in senior positions ashore. The period was a time also of radical change in the freight markets, especially concerning the tanker trades, advances in containerisation and expansion in passenger and reefer traffic. *"The Industry had became more fragmented and specialised. It regrouped around free flags, changed its crews and became more global in outlook."*

Full Membership of the Nautical Institute today conveys both a professional status on the holder – each of whom have qualified as masters in command – and an equally strong professional credibility towards the Institute as a Corporate Body. The designatory letters *"MNI or FNI"* have increasingly, for nearly three decades now, attracted respect within the Industry and with a range of associated external organisations. Members produce a number of authoritative books, and other publications, which are offered also to non-members, thus providing a wider dissemination of professional knowledge. *Indeed, reference has already been made, in Chapter 10, to an extract from one or their local branch Papers.*

NATIONAL UNION OF MARINE, AVIATION AND SHIPPING TRANSPORT OFFICERS – NUMAST

This Organisation is a highly professional Union representing and protecting, amongst others, the interests of all British merchant navy officers – including cadets and officer trainees – as well as foreign officers who serve aboard British ships. NUMAST are active also in preparing representations to shipowners and various governmental committee's on future employment trends and needs for UK officer. The Union has produced a number of authoritative papers on employment matters, particularly, as well as sponsoring a range of appropriate conferences. They have been extremely active over the years in arranging legal representation and there are many officers who have found themselves in diverse situations who have reason to be grateful for NUMAST's involvement and support.

The Association's 36-paged monthly magazine, 'TELEGRAPH', is recognised as an authoritative voice of the shipping industry. Articles cover all aspects of maritime affairs including training, employment, international and governmental policies, and contain often a number of specialised reports dealing with subjects of direct concern to officers on world-wide issues such as piracy, ship flagging and registry, and other subjects coming under the wider umbrella of the International Transport Federation (ITF).

THE OIL COMPANIES INTERNATIONAL MARINE FORUM (OCIMF)

Improving safety and pollution prevention from oil tankers and terminals; these are the prime objectives of OCIMF, formed in 1970 as a result of increasing concern with the problems of pollution in the marine environment. Its formation reflected a need for the oil industry to make its views known and its professional expertise available to governments and inter-governmental bodies. Current membership of OCIMF numbers over 40 companies and groups all over the world, having an interest in the shipment and storage of crude oil and oil products, including gas and petrochemicals. Virtually every important international oil organisation is included, although membership is voluntary.

OCIMF has consultative status at the International Maritime Organization and is a widely recognised and respected authority on matters affecting the marine transportation and terminalling of petroleum and its by-products. OCIMF's area of expertise includes tanker and terminal operation, navigation, marine engineering, training, tanker and equipment design, fire protection, and the bulk carriage of chemicals and gases. OCIMF undertakes a number of research activities and has produced, independently and jointly with other international organisations, over 50 technical publications which are designed to improve further the safe transportation and handling of oil and its by-products, with consequent benefits to the preservation and protection of the environment.

A major OCIMF initiative to improve tanker quality is the Ship Inspection Report (SIRE) programme, which was introduced in 1993. Under this programme, a pool of technical information about the condition and operation of oil tankers is maintained in a computerised database for use by OCIMF members and certain third parties including governmental agencies.

THE ROYAL INSTITUTE OF NAVIGATION

The objects of the Royal Institute of Navigation are to: "unite into one body those who are concerned in navigation and to further its development". Membership is not restricted to those who practice navigation and it does not serve as an institute of **navigators** although, inevitably, it includes a considerable number of practicing professionals within its membership, as well as anyone in essence who can satisfy the Membership Council of their interest in Navigation. It is concerned with both research and development and encompasses virtually anything to do with the theory and/or practice of the subject. They are particularly open to examining new ideas and products on the subject which are scrutinised by Study Groups, Members of Council and Technical forums, conferences and symposia. As a result, the Royal Institute attracts an extremely wide range of members which include nautical college lecturers, scientists and technologists. The 'Journal of Navigation' is a highly technical quarterly publication that publishes papers presented at their meetings as well, also, as contributions from a wide range of sources.

TANKER STRUCTURE CO-OPERATIVE FORUM

The Forum does not exist as a specific institution, but was founded as a consultative body in the 1970's at a time when VLCC's were rapidly increasing in number and tonnage. It consists of an amalgamation of interested parties, including Classification Societies and independent ship owners with interests in the oil industry, each of whom is concerned with awareness of optimising safety at sea, and the prevention of marine pollution from oil tankers. Meetings are convened to share experiences and discuss solutions on a par-industry *'ad hoc'* basis.

The Forum has published two major inspection guidance manuals: one concerning condition assessment of tanker structures, and the other on the maintenance of double hull tanker structures. The manuals, as the Preface to the latter explains: *"are not to impose standards ... but to give guidance in a form which may be adapted to the particular needs of individual organisations. It should be noted that, in some cases, the practices of individual members of the Forum will differ on specific aspects"*.

BIBLIOGRAPHY

Sea Transport: Operation and Economics
Alderton, Captain R.M. – Thomas Reed Ltd 1995

Practical Ship – Handling
Armstrong, Captain M.C. – Brown, Son and Ferguson 1994

Guidelines for Corrosion Protection of Ships
Askheim, N.E. and Others – Det Norske Veritas AS 1994 Revsd

Tanker Handbook for Deck Officers
Baptist, Captain C.-Brown, Son and Ferguson 2000

Tanker Shipping
BES, J. – Barker and Howard 1963

Recommended Practice for the Protection and Painting of Ships
British Ship Research Association – BSRA and CS 1972

The Human Element in Shipping Casualties
Bryant, D. – HMSO 1991

The Tonnage Measurement of Ships
Corkhill, M. – Fairplay Publications 1980

The Mariner's Handbook: NP100
de C. Scott Lt. Cdr. C.J. – HMSO 1979

Running Costs
Downard, J.M. – Fairplay Publications 1981

Bitten by the Officer Gap
Everard, M. – Lloyd's Press 01/97

Ship Construction
Eyres, D.J. – Butterworth-Heinemann 1994

Abbrev: Study of UK's Requirements for Ex-Seafarers'
Gardner, B.M. and Pettit, Dr. S.J. – University of Wales, Cardiff 1996

Double Hull Tankers
Gavin, A.G. – Lloyd's Register 1995

Is 25 Years Really the End of the Road for Tankers?
Gavin, A.G. – Lloyd's Register 1996

Software Advances in Aided Inertial Navigation Systems
Gelb, A. and Sutherland, A.A. Jnr – Analytical Sciences Corp. 1970

Tankcleaning for Dry Dock Work
Grieves, Captain T. W. – ITS Publishing AS 1990

Navigation for Masters
House, D.J. – Witherby and Co 1998

Reduction of Oil Outflows at Collision and Groundings ...
Hysing, T and Torset, O.P. – Det Norske Veritas AB 1993

Bridge Procedures Guide
ICS – Witherby and Co 1998

Guide to Helicopter/Ship Operations
ICS – Witherby and Co 1989

Safety in Oil Tankers
ICS – Witherby and Co 1978

Ship to Ship Transfer Guide Petroleum
ICS OCIMF – Witherby and Co 1997

International Safety Guide for Oil Tankers and Terminals
ICS, OCIMF and IAPH – Witherby and Co 1996

Clean Sea Guide for Oil Tankers
ICS/OCIMF – Witherby and Co 1994

Prevention of Oil Spillages Through Cargo Pumproom Sea Valves
ICS/OCIMF – Witherby and Co 1991

Crude Oil Washing Systems
IMO 1981

International Conference on Load Lines, 1966
IMO 1981

MARPOL 73/78 Consolidated Edition
IMO 1992

STCW 1995
IMO 1995

Guide to Contingency Planning for Oil Spills on Water
IPIECA 1991

Shell Centre Response to Ship Casualties and Oil Spills
Irvine, Captain J.M.-Shell IT and S 1994

Hull Damage in Large Ships
Janzen, S. and Nilsson, O. – Lloyd's of London Press 1972-3

Notes on Cargo Work
Kemp, Professor J. and Young, Captain P. – Butterworth-Heinemann 1996

Ship Stability Notes and Examples
Kemp, Professor J. and Young, Captain P. – Butterworth-Heinemann 1998

Ship Construction Sketches and Notes
Kemp, Professor J. and Young, Captain P. – Butterworth-Heinemann 1997

Seamanship Notes
Kemp, Professor J. and Young, Captain P. – Butterworth-Heinemann 1996

Tanker Practice
King, Captain G.A.B. – Maritime Press Ltd 1965

Discussion on Crude Oil Washing
Lloyd's Register Technical Association – Lloyd's Register 1981

The True Relationship Between Age and Quality in VLCC's
Magelssen, W. – Det Norske Veritas AB 1993

Double Hull – A Political Reality – For and Against
Magelssen, W. – Det Norske Veritas AB 1996

Control of Pollution by Noxious Substances in Bulk
Magill, C.M. – Lloyd's of London Press 1987

Tanker Operations: A Handbook for the Ship's Officer
Marton, G.S. – Cornell Maritime Press 1992

Guidelines for Structural Surveys (Tankers)
Mobil Shipping and Transportation Company – Mobil Ship Management 1996

Prevention of Marine Pollution
Mobil Shipping and Transportation Company – Mobil Ship Management Post-1990

The New Approach to Longitudinal Strength
Murray, MBE., J.M. – Lloyd's of London Press 1961-2

A Seafaring Nation?
NUMAST 1993

Officer Requirements on UK-Flag Ships
NUMAST 1994

Only Human? The Human Element in Safe Shipping
NUMAST Post-1993

Anchoring Systems and Procedures for Large Tankers
OCIMF – Witherby and Co 1982

Disabled Tankers: Report of Studies on Ship Drift and Towage
OCIMF – Witherby and Co 1981

Drift Characteristics of 50,000 to 70,000 DWT Tankers
OCIMF – Witherby and Co 1982

Effective Mooring
OCIMF – Witherby and Co 1989

Hawser Guidelines
OCIMF – Witherby and Co 1987

Inert Flue Gas Safety Guide
OCIMF – ICS Witherby and Co 1978

Mooring Equipment Guidelines
OCIMF – Witherby and Co 1997

Prediction of Wind and Current Loads on VLCC's
OCIMF – Witherby and Co 1994

Recommendations for Equipment Employed in the Mooring of Ships at SPM
OCIMF – Witherby and Co 1993

Recommendations for Oil Tanker Manifolds and Associated Equipment
OCIMF – Witherby and Co 1991

Inerts, Gas and Venting System
Oxford, B.W. – Lloyd's Press 1987

Oil Cargo Losses due to Emission of Cargo Vapours
Riksheim, J.B. and Magelssen, W. – Det Norske Veritas 1990

Piloting VLCC's Through the Straits
Knowles, Captain A.V. – Nautical Institute 1979

Passage Planning Guidelines
Salmon, Captain D.R. – Witherby and Co 1993

Effective Mooring
Shell International Shipping and Transport-Shell International Services 1978

Condition Evaluation and Maintenance of Tanker Structures
Tanker Structure Co-operative Forum – Witherby and Co 1992

Guidance Manual for Tanker Structures
Tanker Structure Co-operative Forum – Witherby and Co 1997

Guidelines for Inspection and Maintenance of Double Hull Tanker Structures
Tanker Structure Co-operative Forum – Witherby and Co 1995

Cargo Work: The Care, Handling and Carriage of Cargoes

Taylor, Captain L.G. — Brown, Son and Ferguson 1992

Merchant Ship Construction
Taylor, D.A. — Butterworths 1993

Tankers: An Introduction to the Transport of Oil by Sea
Valois, Captain Philippe — Witherby and Co 1997

Journals of the Company
Various — HCMM

The MOTOR SHIP Magazine
Various — Reed Business Publishing

Guidelines for the Purchasing and Testing of SPM Hawsers
OCIMF — Witherby & Co 2000

Inspection, Repair and Maintenance of Ship Structures
Caridis, Piero — Witherby & Co 2001

INDEX

ff, indicates – and following pages

A.G. Gavin, Principal Lloyd's Surveyor Quoted 404
A.P. Moller Shipping Company 457, 552
Able Seaman 304, 368
Accommodation – Construction 44, 112, 420
Accommodation – Layout and Planning Officers/Crew 384ff
Acomarit Company 377
Admiralty Cast Type Anchor 90
Admiralty Chart System (ARCS) 453
Admiralty Notices to Mariners 296, 314, 458
Admiralty Publications Computerised 458
Admiralty Sailing Directions-Pilot Books 185, 314
Advanced Ship Entry Form of WNI Oceanroutes 510
Advantages of Using Electronic Charts 458
AFRA – Average Freight Rate Assessment Panel xii, xvii, 198
AFRA DEFINITIONS OF VLCC's xii, 557
Afterpeak Tank 35, 165, 191
After-Peak Tank Inspections 284ff
Ageing of VLCC's 558
Agnew, Captain "Tommy" Cited 290
AKD Stopper 346
Alcohol and Drug Policies 387
Aldis Signalling Lamp 372
Alternative Route Comparisons of WNI Oceanroutes' ORION 511
Aluminium Harness Stretchers 280
American Bureau of Shipping (ABS) 12, 562
American Petroleum Institute – API 171
Analogue Speed Indicators 482
Anchor – Admiralty Cast Type 90
Anchor – Admiralty Solid Stockless Type 90
Anchor – Grapnel Type 346
Anchor – Hall's Patent Type 90
Anchor – Marking 334
Anchor – Spare on Foredeck 90
Anchor – Stoppers 93
Anchor Cable 90ff, 124
Anchor Weights 92
Anchorage's – Designated 333
Anchored VLCC and Dry Cargo Ship Collision 335
Anchoring-Swinging Circle 336
Anchoring Operations 333ff
Annual Report 1996 of Concordia Maritime AB 562
Anodes in Ballast Tanks, 72, 423
Application of On Line Process Computers on Ships 440, 445ff
Approximate Dimensions of a VLCC 5, 159, 428
Arabian Gulf 114, 127, 189, 193, 282, 383, 474
Arc of Visibility over VLCC's Bow 323
Archimedes' Principle 160
Ariake Works of Hitachi Zosen Corporation 409, 414
AROSA – Double-Hulled VLCC Construction Stages 409ff

AROSA – Unique Construction Features in Ballast Tanks 416
ARPA RADAR 304ff, 445, 450/1, 512, 515
Article – Practical Navigation in the Dover Strait 317
Articles – Legally signed-on ship xiii
Atmospheric Affects on DGPS 458
Author's Caution Regarding Electronic Chart Debate 458
Author's Experience of Storm damage at sea 507
Automatic Fire Alarm System 484
Average Day's Run 127, 313

Bachelor of Science (Nautical) degree 364
Back Pressure during Cargo Discharge 202
Bacterial Growth Found in Double Hulled VLCC's 564
Baffle Plates in IGS Scrubbing Tower 250
Ballast System 135
Ballast Tank – Electrolytic Descaling 75
Ballast Tank Access in Double-Hulled VLCC's 423, 565
Ballast Tank Gas Detection in Double-Hulled VLCC's 565
Ballast Tank Protection – Stripe Coating 75
Ballast Tank Protection in Double-Hulled VLCC's 423
Ballast Tank Ventilation in Double-Hulled VLCC's 565
Ballast Tanks – Corrosion Problems 72, 558, 564
Ballast Tanks – Segregated 40, 98, 165, 416
Ballasting Operations 174, 194, 206
Beaufort Wind Scale 13, 179
Beaufort Liferaft 88
Beavington, Mr. Michael Cited 368
Bell Mouth Suction 423
Bending Moments 161/6, 171, 173, 179/80, 441, 447/8, 502, 505/6, 548, 552, 567
Berthing – Alongside 336
Berthing – Approach Speed to Jetty 337
Berthing – Use of Doppler Shift 338, 482
Berthing – Use of Tugs 336
Berthing Incidents 34, 338
Binoculars – Not superseded by Radar or Computerisation 439
Bitten by the Officer Gap, Lloyd's Shipping Manager Article 360, 377
Black Box Recording System 475
Blackout in Sydney Harbour, Author's Article 368
Blind Pilotage by Transas' NAVI-TRAINER 512ff
Bosun 121, 368
Bosun's Store 93, 121, 271, 339
Botlek Stores, Rotterdam 388
Brain Damage Caused by Toxic Gases 218
Breathing Apparatus 273ff
Bridge Control of Ship's Engines 439
Bridge Duties 297ff, 458

Supertankers Anatomy and Operation

Bridge Movement BookLegal Document 323
Bridge Navigation Console 149
Bridge Watchkeeping 295ff, 314ff, 439, 445ff, 513
Bridge Wing Debate 112
British Admiralty Charts 314, 453
British Admiralty Pattern Anchors 90
British India Line 357
British Merchant Shipping Act 1854 8
British Petroleum Shipping 357
British Registered Ships – Demise 358
British Ship Building Yards 559
Brocklebank Line Shipping Company 153, 507
Bulbous Bow 23, 45ff, 160, 183, 438
Bulk/Ore Carriers 403
Bullets in Pressure Vacuum Valve 260
Bunkers Capacity 127
Bunker Market 189
Bureau Veritas 171
Butterworth Locker 112, 195, 217

Cadets – xiv, 57, 146, 260, 335, 357ff, 378, 396, 512
Cadets – Dual Certificate xiv
Canterbury Cathedral 5
Cape Columbine Light 299
Cape Rollers 181, 399
Capers on the Coast, Author's Article 399
Capetown xii, 18, 118, 181, 282, 299, 307, 384, 396
Captain (see also Master) xi, xiii, 273, 289, 296, 313/14, 321, 323, 336, 383, 386, 387, 395, 396, 506
Captain Geoff Cowap, Extra Master and Marine Consultant 171, 440, 552, 559, 567
Captain (now Professor) John Kemp, Ph.D 91, 97, 437
Captain Lyse – Master of mv "KATRINE MAERSK" Cited 457
CAPTER – Sanderson CBT's Training Simulation 525, 567
Cardiff Institute, University of Wales 374, 378
Careers Advisors in Schools 361
Cargo – Pump Room 131
Cargo "Danger Light" 156
Cargo Discharging 201ff, 531ff
Cargo Control Room 136, 169, 174, 202, 250, 535, 552
Cargo Escape due to Grounding 566/7
Cargo Loading 166ff, 190ff, 200ff, 531ff
Cargo Manifolds 109, 160, 195ff, 202ff, 347ff, 351/2
Cargo "Parcels" 190ff, 201
Cargo Pumps 98, 131, 136, 173, 200, 347, 423, 544
Cargo Pump Explosion 282
Cargo Samples 200
Cargo Tank Capacities 124ff
Cargo Tank – Protection 71ff, 564
Cargo Tank Construction 42ff, 416ff
Cargo Tank Construction Nomenclature 32, 406
Cargo Tank Inspections 44, 195, 218, 262, 284, 564/5
Cargo Tanks – Bacterial Growth in Double Hull VLCC's 564
Cargo Tanks – Conditions for an Explosion 211
Cargo Tanks – Gas Freeing 261
Cargo Tanks Used for Ballast 166, 206, 416

Carr, Mr. John – Sperry Marine Systems Quoted 439
Catering aboard VLCC's – High Standard 388
Catering Boy 384, 388, 389
Catering Department 386, 388
CD-ROM for Chart Corrections 458
Centi-Stripping System 131ff
Certificates of Competency – Deck Officers xiii, 361
Chamber of Shipping 358, 361, 374
Chart Correcting – Computerised by CD-ROM 458
Chart Display Unit (CDU) 453
Chart Folios 149, 314, 458
Charter Party Agreement 387
CHARTMASTER integrated with MIRANS 5000 463
CHARTMASTER System of Racal-Decca 456
Charts – Electronic 450ff
Chemical Tankers 553
Chief Cook/Steward 387
Chief Engineering Officer xiv, 199, 270, 383, 386, 393
Chief Officer Mate xiii, 55, 174, 190, 193, 199, 200, 219, 225, 261, 269, 270, 273, 281, 284, 289, 296, 313, 314, 319, 323, 334, 364, 368, 383, 384, 393, 395, 466
CHIEF OFFICER'S LOG Book 319
Chiksans/Manifold 109, 160, 194ff, 198, 202ff, 347, 352ff
Chronometers – Quartz 148
City of London Polytechnic School of Navigation 302, 437
Clan Line Steamers Limited 357
Clarke Chapman Company 145
Clarkson Research Studies, London, Mr. Cliff Tyler 557
Classification Societies/Regulations 8, 10, 12, 23ff, 40, 44, 73, 87, 89, 121, 145, 171, 173, 185, 221, 241, 405, 423, 462, 559, 562, 564/5, 570
Clyde Marine Training 359, 362
CO_2 – Fire Extinguishing Room 139
Coastal Contamination by VLCC "EXXON VALDEZ" 403
Coastguard HM. 374
Coatings of Ballast Tanks 71ff, 423, 559, 564
Cofferdam 40, 121
Collision Avoidance 299ff, 445, 446, 474, 566
Collision Avoidance Training by Transas' NAVI-TRAINER 512
Collision Regulations 89, 153, 156, 321, 329, 330, 333, 366
Combined Wheelhouse/Chartroom 147
Command Certificate – Class One 364
Compact Disc for Small Scale Ocean Chart Folio 456
Compact Discs used to contain World Chart Folio 456
Compact Discs for Chart Corrections 458
Compass Error Book 319
Compressive Stresses 166
Computerisation – Aid to the Ship's Officer 437ff
Computerisation – Cargo Loading, Control/Management Systems 552
Computerisation – Cargo Monitoring and Control 542
Computerisation – Cargo Work on VLCC's 441, 531ff

Computerisation – Consilium Metritape Tank Gauging 531
Computerisation – Effect of non-computerised inputs 479
Computerisation – Gradual Process 445
Computerisation – Guarded Reception at Sea 438
Computerisation – Navigational Applications 440, 445ff
Computerisation – SENTRY 11 of Consilium Metritape 531ff
Computerisation – Tank Temperature and Ullage 531ff
Computerisation – Tidal Streams 440, 495
Computerisation – To Assess Sea/Loading Stresses on Hull 171, 186, 497ff
Computerisation – Use for Clerical Tasks 439
Computerisation – Use in Engine Room 439, 553
Computerisation of Admiralty Publications 458
Computerisation of Chart Correcting 458
Computerisation of World Chart Folio 456
Computerised Weather Routing by Oceanroutes (UK) ORION 506ff
Computerised Assessment of Sea States by Consilium Marine 502
Computerised Barometer WEATHERTREND of Prosser's 496
Computerised Deck Hose Crane SWL by Strainstall Engineering 502
Computerised Docking Log – Consilium Marine's SAL 860 482
Computerised Fire Alarm SALWICO 3000 of Consilium Marine 484ff
Computerised Fleet Monitoring by Transas' NAVI-MANAGER 513
Computerised ISIS 250 Alarm Systems of Racal-Decca 487
Computerised Navigation Training by Transas' Marine NAVI-TRAINER 512ff
Computerised Quick Release Mooring Gear by Strainstall 502
Computerised Safety and Communications System of INMARSAT 490ff
Computerised Speed Log SAL RI of Consilium Marine 482
Computerised Hull Stress Assessment System, STRESSALERT II, of Strainstall Engineering 497ff
Computerised Stress Sensors – Cargo Handling 186, 548ff
Computerised Tidal Stream Predictor TIDESTREAM 495
Computerised Voyage Planning 462ff
CONCORDIA MARITIME – STENA LINE xii, 378/9, 562/3
Consilium Marine AB 225, 446, 458, 462, 482, 497, 502
Consilium Marine's INS 970 System 462
Consilium Marine's MM950 Radar 446
Consilium Marine's SAL 860 Computerised Docking Log 482
Consilium Marine's SAL R1 Log 482, 518
Consilium Marine's SAL Seakeeping Prediction System (SPS) 502

Consilium Marine's SALWICO CS3000 Fire Alarm System 484
Consilium Metritape's SENTRY 11 Displays 531
Consilium Metritape's VANGUARD Control System 534
Consilium SALWICO SW210 Gas Detection System 534
Construction – Swash Bulkheads 44
Construction – Accommodation 44, 112, 420
Construction – Bulbous Bow 45ff, 112
Construction – Cargo Tanks 42ff, 405/6, 416ff
Construction – Cargo Tank Nomenclature 32/4, 406/8
Construction – Cofferdam 40, 121
Construction – Different Types of Stems 46
Construction – Lloyd's Rule 4004 30, 44
Construction – Oil-Tight Transversals 44
Construction – Scantlings 29, 423, 426, 559
Construction – Sequence 26, 30ff, 411ff
Construction – Stem and Fo'c'sle 46, 124, 416
Construction – Stern Frame Area 35, 423/4
Construction – Surprisingly Short Period 24, 409
Construction Complexities of Supertankers 23
Coriolis Force 179, 480
Corrosion Found in Cargo and Ballast Tanks of Double Hull VLCC's 564
Corrugation Effect 179
Coryton – River Thames 18, 20, 189, 193, 201, 202, 206, 283, 345, 474
Course Recorder Sheet 319
COW – See Crude Oil Washing (COW)
Cowap, Captain G. Extra Master and Company Director 171, 440, 552, 559, 567
Cranes Luffing – Safe Working Load (SWL) 428
Cranes Replace Hose Handling Derricks on Double-Hulled VLCC's 428, 502
Creation of Inert Gas from Flue Gas 241ff
Crude Oil Cargo – American measurements 127
Crude Oil Cargo – Parcels 189
Crude Oil Cargo – Advice of Loading 189
Crude Oil Cargo – Chemical Properties 211
Crude Oil Cargo – De-ballasting Operations 194
Crude Oil Cargo – Discharging 166/7, 202ff, 531ff, 542/553
Crude Oil Cargo – Final Loading Ullages 193
Crude Oil Cargo – Load on Top Method 194
Crude Oil Cargo – Loading Duties 195
Crude Oil Cargo – Loading Factors 190ff
Crude Oil Cargo – Nominated Loading Tonnage 191
Crude Oil Cargo – Physical Properties 210
Crude Oil Cargo – Sludge 193, 209ff, 262
Crude Oil Cargo – Specific Gravities 189
Crude Oil Cargo – Temperature 534, 101
Crude Oil Cargoes – Mechanical/LCD Ullaging 101/5, 175, 199/200, 212, 534, 540ff
Crude Oil Discharging – Ballasting Before Sailing 206
Crude Oil Discharging – Chief Officer's Plan 202ff
Crude Oil Discharging – Duties Undertaken 201
Crude Oil Discharging – Oil Pressure Precautions 202ff
Crude Oil Loading – Dipping for Ullages 199
Crude Oil Loading – Topping Off 200
Crude Oil Loading – Cargo Samples 200

Crude Oil Loading – Chief Officer's Responsibility 190
Crude Oil Loading – Correcting List 200
Crude Oil Loading – Gradual Loading Rate 199
Crude Oil Loading – Hull Stress 166/7, 542ff
Crude Oil Loading – Locking of IGS Valves 195
Crude Oil Loading – Mate's Plan 190ff, 199
Crude Oil Loading – Officers' Deck Watches 200
Crude Oil Loading – Rates 199ff
Crude Oil Loading – Specific Voyage Details 193
Crude Oil Loading – Use of Slop Tanks 200
Crude Oil Loading – Weather Precautions 194
Crude Oil Loading – Zone/Load Line Calculations 9, 13/5, 191
Crude Oil Cargowork-Team Operation 200
Crude Oil Washing (COW) xv, 102, 221/242, 262, 559, 564
Crude/Gas Oil cargo experiment ix
CUPROBAN anti-hull-fouling of Jotun-Marine Coatings 68/9
Current – Set and Drift 305, 450
Cyclical Bending – Mr. Rob Drysdale, Company Superintendent Quoted 559

D Type Towing Shackle 87
Damage Reports to Owners 34
Danger of Excessive Trim during Cargo Work 169
Davies, William, Petroleum Tables 171
Day's Steaming 127, 313
Dead Reckoning Position 440
Deadweight Table/Scale of Double-Hulled VLCC's 429ff
Deadweight Tonnage 10ff, Back Cover Plans/Tables
Death Caused by Toxic Gases 218
Deaths to Seafarers Caused by Tank Explosions 217
Daewoo Shipbuilding Yard, South Korea 420
De-Ballasting – Care Exercised Prior to Loading 194
De-Ballasting Methods 194
Debate Concerning Raster/Vector Chart Use 457
Decca "ARKAS" Auto-Pilot 152
Decca Navigator-Reliability 148
Decca Navigator – Use for Anchor Position 335
Decca Navigator Company 147
Decca Navigator Company's DGPS 370, 440, 458, 475
Decca Navigator System 147, 314
Decca Radar RM and TM 149
Deck Apprentices 358
Deck Boy 121, 276, 389
Deck Cadet 145, 277, 295, 309, 321, 361/2, 365ff, 378, 446
Deck Cadet – "A" Level Entrant 359
Deck Cadet – Examinations at sea 371
Deck Cadet – Recruitment Figures 360
Deck Cadet Training – Training Portfolio 365
Deck Cadet Recruitment-Serious Worries 377
Deck Cadet Training – 1980's Policies 357
Deck Cadet Training – Competition 360
Deck Cadet Training – Importance of Deck Work 366
Deck Cadet Training – -In Other Departments 372
Deck Cadet Training – Lloyd's Ship Manager Article 358
Deck Cadet Training – Master's Review 366
Deck Cadet Training – NVQ Input 366
Deck Cadet Training – Objectives 357
Deck Cadet Training – Sponsorship Schemes 361
Deck Cadet Training – Supervision 365
Deck Cadet Training – Theoretical Input 365
Deck Cadet Training – Use of Radar/Simulators 317, 512ff
Deck Cadets – 1997+ Qualifications 364
Deck Cadets – Anecdotal Incidents 368
Deck Cadets – Loss to Industry 364
Deck Cadets – Ship board Duties 309, 366
Deck Man 284, 367
Deck Officer – Serious Shortages 377
Deck Officers (see also Navigating Officer) xiii, 295ff, 360, 374, 386, 393, 437
Deck Officers – Opportunities Ashore 374
Deck Officers – Promotion 364
Deck Officers' Training using Simulators 313, 512ff
Deck Ratings – Recruitment and Training schemes 364
Deck Water Seal in Inert Gas Systems 253
Deckhands 342, 384
Deep Water Anchorage's 334
Defects during Building 284
Demister Unit in IGS 253
Demonstration of Racal-Decca's CHARTMASTER 456
Department of Transport 319, 361, 374
Department of Transport – Eyesight Tests 358
Department of Transport (MAIB) 265
Derricks – 109
Designs of Inert Gas Systems 250
Det Norske Veritas 12
Detection of "Shadow" Areas in Tank Cleaning 230
DGPS 440, 458, 475
Differential GPS use on Electronic Charts 458
Digital Compass – Sperry Marine MK 37 480
Digital Gyropilot – Sperry Marine's ADG Model 482
Dimensions of VLCC's 5ff, 159, 428
Direction Finder 149, 521
Distance-Arabian Gulf/Europe 127
Distribution of Hand Dipping Holes on Deck 239
Docking Operations 83, 336ff, 482
DOCS Development of Certificated Seafarers 361
Doctor of Philosophy degree 364
Domestic Water Tanks 34
Doppler-Radar 338, 482
Doppler-Sonor 338, 482
Doppler Shift in Berthing 338
Double Hull VLCC – Bacterial Growth Encountered 564
Double Hulled VLCC – Computerised Tank Gauging 531
Double Hulled VLCC Construction Examined 406ff
Double-Hull VLCC – Ballast Tank's 416, 428, 564/8
Double-Hulled VLCC – Building forecasts 564
Double-Hulled VLCC Construction Explained 409ff
Double-Hulled VLCC Fuel Consumption 423
Double-Hulled VLCC Hose Handling Cranes 428

Double-Hulled VLCC Plating Thickness 428
Double-Hulled VLCC's – Nomenclature Used in Construction 406ff
Double-Hulled VLCC's Deadweight Table/Scales 429
Double-Hulled/Single Hulled VLCC Construction Differences 428
Douglas, Mr.Graham-Chief Officer Voyage Loading Plan 206
Dover Branch Nautical Institute 317
Dover Strait ix, 114, 298, 314, 317, 457
Dr. John Kemp, Phd 91, 97, 437
Draeger Ltd Breathing Apparatus 274
Draeger Ltd Personal Explosive Gas Monitor MULTIWARN II 221
Draught Marks 11ff, 160, 199, 201
Draught – Loading 199, 201
Drug and Alcohol Policies 387
Dry Cargo Ships 3, 280, 298, 317, 329, 383
Dry Docking of VLCC's 24, 55, 68/9, 217, 426, 562
Drysdale, Rob – Company Superintendent Quoted 559
Duty Mate/Officer 83, 169, 323/5, 335/6, 535
Duty Officer's Loading Duties 200
Dynamic Stresses 159ff, 179ff, 497ff, 553

ECDIS-Electronic Chart Display and Information System 462
ECDIS Integrated with Transas' NAVI-TRAINER 515
Ecology Studies integrated into cadet training 366
Electronic Charts 450
Electronic Charts – Advantages of Use 457/8
Electronic Charts – Dispute on Raster/Vector Use 457
Electronic Charts – IMO/IHO Caution in Ratifying 457
Electronic Charts – Integration of Satellite data 458
Electronic Charts – Practical Voyage Planning 457/8, 474
Electronic Charts – Superimposition of Own Ship 458
Electronic Charts – Use of Appropriate Working Range 458
Electronic Charts – Use of Way Points with GPS 458
Electronic Charts – Use of Zoom Facility 458
Electronic Charts – World Folio on Compact Discs 456/58
Electronic Charts – Zoom Facility 456
Electronics Officer 490
Ellerman Group 357, 358
Elto Domestic Machinery Company 386
Emergency drills 271ff
Emergency Towing System (ETS) Described 118
Energy Marine (International) Ltd 171, 440, 553
Energy Marine's SHIPWRITE/TANKERLOAD Programs 553
Engine Room – Computerisation Mentioned 439, 531
Engine Room – Excluded from Book Examination xv, 531
Engine Room Machinery/Equipment 35, 117/8, 253, 553
Engineer Officers 185, 198, 359/360, 374, 386, 553
Engineer Officers – Serious Shortages 377
Engineering Cadets 357, 374

Engine-Room – Sea-Water Uptake 256
Engines – Steam Turbine 35, 127, 325
Environmental Damage due to Tanker Groundings xi, 403
Environmental Groups/Issues 68, 405, 571, 572, 574
Environmental Protection – Major Policy and Area of Concern xi, 193, 333, 345, 571, 574, 575,
Esso Tanker Company 290
European Ship Building Yards 24, 30, 37, 49, 559
European Waters Pilotage Excellent Standard 321
Europoort 18, 20, 167, 202, 335
Examination Regulations – 361ff
Examinations taken at sea 371, 396
Experiments in UK/USA on Flue Gases 241
Explosimeters – Wheatstone Bridge Principle 218
Explosion Dangers during Tank Cleaning 213
Explosions in VLCC's 213ff
Extra Master's Certificate 364
Extracts from Lloyd's Rules for Construction 30
Exxon Group Shipping Company 409
EXXON VALDEZ. VLCC – Coastal Contamination 403
Eyesight Tests 358

Falls Light Vessel RACON 469
FGI Company Ltd of Leatherhead 250ff
Fibre-Optic Gyro Compass Research 480
Fire Detector – Consilium's SMART/TM Computerised System 484
Fire drill 273/5
Fire Wires 195, 344
First Double-Bottomed VLCC Built (1969) 403
Fleet Movements Monitored by Transas' NAVIMANAGER 513
Flue Gas as a Neutralising Agent Experimentation in UK/USA 241
Foam Monitors 94, 275
Fo'c'sle 45, 88ff, 121
Fog Horn/Siren 89, 329
Fore Peak Tank 45, 123, 165, 191
Fourth/Fifth Special Surveys 64, 175, 558, 562
Frame Numbering used in Damage Reports 34
Freeboard 13, 17
Fuel Consumption of VLCC's 127, 423
FUELSMANAGER System of Whessoe Varec 547
FURUNO FD-177 Direction Finder Integrated with NAVI-TRAINER 521
FYRITE Portable Oxygen Indicator 261

Gardner, Captain Bernard Article 374
Gas Tankers 403, 553
Gavin, Mr.A.G. Principal Lloyd's Surveyor Quoted 404
GCE "A" Level Qualifications 362
GCSE Qualifications 362, 396
General GMDSS Operators' Certificate 522
Germanischer Lloyd 12
Global Fluid Dynamics Laboratory 509
Global Position System (GPS) 370, 440, 453
Glycol/Water Mixture in PV Breaker 261

GMDSS 374, 471, 474, 490ff, 512ff, 521, 553
GMDSS Simulator Integrated with NAVI-TRAINER 512ff
Gotverken Lodicator 171
GPS/DGPS 370, 440, 453, 458, 475
Graham Captain Colin Quoted 200, 277, 296, 342
Gray, Mr. J. of Clyde Marine Training Company 359, 362
Green, Mr. Roger, Operations Manager Quoted 367, 377
Guide to Helicopter/Ship Operations 281
Guildhall University 302 ff
GUNCLEAN Tank Cleaning Machine of Salen and Wicander 223/4
Gyro Compass 145, 313, 370, 445, 479ff
Gyro Compass – Fibre Optic Research 480
Gyro Compass Simulation in Transas' NAVI-TRAINER 521

Hancock, Mr. J. of Lloyd's Register Quoted 440
Harbour Service Launches 330, 346
Harland & Wolff of Belfast 24, 112
Haverton Shipping Company 3
Hawse Pipe 90, 336
Hazardous Area of Inert Gas System 253
Heating Coils in Ballast Tanks 101, 423
Heating Coils in Cargo Tanks 202
Helicopter Landing Area 109
Helicopter Operations 281, 335, 384, 393, 396
High Tensile Steel 29, 559
Hitachi Zosen Corporation's Ariake Works 409
HM Coastguard 374
HMSO 359, 440
HMSO – Marine Directorate Paper 362
HND in Nautical Science 364
Hogging 162ff, 173, 175, 178, 180, 183
Holmes, Mr. Grant – Navigating Officer Quoted 378
Howden FGI (IG) Systems 250ff
Hull Fouling Protection – Cathodic 62ff
Hull Protection – Environmental Concerns 68ff
Hull Stress Detectors 171ff, 471, 497ff, 567
Human Element in Shipping Casualties Report 362
Hydraulic Control Valves 98ff, 133, 174, 200
Hydraulic Pilot Hoists – Dangers 345
Hydrocarbon Gas 211ff, 218ff, 241, 250
Hydrocarbon Gas Detectors 218ff, 565
Hydrocarbon Gases – Flammability Range 211
Hydrocarbon Residue Gas 195
Hydrocarbon Skin Cancers 217
Hydrographic Office, Taunton 453
Hydrostatically Balanced Loading (HBL) 558, 562ff

IACS 8, 29, 87, 121, 185, 405, 462, 564, 570
ICS 211, 281, 405
IGS – See Inert Gas Systems (IGS)
IMCO/IMO xvii
IMO/IHO – International Hydrographic Office 457
IMO Conventions 23

IMO International Electrotechnical Commission 462
International Maritime Organisation (IMO) Conventions/Regulations 8ff, 40, 98, 118, 121, 148, 153, 195, 221, 231, 241, 248, 289, 296, 344, 361, 423, 457, 462, 479, 490, 523, 558, 562/3, 565, 571
IMO Working Parties 405
Incoloy used in IG System 253
Indentured Deck Apprentices 358
Inert Gas Created from Flue Gases 246
Inert Gas Systems (IGS) xv, 102, 149, 167, 194, 200, 202, 206, 241/262, 265, 296, 347, 536, 540, 566
INMARSAT Computerised Safety & Communications System 154, 490ff
INMARSAT Interactive with ORION and WNI 509
INMARSAT-C Integrated with Transas' NAVI-TRAINER 513
Inspections of Cargo Tanks 44, 195, 218, 261/2, 284, 564/5
Integrated Navigation System (INS) 462ff
Integrated System "INS970" of Consilium Marine 462ff
Interlinked Navigation and Voyage Management System "VISION 2100" of Sperry Marine 469ff
Interaction 330
Interaction between ships 330
International Association of Classification Societies (IACS) 8, 29, 87, 121, 185, 405, 462, 564, 570
International Chamber of Shipping – (ICS) 281, 405, 570
International Code of Signal Flags 153, 198, 321
International Electrotechnical Commission of IMO 462
International Hydrographic Office 457, 462
International Maritime Pilots' Association 323
INTERTANKO 405, 572
Iranian Heavy Crude Oil 189
Isherwood, Sir Joseph Cited 24
ISIS 250 Computerised Alarm System of Racal-Decca 487
ISM Code of IMO 558
ITT (Marine) Company 146

Jahre-Wallem Shipping Company 65, 111, 170, 190
Japanese Shipbuilding Yards 24, 29
Jotun-Henry Clark – Marine Coatings 55ff, 63, 75, 77

Kardorama of Potter's Bar 6
Keith Shaw, Sanderson CBT's Sales Director Quoted 527
Kelvin-Hughes "KELAMP" Aldis Lamp 372
Kemp, Captain (now Professor) PhD., John 91, 97, 437
Kent Clear-View Screens 149
Knowles, Captain A – Dover Based VLCC Pilot Quoted 317
Kockum Sonics AB Company 548
Kockum Sonics LEVELMASTER-CALM System 550
Kockum Sonics LOADRITE Cargo Planning System 548
Kockumation's LOADMASTER 542
Kockums AB of Malmo, Sweden 171, 174
Kokumation AB Company 542

Kort-Nozzle Swivelling Rudder 37

Lady Kitty – Motor Barge 283
Launching-Flooding of Dry Dock 49
Launching by Slipway 49
Le Havre 18, 193, 201, 345, 351
Libraries at Sea 396
Life at Sea – Potentially Lonesome Life 359
Lifeboat – Maintenance by Deck Cadets 368
Lifeboats 114
Lifeboats – drills 271
Lightening – Procedures and Operations to a smaller Tanker 347/53
Lindo Shipbuilding Yard, Odense 4, 284, 409
Lloyd's Register of Shipping xi, 12, 24, 377
Lloyd's Register Staff Association 11, 241, 440
Lloyd's Rules for Oil Tanker Construction 29/30
Lloyd's Ship Manager, Article January 1997 358, 360
Lloyd's Surveyors in Haifa 34
Load Line 10ff, 44, 93
Loading 166/178, 189/201, 542/553, 562ff
Loading Control System LCS97 of Transas Marine 548
LOADMASTER Lodicator of Kockum AB 171, 174
LOADMASTER Cargo Handling Computer of Kockumation AB Company 542
LOADSTAR Management Tool of Maersk Data AB 552
Log Book 179, 185, 189, 335, 471, 521
Log Book – Legal Document 319
London School of Navigation 302ff, 437/9
Longitudinal Bending Moments Stress Gauges 160ff, 497ff, 504
Look-Out Duty 298, 304, 319, 527
LORAN C Navigation Aid 147/8, 370
Loss of Ships at Sea 504, 507
Luffing Cranes replace Derricks 112
Lykiardopulo Shipping Company 409

Maersk Data AB's LOADSTAR Management Tool 552
Maersk Line AB (A.P. Moller) 409, 457, 552
Magnetic Compasses 118, 154, 480
Main Deck – Painting 269
Main-Deck – Protection 69
Manholes 44, 284ff
Manifolds 109, 161, 195ff, 200ff, 347, 351/2, 420
Marconi "LODESTONE" D/F 148
Marconi "PREDICTOR" Radar Set 300
Marconi International Marine Company 146, 300
Marine Accident Investigation Branch (MAIB) 265
Marine Navigators – Informal Definition 437
Marine Research Agency 65
Marine Safety Agency 309, 361, 365
Marine Society 371, 396, 397, 573
Mariner's Handbook. NP100 179, 181
Maritime Centre, Warsash 359, 360, 378
Marlowe Ropes Company 339
MARPOL – Pre-1973 VLCC's xvii, 44, 93, 217ff, 284ff

MARPOL 73/78 REGULATIONS 44, 102, 198, 218, 221/2, 230/1, 366, 405/6, 532, 557/8, 564
MARPOL Stripping Line 241, 245
Maryan, Richard (Photographer) 6
Master (See also Captain) xiii, 153, 169, 180, 190, 201, 270, 281, 295, 299, 300, 309, 319, 329, 334, 335, 336, 338, 360, 362, 366, 368, 370, 391, 395, 438, 440, 446, 474, 513
Master Mariner's Certificate 362, 371, 374
Master's Certificate Recognition Ashore 364
Master's University degree 364
Master's Night Order Book 309, 319, 471
Mate/Master Certificate – Class 2 364
Mate's Loading Plan 199
Medical Inspections before sailing 393
Memories of a Supertanker, Author's Article 399
Merchant Navy – 1950's Scenario 358
Merchant Navy – Diminished 357
Merchant Navy- – Recruitment 360
Merchant Navy as a Career in 21st Century 358
Merchant Navy Training Board 361, 364, 365
Merchant Shipping Act 8, 361, 372
Metal Fatigue – Hairline cracks 280/1
Metal Fatigue in Ageing VLCC's 559
Meteorology 179, 506/8
Mike Slavin of Oceanroutes WNI Ltd Quoted 507/8
Mild Steel 29, 559
Mills, John (Photographer) 249/52
MINIGAS MONITOR of Neotronics Ltd 221
MIRANS 5000 VMS of Racal-Decca 462ff
Mitsubishi Heavy Industry Shipyard 24
MM950 Radar of Consilium Marine 446ff
MMC (Europe) Ltd 103
Mobil Shipping Company 75
Moller, A.P. Group 409, 457, 552
Moment to Change Ship's Trim MCT 169
Monitoring of Ship's Passage by VMS 474
Monkey Island 83, 154, 181
Moorings 89, 336ff, 344/5
Mooring Hook Transducers from Strainstall Engineering 502
Morse Code – Retained in Deck Officer's Examinations 372
Motor Ship Magazine 113, 563
Movements of Supertankers 17ff
Mr. Michael Beavington – P&O Deck Cadet 368
Mr. John Carr – Sperry Marine Systems 439
Mr. Rob Drysdale – Company Superintendent 559
Mr. Jim Gray, Clyde Marine Ltd 359, 362
Mr. Roger Green – Maritime Operations Manager 367, 377
Mr. J. Hancock of Lloyd's Register 440
Mr. Grant Holmes – Navigating Officer 378
Mr. Brian Mullan – INMARSAT Management 490
Mr. David Penny – Marine Engineering Superintendent 565/6, 568
Mr. Keith Shaw, Sanderson CBT Company 527
Mr. Mike Slavin 507
MSA Examinations 362ff
Mullan, Mr. Brian-INMARSAT Management 490

Multi-Media training Simulation of Sanderson CBT 523ff
MULTIWARN 11 Personal Monitor 221

Nakashima Propeller Company, Japan 426
National Vocational Qualifications (NVQ) 362
Nautical Almanac – HMSO Publication 309, 440
Nautical Colleges xi, xii, xviii, 359/61, 370, 378, 445, 512
Nautical Institute 317
Navigating Officers (See also Deck Officers) xiiff, 3, 186, 201, 302, 313/4, 365, 371, 439, 451, 456, 471, 490, 505/6, 566, 568
Navigation – Celestial 305ff
Navigation – Coastal Waters 297ff, 313ff, 450ff
Navigation – Computerised 445ff
Navigation – Electronic 305, 370
Navigation – GPS/DGPS 370, 440, 453, 458, 475
Navigation – LORAN C 370
Navigation Ship's Lights 88, 153, 156
Navigation with Electronic Charts 453ff
NAVI-MANAGER Monitoring by Company of Fleet 513
NAVISAILOR System of Transas Marine 457
NAVI-TRAINER of Transas Marine 513ff
Neilson Stretcher and Drill 277
Nelson's Column 5
Neotronics Ltd of Bishop's Stortford 221
New buildings of VLCC's 564
Nicholls's Concise Guide for Navigation Work 371
Nippon Kaizi Kyokaie 12
NKK Corporation of Japan 25
Non-Return Isolating Valve in IGS 256
Norie's Nautical Tables 309, 440
Notices to Mariners 313, 458
Not Under Command (NUC) Lights 156
NUMAST 360, 378, 573
NUMAST Magazine TELEGRAPH 378, 523
Numbers of Single/Double Hulled VLCC's in Service – 2000 557
NVQ – Marine Vessel Operations 362, 364

Oates, Bill (Photographer) 6
Ocean Passage Planning 296, 474, 506ff
Oceanroutes (WNI) Ltd – Mike Slavin Quoted 507
Oceanroutes' ORION On Board Guidance System 506ff
OCIMF 89/90, 195, 241, 339, 346, 351, 387, 405, 574
Odense Shipyard, Denmark 4, 409
Officer's Duties 195, 198, 200, 269ff, 366, 535
Officer-of-the-Watch/Day 153, 295ff, 304ff, 329, 335, 370, 445, 482, 518
Officer's Uniforms Standard Dress 361, 384
Oil Refinery ix, 346, 574
Oil Tankers – Lloyd's Rules for Construction 29/30
Onassis Group xi
One Man Bridge Operation 298
OPA90 118, 404, 532, 557/8
Oral Examinations For Masters' and Mates' 364, 527

Ord's Ltd – Printing Company 319
ORION Weather Forecasting and Guidance System of Oceanroutes (WNI) 506ff
Oxford University xiv
Oxygen Analyser in IGS 261
Oxygen Deficiency during Tank Cleaning 218

P & O Lines Deck Cadet Incident 368
P & O Cruises (UK) Company 524
P & O Line Ferries 3
Paint – Composition and Application 62ff, 289
Paint Manufacturers' and Ship-Owners' Discussions 57, 565
Painting Areas of VLCC's 59
Painting – Shore Gangs onto VLCC's 270
Priming and Preparation for Painting 59ff, 414ff
Passage Planning 296, 323, 438, 457, 474
Passage Planning by Electronic Charts 457
Pawl Type Chain Stopper 87
PDP8 Computer 440
Penny, Mr. David – Marine Engineering Superintendent Quoted 565/6, 568
Personal Explosive/Toxic Gas Monitor MULTIWARN II of Draeger Ltd 221
Personal Radio – Stornophone (VHF) 153, 281, 323, 325, 329, 334
Pettit, Dr. S.J., Cardiff, University of Wales Article 374
Phillip's Personal Address (P/A) Equipment 149
Pilsbury, Mr. P.K. (Photographer) 496
Pilot Books – Admiralty Sailing Directions 296, 314
Pilotage Zone – Need for Caution 321
Pilot and VLCC's 112, 296, 317, 321ff, 335/6, 345, 482
Pipe Systems on VLCC's 95ff, 113, 131/2, 202, 239, 241ff
Pipelines – Flexible 346
Piston Horn Whistle 89, 329, 335
Plating Thickness of Double-Hulled VLCC's 29, 428
Plimsoll Line 10ff
Plymouth Nautical College 512
Pollution Prevention – Major Policy and Area of Concern xi, 68, 193, 333, 345, 404/5, 571/2, 574
Practical Navigation in the Dover Strait Article 221
Port of Registry 11, 23
Ports Visited by VLCC's 18ff
Port Signals 156
Portable Gas Freeing Fans 261
Portable Tank Cleaning Machines 217
Post-graduate University Qualifications 364, 396
Pre-MARPOL REGULATION VLCC'S xvii, 44, 94, 217ff, 284ff
PREDICTOR Radar set of Marconi Marine 300
Pressure Vacuum Valve "Bullets" in IGS 260
Princess Cruises Company 524ff
Principles of Hydrostatically Balanced Loading (HBL) 562ff
Professional Certificates of Officers 361ff
Professor John Kemp, PhD 91, 97, 437
Programmable Tank Cleaning Machines 224ff
Propeller 23, 39, 64ff, 160, 325, 426

Propeller Thrust to Assist Turning 325
Propeller Weight 39, 426
Prosser Scientific Instruments of Ipswich 495
Prosser Scientific's WEATHERTREND Digital Barometer 496
Pump Room 24, 30, 40, 98, 113, 131ff, 149, 200, 219, 239, 282, 423
Pumps – Ballast 98, 131, 135, 174/5, 544
Pumps – Cargo 98, 131, 136, 200, 282, 347, 423, 544
Pumps – Stripping 131, 423, 544
Pumps Capacities 131ff
Pump Explosion 282
Purge Pipe in IGS 260
Pusnes Company of Norway 118
PV Breaker in IGS 260
PYRATE, Victor Tank Cleaning Machines 224/230

Qualifications – First Watchkeeping Certificate 365
Quartermaster 335
Quick Release Mooring Equipment by Strainstall Engineering 502

Racal-Decca Company 149, 456ff, 463ff, 515
Racal-Decca's Alarm System "ISIS" 2500 487ff
Racal-Decca's BRIDGEMASTER and Transas Marine's NAVI-TRAINER 515
Racal-Decca's CHARTMASTER System 456ff
Racal-Decca's Live Situation Report (LSR) 463ff
Racal-Decca's MIRANS 5000 Voyage Management System (VMS) 463ff
RACON Used on Radar for Identification 321, 450, 469
Radar – "Assisted Collisions" in early days 300, 438
Radar – Action taken on Insufficient Information 438
Radar – ARPA Displays 304ff, 445, 450, 512, 515
Radar – Daylight Viewing Screen 446
Radar – Fast Moving Target Plotting 450
Radar – Guard Zones 450
Radar – Height of Scanner on VLCC 299
Radar – Introduction of Video graphics 446
Radar – Land Shadowing 323
Radar – Mis-interpretation of PPI 300, 438
Radar – MM950 of Consilium Marine Model Described 446ff
Radar – Parallel Indexing techniques 305
Radar – Plotting Essential in Collision Avoidance 300
Radar – Plan Position Indicator (PPI) 300, 445
Radar – PREDICTOR Model of Marconi Marine 300
Radar – Projected Danger Zone 450
Radar – Relative Motion Display (RM) 149, 300, 450ff
Radar – Spurious Echoes 527
Radar – Target Return 299, 300
Radar – Tendency to Over-Rely 300, 304
Radar – True Motion Display(TM) 149, 300, 302, 450ff
Radar – Unable to Determine Targets' Initial Movements 300
Radar – Use for Anchor Position 334
Radar – Use for Collision Avoidance 299, 450
Radar – Use in Coastal Navigation 450ff

Radar – Use of "Windows" 451
Radar – Use of in Restricted Visibility 329
Radar – Use with Electronic Charts 450ff
Radar – Value of Simulator Training 317, 512ff
Radar – Technology Developed since Second World War 446
Radio Officers 146, 374, 386, 397, 490
Radio-Telephony (R/T) Watch 145, 335
RASCAR VT Radar of Sperry Marine 471
Raster Chart Display System (RCDS) 453ff
Raster/Vector Chart Use and IMO/IHO Regulations 457
Ratholes 44, 288
Ratification of Electronic Charts by IMO Essential 462
Red Ensign Fleet Diminished 357
Reed's Sight Reduction Tables 440
Refinery ix, 346, 547
Restricted GMDSS Operators' Certificate 523
Restricted Visibility 300, 329
River Thames – Coryton 18, 20, 189, 193, 201/2, 206, 283, 345, 474
Rodger, Mr. David of Acomarit Company Cited 377
Ropes – Synthetic 95, 339ff
Royal Mail Line 357
Royal National Lifeboat Institution (RNLI) 374
Royal Naval Officers – Recruitment into MN 378
Rudder 23, 37, 160, 289, 426
Rudder Cycling to Assist Stopping of VLCC 329
Rudder Limits 145
Rudder Types 38
Rust – Composition of 55ff
Rust – Main Deck 69/70, 269/70

Saab Marine Electronics Company 535ff
Saab TANKRADAR G3 Gauging System 535
Saab TANKRADAR Mac Cargo Monitoring and Control System 535
Safe Working Load (SWL) of Derricks 112
Safe Working Load (SWL) of Luffing Cranes 428
Sagging Stress 162, 165, 173, 175, 179, 183
SAL R1 Log of Consilium Marine 482, 512
SAL Seakeeping Prediction Systems (SPS) of Consilium Marine 502ff
Salen and Wicander 223
SALWICO CS3000 Fire Alarm of Consilium Marine 484
SALWICO SW2010 Gas Detection System of Consilium Marine 534
Sanderson CBT Multi-Media Simulator Training CAPTER 523ff, 567
Satellite Communications System – INMARSAT 154, 409ff, 509, 513
Satellite Data Integrated into Electronic Chart Systems 458
SBM – Single Buoy Mooring 201, 345ff
SBM – OCIMF Recommendations 346
Scantlings 29, 428, 559
Scrubber Unit in IGS 250
Sea breezes Magazine, Author's Articles, 368, 399
Sea Technology – American Shipping Magazine 314
Seafarers' Problems Experienced Working Ashore 399

Supertankers Anatomy and Operation

Seakeeping Prediction System (SPS) of Consilium Marine 502
Sea-Time Completion – Controversy 362
Second Engineering Officer 270, 383, 386, 393
Second Mate/Officer 57, 145, 296, 313, 365, 371, 393, 458, 474
Second World War 24, 241, 446
Segregated Ballast Tanks SBT 40, 98, 166, 406
SELESMAR Radars of Consilium Marine 446
Senior Officers Jailing Incident 191
SENTRY 11 Sensor Displays Consilium Marine 531
Sextants – Errors Introduced due to Vibration 185
Shaw, Keith – Sanderson CBT's Sales Director 527
Shearing Forces 161/6, 171, 173, 178, 183, 441, 548, 552, 567
Shell International Trading & Shipping Group xi
Shell Plating Thickness 44, 428
Ship Building Yards 23ff, 409ff
Ship Charterers xvi, 191
Ship Losses at Sea 504, 507
Ship Performance Weather Studies of WNI Oceanroutes 506ff
Ship Stresses – Assessment by Computer 171ff, 497ff, 505
Ship Stresses – Bending Moments 161/6, 171, 173, 179/80, 441, 497/8, 502, 505/6, 548, 552, 567
Ship Stresses – Dynamic Stresses 159, 179, 497ff, 553
Ship Stresses – Shearing Forces 161/6, 171, 173, 178, 183, 548, 552, 567
Ship Stresses – Static 159ff, 179ff, 497ff, 553
Shipping Federation 358
Ship's Draught 11ff, 83, 199, 201, 206, 336, 563
Ship's Log Book 179, 185, 189, 319, 335, 471, 521
Ship's Rate of Turn 325, 484
Ships: AKADEMIC KRYLOV.mv 317
Ships: AMOCO CADIZ. VLCC 403
Ships: AROSA. VLCC 409ff
Ships: BRAER.mt xi
Ships: BRITISH AVIATOR.mt 438
Ships: BURMAH ENDEAVOUR. ULCC xii
Ships: CRYSTAL JEWEL.mv 438
Ships: DERBYSHIRE. mv 507
Ships: EAGLE. VLCC 429
Ships: ELEO MAERSK. VLCC 409
Ships: ESSO ATLANTIC. ULCC 409
Ships: ESSO CARDIFF.mt 347, 351
Ships: ESSO PACIFIC. ULCC 409
Ships: EXXON VALDEZ.VLCC 387, 403, 566
Ships: GLOBTIK LONDON. ULCC xii
Ships: GLOBTIK TOKYO. ULCC xii
Ships: GOLD VARDA. mv 3, 4, 34, 46, 217
Ships: HELOS FOS.ULCC 39
Ships: JAHRE VIKING.ULCC xii, 7, 16, 20, 39, 63, 113, 127, 169, 190, 429, 566
Ships: KAPETAN GLANNIS. ULCC 409
Ships: KAPETAN MICHALIS. ULCC 409
Ships: KATRINE MAERSK.mv 457
Ships: MATRA.mv 153, 507
Ships: MOUNTAIN CLOUD.VLCC 95
Ships: PRIDE OF BILBAO.mv 3

Ships: QUEEN ELIZABETH 11. mv 440, 490
Ships: RAMLAH. ULCC 463
Ships: RANIA CHANDRIS.VLCC 4, 5, 7, 83, 106, 125, 140
Ships: SAUDI SPLENDOUR.VLCC 75, 113
Ships: SEA EMPRESS.mt xi
Ships: STENA CONGRESS. VLCC 562
Ships: STENA CONVOY. VLCC 559/563
Ships: STENA KING. ULCC 379
Ships: STENA QUEEN. ULCC xii
Ships: TINA ONASSIS.mt v, xi
Ships: TORREY CANYON. mt 403
Ships: UNIVERSE APOLLO.VLCC xi/xii
Ships: UNIVERSE IRELAND.ULCC xii
Ships: VARDA. mv 4
Ships: WORLD UNICORN.VLCC 7, 46/9
Ships: ZATHON.mt xi
Ships: ZENATIA.mt xi
SHIPWRITE/TANKERLOAD Programs of Energy Marine Company 553
Shore Leave 383
Shore Maintenance Crews 270
Sight Reductions 309
Simpson's Waterplane Coefficients 169
Simulation of Sea/Air Rescue (SAR) Work 523
Simultaneous use of Raster/Vector Charts 457
Single Buoy Mooring (SBM) 201, 345ff
Single Pipe/Point Mooring (SPM) 346/7
Single-Hulled VLCC's – Instances of Neglected Maintenance 558
Single-Nozzle Tank Cleaning Machines 222ff
Skin Cancers 217
Slamming Stresses 179ff, 498ff, 568
Slipway Launching 49
Slop Tanks 40, 194, 200/1, 239, 421, 535
Sludge – Formation and Difficulties of Removal 209ff, 262
Small Scale Ocean Charts on Compact Disc 456
SMART/TM Fire Detector of Consilium Marine 484
SMIT Towing Bracket 87
SOLAS Regulations xi, 68, 114, 193, 221, 241, 248, 273, 289, 319, 333, 366, 456, 458, 490, 532, 557
Solly, Dr. R.J. Author's Article 368, 399
South Korean Daewoo Yard 420
Southampton Maritime Centre, Warsash xii, 359/60, 378
Special Fourth/Fifth Surveys 64, 175, 558, 562
Specific Gravities of Crude Oil 189
Speed of VLCC's – Loaded and in Ballast 39, 127, 423
Sperry Marine Company 145, 314, 439, 463, 469ff
Sperry Marine MK 37 Digital Compass 480
Sperry Marine's Adaptive Digital Gyropilot (ADG) 480
Sperry Marine's Automatic Navigation and Track Keeping System (ANTS) 463
Sperry Marine's RASCAR VT Radar 471
Sperry Marine's Voyage Management System VMS/VISION 2100 314, 457, 469ff
Spurling Pipe 91, 124
Squat 329

Srubbing Tower — Function in IG System 250
Stagecoach South Bus Company 6
Standard Dress Replaces Uniform 361
Standard Tidal Ports 495
Static Stresses 159ff, 179ff, 497ff, 553
Statistics of VLCC's 2000 557
STCW95 Convention 360/1, 512, 523
Steel — High Tensile 29, 559
Steel Replacement on VLCC's due to Ageing 559
Steering Compartment 145ff, 339
Steering Console 152ff, 482
Stem Construction — U or V-shaped 46
Stern Frame Construction 35, 423
Stone Manganese Marine Company 39
Stornophone VHF sets 153, 281, 323, 325, 329, 334
Strainstall Engineering Quick Release Mooring Equipment 502
Strainstall Engineering Services Stress Computers 498
Strainstall Engineering's Computerised Deck Crane SWL's 502
Strainstall Engineering's Range of Marine Equipment 502
Strainstall's STRESSALERT II Monitoring Equipment 497ff
Stress affecting VLCC officers 527
STRESSALERT II Monitoring Equipment from Strainstall's 497ff
Stretcher Party Drill 277
Stripping Line — MARPOL Small Diameter 241
Stripping Lines/Pumps 113, 131ff, 194, 209, 241, 245, 421/5
Study at Sea 396
Suez Canal xi, 63, 474
Suez Canal — Searchlight 123
Sumitomo Heavy Industry Shipbuilding Yard, Japan 429
Swan Hunter's Shipyard 7, 24, 28, 45/9, 124
Swedish Corrosion Institute 55/6
Swell — Extract from NP100 179
Swettenham Fenders — Use in Lightening 351
Synthetic Ropes — Breaking Strain 339

Tank Capacities 126, 428
Tank Cleaning 133, 195, 207/267, 409
Tank Cleaning — Shadow Areas and Drawings 230ff
Tank Cleaning — Use of Portable Machines 217/8
Tank Cleaning — Wash Patterns 224
Tank Cleaning Explosions 213ff
Tank Cleaning Machines — Deck Mounted 224
Tank Cleaning Machines — Fixed 222
Tank Cleaning Machines — Numbers Used 236
Tank Cleaning Machines — Programmable 224
Tank Cleaning Machines — Single-Nozzle 222
Tank Cleaning Machines — Submerged 230
Tank Cleaning Machines — Twin-Nozzle 226
Tank Dimensions 46, 428
Tank Gauging — Alarm Systems 534
Tank Gauging — Consilium Metritape Sensors 531
Tank Gauging — Temperatures 531
Tank Rescue Drills 277

Tanker Officer's Turmoil on a Dry Cargo Ship, Author's Article 399
Tanker Structure Co-Operative Forum 71, 406, 575
TANKRADAR G3 Gauging System of Saab Marine Electronics 535ff
TANKRADAR MaC — Specification 535
Tavistock Institute of Human Relations 362
TCIM/M Tank Control Management System of Whessoe Varec 546
TELEGRAPH-Magazine of NUMAST 378, 523
There's no Smoke Without Fire … Or is There?, Author's Article 368
Third Mate/Officer 323, 361, 365, 368, 374, 393
Third Watchkeeping Certificate 364
Thirty Month Survey for VLCC's 69, 261
Tidal Atlases Computerised 458, 495
Tidal Predictions 474, 495
Tidal Rip Experienced 317, 469
TIDECLOCK 3 of Prosser Scientific Instruments 496
TIDESTREAM 3 of Prosser Scientific Instruments 496
Tonnage's — Defined 8ff
Topping Off during Loading 200
Toxic Gases — Detected by Meters 218
Traffic Separation Zones 314, 345, 438, 469, 566
Training Portfolio 365ff
Transas Marine Company, Southampton 457/8, 474, 513ff, 548
Transas Marine's Development of LOADRITE System 548
Transas Marine's NAVISAILOR System 457
Transas Marine's NAVI-TRAINER 513
TRANSIT Satellite System 440
Transversal Stresses 160ff
Trinity House 374
Tugs 89
Tugs — Bollard Pull 336/7
Tugs — Disposition During Berthing 337, 344
Tugs — In Attendance 195, 323
Tugs used in Launching 49
Tunnard's Tanker Tables 171
Turning Circle of VLCC's 325ff, 438
Twin-Nozzle Tank Cleaning Machines/Wash Times 226ff
Twin-Nozzle Wash Patterns 230
Tyler, Mr Cliff, Clarkson Research Studies, London 557

ULCC — Forepeak Damage 72
Ullage/Ullaging 10, 102/5, 175, 193, 199/200, 212, 534, 540ff
Uniform — Protection by Merchant Shipping Act 361
United Kingdom Hydrographic Office 453, 457
United Kingdom Merchant Navy Diminished 357/8
United States Ship Building Yards 559
Universe Tanker Company Inc xi
University of Wales, Cardiff Institute Survey Report 374
University Qualifications 364, 396
U-Type Stem Construction 46

Supertankers Anatomy and Operation

Vacancies for Deck Officers Ashore and at Sea 360
Vacuum Breaker (Bullet) on IG System 260ff
Value of Practical Deckwork – Incident 366/8
Valves – Different Types/Uses 93, 98ff, 133ff, 239ff, 250ff, 535ff, 550
VANGUARD Shipboard Control System of Consilium Marine 534
Varne Light Vessel 314, 456
Vector Charts – Advantages 456
Vector Charts – "Layered" Information 456, 467
Vector/Raster Chart Use and IMO/IHO Regulations 456
Vector/Raster Charts used Simultaneously 457
Vent Riser in Inert Gas Systems 261
Vibration 37, 185/6
Victor Pyrate Ltd 224/238
VISION 2100 Navigation System of Sperry Marine 314, 457, 469ff
Visual Look-out 298, 304, 319, 439, 527
VLCC – Conform to Collision Rules Always 321
VLCC – "First" Double-Bottomed Built in 1969 403
VLCC – 2000 New Buildings 564
VLCC – 2000 Statistics 557
VLCC – Officers/Crew Accommodation 384ff
VLCC – Accommodation Construction 44, 112, 420
VLCC – Afterpeak Tank 35, 165, 191, 284ff
VLCC – Alcohol and Drug Policies 387
VLCC – Anchor Equipment 90ff, 124
VLCC – Anchoring 333ff
VLCC – Approximate Dimensions 5ff, 159, 428
VLCC – Arc of Visibility over Bow 323
VLCC – Average Day's Run 127, 313
VLCC – Ballast Tanks Segregated 40, 98, 165, 416
VLCC – Ballast System Arrangements 135
VLCC – Ballast Tank Coatings 72ff, 559, 564
VLCC – Bending Moments 161/6, 171, 173, 179/80, 441, 497/8, 502, 506, 548, 552, 567
VLCC – Berthing 336ff, 482
VLCC – Bridge/Wheelhouse Arrangements 147
VLCC – Bridge Wing Debate 112
VLCC – Bulbous Bow Construction Function 23, 45ff, 160, 183, 438
VLCC – Bunker Capacity 127
VLCC – Cargo "Parcels" 190ff, 201
VLCC – Cargo Control Room 136, 169, 174, 202, 250, 535, 552
VLCC – Cargo Discharging 201ff, 531ff
VLCC – Cargo Loading 166ff, 190ff, 200ff, 531ff, 200ff
VLCC – Cargo Pumps 98, 131, 136, 173, 200, 347, 423, 544
VLCC – Cargo Samples 200
VLCC – Cargo Tank Capacities 124ff, 428
VLCC – Cargo Tank Construction Terms 32, 406
VLCC – Cargo Tank Dimensions 46, 428
VLCC – Cargo Tank Inspections 44, 195, 218, 262, 284ff, 562ff, 565
VLCC – Catwalk 93, 113
VLCC – Classification Societies 8, 10, 12, 23ff, 40ff, 72, 87ff, 121, 145, 171, 173, 185, 221, 241, 405, 423, 462ff, 559, 562ff, 570

VLCC – CO_2 Fire Extinguishing Room 139
VLCC – Cofferdam 40, 121
VLCC – Comparisons with Familiar Objects 5
VLCC – Complexities of Loading 190
VLCC – Compressive Stresses 166
VLCC – Computerisation INMARSAT 490ff
VLCC – Computerised Weather Routing 506ff
VLCC – Computerised Tidal Predictions 495
VLCC – Computerised Engine Room 439, 553
VLCC – Computerised Cargo Monitoring 531ff
VLCC – Computerised Hull Sensors 497ff
VLCC – Computerised Navigation 440ff
VLCC – Computerised Navigation Training 512ff
VLCC – Crude Oil Washing (COW) Techniques, Procedures and Operations xv, 102, 221/242, 262, 559, 564
VLCC – Day's steaming 127, 313
VLCC – Deep draft Signal 330
VLCC – Designated Deep Water Anchorage's 334
VLCC – Direct Bridge Control 440
VLCC – Double Hull Building Forecasts 564
VLCC – Double-Hulled Construction 403/428
VLCC – Dry-docking 24, 55, 68/69, 217, 426, 562
VLCC – Dynamic Stresses 155ff, 179ff, 497ff, 553
VLCC – Emergency Towing System (ETS) 118
VLCC – Engine Room Machinery 35, 117/8, 253, 553
VLCC – Engine Speed 39, 127, 423
VLCC – Environmental Protection-Major Area of Concern xi, 68, 193, 333, 345, 404/5, 571/2, 574/5
VLCC – Explosions in December 1969 213ff
VLCC – Fire Drill 273/5
VLCC – Fire Monitors 94
VLCC – Foam Monitor Drills 94, 275
VLCC – Fo'c'sle 45, 88ff, 121ff
VLCC – Forepeak Tank 45, 123, 165, 191
VLCC – Forward Observation Platform 83
VLCC – Fourth/Fifth Special Surveys 64, 175, 558, 562
VLCC – Fuel Consumption 127, 423
VLCC – Heating Coils in Ballast Tanks 101, 423
VLCC – Heating Coils in Cargo Tanks 202
VLCC – Height of Radar Scanner 299
VLCC – Helicopter Operations 281, 335, 384, 393, 396
VLCC – High Standard of Catering 388
VLCC – Hogging Stress 162/5, 175, 178, 180, 183
VLCC – Hull Stress and Detectors 171ff, 497ff, 567
VLCC – Hydraulic Pilot Hoist Dangers 345
VLCC – Hydrostatically Balanced Loading (HBL) 558, 562ff
VLCC – IMO Conventions/Regulations 8ff, 40, 98, 118, 121, 148, 153, 195, 221, 231, 241, 248, 289, 296, 344, 361, 423, 457, 462, 479, 490, 523, 558, 562/3, 565, 571
VLCC – Inert Gas System xv, 102, 149, 167, 194, 200, 202, 206, 241/262, 265, 296, 347, 536, 540, 566
VLCC – Interaction 330
VLCC – Large Turning Circle 325ff, 438
VLCC – Launched by Flooded Dock 49
VLCC – Lifeboat Drills 271
VLCC – Lifeboats 114
VLCC – Lightening 347/353

VLCC – Loading 166/178, 189/201, 542/553, 562ff
VLCC – Loadline 10ff, 44, 93
VLCC – Log Book Important Legal Document 319
VLCC – Longitudinal Stresses 160ff, 497ff, 504
VLCC – Luffing Cranes replace Derricks 112
VLCC – Manifold 109, 161, 195ff, 347, 351/2, 420
VLCC – Manoeuvring Data 324ff
VLCC – MARPOL References xvii, 44, 93, 102, 198, 218, 221/2, 230/1, 289, 366, 405/6, 532, 557/8, 564
VLCC – Master's Night Order Book 309, 319, 471
VLCC – Mild and HT Steel Use 29, 559
VLCC – Monkey Island 83, 154, 181
VLCC – Mooring and Equipment 87/90, 336/345, 502
VLCC – Navigation Lights 88, 153, 156
VLCC – Numbers of Single/Doubled Hulled in Service 2000 557
VLCC – Officers affected by Stress 527
VLCC – Paint Layer Thickness Testing 289
VLCC – Painting Area 59
VLCC – Passage Planning 296, 323, 336, 438, 457, 474
VLCC – Pilot's Duties 112, 296, 317, 321ff, 335/6, 345, 482
VLCC – Pipe Systems 95ff, 113, 131/2, 202, 239, 241ff
VLCC – Ports Visited 17ff
VLCC – Practical Lightening Operation 347/53
VLCC – Pre-Handing-Over Inspections 283ff
VLCC – Procedures in Restricted Visibility 300, 329
VLCC – Propeller 23, 39, 64ff, 160, 325, 426
VLCC – Pumps 98, 131, 135/6, 175, 200, 282, 347, 423, 544
VLCC – Pump Explosion 282
VLCC – Pump Room 24, 30, 40, 98, 113, 131ff, 149, 200, 219, 239, 282, 423
VLCC – Radio 145/6, 335, 374, 386, 397, 490
VLCC – Ratholes and Manholes 44, 284ff
VLCC – Recreational Activities 396
VLCC – Rudder 23, 37/38, 145, 160, 289, 426
VLCC – Rudder Cycling to Assist Stopping 329
VLCC – Sagging Stress 162/5, 173, 175, 179, 183
VLCC – Scantlings 29, 428, 559
VLCC – Seafarers Deaths by Explosions 213ff
VLCC – Segregated Ballast Tanks SBT 40, 98, 165, 416
VLCC – Shearing Forces 161/6, 171, 173, 178, 183, 441, 548, 552, 567
VLCC – Shell Plating Thickness 44, 428
VLCC – Ship's Library 396
VLCC – Shore Leave 383
VLCC – Signal Mast Light Arrangements 156
VLCC – Single Buoy Mooring SBM 201, 345ff
VLCC – Single/Double Hulled Construction Differences 428
VLCC – Slamming Stresses 179ff, 498ff, 568
VLCC – Slipway Launching 49
VLCC – Slop Tanks 40, 194, 200/1, 239, 421, 535
VLCC – Slow Helm Response Time 325, 438
VLCC – Sludge 209ff, 262
VLCC – Social Conditions 396
VLCC – SOLAS Regulations xi, 68, 193, 221, 241, 248, 273, 289, 319, 333, 366, 456, 458, 490, 532, 557

VLCC – Special Fourth/Fifth Surveys 64, 175, 558, 562
VLCC – Squat 329
VLCC – Static Stresses 159ff, 179ff, 497ff, 553
VLCC – Static Electricity Dangers 213
VLCC – Steel Replacement with Ageing 559
VLCC – Steering 145ff, 152ff, 339, 482
VLCC – Stem and Foc'sle Construction 45
VLCC – Stern Construction 35, 46, 423
VLCC – Stopping Distances 325
VLCC – Stripping Lines and Pumps 113, 131ff, 194, 209, 241, 423
VLCC – Suez Canal Searchlight 123
VLCC – Surprisingly Short Period of Building 24, 409
VLCC – SWL Cranes and Derricks 112, 428
VLCC – Tank Cleaning 133, 195, 207/267, 409
VLCC – Tank Inspections 44, 195, 218, 261/2, 284, 564/5
VLCC – Tank Rescue Drill 277
VLCC – Thirty Month Survey 69, 261
VLCC – Trading Ports 17ff
VLCC – Transversal Stresses 160ff
VLCC – Tugs 49, 89, 195, 323, 336/7, 344
VLCC – Turning Circle 325ff, 428
VLCC – Typical Voyages 17ff
VLCC – Ullages/Ullaging 102/5, 175, 193, 199/200, 212, 534, 540ff
VLCC – Use of AKD Stopper 346
VLCC – Use of Doppler to Berth 338, 482
VLCC – Use of Explosimeters 218
VLCC – Use of Fog Horns/Sirens 89, 329
VLCC – Vibration 37, 185/6
VLCC – Lookout Duty Essential 298, 304, 319, 527
VLCC – Ports Visited 17ff
VLCC – Sample Voyages 20
Voith Schneider Rudder 37
Voyage Lengths and Leave Patterns 383
Voyage Management System VISION 2100 (VMS) of Sperry Marine 314, 457, 469ff
Voyage Planning by Consilium's INS 970 System 462
Voyage Planning by Racal-Decca's MIRANS 5000 (LSR) 463ff
Voyage Planning at Transas Marine 474
VP MAGNA Twin-Nozzle Cleaning Machine of Victor Pyrate 230
VP MONOMATIC 2 Tank Cleaning Machine of Victor Pyrate 224
VTMS of Port Authorities by Transas' NAVI-TRAINER 513
V-Type Stem Construction 46

Walport Video Company, London 397
Warsash Maritime Centre xii, 359, 360, 378
Watchkeeping on Bridge at Sea 216, 295ff, 439, 445ff, 513
Wave Motion Measurement by STRESSALERT II of Strainstall Engineering Company 497ff
Way Points on Electronic Charts Using DGPS 458
Weather Facsimile Map 445, 508
Weather Forecasting – A Vital Service to Mariners 506ff

Weather Routing Using On-Board Guidance System of Oceanroutes (WNI) 506
Weekly Notices to Mariners – Computerised 458
Weekly Officer's Meetings 270, 395
Wheelhouse 147, 181, 185, 420, 445, 553
WHESSMATIC 550M System and TCIM/M of Whessoe Varec 546/7
Whessoe Gauges 102, 199, 200, 281
Whessoe Varec's FUELSMANAGER System 547
Whessoe Varec's RADAR LEVEL GAUGE 7600 RLG 548
Wives at Sea 393
WNI (Weathernews) Group Oceanroutes 506
WNI Interactive with ORION and INMARSAT 510
World's Charts Produced on Compact Discs 456
World-Wide Shipping Company 7, 49

Yachtsmen 456, 495
Young, Captain Paul 91, 97

Zener Barrier 498, 538, 544
Zone/Load Line Importance 13ff, 190
Zoom Facility on Electronic Charts 458
Zoom Facility on Radars 451